Foundations of Quantitative Finance

Chapman & Hall/CRC Financial Mathematics Series

Recently Published Titles

Geometry of Derivation with Applications
Norman L. Johnson

Foundations of Quantitative Finance
Book I: Measure Spaces and Measurable Functions
Robert R. Reitano

Foundations of Quantitative Finance
Book II: Probability Spaces and Random Variables
Robert R. Reitano

Foundations of Quantitative Finance
Book III: The Integrals of Riemann, Lebesgue and (Riemann-)Stieltjes
Robert R. Reitano

Foundations of Quantitative Finance
Book IV: Distribution Functions and Expectations
Robert R. Reitano

Foundations of Quantitative Finance
Book V: General Measure and Integration Theory
Robert R. Reitano

Computational Methods in Finance, Second Edition
Ali Hirsa

Interest Rate Modeling
Theory and Practice, Third Edition
Lixin Wu

Data Science and Risk Analytics in Finance and Insurance
Tze Leung Lai and Haipeng Xing

Foundations of Quantitative Finance
Book VI: Densities, Transformed Distributions, and Limit Theorems
Robert R. Reitano

For more information about this series please visit: https://www.routledge.com/Chapman-and-HallCRC-Financial-Mathematics-Series/book-series/CHFINANCMTH

Foundations of Quantitative Finance

Book VI: Densities, Transformed Distributions, and Limit Theorems

Robert R. Reitano
Brandeis International Business School
Waltham, MA

CRC Press is an imprint of the
Taylor & Francis Group, an **informa** business
A CHAPMAN & HALL BOOK

First edition published 2025
by CRC Press
2385 Executive Center Drive, Suite 320, Boca Raton, FL 33431

and by CRC Press
4 Park Square, Milton Park, Abingdon, Oxon, OX14 4RN

CRC Press is an imprint of Taylor & Francis Group, LLC

Library of Congress Cataloging-in-Publication Data
Names: Reitano, Robert R., 1950- author.
Title: Foundations of quantitative finance. Book VI, Densities, transformed distributions, and limit theorems / Robert R. Reitano.
Other titles: Densities, transformed distributions and limit theorems
Description: Boca Raton, FL : CRC Press, 2025. | Includes bibliographical references and index.
Identifiers: LCCN 2024020819 | ISBN 9781032231167 (hardback) | ISBN 9781032229492 (paperback) | ISBN 9781003275770 (ebook)
Subjects: LCSH: Finance--Mathematical models. | Distribution (Probability theory) | Limit theorems (Probability theory)
Classification: LCC HG106 .R4486 2025 | DDC 332.01/5195--dc23/eng/20240628
LC record available at https://lccn.loc.gov/2024020819

ISBN: 978-1-032-23116-7 (hbk)
ISBN: 978-1-032-22949-2 (pbk)
ISBN: 978-1-003-27577-0 (ebk)

DOI: 10.1201/9781003275770

Typeset in Latin Modern
by KnowledgeWorks Global Ltd.

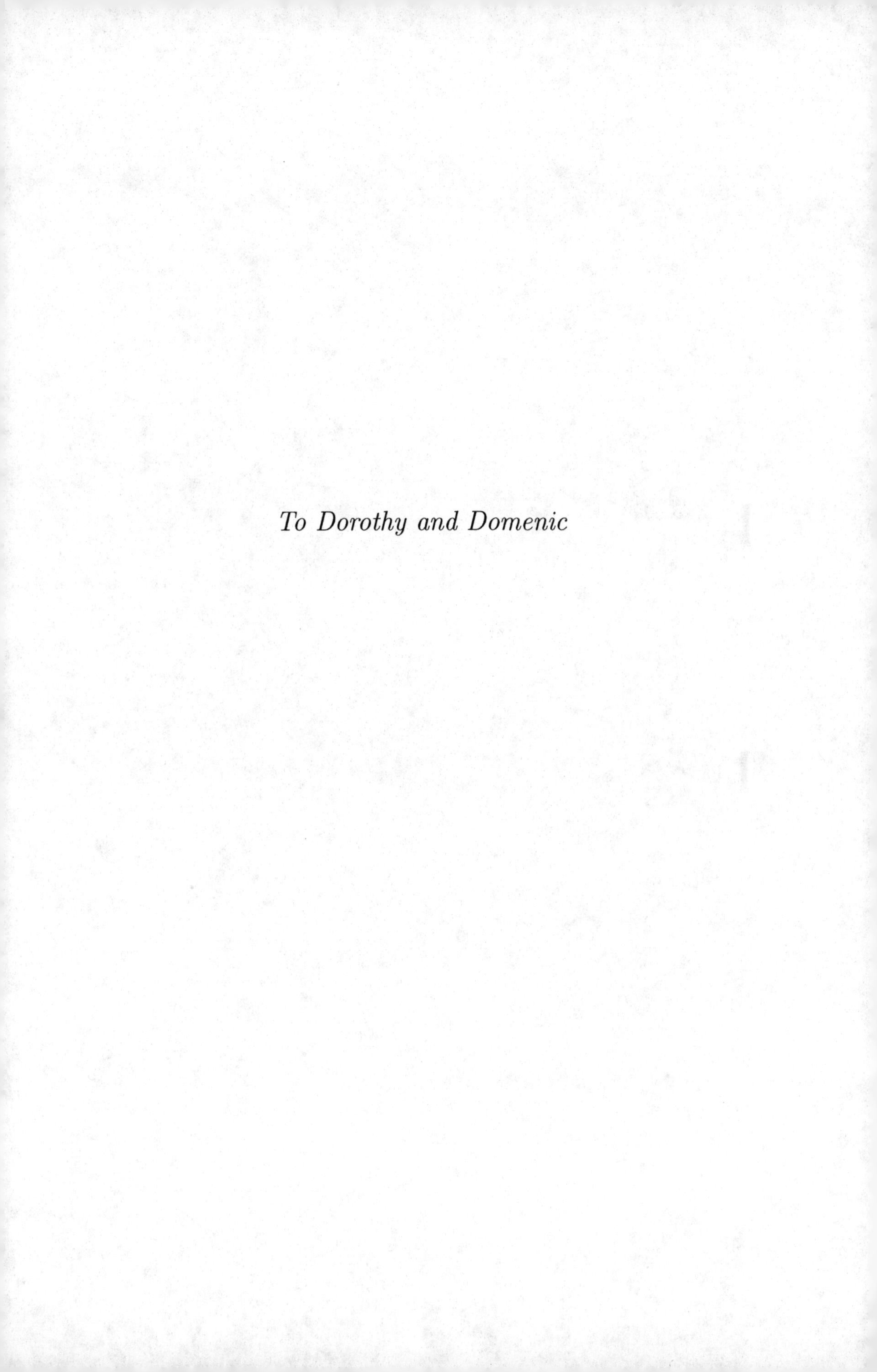

To Dorothy and Domenic

Contents

Preface

The idea for a reference book on the mathematical foundations of quantitative finance has been with me throughout my professional and academic careers in this field, but the commitment to finally write one didn't materialize until completing my introductory quantitative finance book in 2010.

My original academic studies were in pure mathematics in a field of mathematical analysis, and neither applications generally nor finance in particular were then even on my mind. But on completion of my degree, I decided to temporarily investigate a career in applied math, becoming an actuary, and in short order became enamored with mathematical applications in finance.

One of my first inquiries was into better understanding yield curve risk management, ultimately introducing the notion of partial durations and related immunization strategies. This experience led me to recognize the power of greater precision in the mathematical specification and solution of even an age-old problem. From there my commitment to mathematical finance was complete, and my temporary investigation into this field became permanent.

In my personal studies, I found that there were a great many books in finance that focused on markets, instruments, models, and strategies, and which typically provided an informal acknowledgement of the background mathematics. There were also many books in mathematical finance focusing on more advanced mathematical models and methods, and typically written at a level of mathematical sophistication requiring a reader to have significant formal training and the time and motivation to derive omitted details.

The challenge of acquiring expertise is compounded by the fact that the field of quantitative finance utilizes advanced mathematical theories and models from a number of fields. While there are many good references on any of these topics, most are again written at a level beyond many students, practitioners, and even researchers of quantitative finance. Such books develop materials with an eye to comprehensiveness in the given subject matter, rather than with an eye toward efficiently curating and developing the theories needed for applications in quantitative finance.

Thus the overriding goal I have for this collection of books is to provide a complete and detailed development of the many foundational mathematical theories and results one finds referenced in popular resources in finance and quantitative finance. The included topics have been curated from a vast mathematics and finance literature for the express purpose of supporting applications in quantitative finance.

I originally budgeted 700 pages per book, in two volumes. It soon became obvious this was too limiting, and two volumes ultimately turned into ten. In the end, each book was dedicated to a specific area of mathematics or probability theory, with a variety of applications to finance that are relevant to the needs of financial mathematicians.

My target readers are students, practitioners, and researchers in finance who are quantitatively literate, and recognize the need for the materials and formal developments presented. My hope is that the approach taken in these books will motivate readers to navigate these details and master these materials.

Most importantly for a reference work, all ten volumes are extensively self-referenced. The reader can enter the collection at any point of interest, and then using the references cited, work backwards to prior books to fill in needed details. This approach also works for a course on a given volume's subject matter, with earlier books used for reference, and for both course-based and self-study approaches to sequential studies.

The reader will find that the developments herein are presented at a much greater level of detail than most advanced quantitative finance books. Such developments are of necessity typically longer, more meticulously reasoned, and therefore can be more demanding on the reader. Thus before committing to a detailed line-by-line study of a given result, it can be more efficient to first scan the derivation once or twice to better understand the overall logic flow.

I hope the scope of the materials, and the additional details presented, will support your journey to better understanding.

I am grateful for the support of my family: Lisa, Michael, David, and Jeffrey, as well as the support of friends and colleagues at Brandeis International Business School.

Robert R. Reitano
Brandeis International Business School

Author

Robert R. Reitano is Professor of the Practice of Finance at the Brandeis International Business School where he specializes in risk management and quantitative finance, and where he previously served as MSF Program Director and Senior Academic Director. He has a PhD in mathematics from MIT, is a fellow of the Society of Actuaries, and a Chartered Enterprise Risk Analyst. He has taught as Visiting Professor at Wuhan University of Technology School of Economics and Reykjavik University School of Business, and as Adjunct Professor in Boston University's master's degree program in mathematical finance. Dr. Reitano consults in investment strategy and asset/liability risk management, and previously had a 29-year career at John Hancock/Manulife in investment strategy and asset/liability management, advancing to Executive Vice President & Chief Investment Strategist. His research papers have appeared in a number of journals and have won an Annual Prize of the Society of Actuaries and two F.M. Redington Prizes of the Investment Section of the Society of Actuaries. Dr. Reitano serves on various not-for-profit boards and investment committees.

Author

[illegible]

Introduction

Foundations of Quantitative Finance is structured as follows:

Book I: *Measure Spaces and Measurable Functions*
Book II: *Probability Spaces and Random Variables*
Book III: *The Integrals of Riemann, Lebesgue and (Riemann-)Stieltjes*
Book IV: *Distribution Functions and Expectations*
Book V: *General Measure and Integration Theory*
Book VI: *Densities, Transformed Distributions, and Limit Theorems*
Book VII: *Brownian Motion and Other Stochastic Processes*
Book VIII: *Itô Integration and Stochastic Calculus 1*
Book IX: *Stochastic Calculus 2 and Stochastic Differential Equations*
Book X: *Classical Models and Applications in Finance*

The series is logically sequential. Books I, III, and V develop foundational mathematical results needed for the probability theory and finance applications of Books II, IV, and VI, respectively. Books VII, VIII, and IX then develop results in the theory of stochastic processes. While these latter three books introduce ideas from finance as appropriate, the final realization of the applications of these stochastic and other models to finance is deferred to Book X.

This Book VI, *Densities, Transformed Distributions, and Limit Theorems*, develops a number of more advanced topics in probability theory, and then turns to applications in the modeling and pricing of financial derivatives.

Chapter 1 investigates density functions. The first section reviews Book V results on defining a measure in terms of an integral of a density function, and the famous existence result of Radon-Nikodým which characterizes when a given measure on a measure space can be so characterized. The next section applies these results to distribution functions of random variables and vectors and the associated probability measures. The density function for a given distribution is defined, when it exists, as the density function of the induced probability measure. Results are then derived on existence and properties of density functions for joint, marginal, and conditional distribution functions, as well as for density functions for random vectors with independent components.

Transformations of random vectors are investigated in Chapter 2, starting with a useful technical result known as Cavalieri's principle. Though formulated centuries earlier, it is now a corollary of Fubini's theorem of Book V, and provides an elegant tool for generalizing the Book II studies on the distribution functions of sums of random vectors, as well as the associated density functions when they exist. These results then provide for an important generalization of the Book V result on the integrability of the convolution of integrable functions. The final sections of this chapter derive the distribution and density functions for a ratio of independent random variables, and then address distribution and density function results on general measurable transformations of random vectors.

With the integration theory of Book V, Chapter 3 derives deeper results on the weak convergence of random vectors, extending the studies and applications of Books II and IV. Starting with a brief version of the portmanteau theorem on $\mathbb{R}$ applicable to random variables, a general version of this result is derived on $\mathbb{R}^m$. This theorem provides a host of useful, equivalent formulations for the weak convergence of a sequence of probability measures, or equivalently, the convergence in distribution of a sequence of random vectors. Turning next to applications, the chapter generalizes Book II results from $\mathbb{R}$ to $\mathbb{R}^m$ for the mapping theorem, Mann-Wald theorem, Slutsky's theorem, the Delta method, and Prokhorov's theorem, and then introduces Scheffé's theorem and part one of a very useful result called the Cramér-Wold device.

Chapter 4 completes and extends the introductory studies of Book IV on expectations of random variables. With the integration theory of Book V, expectations can be defined in terms of integrals on the underlying probability space. Using various transformations and change of variable results, this definition is seen to have various guises, including the familiar formulas from continuous and discrete probability theory. Moments and the moment generating functions follow as examples of expectations, and various properties are developed, including an investigation into weak convergence of distribution functions and the implications for convergence of moments and moment generating functions. The final sections study the general notions of conditional probabilities and conditional expectations. After introducing the ideas from the perspective of elementary probability theory, these notions are formally defined, and then proved to exist with the aid of the Radon-Nikodým theorem. Various foundational properties of conditional expectations are then derived, as is Jensen's inequality and $L_p(\mathcal{S})$-space properties. The final section develops various results related to integration to the limit.

Characteristic functions of distribution functions are the focus of Chapter 5, developing results on $\mathbb{R}$ and $\mathbb{R}^n$. As these results generalize those of moment generating functions, for which the 1-dimensional theory was developed in Book IV, the first section investigates results on multivariate moment generating functions on $\mathbb{R}^n$. Then after a short section on integration of complex-valued functions, characteristic functions are formally defined in terms of probability space integrals. This definition is then transformed into various guises using the Book V results on change of variables, and this obtains the more familiar definitions from discrete and continuous probability theory. Adding to the examples of Books II and IV, the next section illustrates the characteristic function for a variety of familiar distribution functions. Properties of characteristic functions are derived next, first on $\mathbb{R}$ and requiring only a notational translation of the Book V results on the Fourier transform, and then generalized to characteristic functions on $\mathbb{R}^n$. The final two sections derive Bochner's theorem, which identifies necessary and sufficient conditions for a function to be a characteristic function, and then prove a result on uniqueness of moments that was quoted and used in the Book IV investigation.

Chapter 6 investigates the multivariate normal distribution, first motivating a definition of this distribution function by investigating affine transformations of random vectors of independent normal variates, and then generalizing. While the initial construction obtained a density function for this distribution, the general definition, framed in terms of characteristic or moment generating functions, encompasses distributions without densities. Existence of density functions within the general framework is studied next, as is how such existence is affected by affine transformations. Continuing the discussions of Book IV, simulation of multivariate normal vectors is discussed, for which the Cholesky decomposition of a positive definite matrix plays a central role. The final section derives properties of the multivariate normal distribution function, starting with general moment formulas, and then derives the interesting and oft-misquoted property that the component variates of a multivariate normal are independent if and only if they are uncorrelated. Samples of normal variates are then investigated, proving that the sample mean and sample variance are independent random variables, with distribution functions that are derived.

A variety of applications of characteristic functions to topics of interest are provided in Chapter 7. First, the central limit theorem of Book IV can now be significantly generalized from the earlier context of distribution functions with moment generating functions, to distribution functions with only two moments, and then to results on independent but not identically distributed variates, and finally to a central limit theorem in $\mathbb{R}^n$. The second investigation focuses on distribution functions of sums of random variables, and in particular, distribution families related under addition. The approach used is similar to that in Book IV, but replacing moment generating functions with characteristic functions provides far more general results. These methods are then applied to investigate infinitely divisible distributions and to derive some of their properties. The final section considers the distribution function of products of random variables.

Chapter 8 investigates asset price models in discrete time. These models are envisioned as two dimensional, having both spatial and temporal distributional specifications, and various earlier results are applied to investigate models that have both useful and mathematically tractable characteristics. In the first section, asset price models are introduced in terms of both additive and multiplicative temporal specifications, and the induced spatial distributions are derived. These model specifications are refined in the next section by introducing and illustrating various notions of scalability within a model, where this term identifies how models are related as the time-step $\Delta t \to 0$. The final section investigates limiting distributions associated with scalable models, for which the earlier central limit theorems play a prominent role. While this chapter uses the terminology of "asset prices" to simplify the language, the results discussed apply equally well to the modeling of other financial variables.

The discrete-time pricing of financial derivatives on various underlying assets is the subject of Chapter 9, where it is assumed that such assets can be modeled in the real world according to the multiplicative binomial temporal models of the prior chapter. The first section develops the binomial lattice framework for pricing European and American versions of such derivatives based on the notion of a replicating portfolio. The pricing formulas for European derivatives are seen to reduce to the expected value of discounted settlements, with expectations defined relative to new "risk-neutral" probabilities for the binomial lattice.

The next section then investigates limiting distributions of asset prices as $\Delta t \to 0$ under this risk-neutral probability structure, as well as under an alternative probability structure for real-world modeling. Then limits of the binomial lattice prices for general European derivatives are obtained as $\Delta t \to 0$, and for European puts and calls, these limiting results are the famous Black-Scholes-Merton pricing formulas. Various results related to these limiting prices are then derived, such as derivative price convergence to settlement values, and that European derivative prices satisfy the Black-Scholes-Merton partial differential equation.

Path-dependent financial derivatives are then investigated, and initially modeled within a non-recombining binomial lattice framework to obtain replicating portfolio pricing formulas. Monte Carlo pricing, meaning pricing based on samples of asset price paths, is considered next, as are convergence results to lattice prices as the sample size increases. The final section investigates lognormal model pricing of financial derivatives, exact and by Monte Carlo methods, and extends the Black-Scholes-Merton framework to path-dependent European derivatives.

While Chapters 8-9 studied the limiting distributions of the asset price X_t for $t = kT/n$ for $1 \leq k \leq n$ for models under various probability structures, Chapter 10 investigates such models in continuous time, introducing ideas to be explored in Books VII and beyond. Beginning with binomial paths, which are defined to interpolate binomial sums defined with $p = \frac{1}{2}$ between successive time points $k\Delta t$ and $(k+1)\,\Delta t$, we investigate two types of limits as $n \to \infty$. Pathwise binomial motion is first considered as the uniform limits of these binomial paths, but while continuous, the distributional properties of binomial paths are lost. A more rewarding investigation into the distributional limits of binomial paths is studied next, defining such limits to be standard binomial motion.

After deriving a variety of results on properties of such distributional limits, nonstandard binomial motion is introduced. The first inquiry is into binomial paths with $p \neq \frac{1}{2}$, and then where $p = p(\Delta t)$ is given by the risk neutral probabilities or the alternative real-world probabilities of Chapter 9. The final section considers distributional limits of real-world and risk-neutral asset models defined with respect to the respective binomial paths, extending earlier results.

I hope this book and the other books in the collection serve you well.

Notation 0.1 (Referencing within FQF Series) *To simplify the referencing of results from other books in this series, we use the following convention.*

A reference to "Proposition I.3.33" is a reference to Proposition 3.33 of Book I, while "Chapter III.4" is a reference to Chapter 4 of Book III, and "II.(8.5)" is a reference to formula (8.5) of Book II, and so forth.

1

Density Functions

Distribution functions and their associated Borel measures were studied in Books II and IV, and with the aid of the general integration theory of Book V, we are now in the position to formally introduce the associated density functions when they exist.

The first section reviews Book V results on defining a measure in terms of an integral of a density function, and the famous existence result of Radon-Nikodým, which characterizes when a given measure on a measure space can be so characterized. The next section applies these results to distribution functions of random variables and vectors and the associated probability measures.

The density function for a given distribution is defined, when it exists, as the density function of the induced probability measure. Results are then derived on existence and properties of density functions for joint, marginal, and conditional distribution functions, as well as for density functions for random vectors with independent components.

1.1 Density Functions of Measures

Books I and V contain various background materials on measures and measure spaces, and the reader can find a summary of some Book I results in Chapter V.1. Of particular interest for the current discussion is the Book V investigation into measures induced by integrable functions, and for this we recall the notion of a sigma finite or σ-finite measure space.

Definition 1.1 (σ-finite measure space) *A measure space $(X, \sigma(X), \mu)$ is said to be* ***sigma finite,*** *or* ***σ-finite****, if there exists a countable collection $\{B_j\}_{j=1}^{\infty} \subset \sigma(X)$ with $\mu(B_j) < \infty$ for all j, and $X = \bigcup_{j=1}^{\infty} B_j$. In this case, it is also said that the measure μ is σ-finite.*

If $(X, \sigma(X), \mu)$ is a **probability space**, meaning $\mu(X) = 1$, or more generally a **finite measure space** with $\mu(X) < \infty$, then $(X, \sigma(X), \mu)$ is σ-finite, but there are many other examples. See Exercise 1.3.

The $\sigma(X)$-measurability of a function defined on a measure space $(X, \sigma(X), \mu)$ has been a fundamental notion in the measure theory and probability theory of earlier books. For this definition, the Borel sigma algebras of Definition 1.2, named for **Émile Borel** (1871–1956), played a prominent role. For a discussion of open sets and topologies, see Section I.2.1. See also Remark 1.54.

Definition 1.2 (Borel sigma algebras $\mathcal{B}(\mathbb{R}^n)$ and $\mathcal{B}(X)$; Borel measures)
The ***Borel sigma algebra*** *$\mathcal{B}(\mathbb{R}^n)$ is the smallest sigma algebra that contains the open sets of $\mathbb{R}^n$.*

More generally, if X is a topological space, the Borel sigma algebra $\mathcal{B}(X)$ is defined as the smallest sigma algebra on X that contains the open sets defined by the topology.

DOI: 10.1201/9781003275770-1

A **Borel measure** *λ is a measure defined on $\mathcal{B}(\mathbb{R}^n)$, respectively $\mathcal{B}(X)$, with the added property that $\lambda(A) < \infty$ for all compact $A \in \mathcal{B}(\mathbb{R}^n)$, respectively compact $A \in \mathcal{B}(X)$. When λ is a Borel measure, we say that $(\mathbb{R}^n, \mathcal{B}(\mathbb{R}^n), \lambda)$ or $(X, \mathcal{B}(X), \lambda)$, respectively, is a* **Borel measure space.**

Lebesgue measure m on $\mathbb{R}^n$ is a measure on the Borel sigma algebra $\mathcal{B}(\mathbb{R}^n)$ since $\mathcal{B}(\mathbb{R}^n) \subset \mathcal{M}_L(\mathbb{R}^n)$, where $\mathcal{M}_L(\mathbb{R}^n)$ denotes the complete Lebesgue measure space. See Proposition I.2.38 and Section I.7.6.1.

Exercise 1.3 (Borel measures are σ-finite) *Check that Lebesgue measure m defined on $\mathcal{B}(\mathbb{R}^n)$ is σ-finite, as is every Borel measure on $\mathcal{B}(\mathbb{R}^n)$. Hint: How does $\lambda(A) < \infty$ for all compact $A \in \mathcal{B}(\mathbb{R}^n)$ imply this?*

Turning next to $\sigma(X)$-measurability of a function defined on a measure space $(X, \sigma(X), \mu)$, recall Definition V.2.1.

Definition 1.4 ($\sigma(X)$-measurable function; $f = g$, μ-a.e.) *A function $f : X \to \mathbb{R}$ defined on a measure space $(X, \sigma(X), \mu)$ is said to be $\sigma(X)$-***measurable function** *if $f^{-1}(A) \in \sigma(X)$ for all $A \in \mathcal{B}(\mathbb{R})$. Stated more compactly:*

$$f^{-1}(\mathcal{B}(\mathbb{R})) \subset \sigma(X),$$

where $f^{-1}(\mathcal{B}(\mathbb{R})) \equiv \{f^{-1}(A) | A \in \mathcal{B}(\mathbb{R})\}$ is a sigma algebra by Corollary I.3.27.

Given $\sigma(X)$-measurable functions f and g defined on $(X, \sigma(X), \mu)$, we write $f = g$, ***μ-a.e.****, which is read $f = g$,* ***μ-almost everywhere,*** *to mean that:*

$$\mu\left[\{x | f(x) \neq g(x)\}\right] = 0. \tag{1.1}$$

Remark 1.5 ($f = g$, μ-a.e.) *Perhaps the most significant consequence of knowing that $f = g$, μ-a.e. for $\sigma(X)$-measurable functions f and g defined on $(X, \sigma(X), \mu)$, is that by item 3 of Proposition V.3.42, either both or neither of f and g are integrable over any given set $E \in \sigma(X)$, and when both are integrable, the values of these integrals agree. Hence for integration purposes, f and g are indistinguishable.*

As stated in Proposition V.4.5, every **nonnegative** $\sigma(X)$-measurable function $f(x)$ induces a new measure ν on σ-finite $(X, \sigma(X))$. This result applied the integration theory of Chapter V.3, and readily generalizes to $\sigma(X)$-measurable functions that are **nonnegative, μ-a.e.**

Proposition 1.6 ($f(x)$-induced measure on $\sigma(X)$) *If $f(x)$ is a $\sigma(X)$-measurable function on σ-finite $(X, \sigma(X), \mu)$ with $f \geq 0$ μ-a.e., then the set function ν defined on $A \in \sigma(X)$ by:*

$$\nu(A) \equiv \int_A f(x) d\mu, \tag{1.2}$$

is a measure on $\sigma(X)$.

If $f(x)$ is bounded μ-a.e., then ν is a σ-finite measure. If $f(x)$ is μ-integrable, then ν is a finite measure.

Proof. *Proposition V.4.5 obtains this result for nonnegative $\sigma(X)$-measurable function $f(x)$, noting the change of notation.*

If $g(x)$ is $\sigma(X)$-measurable with $g \geq 0$, μ-a.e., define $f(x) = \max(0, g(x))$. Then $f(x)$ is $\sigma(X)$-measurable by Proposition V.2.19, and $f = g$, μ-a.e.

Then by item 3 of Proposition V.3.42:

$$\int_A f(x)d\mu = \int_A g(x)d\mu,$$

for all $A \in \sigma(X)$, though such integrals need not be finite. In any case, this obtains that $\nu(A) \equiv \int_A g(x)d\mu$ is a measure by Proposition V.4.5.

We leave the remaining details of the last sentence as an exercise. ■

While boundedness of measurable $f(x)$ is sufficient to obtain a measure ν that is σ-finite, it is not necessary. This result can be generalized to **locally integrable** $f(x)$. This is a generalization because bounded functions are locally integrable by item 5 of Proposition V.3.42, as are functions bounded μ-a.e.

Definition 1.7 (Locally integrable) *A $\sigma(X)$-measurable function $f(x)$ defined on $(X, \sigma(X), \mu)$ is said to be **locally integrable** if for all $A \in \sigma(X)$ with $\mu(A) < \infty$:*

$$\int_A |f(x)|\, d\mu < \infty. \tag{1.3}$$

Corollary 1.8 ($f(x)$-induced measure on $\sigma(X)$) *If $f(x)$ is a $\sigma(X)$-measurable function on σ-finite $(X, \sigma(X), \mu)$ that is nonnegative μ-a.e. and locally integrable, then the set function ν of (1.2) is a σ-finite measure.*

Proof. *Let $\{B_j\}_{j=1}^{\infty} \subset \sigma(X)$ be as in Definition 1.1. Then $\mu(B_j) < \infty$ assures that $\nu(B_j) < \infty$ by (1.3), and the proof is complete.* ■

Of interest generally, and especially in probability theory, is the following question:

When can a measure ν defined on $(X, \sigma(X), \mu)$ be represented as in (1.2) for some $\sigma(X)$-measurable, nonnegative μ-a.e. $f(x)$?

In such cases, we will call $f(x)$ the density function of ν with respect to μ.

Definition 1.9 (Density function of ν w.r.t. μ) *If ν is a measure on a measure space $(X, \sigma(X), \mu)$ that is given as in (1.2) for a $\sigma(X)$-measurable function $f(x)$ that is nonnegative μ-a.e., we will say that $f(x)$ **is a density function of ν with respect to** μ, or when μ is clear from the context, $f(x)$ **is a density function of** ν.*

It should be emphasized that in Definition 1.9, $f(x)$ was called **a density function,** because density functions, when they exist, cannot be unique.

Proposition 1.10 (On uniqueness of density functions, μ-a.e.) *If $f(x)$ **is a density function of ν with respect to** μ and $g(x)$ is a $\sigma(X)$-measurable function with $f = g$, μ-a.e., then $g(x)$ **is a density function of ν with respect to** μ.*

Proof. *This follows from item 3 of Proposition V.3.42, recalling the introductory comments to this result.* ■

Returning to the above question, we first note that not every measure ν on $(X, \sigma(X), \mu)$ will have a density function. By item 3 of Proposition V.3.29, (1.2) requires that $\nu(A) = 0$ for all $A \in \sigma(X)$ with $\mu(A) = 0$. This special property was identified in Definition V.8.3 as absolute continuity.

Definition 1.11 (Absolutely continuous; Equivalent) *Given a measure space $(X, \sigma(X), \mu)$ and measure ν defined on $\sigma(X)$:*

1. *ν is* ***absolutely continuous with respect to*** *μ, denoted:*

$$\nu \ll \mu,$$

if $\nu(A) = 0$ for all $A \in \sigma(X)$ with $\mu(A) = 0$.

2. *ν is* ***equivalent to*** *μ, denoted:*

$$\nu \sim \mu,$$

if:

$$\nu \ll \mu, \text{ and, } \mu \ll \nu.$$

Thus as noted above, if a measure ν on $(X, \sigma(X), \mu)$ has a density function $f(x)$, it is **necessary** that $\nu \ll \mu$. Remarkably, $\nu \ll \mu$ is also **sufficient** for the existence of a density function for σ-finite measures ν defined on a σ-finite measure space $(X, \sigma(X), \mu)$.

The **Radon-Nikodým theorem** is named for **Johann Radon** (1887–1956), who proved this result for $X = \mathbb{R}^n$, and **Otto Nikodým** (1887–1974), who generalized Radon's result to all σ-finite measure spaces. To better understand the need for σ-finiteness of the measure space $(X, \sigma(X), \mu)$, see Example V.8.19.

While $f(x)$ in this theorem is conventionally stated as nonnegative, the last line of the result makes it clear that we can only assert that $f \geq 0$, μ-a.e.

Proposition 1.12 (Radon-Nikodým theorem) *Let $(X, \sigma(X), \mu)$ be a σ-finite measure space and ν a σ-finite measure on $\sigma(X)$ which is absolutely continuous with respect to μ, so $\nu \ll \mu$.*

Then there exists a nonnegative $\sigma(X)$-measurable function $f : X \to \mathbb{R}$, also denoted $f \equiv \frac{\partial \nu}{\partial \mu}$ and called the ***Radon–Nikodym derivative of*** *ν* ***with respect to*** *μ, so that for all $A \in \sigma(X)$:*

$$\nu(A) = \int_A f d\mu. \tag{1.4}$$

Further, f is unique μ-a.e. If g is a $\sigma(X)$-measurable function so that (1.4) is satisfied with g, then $g = f$, μ-a.e.
Proof. *Proposition V.8.23.* ■

We now summarize the above into our main result for the existence of density functions of measures.

Proposition 1.13 (Existence of density functions of measures) *Let $(X, \sigma(X), \mu)$ be a σ-finite measure space and ν a σ-finite measure on $\sigma(X)$. Then ν has a density function $f(x)$ with respect to μ if and only if $\nu \ll \mu$.*
Proof. *Necessity of $\nu \ll \mu$ is item 3 of Proposition V.3.29, while sufficiency is the Radon-Nikodým theorem.* ■

Remark 1.14 (On density functions and $f \geq 0$, μ-a.e.) *From this point forward, we will refer to a $\sigma(X)$-measurable function f that satisfies $f \geq 0$, μ-a.e., as a $\sigma(X)$-measurable, nonnegative function. This is conventional and rarely creates ambiguity by Remark 1.5.*

If we are given $f \geq 0$, μ-a.e., we can if needed define a truly nonnegative $\tilde{f} = \max(0, f(x))$, which is then also $\sigma(X)$-measurable by Proposition V.2.19. Further, the integrals of $\tilde{f}$ and f over any measurable set A will have the same value by item 3 of Proposition V.3.42, noting the introductory comments to this result.

1.2 Density Functions of Distributions

Density functions of distribution functions of random variables and random vectors defined on a probability space $(\mathcal{S}, \mathcal{E}, \mu)$ were informally introduced in Sections IV.1.3 and IV.3.5. In this section, we investigate the existence of density functions associated with distribution functions, where such density functions are defined in terms of the associated probability measures on $\mathbb{R}$ or $\mathbb{R}^n$.

Hence we begin with a review of random vectors, distribution functions, and induced probability measures, and then turn to the subject at hand.

1.2.1 Distribution Functions and Random Vectors

Recall the notion of a random variable or random vector X defined on a probability space $(\mathcal{S}, \mathcal{E}, \mu)$. As in Definition 1.4, measurability is defined in terms of the Borel sigma algebras.

Definition 1.15 (Random variable; Random vector) *A* ***random variable*** *X defined on a probability space $(\mathcal{S}, \mathcal{E}, \mu)$ is an $\mathcal{E}$-measurable function, so $X : \mathcal{S} \to \mathbb{R}$ and $X^{-1}(\mathcal{B}(\mathbb{R})) \subset \mathcal{E}$.*

A ***random vector*** *X defined on a probability space $(\mathcal{S}, \mathcal{E}, \mu)$ is an $\mathcal{E}$-measurable vector-valued function, so $X : \mathcal{S} \to \mathbb{R}^n$ and $X^{-1}(\mathcal{B}(\mathbb{R}^n)) \subset \mathcal{E}$.*

Exercise 1.16 ($X^{-1}(\mathcal{B}(\mathbb{R}^n))$ is a sigma algebra) *As noted in the previous section, $X^{-1}(\mathcal{B}(\mathbb{R}))$ is a sigma algebra by Corollary I.3.27. Generalize this and prove that if $X : \mathcal{S} \to \mathbb{R}^n$ is $\mathcal{E}$-measurable, then $X^{-1}(\mathcal{B}(\mathbb{R}^n))$ is a sigma algebra on X.*

If $X : \mathcal{S} \to \mathbb{R}^n$, then $X \equiv (X_1, ..., X_n)$ is defined pointwise by:

$$X(s) \equiv (X_1(s), X_2(s), ..., X_n(s)).$$

As noted in Book II, the measurability requirements for the random vector X and random variables $\{X_j\}_{j=1}^n$ are related.

Proposition 1.17 (Random vectors and Random variable components) *Let $X : \mathcal{S} \to \mathbb{R}^n$ be defined on a probability space $(\mathcal{S}, \mathcal{E}, \mu)$, with $X \equiv (X_1, ..., X_n)$.*

Then X is a random vector if and only if X_j is a random variable for all j. That is, $X^{-1}(\mathcal{B}(\mathbb{R}^n)) \subset \mathcal{E}$ if and only if $X_j^{-1}(\mathcal{B}(\mathbb{R})) \subset \mathcal{E}$ for all j.

Proof. *Proposition II.3.32.* ■

Random variables and vectors give rise to distribution functions defined on the range space $\mathbb{R}$ or $\mathbb{R}^n$. Recalling Chapter II.3:

Definition 1.18 (Distribution functions) *If $X : \mathcal{S} \longrightarrow \mathbb{R}$ is a random variable defined on $(\mathcal{S}, \mathcal{E}, \mu)$, the* ***distribution function (d.f.)****, or* ***cumulative distribution function (c.d.f.)*** *of X, denoted F or F_X, is defined on $x \in \mathbb{R}$ by:*

$$F(x) = \mu \left[X^{-1}(-\infty, x]\right]. \tag{1.5}$$

If $X : \mathcal{S} \longrightarrow \mathbb{R}^n$ is a random vector defined on $(\mathcal{S}, \mathcal{E}, \mu)$, $X = (X_1, X_2, ..., X_n)$, the ***joint distribution function (d.f.)****, or* ***joint cumulative distribution function (c.d.f.)*** *associated with X, denoted F or F_X, is defined on $(x_1, x_2, ..., x_n) \in \mathbb{R}^n$ by:*

$$F(x_1, x_2, ..., x_n) = \mu \left[\bigcap\nolimits_{j=1}^n X_j^{-1}(-\infty, x_j]\right]. \tag{1.6}$$

It is common in probability theory to say that $F(x_1, x_2, ..., x_n)$ is the **probability** that $X \in A_{(x_1,x_2,...,x_n)}$:

$$F(x_1, x_2, ..., x_n) = \Pr\left[X \in A_{(x_1,x_2,...,x_n)}\right],$$

where $A_{(x_1,x_2,...,x_n)} \subset \mathbb{R}^n$ is defined as the **unbounded right semi-closed rectangle:**

$$A_{(x_1,x_2,...,x_n)} \equiv \prod_{j=1}^{n} (-\infty, x_j]. \tag{1.7}$$

This follows because $X(s) \in A_{(x_1,x_2,...,x_n)}$ if and only if $s \in \bigcap_{j=1}^{n} X_j^{-1}(-\infty, x_j]$, and so the probability of this X-event is by definition the μ-measure or μ-probability of this s-event as given in (1.6). Thus $F(x_1, x_2, ..., x_n)$ can also be expressed:

$$F(x_1, x_2, ..., x_n) = \mu\left[X^{-1}\left(\prod_{j=1}^{n}(-\infty, x_j]\right)\right]. \tag{1.8}$$

Properties of distribution functions of random variables can be summarized as follows:

Proposition 1.19 (Properties of a d.f. $F(x)$) *Given a random variable X defined on a probability space $(\mathcal{S}, \mathcal{E}, \mu)$, the distribution function $F(x)$ in (1.5) has the following properties:*

1. *$F(x)$ is a nonnegative, increasing function on $\mathbb{R}$ which is Borel, and hence, Lebesgue measurable.*
2. *$F(x)$ is right continuous:*
$$\lim_{y \to x+} F(y) = F(x), \tag{1.9}$$
and has left limits:
$$\lim_{y \to x-} F(y) = F(x) - \mu(\{X(s) = x\}). \tag{1.10}$$
Hence, $F(x)$ is continuous at x if and only if $\mu(\{X(s) = x\}) = 0$.
3. *$F(x)$ has at most countably many discontinuities.*
4. *The limits of $F(x)$ exist as $x \to \pm\infty$:*
$$\lim_{x \to -\infty} F(x) = 0. \tag{1.11}$$
$$\lim_{x \to \infty} F(x) = 1. \tag{1.12}$$

Proof. *Proposition II.6.1.* ■

The above result also has a converse.

Proposition 1.20 (Identifying a distribution function) *Let $F(x)$ be an increasing function that is right continuous and satisfies $F(-\infty) = 0$ and $F(\infty) = 1$, defined as limits.*

Then there exists a probability space $(\mathcal{S}, \mathcal{E}, \mu)$ and random variable X so that $F(x) = \mu[X^{-1}(-\infty, x]]$. In other words, every such function is the distribution function of a random variable.

Proof. *Proposition II.3.6.* ■

For joint distribution functions, properties can again be summarized, but first we recall the definition of **continuous from above** and **n-increasing.** Note that for $n = 1$, continuous from above reduces to right continuous, and n-increasing reduces to increasing.

Definition 1.21 (Continuous from above; n-increasing) *A function $F:\mathbb{R}^n\rightarrow\mathbb{R}$ is said to be* **continuous from above** *if given $x=(x_1,...,x_n)$ and a sequence $x^{(m)}=(x_1^{(m)},...,x_n^{(m)})$ with $x_i^{(m)}\geq x_i$ for all i and m, and $x^{(m)}\rightarrow x$ as $m\rightarrow\infty$:*

$$F(x)=\lim_{m\rightarrow\infty}F(x^{(m)}). \tag{1.13}$$

Further, F is said to be **n-increasing** *or to satisfy the* **n-increasing condition,** *if for any bounded right semi-closed rectangle $\prod_{i=1}^{n}(a_i,b_i]$:*

$$\sum\nolimits_x sgn(x)F(x)\geq 0. \tag{1.14}$$

Each $x=(x_1,...,x_n)$ in this summation is one of the 2^n vertices of this rectangle, so each $x_i=a_i$ or $x_i=b_i$, and $sgn(x)$ is defined as -1 if the number of a_i-components of x is odd, and $+1$ otherwise.

Exercise 1.22 ($F(x)=\prod_{j=1}^{n}F_j(x_j)$) *If $F:\mathbb{R}^n\rightarrow\mathbb{R}$ is a product function, $F(x)=\prod_{j=1}^{n}F_j(x_j)$, check that $F(x)$ is continuous from above if all $\{F_j\}_{j=1}^{n}$ are right continuous, and is n-increasing if all $\{F_j\}_{j=1}^{n}$ are increasing. Then show that:*

$$\sum\nolimits_x sgn(x)F(x)=\prod\nolimits_{j=1}^{n}\left[F_j(b_j)-F_j(a_j)\right].$$

Properties of joint distribution functions are summarized as follows. By Proposition II.3.53, the joint distribution function $F(x)$ in (1.6) is a product function of Exercise 1.22 if and only if $\{X_j\}_{j=1}^{n}$ are independent random variables (Definition 1.53).

Proposition 1.23 (Properties of a joint d.f. $F(x)$) *Given a random vector $X\equiv(X_1,...,X_n)$ defined on a probability space $(\mathcal{S},\mathcal{E},\mu)$, the joint distribution function $F(x)$ in (1.6) has the following properties:*

1. *$F(x)$ is continuous from above.*
2. *$F(x)$ satisfies the n-increasing condition.*
3. *With $x=(x_1,...,x_n)$, the limits of $F(x)$ exist as $x\rightarrow\pm\infty$:*

$$\lim_{x\rightarrow-\infty}F(x)=0,\quad \lim_{x\rightarrow\infty}F(x)=1, \tag{1.15}$$

where the notation $x\rightarrow\pm\infty$ means that $x_i\rightarrow\pm\infty$ for all i.

Proof. *Items 1 and 2 are Proposition II.6.9, while item 3 is left as an exercise. Hint: Use (1.8) and continuity of μ (Proposition I.2.45).* ■

This result again has a converse.

Proposition 1.24 (Identifying a joint distribution function) *Let $F:\mathbb{R}^n\rightarrow\mathbb{R}$ be n-increasing and continuous from above, with $F(-\infty)=0$ and $F(\infty)=1$ defined as limits.*

Then there exists a probability space $(\mathcal{S},\mathcal{E},\mu)$ and a random vector X so that F is the joint distribution function of X.

Proof. *Proposition II.6.10.* ■

1.2.2 Distribution Functions and Probability Measures

A random variable or vector X defined on a probability space $(\mathcal{S}, \mathcal{E}, \mu)$ induces a probability measure on $(\mathbb{R}, \mathcal{B}(\mathbb{R}))$, respectively $(\mathbb{R}^n, \mathcal{B}(\mathbb{R}^n))$. This can be developed in two ways, with results which prove to be equivalent. The first approach reflects the constructions of Book I and utilizes the associated distribution functions, while the second reflects the Book V notion of measures induced by measurable transformations.

To begin with the Book I construction, we first recall the notions of semi-algebra and algebra of sets.

Definition 1.25 (Semi-algebra, algebra of sets) *A collection $\mathcal{A}'$ of sets from a space X is called a* ***semi-algebra of sets on*** *X:*

1. *If $A_1', A_2' \in \mathcal{A}'$, then $A_1' \bigcap A_2' \in \mathcal{A}'$, and thus this holds by induction for all finite intersections.*

2. *If $A' \in \mathcal{A}'$, then there exists disjoint $\{A_j'\}_{j=1}^n \subset \mathcal{A}'$ so that $\widetilde{A'} = \bigcup_{j=1}^n A_j'$.*

   The collection $\mathcal{A}' \equiv \mathcal{A}$ is an ***algebra*** *if in place of item 2:*

2'. *If $A \in \mathcal{A}$, then $\widetilde{A} \in \mathcal{A}$.*

These collections of sets are the building blocks in measure theory:

- **Exercise I.6.10** proved that given a semi-algebra $\mathcal{A}'$, the collection of all finite disjoint unions of $\mathcal{A}'$-sets, including the empty set $\emptyset$ if not already in $\mathcal{A}'$, is an algebra of sets, called the **algebra generated by** $\mathcal{A}'$.

- **Exercise I.2.8** proved that given any algebra $\mathcal{A}$, the smallest sigma algebra that contains $\mathcal{A}$, denoted $\sigma(\mathcal{A})$, is well-defined. First, the power set $\mathcal{P}(X)$ of all subsets of X is one sigma algebra that contains $\mathcal{A}$, and this exercise derives that the intersection of sigma algebras is a sigma algebra. Thus $\sigma(\mathcal{A})$ is the intersection of all sigma algebras that contain $\mathcal{A}$.

- Given a semi-algebra $\mathcal{A}'$, $\sigma(\mathcal{A}') = \sigma(\mathcal{A})$, where $\mathcal{A}$ is the algebra generated by $\mathcal{A}'$, and $\sigma(\mathcal{A}')$ is defined as the smallest sigma algebra that contains $\mathcal{A}'$. This follows because $\mathcal{A}' \subset \mathcal{A} \subset \sigma(\mathcal{A}')$ by definition.

Example 1.26 (Semi-algebras on $\mathbb{R}$, $\mathbb{R}^n$ and $\mathcal{B}(\mathbb{R}^n)$) *The collection $\mathcal{A}' \equiv \{(a, b]\}$ of all right semi-closed intervals is a semi-algebra on $\mathbb{R}$ by Example I.6.12.*

The collection $\mathcal{A}' \equiv \{\prod_{i=1}^n (a_i, b_i]\}$ of all right semi-closed rectangles is a semi-algebra on $\mathbb{R}^n$ by Corollary I.7.3 and Example I.7.4.

By Proposition I.8.1, $\sigma(\mathcal{A}') = \mathcal{B}(\mathbb{R})$, respectively $\sigma(\mathcal{A}') = \mathcal{B}(\mathbb{R}^n)$, for these semi-algebras.

To understand the linkage between distribution functions of random variables or vectors and the induced probability measures we seek, we begin with a summary of two Book I investigations that connect these notions.

Given a Borel measure λ on $(\mathbb{R}, \mathcal{B}(\mathbb{R}))$ or $(\mathbb{R}^n, \mathcal{B}(\mathbb{R}^n))$, which includes probability measures, one can identify a function $F_\lambda(x)$ on $\mathbb{R}$ or $\mathbb{R}^n$ so that:

1. On $\mathbb{R}$, for all $(a, b] \in \mathcal{A}'$:
$$\lambda[(a, b]] = F_\lambda(b) - F_\lambda(a), \tag{1.16}$$
   and the function $F_\lambda(x)$ is increasing and right continuous.

2. On $\mathbb{R}^n$, for all $\prod_{i=1}^n (a_i, b_i] \in \mathcal{A}'$:

$$\lambda\left(\prod\nolimits_{i=1}^{n}(a_i, b_i]\right) = \sum\nolimits_x sgn(x) F_\lambda(x), \tag{1.17}$$

and the function $F_\lambda(x)$ is n-increasing and continuous from above. The summation in (1.17) is defined as in (1.14).

The general constructions of $F_\lambda(x)$ are found in I.(5.1) and I.(8.12), but for finite Borel measures such as probability measures, these definitions simplify in I.(5.3) and I.(8.2) to:

$$\begin{aligned} F_\lambda(x) &= \lambda\left[(-\infty, x]\right], \\ F_\lambda(x) &= \lambda\left[\prod\nolimits_{j=1}^{n}(-\infty, x_j]\right]. \end{aligned} \tag{1.18}$$

And when λ is a probability measure, the function $F_\lambda(x)$ has limits $F_\lambda(-\infty) = 0$ and $F_\lambda(\infty) = 1$ as in Propositions 1.19 and 1.23, and thus has all the requisite properties of a distribution function of a random variable or random vector.

Unsurprisingly, $F_\lambda(x)$ is therefore called **a distribution function associated with** λ.

The second half of Chapters I.5 and I.8 show that the above results are reversible, that from distribution functions, one obtains measures.

With a small change of notation:

1. **Proposition I.5.23:** Given an increasing and right continuous function $F(x)$ defined on $\mathbb{R}$, there exists a Borel measure λ_F defined on $(\mathbb{R}, \mathcal{B}(\mathbb{R}))$ so that (1.16) is satisfied.

   Further, by the discussion in Section I.5.3, $F(x)$ is a distribution function associated with λ_F. This proposition also obtains that λ_F is defined on a complete sigma algebra $\mathcal{M}_F(\mathbb{R})$ with $\mathcal{B}(\mathbb{R}) \subset \mathcal{M}_F(\mathbb{R})$.

2. **Proposition I.8.15:** Given an n-increasing and continuous from above function $F(x)$ defined on $\mathbb{R}^n$, there exists a Borel measure λ_F defined on $(\mathbb{R}^n, \mathcal{B}(\mathbb{R}^n))$ so that (1.17) is satisfied.

   Further, by Corollary I.8.17, $F(x)$ is a distribution function associated with λ_F. This proposition also obtains that λ_F is defined on a complete sigma algebra $\mathcal{M}_F(\mathbb{R}^n)$ with $\mathcal{B}(\mathbb{R}^n) \subset \mathcal{M}_F(\mathbb{R}^n)$.

3. **In both cases,** the above constructions obtain that if $F(x)$ has limits $F(-\infty) = 0$ and $F(\infty) = 1$, the measure λ_F is in fact a probability measure.

Returning to the subject of this section, these Book I constructions obtain **probability measures induced by the distribution functions of random variables and random vectors.**

Proposition 1.27 (Probability measures induced by an RV) *If X is a random variable on $(\mathcal{S}, \mathcal{E}, \mu)$ with distribution function $F(x)$, there exists a probability measure λ_F defined on $(\mathbb{R}, \mathcal{B}(\mathbb{R}))$ that satisfies (1.16) for all $(a, b] \in \mathcal{A}'$.*

If X is a random vector defined on $(\mathcal{S}, \mathcal{E}, \mu)$ with joint distribution function $F(x)$, there exists a probability measure λ_F defined on $(\mathbb{R}^n, \mathcal{B}(\mathbb{R}^n))$ that satisfies (1.17) for all $\prod_{i=1}^n (a_i, b_i] \in \mathcal{A}'$.

In both cases, λ_F is the unique extension of (1.16) and (1.17) from the respective semi-algebras $\mathcal{A}'$ to $\mathcal{B}(\mathbb{R})$ and $\mathcal{B}(\mathbb{R}^n)$.

Proof. *The distribution function $F(x)$ of a random variable X is increasing and right continuous by Proposition 1.19, and thus Proposition I.5.23 obtains the existence of a Borel measure λ_F on $\mathcal{B}(\mathbb{R})$ that satisfies (1.16). Since $\lim_{x \to -\infty} F(x) = 0$, (1.16) yields that*

$F(x) = \lambda_F[(-\infty, x]]$, and λ_F is a probability measure since $\lim_{x\to\infty} F(x) = 1$. The same logic applies in the case of a random vector using Proposition I.8.15, with details left as an exercise.

Let λ'_F be another extension of the set functions λ_F in (1.16) and (1.17) from the respective semi-algebra $\mathcal{A}'$ to a sigma algebra containing $\mathcal{A}'$. This sigma algebra then also contains the associated algebra $\mathcal{A}$ of all finite unions of $\mathcal{A}'$-sets (Exercise I.6.10), so by finite additivity, $\lambda'_F = \lambda_F$ on $\mathcal{A}$.

Since these measures are finite and so too σ-finite, Propositions 1.23 and I.6.14 on uniqueness of extensions obtain that $\lambda'_F = \lambda_F$ on $\sigma(\mathcal{A})$, the smallest sigma algebra that contains $\mathcal{A}$. Then Proposition I.8.1 proves that $\sigma(\mathcal{A}) = \mathcal{B}(\mathbb{R})$, respectively $\sigma(\mathcal{A}) = \mathcal{B}(\mathbb{R}^n)$, and the proof is complete. ■

Thus from a random variable or random vector X, one can construct a unique probability measure λ_F on $\mathcal{B}(\mathbb{R})$, respectively $\mathcal{B}(\mathbb{R}^n)$, using he associated distribution function, respectively, joint distribution function, $F(x)$.

As noted in the introduction, there is an alternative approach to such probability measures based on the Book V investigation of **measures induced by measurable transformations**. Recall Definition V.4.12 and the comment there that while μ_T is called a "measure" in this definition, it is only a set function until the properties of a measure are proved.

Definition 1.28 (Measurable transformations; Induced measures) *Given measure spaces $(X, \sigma(X), \mu)$ and $(X', \sigma(X'), \mu')$, a transformation $T : X \to X'$ is* ***measurable,*** *and sometimes $\sigma(X)/\sigma(X')$-**measurable**, if:*

$$T^{-1}(A') \in \sigma(X) \text{ for all } A' \in \sigma(X'), \tag{1.19}$$

or more economically:

$$T^{-1}[\sigma(X')] \subset \sigma(X).$$

*If $(X', \sigma(X'), \mu') = (\mathbb{R}, \mathcal{B}(\mathbb{R}), \mu')$ for a Borel measure μ', then T is called a $\sigma(X)$-**measurable function**.*

A measurable transformation/function induces a measure μ_T on the range space $(X', \sigma(X'))$, and hence induces a new measure space denoted $(X', \sigma(X'), \mu_T)$. The measure μ_T is called ***the measure induced by*** *T and defined on $A' \in \sigma(X')$ by:*

$$\mu_T(A') = \mu\left[T^{-1}(A')\right]. \tag{1.20}$$

The set function μ_T is well-defined since T is measurable. Thus $T^{-1}(A) \in \sigma(X)$ for all $A \in \sigma(X')$ and so $\mu\left[T^{-1}(A)\right]$ is well-defined. Further, μ_T is indeed a measure on $\sigma(X')$ by Proposition V.4.13.

Example 1.29 ($T = X$, a random variable/vector) *If X is a random variable defined on $(\mathcal{S}, \mathcal{E}, \mu)$:*

$$X : (\mathcal{S}, \mathcal{E}, \mu) \to (\mathbb{R}, \mathcal{B}(\mathbb{R})),$$

the measure μ_X of (1.20) is denoted λ_X and defined on $A \in \mathcal{B}(\mathbb{R})$ by:

$$\lambda_X(A) \equiv \mu\left[X^{-1}(A)\right]. \tag{1.21}$$

The same definition applies to $A \in \mathcal{B}(\mathbb{R}^n)$ if X is a random vector defined on $(\mathcal{S}, \mathcal{E}, \mu)$:

$$X : (\mathcal{S}, \mathcal{E}, \mu) \to (\mathbb{R}^n, \mathcal{B}(\mathbb{R}^n)).$$

Since λ_X is indeed a measure on $\mathcal{B}(\mathbb{R})$ or $\mathcal{B}(\mathbb{R}^n)$ as noted above, and it is natural to wonder how λ_X compares to λ_F of Proposition 1.27, the measure induced by the distribution function $F(x)$ of X.

Proposition 1.30 (Probability measures induced by an RV) *If X is a random variable/vector defined on $(\mathcal{S}, \mathcal{E}, \mu)$ with distribution function $F(x)$, the probability measure λ_F of Proposition 1.27 is given on $(\mathbb{R}, \mathcal{B}(\mathbb{R}))$ / $(\mathbb{R}^n, \mathcal{B}(\mathbb{R}^n))$ satisfies:*

$$\lambda_F(A) \equiv \mu\left[X^{-1}(A)\right],$$

and thus:

$$\lambda_F = \lambda_X, \text{ on } \mathcal{B}(\mathbb{R}) \text{ or } \mathcal{B}(\mathbb{R}^n). \tag{1.22}$$

Proof. *Example V.4.14 and Proposition V.4.15* ■

Remark 1.31 (λ_X **vs.** λ_F) *It should be noted that λ_F as constructed in Proposition 1.27 is a measure both on the Borel sigma algebras $\mathcal{B}(\mathbb{R})/\mathcal{B}(\mathbb{R}^n)$, as well as on larger complete sigma algebras $\mathcal{M}_F(\mathbb{R})/\mathcal{M}_F(\mathbb{R}^n)$ as seen in the Book II references.*

Formally, λ_X as constructed in (1.21), is only definable on $\mathcal{B}(\mathbb{R})/\mathcal{B}(\mathbb{R}^n)$. It cannot be defined on the larger sigma algebras $\mathcal{M}_F(\mathbb{R})/\mathcal{M}_F(\mathbb{R}^n)$, or on the larger Lebesgue counterparts, $\mathcal{M}_L(\mathbb{R})/\mathcal{M}_L(\mathbb{R}^n)$, for the very simple reason that a random variable is only required by Definition 1.15 to satisfy $X^{-1}(\mathcal{B}(\mathbb{R})) \subset \mathcal{E}$, and similarly for a random vector. Thus if $A \in \mathcal{M}_L(\mathbb{R}) - \mathcal{B}(\mathbb{R})$, it need not be true that $X^{-1}(A) \in \mathcal{E}$, and thus $\lambda_X(A) \equiv \mu\left[X^{-1}(A)\right]$ need not be defined.

Thus the statement in (1.22), that $\lambda_F = \lambda_X$ on $\mathcal{B}(\mathbb{R})/\mathcal{B}(\mathbb{R}^n)$, is to be understood in this context.

But we will see in Proposition 1.35 that if $\lambda_X \ll m$, the Borel measure λ_X on $\mathcal{B}(\mathbb{R})/\mathcal{B}(\mathbb{R}^n)$ can always be extended to a probability measure $\tilde{\lambda}_X$ on the complete Lebesgue sigma algebras $\mathcal{M}_L(\mathbb{R})/\mathcal{M}_L(\mathbb{R}^n)$. By extended we mean that $\lambda_X(A) = \tilde{\lambda}_X(A)$ for all Borel sets A. We will also see in Remark 1.38 that there is good reason to work with this extension and its associated Lebesgue measurable density function.

Definition 1.32 (Probability measures induced by X, or $F(x)$) *If X is a random variable, respectively random vector, defined on $(\mathcal{S}, \mathcal{E}, \mu)$, the probability measure λ_X defined by (1.21) on $(\mathbb{R}, \mathcal{B}(\mathbb{R}))$, respectively $(\mathbb{R}^n, \mathcal{B}(\mathbb{R}^n))$, is called the* ***Borel probability measure induced by*** *X, or, the* ***Borel probability measure induced by*** *$F(x)$, where $F(x)$ is the distribution function, respectively, joint distribution function, of X.*

Similarly, the extended probability measure $\tilde{\lambda}_X$ defined in Proposition 1.35 on $(\mathbb{R}, \mathcal{M}_L(\mathbb{R}))$, respectively $(\mathbb{R}^n, \mathcal{M}_L(\mathbb{R}^n))$, is called the ***Lebesgue probability measure induced by*** *X, or, the* ***Lebesgue probability measure induced by*** *$F(x)$, where $F(x)$ is the distribution function, respectively, joint distribution function, of X.*

1.2.3 Existence of Density Functions

To simplify the presentation, we now largely suppress the distinction between random variables ($n = 1$) and random vectors ($n > 1$), and use the terminology of the latter. Simply setting $n = 1$ in the following discussion will obtain results for random variables without ambiguity.

If X is a random vector defined on a probability space $(\mathcal{S}, \mathcal{E}, \mu)$ with joint distribution function $F(x)$, we define the density function associated with X, when it exists, as follows. But see Notation 1.39.

Definition 1.33 (Density functions of X, or of $F(x)$) *Let X be a random vector defined on a probability space $(\mathcal{S},\mathcal{E},\mu)$ with joint distribution function $F(x)$ defined on $\mathbb{R}^n$, and Borel probability measure λ_X defined on $\mathcal{B}(\mathbb{R}^n)$. If λ_X has a density function $f(x)$ relative to Lebesgue measure m defined on $\mathcal{B}(\mathbb{R}^n)$ by Definition 1.9, we call $f(x)$* ***a Borel density function of** X, and also,* ***a Borel density function associated with** $F(x)$.*

Similarly, if the extended Lebesgue probability measure $\tilde{\lambda}_X$ has a density function $\tilde{f}(x)$ relative to Lebesgue measure m defined on $\mathcal{M}_L(\mathbb{R}^n)$ by Definition 1.9, we call $\tilde{f}(x)$ ***a Lebesgue density function of** X, and also,* ***a Lebesgue density function associated with** $F(x)$.*

For example, given:

$$X:(\mathcal{S},\mathcal{E},\mu)\to(\mathbb{R}^n,\mathcal{B}(\mathbb{R}^n),m),$$

and λ_X, the induced probability measure on $\mathcal{B}(\mathbb{R}^n)$ in (1.21), the Borel density function $f(x)$ of X is defined as the (Borel measurable) density function of λ_X with respect to m, when this density exists.

Such $f(x)$ is nonnegative and Borel measurable by definition, so $f^{-1}(\mathcal{B}(\mathbb{R}))\subset\mathcal{B}(\mathbb{R}^n)$. Since $\mathcal{B}(\mathbb{R}^n)\subset\mathcal{M}_L(\mathbb{R}^n)$, the Lebesgue measure space, it follows that $f^{-1}(\mathcal{B}(\mathbb{R}^n))\subset\mathcal{M}_L(\mathbb{R}^n)$, and thus density functions of random vectors, when they exist, are Lebesgue measurable. Hence by (1.2), $\lambda_X(A)$ is defined on $A\in\mathcal{B}(\mathbb{R}^n)$ as the Lebesgue integral:

$$\lambda_X(A)=\int_A f(x)dm. \tag{1.23}$$

If $A=A_x\equiv\prod_{j=1}^n(-\infty,x_j]$ for $x=(x_1,...,x_n)$, then by (1.8) and (1.22):

$$F(x_1,x_2,...,x_n)=\lambda_X(A_x)=\int_{A_x}f(x)dm.$$

Since $\mathcal{M}_L(\mathbb{R}^n)$ is complete and σ-finite, $\lambda_X(\mathbb{R}^n)=1$ assures that nonnegative $f(x)$ is Lebesgue integrable. It then follows from Fubini's theorem of Proposition V.5.16 that this integral can be evaluated as an iterated Lebesgue integral, and for all $(x_1,x_2,...,x_n)\in\mathbb{R}^n$:

$$F(x_1,x_2,...,x_n)=\int_{-\infty}^{x_n}\cdots\int_{-\infty}^{x_1}f(y_1,y_2,...,y_n)dy_1...dy_n. \tag{1.24}$$

This formulation is likely familiar to the reader, as density functions of random variables and random vectors are usually **defined** in terms of an identity as in (1.24).

The following result states that (1.23) and (1.24) are equivalent formulations. In detail, if there exists a nonnegative Lebesgue integrable function $f(x)$ that satisfies either of these identities, then the same function satisfies the other.

Proposition 1.34 ((1.23)⇔(1.24)) *Let X be a random vector defined on a probability space $(\mathcal{S},\mathcal{E},\mu)$ with joint distribution function $F(x)$ defined on $\mathbb{R}^n$, and Borel probability measure λ_X defined on $\mathcal{B}(\mathbb{R}^n)$.*

Then there exists a nonnegative, Borel measurable, Lebesgue integrable function $f_1(x)$ that satisfies (1.23) for all $A\in\mathcal{B}(\mathbb{R}^n)$, if and only if there exists a nonnegative, Borel measurable, Lebesgue integrable function $f_2(x)$ that satisfies (1.24) for all $x\in\mathbb{R}^n$. Further, in each case one can choose $f_1(x)=f_2(x)$.

Proof. *If such $f_1(x)$ exists and satisfies (1.23) for all $A\in\mathcal{B}(\mathbb{R}^n)$, then as noted above, (1.24) is satisfied for all $x\in\mathbb{R}^n$ with $f_2(x)=f_1(x)$.*

Conversely, assume such $f_2(x)$ exists so that (1.24) is satisfied for all $x\in\mathbb{R}^n$. We prove that (1.23) is satisfied for all $A\in\mathcal{B}(\mathbb{R}^n)$ with $f_1(x)=f_2(x)$.

To this end, we first prove that for every bounded right semi-closed rectangle $A \equiv \prod_{j=1}^{n}(a_j, b_j]$:

$$\int_A f_2(y)dy = \lambda_X[A]. \tag{1.25}$$

Since $f_2(x)$ *is integrable, and* $\mathcal{M}_L(\mathbb{R}^n)$ *is complete and* σ*-finite, Fubini's theorem of Proposition V.5.16 obtains that the integral in (1.25) can be expressed as an iterated integral. Thus by Lebesgue integral properties of Proposition III.2.49:*

$$\begin{aligned}
&\int_{a_n}^{b_n} \cdots \int_{a_1}^{b_1} f_2(y_1, y_2, ..., y_n)dy_1...dy_n \\
&= \int_{a_n}^{b_n} \cdots \int_{a_2}^{b_2} \left[\int_{-\infty}^{b_1} f_2(y_1, y_2, ..., y_n)dy_1 - \int_{-\infty}^{a_1} f_2(y_1, y_2, ..., y_n)dy_1\right] dy_2...dy_n \\
&= \int_{a_n}^{b_n} \cdots \int_{a_2}^{b_2} \int_{-\infty}^{b_1} f_2(y_1, y_2, ..., y_n)dy_1dy_2...dy_n \\
&\quad - \int_{a_n}^{b_n} \cdots \int_{a_2}^{b_2} \int_{-\infty}^{a_1} f_2(y_1, y_2, ..., y_n)dy_1dy_2...dy_n.
\end{aligned}$$

Each of these integrals can be similarly expressed, for example:

$$\begin{aligned}
&\int_{a_n}^{b_n} \cdots \int_{a_2}^{b_2} \int_{-\infty}^{b_1} f_2(y_1, y_2, ..., y_n)dy_1dy_2...dy_n \\
&= \int_{a_n}^{b_n} \cdots \int_{a_3}^{b_3} \int_{-\infty}^{b_2} \int_{-\infty}^{b_1} f_2(y_1, y_2, ..., y_n)dy_1dy_2...dy_n \\
&\quad - \int_{a_n}^{b_n} \cdots \int_{a_3}^{b_3} \int_{-\infty}^{a_2} \int_{-\infty}^{b_1} f_2(y_1, y_2, ..., y_n)dy_1dy_2...dy_n.
\end{aligned}$$

It then follows by induction that:

$$\begin{aligned}
&\int_{a_n}^{b_n} \cdots \int_{a_1}^{b_1} f_2(y_1, y_2, ..., y_n)dy_1...dy_n \\
&= \sum\nolimits_x sgn(x) \int_{-\infty}^{x_n} \cdots \int_{-\infty}^{x_1} f_2(y_1, y_2, ..., y_n)dy_1...dy_n \\
&= \sum\nolimits_x sgn(x)F(x_1, x_2, ..., x_n).
\end{aligned}$$

Here each $x = (x_1, ..., x_n)$ *is one of the* 2^n *vertices of* $\prod_{i=1}^{n}(a_i, b_i]$, *so* $x_i = a_i$ *or* $x_i = b_i$, *and* $sgn(x)$ *is defined as* -1 *if the number of components of* x *that equal* a_i *is odd, and* $+1$ *otherwise. Thus the integral in (1.25) equals* $\lambda_X[A]$ *by (1.17) and (1.22).*

To prove (1.23), define the set function λ' *on* $A \in \mathcal{B}(\mathbb{R}^n)$ *by:*

$$\lambda'[A] = \int_A f_2 dm. \tag{1.26}$$

Since f_2 *is Lebesgue integrable,* λ' *defines a measure on* $\mathcal{B}(\mathbb{R}^n)$ *by Proposition V.4.3. Further,* $\lambda' = \lambda_X$ *on the semi-algebra* $\mathcal{A}'$ *of right semi-closed rectangles by (1.25).*

The identity $\lambda' = \lambda_X$ *then extends additively to the algebra* $\mathcal{A}$ *of finite disjoint unions of* $\mathcal{A}'$*-sets, and thus by the uniqueness theorem of Proposition I.6.14,* $\lambda' = \lambda_X$ *on* $\sigma(\mathcal{A})$, *the smallest sigma algebra generated by* $\mathcal{A}$. *Since* $\sigma(\mathcal{A}) = \mathcal{B}(\mathbb{R}^n)$ *by Proposition I.8.1, (1.23) is satisfied for all* $A \in \mathcal{B}(\mathbb{R}^n)$ *with* $f_1(x) = f_2(x)$. ■

While Borel measurability of density functions for λ_X is adequate for the equivalence of (1.23) and (1.24), this measurability criterion will prove to be overly restrictive for results to come. Allowing λ_X to have Lebesgue measurable density functions would provide more generality below.

To be true to Definition 1.33, we next prove that if $\lambda_X \ll m$ by Definition 1.11, then the Borel measure λ_X on $\mathcal{B}(\mathbb{R}^n)$ can be extended to a measure $\tilde{\lambda}_X$ on $\mathcal{M}_L(\mathbb{R}^n)$ with $\tilde{\lambda}_X \ll m$. It will then follow below that any Lebesgue measurable density $\tilde{f}(x)$ for $\tilde{\lambda}_X$ serves equally well as a density for λ_X.

Proposition 1.35 (Extension of λ_X to $\tilde{\lambda}_X$ on $\mathcal{M}_L(\mathbb{R}^n)$) *Given a random vector X defined on a probability space $(\mathcal{S}, \mathcal{E}, \mu)$, let λ_X be the Borel probability measure defined on $\mathcal{B}(\mathbb{R}^n)$ by (1.21).*

If $\lambda_X \ll m$, then λ_X can be extended to a Lebesgue probability measure $\tilde{\lambda}_X$ on $\mathcal{M}_L(\mathbb{R}^n)$ with $\tilde{\lambda}_X \ll m$.

Proof. *Given $E \in \mathcal{M}_L(\mathbb{R}^n)$, Proposition I.2.42 obtains a Borel set $A \in \mathcal{B}(\mathbb{R}^n)$ with $A \subset E$ and $m(E - A) = 0$, and we define:*

$$\tilde{\lambda}_X(E) = \lambda_X(A). \tag{1.27}$$

To check that (1) is well-defined, assume that for $j = 1, 2$ that there exists $A_j \in \mathcal{B}(\mathbb{R}^n)$ with $A_j \subset E$ and $m(E - A_j) = 0$. Then $A_1 \bigcup (E - A_1) = A_2 \bigcup (E - A_2)$ and $A_1 = A_2 \bigcup (A_1 - A_2)$ obtain:

$$A_2 \bigcup (A_1 - A_2) \bigcup (E - A_1) = A_2 \bigcup (E - A_2).$$

Hence $A_1 - A_2 \subset E - A_2$ and $m(A_1 - A_2) = 0$. Similarly $m(A_2 - A_1) = 0$.

As $A_1 - A_2, A_2 - A_1 \in \mathcal{B}(\mathbb{R}^n)$, absolute continuity obtains that $\lambda_X(A_1 - A_2) = \lambda_X(A_2 - A_1) = 0$, and thus $\lambda_X(A_1) = \lambda_X(A_2)$. Hence $\tilde{\lambda}_X$ is well-defined, and then it follows that $\tilde{\lambda}_X = \lambda_X$ on $\mathcal{B}(\mathbb{R}^n)$.

We leave as an exercise to check that $\tilde{\lambda}_X$ satisfies countable additivity of Definition V.1.7 and is therefore a measure.

To see that $\tilde{\lambda}_X \ll m$, let $E \in \mathcal{M}_L(\mathbb{R}^n)$ with $m(E) = 0$. Then again there exists Borel $A \subset E$, and with $m(A) = 0$. Then $\lambda_X \ll m$ obtains $\lambda_X(A) = 0$, so $\tilde{\lambda}_X(E) = 0$ by (1.27) and $\tilde{\lambda}_X \ll m$. ■

Exercise 1.36 ($\lambda_X \ll m$ is needed for existence of $\tilde{\lambda}_X$) *The above result derived an extension of λ_X as defined on $\mathcal{B}(\mathbb{R}^n)$ to $\tilde{\lambda}_X$ defined on $\mathcal{M}_L(\mathbb{R}^n)$ under the assumption that $\lambda_X \ll m$. Prove that given λ_X for which $\lambda_X \ll m$ is false, no extension of $\tilde{\lambda}_X$ can exist. Hint: What would happen, for example, if $\lambda_X(x_0) > 0$ for some $x_0 \in \mathbb{R}^n$? Now generalize using Definition 1.11.*

We are now ready for an existence result on density functions. With the aid of studies from Books III and V, this result is enhanced in the case of $n = 1$ below.

Proposition 1.37 (Existence of Density Functions) *Let X be a random vector defined on a probability space $(\mathcal{S}, \mathcal{E}, \mu)$ with joint distribution function $F(x)$ defined on $\mathbb{R}^n$, and Borel probability measure λ_X defined on $\mathcal{B}(\mathbb{R}^n)$.*

1. *Then λ_X has a Borel density function $f(x)$ if and only if $\lambda_X \ll m$.*

2. *If $\lambda_X \ll m$, then $\tilde{\lambda}_X$ has a Lebesgue density function $\tilde{f}(x)$.*

3. $f(x) = \tilde{f}(x)$, *m-a.e.*

4. The density $\tilde{f}(x)$ *can also be used in (1.23) to define* λ_X*, and in (1.24) to define* $F(x)$*.*

Proof. *As a probability measure on* $(\mathbb{R}^n, \mathcal{B}(\mathbb{R}^n), m)$, λ_X *has a Borel density function* $f(x)$ *if and only if* $\lambda_X \ll m$ *by Proposition 1.13. If* $\lambda_X \ll m$ *then* $\tilde{\lambda}_X \ll m$ *by Proposition 1.35, and thus as a probability measure on* $(\mathbb{R}^n, \mathcal{M}_L(\mathbb{R}^n), m)$, $\tilde{\lambda}_X$ *has a Lebesgue density function* $\tilde{f}(x)$ *by the same result.*

To see that $f(x) = \tilde{f}(x)$, *m-a.e., let* $E \in \mathcal{M}_L(\mathbb{R}^n)$*. By Proposition I.2.42, there is a Borel set* $A \in \mathcal{B}(\mathbb{R}^n)$ *with* $A \subset E$ *and* $m(E - A) = 0$*. By Lebesgue measurability of* $f(x)$, $\int_E f(x)dm$ *is well-defined, and by item 6 of Proposition III.2.49:*

$$\int_E f(x)dm = \int_A f(x)dm + \int_{E-A} f(x)dm.$$

Now the first integral equals $\lambda_X(A)$ *by definition, while the second is 0 by item 9 of Proposition V.3.42, and so:*

$$\int_E f(x)dm = \lambda_X(A).$$

On the other hand, by definition of $\tilde{\lambda}_X$*:*

$$\int_E \tilde{f}(x)dm = \tilde{\lambda}_X(E) \equiv \lambda_X(A).$$

Thus for all $E \in \mathcal{M}_L(\mathbb{R}^n)$*:*

$$\int_E \left[\tilde{f}(x) - f(x)\right] dm = 0,$$

and $f(x) = \tilde{f}(x)$, *m-a.e. follows from item 8 of Proposition III.2.49.*

For item 4, if $E \in \mathcal{M}_L(\mathbb{R}^n)$ *then:*

$$\tilde{\lambda}_X(E) = \int_E \tilde{f}(x)dm. \tag{1.28}$$

Thus if $A \in \mathcal{B}(\mathbb{R}^n) \subset \mathcal{M}_L(\mathbb{R}^n)$*, then:*

$$\int_A \tilde{f}(x)dm = \tilde{\lambda}_X(A) \equiv \lambda_X(A), \tag{1}$$

and so $\tilde{f}(x)$ *can be used in (1.23) to define* λ_X*.*

The statement on (1.24) and $F(x)$ *follows by letting* $A = A_x \equiv \prod_{j=1}^n (-\infty, x_j]$ *in (1) and applying Fubini's theorem of Proposition V.5.16.* ■

Remark 1.38 (Measurability of densities) *By the Radon-Nikodým theorem that drives Proposition 1.13, if* λ_X *has a density function* $f(x)$*, then it has infinitely many defined by:*

$$D_{\lambda_X} = \{g(x) | g \text{ is Borel measurable and } g = f, \ m\text{-a.e.}\}$$

In this case, $\tilde{\lambda}_X$ *also has a density function* $\tilde{f}(x)$*, and indeed infinitely many defined by:*

$$D_{\tilde{\lambda}_X} = \{g(x) | g \text{ is Lebesgue measurable and } g = \tilde{f}, \ m\text{-a.e.}\}$$

Further, by Lebesgue measurability of $f(x)$ *and item 3 of Proposition 1.37,* $f \in D_{\tilde{\lambda}_X}$ *and thus* $D_{\lambda_X} \subset D_{\tilde{\lambda}_X}$*. Specifically,* D_{λ_X} *is the collection of functions in* $D_{\tilde{\lambda}_X}$ *that are Borel measurable.*

But there is an important distinction in identifying functions in either collection:

- *If $g(x) = \tilde{f}(x)$ m-a.e., then $g(x)$ is **always** an element of $D_{\tilde{\lambda}_X}$ and thus always a Lebesgue density for $\tilde{\lambda}_X$:*

  The function $g(x)$ is Lebesgue measurable by completeness of $\mathcal{M}_L(\mathbb{R}^n)$ and Proposition V.2.14, and obtains the same values in (1.28) for all $E \in \mathcal{M}_L(\mathbb{R}^n)$ by item 3 of Proposition III.2.49.

- *If $g(x) = f(x)$ m-a.e., then $g(x)$ is an element of D_{λ_X} and thus a Borel density for λ_X **only if Borel measurable:***

  While such $g(x)$ is Lebesgue measurable, it need not be Borel measurable, and so this must be independently established. However, given measurability, the same values are again obtained in (1.28) for all $E \in \mathcal{B}(\mathbb{R}^n)$ as above.

Thus there is a small but important advantage in working with Lebesgue measurable density functions, even for the Borel measure λ_X. Lebesgue measurable densities are always indifferent to redefinitions on sets of Lebesgue measure 0, *while Borel measurable densities are not.*

In addition to this advantage, nothing is lost by using Lebesgue measurable density functions, since by Proposition 1.37, they integrate equally well as densities for the Borel measure λ_X.

Notation 1.39 (Measurability of density functions of X, or of $F(x)$) *Given a random vector X on $(\mathcal{S}, \mathcal{E}, \mu)$, if the induced probability measure λ_X on $\mathcal{B}(\mathbb{R}^n)$ and defined by (1.21) has a density function by Radon-Nikodým theorem, then this function must be Borel measurable, by definition. However, given Proposition 1.37 and Remark 1.38, there are advantages to expanding the class of such densities to include Lebesgue measurable functions, and indeed, there is no apparent disadvantage.*

*Thus when λ_X has a density function, we will always assume that the class of density functions of λ_X **includes Lebesgue measurable functions.** The theoretical justification related to Radon-Nikodým is provided by Proposition 1.37.*

To derive additional insights on existence of density functions in the special case where $n = 1$, we first recall Definition III.3.54 on the notion of an absolutely continuous function. By Definition 1.40 with $m = 1$, such functions are continuous, and thus Borel and Lebesgue measurable by Proposition I.3.11. But there are continuous functions that are not absolutely continuous, recalling Example III.3.57 of the Cantor function.

Definition 1.40 (Absolute continuity) *A real-valued function $F(x)$ is **absolutely continuous** on $[a, b]$ if for any $\epsilon > 0$, there is a δ so that:*

$$\sum\nolimits_{i=1}^{m} |F(x_i) - F(x_i')| < \epsilon, \tag{1}$$

for any finite collection of disjoint subintervals $\{(x_i', x_i)\}_{i=1}^m \subset [a, b]$ with:

$$\sum\nolimits_{i=1}^{m} |x_i - x_i'| < \delta. \tag{2}$$

*More generally, $F(x)$ is **absolutely continuous** if* (1) *holds for any finite collection of disjoint subintervals $\{(x_i', x_i)\}_{i=1}^m \subset \mathbb{R}$ that satisfies* (2).

The next result characterizes the existence of density functions in terms of the absolute continuity of the distribution function $F(x)$.

Proposition 1.41 (Existence of Density Functions, $n = 1$) *Let X be a random variable defined on a probability space $(\mathcal{S}, \mathcal{E}, \mu)$ with distribution function $F(x)$ defined on $\mathbb{R}$, and probability measure λ_X defined on $\mathcal{B}(\mathbb{R})$.*

Then X has a density function $f(x)$ if and only if $F(x)$ is absolutely continuous, and then:

$$f(x) = F'(x), \ m\text{-a.e.} \tag{1.29}$$

Proof. *By Propositions 1.30 and V.8.18, $\lambda_X = \lambda_F \ll m$ on $(\mathbb{R}, \mathcal{B}(\mathbb{R}), m)$ if and only if F is absolutely continuous, so existence of $f(x)$ follows from Proposition 1.13.*

If $f(x)$ is such a density function for X, then by (1.24):

$$F(x) = \int_{-\infty}^{x} f(y)dm.$$

For arbitrary fixed $a < x$, it follows from item 6 of Proposition III.2.49 that:

$$F(x) = F(a) + \int_{a}^{x} f(y)dm. \tag{1}$$

Since $f(x)$ is Lebesgue measurable, Proposition III.3.39 obtains that $F(x)$ is differentiable m-a.e., and $f(x) = F'(x)$, m-a.e.

Conversely, assume that $F(x)$ is absolutely continuous. Then $F(x)$ is differentiable m-a.e. and $F'(x)$ is Lebesgue measurable and integrable on every bounded interval by Proposition III.3.59.

For arbitrary a and $x > a$, define:

$$G(x) = \int_{a}^{x} F'(y)dm,$$

and so by Proposition III.3.62:

$$G(x) = F(x) - F(a).$$

Letting $a \to -\infty$, and recalling that $F(-\infty) = 0$ in the limit, obtains $G(x) = F(x)$, and thus:

$$F(x) = \int_{-\infty}^{x} F'(y)dm. \tag{2}$$

Hence $f(x) = F'(x)$ is a (Lebesgue measurable) density for X.

If $f(x)$ is any other density function for X, then for all x:

$$\int_{-\infty}^{x} [F'(y) - f(y)]\, dm = 0.$$

Proposition III.3.36 now applies to obtain $f(x) = F'(x)$, m-a.e. ■

Remark 1.42 (On measurability of $f(x)$) *The statement and proof of Proposition 1.41 were simplified by the Notation 1.39 convention that density functions of λ_X include Lebesgue measurable functions.*

If we insisted that such densities be Borel measurable, the existence of Borel measurable $f(x)$ would again obtain that $f(x) = F'(x)$, m-a.e., though in general, $F'(x)$ need only be Lebesgue measurable.

But we would fall short of deriving that all absolutely continuous distribution functions $F(x)$ have Borel measurable density functions. The problem here is that if $F(x)$ is absolutely *continuous, we can only assert the existence of $F'(x)$ m-a.e., and then the theory only assures the Lebesgue measurability of such $F'(x)$, not Borel measurability.*

If we assumed more, that $F(x)$ was differentiable and thus that $F'(x)$ existed everywhere, then $f(x) = F'(x)$ is Borel measurable. Differentiability of $F(x)$ assures continuity and hence Borel measurability. Thus:

$$g_n(x) \equiv n\left[F(x+1/n) - F(x)\right],$$

is Borel measurable for all n, and the Borel measurability of $F'(x) = \lim g_n(x)$ follows from item 7 of Proposition V.2.19. That $f(x) = F'(x)$ is a density then follows as above.

While compelled to provide an example of the utility of Notation 1.39, we will not return to this discussion again.

When the density function of Proposition 1.41 is continuous, the result in (1.29) can be sharpened.

Proposition 1.43 (Continuous density functions: $n = 1$) *Let X be a random variable defined on a probability space $(\mathcal{S}, \mathcal{E}, \mu)$ with distribution function $F(x)$ defined on $\mathbb{R}$, and probability measure λ_X defined on $\mathcal{B}(\mathbb{R})$.*

Then X has a continuous density function $f(x)$ if and only if $F(x)$ is continuously differentiable, and then:

$$f(x) = F'(x). \tag{1.30}$$

Proof. *If the density $f(x)$ is continuous, it is Riemann integrable on every interval $[a, b]$ by Proposition III.1.15, and this integral equals the associated Lebesgue integral by Proposition III.2.18. Letting $a \to -\infty$ and $b \to \infty$ obtains that $f(x)$ is Riemann integrable on $\mathbb{R}$.*

Thus $F(x)$ in (1.24) can be expressed as a Riemann integral:

$$F(x) = \int_{-\infty}^{x} f(y)dy = F(a) + \int_{a}^{x} f(y)dy.$$

Then $F'(x) = f(x)$ follows from Proposition III.1.33, and $F(x)$ is continuously differentiable by definition.

Conversely, if $F(x)$ is continuously differentiable then $F'(x)$ is continuous and thus Riemann integrable over every bounded interval by Proposition III.1.15. By Proposition III.1.30:

$$\int_{a}^{x} F'(y)dy = F(x) - F(a),$$

and letting $a \to -\infty$ obtains:

$$F(x) = \int_{-\infty}^{x} F'(y)dy.$$

Thus $F'(y)$ is a continuous density for X. ■

It is natural to wonder if the result in (1.30) of Proposition 1.43 generalizes to joint distribution functions. When $n > 1$, the next result generalizes the relationship between $F(x)$ and continuous $f(x)$ subject to some additional integrability assumptions in (1.31).

Proposition 1.44 (Continuous density functions: $n > 1$) *Let X be a random vector defined on a probability space $(\mathcal{S}, \mathcal{E}, \mu)$ with distribution function $F(x)$, and a continuous density function $f(x)$. For $1 \leq j < n$, denote $(x_1, x_2, ..., x_n) \equiv (x^{(j)}, x^{(n-j)})$ where $x^{(j)} = (x_1, x_2, ..., x_j)$ and $x^{(n-j)} = (x_{j+1}, ..., x_n)$.*

Assume that for some ordering and relabelling of variates, that there exists Lebesgue integrable functions $g_j(x^{(n-j)})$ on $\mathbb{R}^{n-j}$ for $1 \leq j < n$ so that:

$$f(x^{(j)}, x^{(n-j)}) \leq g_j(x^{(n-j)}), \text{ for all } x^{(j)}. \tag{1.31}$$

Then:

$$f(x_1, x_2, ..., x_n) = \frac{\partial^n F}{\partial x_1 ... \partial x_n}. \tag{1.32}$$

Proof. *First, by continuity of $f(x)$, we can as in the proof of Proposition 1.43 interpret the integrals below as Riemann or Lebesgue. In detail, continuous $f(x)$ is Riemann integrable on every rectangle $\prod_{i=1}^n [a_i, b_i]$ by Proposition III.1.63, and this integral equals the associated Lebesgue integral by Proposition III.2.18. Letting all $a_i \to -\infty$ and $b_i \to \infty$ obtains that $f(x)$ is Riemann integrable on $\mathbb{R}^n$, and by Proposition III.2.56, $F(x)$ in (1.24) can be expressed as a Riemann integral. This applies also to other integrals below.*

Letting $(x_1, x_2, ..., x_n)$ reflect the given relabelling of variates, it follows as in (1.24):

$$F(x_1, x_2, ..., x_n) = \int_{-\infty}^{x_n} \cdots \int_{-\infty}^{x_2} \int_{-\infty}^{x_1} f(y_1, y_2, ..., y_n) dy_1 dy_2 ... dy_n.$$

We first prove that:

$$\frac{\partial F}{\partial x_1}(x_1, x_2, ..., x_n) = \int_{-\infty}^{x_n} \cdots \int_{-\infty}^{x_2} f(x_1, y_2, ..., y_n) dy_2 ... dy_n. \tag{1}$$

By linearity of the integrals, and using the above notation:

$$\frac{F(x_1 + \Delta x_1, x^{(n-1)}) - F(x_1, x^{(n-1)})}{\Delta x_1} = \int_{-\infty}^{x^{(n-1)}} \left[\frac{1}{\Delta x_1} \int_{x_1}^{x_1 + \Delta x_1} f(y_1, y^{(n-1)}) dy_1 \right] dy^{(n-1)}.$$

Then by linearity and the triangle inequality of item 7 of Proposition III.2.49:

$$\left| \frac{F(x_1 + \Delta x_1, x^{(n-1)}) - F(x_1, x^{(n-1)})}{\Delta x_1} - \int_{-\infty}^{x^{(n-1)}} f(x_1, y^{(n-1)}) dy^{(n-1)} \right|$$
$$\leq \int_{-\infty}^{x^{(n-1)}} \left| \frac{1}{\Delta x_1} \int_{x_1}^{x_1 + \Delta x_1} f(y_1, y^{(n-1)}) dy_1 - f(x_1, y^{(n-1)}) \right| dy^{(n-1)}. \tag{2}$$

By the mean value theorem for integrals of Corollary III.1.24, the integrand of the $dy^{(n-1)}$-integral in (2) *converges pointwise to* 0 *as $\Delta x_1 \to 0$ for every $y^{(n-1)}$. Also, this integrand is dominated by an integrable function of $y^{(n-1)}$, since $f(y_1, y^{(n-1)}) \leq g_1(y^{(n-1)})$ for all y_1 by (1.31) implies that:*

$$\begin{aligned} &\left| \frac{1}{\Delta x_1} \int_{x_1}^{x_1 + \Delta x_1} f(y_1, y^{(n-1)}) dy_1 - f(x_1, y^{(n-1)}) \right| \\ \leq\; &\frac{1}{\Delta x_1} \int_{x_1}^{x_1 + \Delta x_1} \left| f(y_1, y^{(n-1)}) - f(x_1, y^{(n-1)}) \right| dy_1 \\ \leq\; &2 g_1(y^{(n-1)}). \end{aligned}$$

An application of Lebesgue's dominated convergence theorem of Proposition III.2.52 to (2) *then obtains:*

$$\left| \frac{F(x_1 + \Delta x_1, x^{(n-1)}) - F(x_1, x^{(n-1)})}{\Delta x_1} - \int_{-\infty}^{x^{(n-1)}} f(x_1, y^{(n-1)}) dy^{(n-1)} \right| \to 0,$$

proving (1).

Denoting this partial derivative by F_1*, the above steps can be repeated to produce with* $x^{(n-2)} = (x_3, ..., x_n)$*:*

$$\left| \frac{F_1(x_1, x_2 + \Delta x_2, x^{(n-2)}) - F_1(x_1, x_2, x^{(n-2)})}{\Delta x_2} - \int_{-\infty}^{x^{(n-2)}} f(x_1, x_2, y^{(n-2)}) dy^{(n-2)} \right|$$

$$\leq \int_{-\infty}^{x^{(n-2)}} \left| \frac{1}{\Delta x_2} \int_{x_2}^{x_2 + \Delta x_2} f(x_1, y_2, y^{(n-2)} y) dy_2 - f(x_1, x_2, y^{(n-2)}) \right| dy^{(n-2)}.$$

Pointwise convergence of the integrand to zero again follows from continuity, and the integrand is again dominated by an integrable function:

$$\left| \frac{1}{\Delta x_2} \int_{x_2}^{x_2 + \Delta x_2} f(x_1, y_2, y^{(n-2)}) dy_2 - f(x_1, x_2, y^{(n-2)}) \right| \leq 2g_2(y^{(n-2)}).$$

Hence by Lebesgue's dominated convergence theorem again, the second partial satisfies:

$$\frac{\partial^2 F}{\partial x_1 \partial x_2}(x_1, x_2, ..., x_n) = \int_{-\infty}^{x^{(n-2)}} f(x_1, x_2, y^{(n-2)}) dy^{(n-2)},$$

and the result follows by induction. ■

1.3 Marginal Density Functions

In this section we investigate results on the density functions of marginal distribution functions. Recall that these distribution functions were introduced in Definition II.3.34.

Definition 1.45 (Marginal distribution functions) *Let* $X = (X_1, X_2, ..., X_n)$ *be a random vector defined on a probability space* $(\mathcal{S}, \mathcal{E}, \mu)$*,* $X : \mathcal{S} \longrightarrow \mathbb{R}^n$*, with joint distribution function* $F(x_1, x_2, ..., x_n)$*.*

1. ***Special Case*** $n = 2$ *: Given* $F(x_1, x_2)$*, define two* ***marginal distribution functions*** $F_1(x_1)$ *and* $F_2(x_2)$ *on* $\mathbb{R}$ *by:*

$$F_1(x_1) \equiv \lim_{x_2 \to \infty} F(x_1, x_2), \quad F_2(x_2) \equiv \lim_{x_1 \to \infty} F(x_1, x_2). \tag{1.33}$$

2. ***General Case*** $n > 2$*: Given* $F(x_1, x_2, ..., x_n)$ *and* $I = \{i_1, ..., i_m\} \subset \{1, 2, ..., n\}$*, let* $x_J \equiv (x_{j_1}, x_{j_2}, ..., x_{j_{n-m}})$ *for* $j_k \in J \equiv \widetilde{I}$*. The* ***marginal distribution function*** $F_I(x_I) \equiv F_I(x_{i_1}, x_{i_2}, ..., x_{i_m})$ *is defined on* $\mathbb{R}^m$ *by:*

$$F_I(x_I) \equiv \lim_{x_J \to \infty} F(x_1, x_2, ..., x_n). \tag{1.34}$$

Remark 1.46 (Marginal combinatorics) *Given* $F(x_1, x_2, ..., x_n)$*, there are* $2^n - 2$ ***proper*** *marginal distribution functions defined by the* $2^n - 2$ *proper subsets of* $\{1, 2, ..., n\}$*. Of these, the most important in many applications are the n marginal distributions* $\{F_i(x_i)\}_{i=1}^n$*, respectively defined with* $I = \{i\}$ *and the notational simplification,* $F_{\{i\}}(x_i) \equiv F_i(x_i)$*. See the Chapter II.7 on Copulas and Sklar's Theorem.*

Proposition II.3.36 proved that the marginal "distribution function" $F_I(x_I)$ defined in (1.34) is indeed a distribution function. Denoting $X_I \equiv (X_{i_1}, X_{i_2}, ..., X_{i_m})$:

$$F_I(x_I) = \mu\left[X_I^{-1}\left(\prod\nolimits_{k=1}^{m}(-\infty, x_{i_k}]\right)\right], \tag{1.35}$$

and consequently, $F_I(x_I)$ is the joint distribution function of X_I:

$$F_I(x_{i_1}, x_{i_2}, ..., x_{i_m}) = F_{X_I}(x_{i_1}, x_{i_2}, ..., x_{i_m}). \tag{1.36}$$

Thus a marginal distribution function of X_I retains no "memory" of the random variables X_{j_k} for $j_k \in \widetilde{I}$ which were originally reflected in the definition of $F(x_1, x_2, ..., x_n)$.

Another way to think about X_I is to introduce a **projection mapping** $\pi_I : \mathbb{R}^n \to \mathbb{R}^m$ indexed by $I = \{i_1, ..., i_m\}$ and defined by:

$$\pi_I(x_1, x_2, ..., x_n) = (x_{i_1}, x_{i_2}, ..., x_{i_m}). \tag{1.37}$$

It is an exercise to verify that projection mappings are continuous by Definition V.2.9.

The random vector $X_I : \mathcal{S} \to \mathbb{R}^m$ is then given by:

$$X_I = \pi_I(X). \tag{1.38}$$

Diagrammatically:

$$(\mathcal{S}, \mathcal{E}, \mu) \quad \xrightarrow{X} \quad (\mathbb{R}^n, \mathcal{B}(\mathbb{R}^n), m) \quad \xrightarrow{\pi_I} \quad (\mathbb{R}^m, \mathcal{B}(\mathbb{R}^m), m).$$

$$(\mathcal{S}, \mathcal{E}, \mu) \quad \xrightarrow{X_I} \quad (\mathbb{R}^m, \mathcal{B}(\mathbb{R}^m), m).$$

By Proposition 1.23, the distribution function $F_I(x_I)$ is continuous from above and satisfies the n-increasing condition of Definition 1.21 for any index set $I \subset \{1, 2, ..., n\}$. Hence, each such $F_I(x_I)$ induces a probability measure $\lambda_{F_I} = \lambda_{X_I}$ on $\mathcal{B}(\mathbb{R}^m)$ by Proposition 1.27.

Proposition 1.47 (Marginal probability measure λ_{X_I}) *Let $X = (X_1, X_2, ..., X_n)$ be a random vector defined on $(\mathcal{S}, \mathcal{E}, \mu)$, $X : \mathcal{S} \longrightarrow \mathbb{R}^n$, with joint distribution function $F(x_1, x_2, ..., x_n)$ and induced probability measure λ_X, and for $I = \{i_1, ..., i_m\} \subset \{1, 2, ..., n\}$, let X_I given by (1.38).*

Then the probability measure λ_{X_I} induced by X_I is defined on $A \in \mathcal{B}(\mathbb{R}^m)$ by:

$$\lambda_{X_I}(A) = \lambda_X\left[\pi_I^{-1}(A)\right]. \tag{1.39}$$

Proof. *Recall that for $A \in \mathcal{B}(\mathbb{R}^m)$:*

$$\pi_I^{-1}(A) \equiv \{x \in \mathbb{R}^n | \pi_I(x) \in A\}.$$

Since projection mappings are continuous, π_I^{-1} maps open sets in $\mathbb{R}^m$ to open sets in $\mathbb{R}^n$. This requires a modest generalization of the proof of Proposition V.2.10, which is left as an exercise.

Now the open sets generate the Borel sigma algebras by Definition 1.2, so by Exercise 1.48, $\pi_I^{-1}(A) \in \mathcal{B}(\mathbb{R}^n)$ for $A \in \mathcal{B}(\mathbb{R}^m)$. That is:

$$\pi_I^{-1}\left[\mathcal{B}(\mathbb{R}^m)\right] \subset \mathcal{B}(\mathbb{R}^n), \tag{1}$$

and hence $\lambda_X\left[\pi_I^{-1}(A)\right]$ is well-defined for all $A \in \mathcal{B}(\mathbb{R}^m)$.

Finally from (1.21) and (1.38), if $A \in \mathcal{B}(\mathbb{R}^m)$:

$$\lambda_{X_I}(A) \equiv \mu\left[X_I^{-1}(A)\right] = \mu\left[X^{-1}\left[\pi_I^{-1}(A)\right]\right] = \lambda_X\left[\pi_I^{-1}(A)\right].$$

■

Exercise 1.48 ($\pi_I^{-1}[\mathcal{B}(\mathbb{R}^m)] \subset \mathcal{B}(\mathbb{R}^n)$) *Complete the proof that* $\pi_I^{-1}[\mathcal{B}(\mathbb{R}^m)] \subset \mathcal{B}(\mathbb{R}^n)$. *Hint: Verify that* π_I *is continuous and generalize the proof of Proposition V.2.10 to obtain that* π_I^{-1} *maps open sets in* $\mathbb{R}^m$ *to open sets in* $\mathbb{R}^n$. *Then check that* $\pi_I^{-1}[\bigcup A_i] = \bigcup \pi_I^{-1} A_i$, $\pi_I^{-1}[\bigcap A_i] = \bigcap \pi_I^{-1} A_i$ *and* $\pi_I^{-1}\left[\widetilde{A}\right] = \widetilde{\pi_I^{-1}[A]}$. *Finally, recalling the definition of the Borel sigma algebras, show that these results assure that* $\pi_I^{-1}(A) \in \mathcal{B}(\mathbb{R}^n)$ *for* $A \in \mathcal{B}(\mathbb{R}^m)$.

The following existence result states that if the joint distribution function $F(x)$ has an associated density function $f(x)$, then every marginal distribution F_I also has an associated density function f_I definable by (1.40).

Proposition 1.49 (Existence of marginal density functions) *Let* $X = (X_1, X_2, ..., X_n)$ *be a random vector defined on a probability space* $(\mathcal{S}, \mathcal{E}, \mu)$ *with joint distribution function* $F(x)$ *and a density function* $f(x)$.

Then for any $I = \{i_1, ..., i_m\} \subset \{1, 2, ..., n\}$, *the marginal distribution function* $F_I(x_I)$ *has a density function* $f_I(x_I)$ *defined by:*

$$f_I(x_{i_1}, x_{i_2}, ..., x_{i_m}) = \int_{-\infty}^{\infty} \cdots \int_{-\infty}^{\infty} f(x_1, x_2, ..., x_n) dx_{j_1} ... dx_{j_{n-m}}. \tag{1.40}$$

As in Definition 1.45, $\{x_{j_1}, x_{j_2}, ..., x_{j_{n-m}}\}$ *are defined by* $j_k \in J \equiv \widetilde{I}$.

Proof. *To simplify notation, assume that* $I = \{1, 2, ..., m\}$ *and thus* $J = \{m+1, ..., n\}$.

By Fubini's theorem of Proposition V.5.16, the iterated integral defining F *can be reordered, and with* $x_I = (x_1, x_2, ..., x_m)$ *and* $x_J = (x_{m+1}, x_{m+2}, ..., x_n)$, *(1.34) can be expressed as:*

$$\begin{aligned} F_I(x_1, x_2, ..., x_m) &\equiv \lim_{x_J \to \infty} \int_{-\infty}^{x_I} \left[\int_{-\infty}^{x_J} f(y_I, y_J) dy_J \right] dy_I \\ &= \lim_{N \to \infty} \int_{-\infty}^{x_I} \left[\int_{-\infty}^{N} f(y_I, y_J) dy_J \right] dy_I, \end{aligned} \tag{1}$$

where $\int_{-\infty}^{N}$ *denotes* $\int_{-\infty}^{x_J}$ *with* $x_J = (N, N, ..., N)$ *for integer* N.

This substitution in (1) *is justified by noting that since* $f(y_I, y_J) \geq 0$*:*

$$\int_{-\infty}^{x_J^{\min}} f(y_I, y_J) dy_J \leq \int_{-\infty}^{x_J} f(y_I, y_J) dy_J \leq \int_{-\infty}^{x_J^{\max}} f(y_I, y_J) dy_J,$$

where $x_J^{\min}$ *is defined with all components equal to* $\min\{x_{m+1}, x_{m+2}, ..., x_n\}$, *and similarly for* $x_J^{\max}$.

To justify taking the limiting operation inside the dy_I*-integral in* (1), *let:*

$$g_N(y_I) \equiv \int_{-\infty}^{N} f(y_I, y_J) dy_J.$$

By Fubini's theorem, each $g_N(y_I)$ *is well-defined m-a.e and Lebesgue measurable, and* $\{g_N(y_I)\}_{N=1}^{\infty}$ *is an increasing sequence of nonnegative functions since* $f(y_I, y_J) \geq 0$. *Thus Lebesgue's monotone convergence theorem of Proposition V.3.23 obtains:*

$$\lim_{N \to \infty} \int_{-\infty}^{x_I} \left[\int_{-\infty}^{N} g_N(y_I) dy_J \right] dy_I = \int_{-\infty}^{x_I} \left[\int_{-\infty}^{\infty} f(y_I, y_J) dy_J \right] dy_I. \tag{2}$$

Hence from (1)*:*

$$F_I(x_I) = \int_{-\infty}^{x_I} \left[\int_{-\infty}^{\infty} f(y_I, y_J) dy_J \right] dy_I.$$

The integrand of this expression is nonnegative, and is Lebesgue measurable in y_I by Fubini's theorem. Hence:

$$f_I(x_I) \equiv \int_{-\infty}^{\infty} f(x_I, y_J) dy_J,$$

is a density function associated with $F_I(x_I)$, and this proves (1.40) with another application of Fubini's theorem. ■

Of all $2^n - 2$ proper marginal distribution functions, by far the most important are the n marginal distribution functions associated with the index sets $I = \{i\}$ for $i = 1, 2, ..., n$. This is reinforced by the copula theory of Chapter II.7. These special marginal distribution functions $F_{\{i\}}(x_i)$ can be denoted $F_i(x_i)$ as noted in Remark 1.46, and are the distribution functions of X_i. Thus:

$$F_i(x_i) = \int_{-\infty}^{x_i} f_i(y_i) dy_i, \tag{1.41}$$

where by (1.40):

$$f_i(y_i) = \int_{\mathbb{R}^{n-1}} f(y_1, y_2, ..., y_n) dy_1 ... \widehat{dy_i} ... dy_n. \tag{1.42}$$

Here $\widehat{dy_i}$ denotes that this integration variable is omitted.

Proposition 1.50 (Marginal density functions: $n = 1$) *Let $X = (X_1, X_2, ..., X_n)$ be a random vector defined on a probability space $(\mathcal{S}, \mathcal{E}, \mu)$ with joint distribution function $F(x)$ and a density function $f(x)$.*

Then the density function $f_i(x_i)$ of the marginal distribution function $F_i(x_i)$ is given by:

$$f_i(x_i) = F_i'(x_i), \ m\text{-a.e.} \tag{1.43}$$

Proof. *Since $f_i(y_i)$ in (1.42) is a Lebesgue integrable function by Fubini's theorem of Proposition V.5.16, the marginal distribution $F_i(x_i)$ in (1.41) can be expressed for arbitrary a and $x_i > a$:*

$$F_i(x_i) \equiv \int_{-\infty}^{x_i} f_i(y_i) dy_i = F_i(a) + \int_a^{x_i} f_i(y_i) dy_i. \tag{1}$$

Hence $F_i(x_i)$ is absolutely continuous on every interval $[a, b]$ by Proposition III.3.58.

Then by Proposition III.3.59, $F_i'(x_i)$ exists m-a.e. and is Lebesgue integrable on every bounded interval, while by Proposition III.3.62:

$$\int_a^{x_i} F_i'(y_i) dy_i = F_i(x_i) - F_i(a).$$

This obtains with (1) *for all $x_i > a$:*

$$\int_a^{x_i} \left[F_i'(y_i) - f_i(y_i) \right] dy_i = 0. \tag{2}$$

Finally, (2) *and Proposition III.3.36 yield that $f_i(y_i) = F_i'(y_i)$ m-a.e. on any interval $[a, b]$, and since a and b are arbitrary, this is (1.43).* ■

In the general case of distribution functions with continuous densities, we have the following result on the density functions of the associated marginal distribution functions. In contrast to the assumption of Proposition 1.44 on the joint density functions, for marginal density functions, we require that (1.31) is satisfied for any ordering of variates.

Proposition 1.51 (Marginal densities for continuous $f(x)$) *Let $X = (X_1, X_2, ..., X_n)$ be a random vector defined on a probability space $(\mathcal{S}, \mathcal{E}, \mu)$ with distribution function $F(x)$ and a continuous density function $f(x)$. For $1 \leq j < n$, let $(x_1, x_2, ..., x_n) \equiv (x^{(j)}, x^{(n-j)})$ where $x^{(j)} = (x_1, x_2, ..., x_j)$ and $x^{(n-j)} = (x_{j+1}, ..., x_n)$.*

Assume that for any ordering and relabelling of variates, that there exists integrable functions $g_j(x^{(n-j)})$ on $\mathbb{R}^{n-j}$ for $1 \leq j < n$ so that as in (1.31):

$$f(x^{(j)}, x^{(n-j)}) \leq g_j(x^{(n-j)}), \text{ for all } x^{(j)}.$$

Then for any $I = \{i_1, ..., i_m\} \subset \{1, 2, ..., n\}$, the density function $f_I(x_{i_1}, x_{i_2}, ..., x_{i_m})$ defined in (1.40) is continuous and:

$$\frac{\partial^m F_I}{\partial x_{i_1} ... \partial x_{i_m}} = f_I(x_{i_1}, x_{i_2}, ..., x_{i_m}). \tag{1.44}$$

Proof. *The result in (1.44) follows from Proposition 1.44 if it is proved that $f_I(x_{i_1}, x_{i_2}, ..., x_{i_m})$ is continuous and satisfies the conditions of (1.31) for any I.*

Since $f(x)$ satisfies (1.31) for any ordering of variates, given I we can reorder so that $I = \{1, ..., m\}$. Choose k with $1 \leq k < m$, and reorder variates:

$$(x_{k+1}, ..., x_m, x_1, ..., x_k, x_{m+1}, ..., x_n).$$

By assumption there is an integrable function $g_{m-k}(x_1, ..., x_k, x_{m+1}, ..., x_n)$ so that for all $(x_{k+1}, ..., x_m)$:

$$f(x_1, x_2, ..., x_n) \leq g_{m-k}(x_1, ..., x_k, x_{m+1}, ..., x_n). \tag{1}$$

Integrating (1) *with respect to $(x_{m+1}, ..., x_n)$ and applying (1.40) obtains that for all $(x_{k+1}, ..., x_m)$:*

$$\begin{aligned} f_I(x_1, ..., x_m) &\leq \int_{\mathbb{R}^{n-m}} g_{m-k}(x_1, ..., x_k, x_{m+1}, ..., x_n)\, dx_{m+1} ... dx_n \\ &\equiv \widetilde{g}_{m-k}(x_1, ..., x_k). \end{aligned}$$

By Fubini's theorem of Proposition V.5.16, integrability of $g_{m-k}(x_1, ..., x_k, x_{m+1}, ..., x_n)$ assures integrability of $\widetilde{g}_{m-k}(x_1, ..., x_k)$, and thus f_I satisfies the conditions of (1.31).

For continuity of $f_I(x_1, ..., x_m)$, we prove sequential continuity and apply Exercise 1.52.

Given $(x_1, ..., x_m)$, let $\{(x_1^{(j)}, ..., x_m^{(j)})\}_{j=1}^{\infty} \subset \mathbb{R}^m$ be a sequence of points with $(x_1^{(j)}, ..., x_m^{(j)}) \to (x_1, ..., x_m)$. Define h, $\{h_j\}_{j=1}^{\infty}$ on $\mathbb{R}^{n-m}$ by:

$$\begin{aligned} h_j(x_{m+1}, ..., x_n) &\equiv f(x_1^{(j)}, ..., x_m^{(j)}, x_{m+1}, ..., x_n), \\ h(x_{m+1}, ..., x_n) &\equiv f(x_1, ..., x_m, x_{m+1}, ..., x_n). \end{aligned}$$

Then $h_j \to h$ for all $(x_{m+1}, ..., x_n)$ by continuity of f, while by (1.31):

$$h_j(x_{m+1}, ..., x_n) \leq g_m(x^{(n-m)}),$$

where $g_m(x^{(n-m)}) \equiv g_m(x_{m+1}, ..., x_n)$ is integrable on $\mathbb{R}^{n-m}$.

Hence by Lebesgue's dominated convergence theorem of Proposition III.2.52, and (1.40):

$$\begin{aligned} f_I(x_1, ..., x_m) &\equiv \int_{\mathbb{R}^{n-m}} f(x_1, ..., x_m, x_{m+1}, ..., x_n) dx_{m+1} ... dx_n \\ &= \int_{\mathbb{R}^{n-m}} h(x_{m+1}, ..., x_n) dx_{m+1} ... dx_n \end{aligned}$$

$$= \lim_{j\to\infty} \int_{\mathbb{R}^{n-m}} h_j(x_{m+1},...,x_n)dx_{m+1}...dx_n$$
$$\equiv \lim_{j\to\infty} f_I(x_1^{(j)},...,x_m^{(j)}).$$

Thus f_I is sequentially continuous, and also continuous by Exercise 1.52.

The proof is completed by noting that (1.44) follows from (1.32). ■

Exercise 1.52 (Continuity; Sequential continuity) *Prove that a function $f(x)$ defined on $\mathbb{R}^m$ is continuous at x if and only if it is sequentially continuous at x. Hint: See Proposition III.4.58 on linear transformations, and confirm that this proof works in general.*

1.4 Densities and Independent RVs

As was the case for joint distribution functions, the density functions related to independent random variables and vectors have an especially simple form which we investigate in this section.

The notion of independence was introduced in Definition II.3.47.

Definition 1.53 (Independent random variables/vectors) *If $\{X_j\}_{j=1}^n : \mathcal{S} \longrightarrow \mathbb{R}$ are random variables on $(\mathcal{S},\mathcal{E},\mu)$, we say that $\{X_j\}_{j=1}^n$ are* ***independent random variables*** *if given $\{A_j\}_{j=1}^n \subset \mathcal{B}(\mathbb{R})$:*

$$\mu\left(\bigcap\nolimits_{j=1}^n X_j^{-1}(A_j)\right) = \prod\nolimits_{j=1}^n \mu\left(X_j^{-1}(A_j)\right). \tag{1.45}$$

Equivalently, if $X \equiv (X_1, X_2, ..., X_n)$ is the associated random vector by Proposition 1.17, then by (1.8) and (1.6):

$$\mu\left[X^{-1}\left(\prod\nolimits_{j=1}^n A_j\right)\right] = \prod\nolimits_{j=1}^n \mu\left(X_j^{-1}(A_j)\right). \tag{1.46}$$

If $\{X_j\}_{j=1}^n$ are random vectors on $(\mathcal{S},\mathcal{E},\mu)$, $X_j : \mathcal{S} \longrightarrow \mathbb{R}^{n_j}$, we say that $\{X_j\}_{j=1}^n$ are ***independent random vectors*** *if given $\{A_j\}_{j=1}^n$ with $A_j \in \mathcal{B}(\mathbb{R}^{n_j})$:*

$$\mu\left(\bigcap\nolimits_{j=1}^n X_j^{-1}(A_j)\right) = \prod\nolimits_{j=1}^n \mu\left(X_j^{-1}(A_j)\right). \tag{1.47}$$

In either case, an infinite collection of random variables/vectors $\{X_j\}_{j=1}^\infty$ is independent if every ***finite*** *subcollection is independent.*

Proposition II.3.53 proved that $\{X_j\}_{j=1}^n$ are independent random variables or vectors if and only if:

$$F(x_1, x_2, ..., x_n) = \prod\nolimits_{j=1}^n F_j(x_j), \tag{1.48}$$

where F and F_j denote the respective distribution functions of $(X_1, X_2, ..., X_n)$ and X_j. For independent random vectors of Definition 1.53, $X_j : \mathcal{S} \longrightarrow \mathbb{R}^{n_j}$, so $x_j \in \mathbb{R}^{n_j}$ and all distribution functions in (1.48) are joint distribution functions.

This result can also be framed in the context of the associated probability measures, but first a remark on so-called product topologies and the associated Borel sigma algebras.

Remark 1.54 (Product topologies and $\mathcal{B}\left(\prod_{j=1}^{n}\mathbb{R}^{n_j}\right)=\mathcal{B}\left(\mathbb{R}^N\right)$) *If X is a topological space, then by Definition 1.2, $\mathcal{B}(X)$ is the smallest sigma algebra that contains the open sets of X as defined by this topology. By Definition I.2.15, a topology defines the open sets of X, and any such collection must contain $\emptyset$ and X, and be closed under finite intersections and arbitrary unions.*

In the space $\mathbb{R}^n$, a set $A \subset \mathbb{R}^n$ is open by Definition I.2.10 if given $x \in A$, there exists $r > 0$ so that the open ball $B_r(x)$ about x of radius r is contained in A:

$$B_r(x) \equiv \{y \mid |x-y| < r\} \subset A.$$

It is an exercise to verify that the collection of open sets is indeed a topology on $\mathbb{R}^n$, meaning that this collection satisfies the above criteria.

As a product space, the ***product topology*** *on $\prod_{j=1}^{n}\mathbb{R}^{n_j}$ is obtained by defining the open sets of $\prod_{j=1}^{n}\mathbb{R}^{n_j}$ as the collection of sets formed by arbitrary unions of products of open sets, $\prod_{j=1}^{n}A_j$, with A_j open in $\mathbb{R}^{n_j}$. This collection turns out to be "equivalent" to the open sets in $\mathbb{R}^N$, directly defined, where $N=\sum_{j=1}^{n}n_j$. For this equivalency, we agree to identify $x=(x_1,...,x_n)\in\prod_{j=1}^{n}\mathbb{R}^{n_j}$ with $x_j\in\mathbb{R}^{n_j}$ and $x=(y_1,...,y_N)$ with $y_k\in\mathbb{R}$, where the y_k-coordinates are defined with the coordinates of the x_j-vectors.*

For this equivalency, certainly the $\prod_{j=1}^{n}A_j$-sets are open in $\mathbb{R}^N$ by the above definition. The reverse conclusion reflects the observation that an open ball $B_r(x)$ in $\mathbb{R}^N$ with $x=(x_1,...,x_n)$ and $x_j\in\mathbb{R}^{n_j}$ contains a product of open balls $\prod_{j=1}^{n}B_{r_j}(x_j)$. Details are left as an exercise.

This product topology also has the property that the projection mappings $\{\pi_k\}_{k=1}^{n}$ are continuous, where $\pi_k:\prod_{j=1}^{n}\mathbb{R}^{n_j}\to\mathbb{R}^{n_k}$ is defined analogously with (1.37) by:

$$\pi_k(x_1,...,x_n)=x_k,$$

where $x_j\in\mathbb{R}^{n_j}$. Recalling Proposition I.3.12, a function $f:\mathbb{R}^n\to\mathbb{R}$ is continuous if and only if $f^{-1}(A)$ is open in $\mathbb{R}^n$ for all A open in $\mathbb{R}$.

This result generalizes to transformations $\pi_k:\prod_{j=1}^{n}\mathbb{R}^{n_j}\to\mathbb{R}^{n_k}$, so continuity of π_k means that $\pi_k^{-1}(A_k)$ is open in $\prod_{j=1}^{n}\mathbb{R}^{n_j}$ for all A_k open in $\mathbb{R}^{n_k}$. And indeed this is the case with the product topology by definition since:

$$\pi_k^{-1}(A_k)=\prod\nolimits_{j=1}^{k-1}\mathbb{R}^{n_j}\times A_k\times\prod\nolimits_{j=k+1}^{n}\mathbb{R}^{n_j}.$$

The Borel sigma algebra $\mathcal{B}\left(\prod_{j=1}^{n}\mathbb{R}^{n_j}\right)$ is then the smallest sigma algebra that contains the open sets of $\prod_{j=1}^{n}\mathbb{R}^{n_j}$, and by the above observation, this is "equivalent" to $\mathcal{B}\left(\mathbb{R}^N\right)$:

$$\mathcal{B}\left(\prod\nolimits_{j=1}^{n}\mathbb{R}^{n_j}\right)=\mathcal{B}\left(\mathbb{R}^N\right). \tag{1.49}$$

Proposition 1.55 (Probability measures λ_X for independent RVs) *Let $\{X_j\}_{j=1}^{n}:\mathcal{S}\longrightarrow\mathbb{R}$ be random variables on $(\mathcal{S},\mathcal{E},\mu)$, $X\equiv(X_1,X_2,...,X_n)$ the associated random vector, and λ_X, $\{\lambda_{X_j}\}_{j=1}^{n}$ the associated probability measures defined on $\mathcal{B}(\mathbb{R}^n)$ and $\mathcal{B}(\mathbb{R})$, respectively.*

Then $\{X_j\}_{j=1}^{n}$ are independent random variables if and only if λ_X is a product measure on $\mathcal{B}(\mathbb{R}^n)$ in the sense of Proposition I.7.20, from the measure spaces $\{(\mathbb{R},\mathcal{B}(\mathbb{R}),\lambda_{X_j})\}_{j=1}^{n}$:

$$\lambda_X=\prod\nolimits_{j=1}^{n}\lambda_{X_j}. \tag{1.50}$$

Similarly, $\{X_j\}_{j=1}^n$ are independent random vectors if and only if λ_X is a product measure on $\mathcal{B}\left(\prod_{j=1}^n \mathbb{R}^{n_j}\right)$ from the measure spaces $\{(\mathbb{R}^{n_j}, \mathcal{B}(\mathbb{R}^{n_j}), \lambda_{X_j})\}_{j=1}^n$.

Proof. *For the result on random variables, assume that (1.50) is satisfied. Then by Proposition I.7.20, if $A = \prod_{j=1}^n A_j \in \mathcal{B}(\mathbb{R}^n)$ is a measurable rectangle with $A_j \in \mathcal{B}(\mathbb{R})$:*

$$\lambda_X(A) = \prod_{j=1}^n \lambda_{X_j}(A_j). \tag{1}$$

This is (1.46) by the definition of probability measure in (1.21), and so $\{X_j\}_{j=1}^n$ are independent random variables by Definition 1.53. The same logic applies for random vectors.

Conversely by reversing this logic, if $\{X_j\}_{j=1}^n$ are independent random variables/vectors, then (1) is satisfied for all measurable rectangles $A = \prod_{j=1}^n A_j$ with $A_j \in \mathcal{B}(\mathbb{R})$, respectively $A_j \in \mathcal{B}(\mathbb{R}^{n_j})$, recalling that these collections of sets are semi-algebras $\mathcal{A}'$ by Proposition I.7.2.

Let $\lambda' \equiv \prod_{j=1}^n \lambda_{X_j}$ be the product measure of Proposition I.7.20 obtained from $\{(\mathbb{R}, \mathcal{B}(\mathbb{R}), \lambda_{X_j})\}_{j=1}^n$, respectively $\{(\mathbb{R}^{n_j}, \mathcal{B}(\mathbb{R}^{n_j}), \lambda_{X_j})\}_{j=1}^n$. It then follows from this result and (1) that for all $A = \prod_{j=1}^n A_j \in \mathcal{A}'$:

$$\lambda_X(A) = \lambda'(A), \tag{2}$$

and (2) extends additively to the associated algebra $\mathcal{A}$ of all finite disjoint unions of $\mathcal{A}'$-sets (recall Exercise I.6.10). By the uniqueness of extensions theorem of Proposition I.6.14, (2) is satisfied for all $A \in \sigma(\mathcal{A})$, the smallest sigma algebra that contains $\mathcal{A}$.

The proof is complete for random variables by Proposition I.8.1 which obtains that $\sigma(\mathcal{A}) = \mathcal{B}(\mathbb{R}^n)$, and by Remark 1.54, this result generalizes to $\sigma(\mathcal{A}) = \mathcal{B}\left(\prod_{j=1}^n \mathbb{R}^{n_j}\right)$ in the case of random vectors. ■

Remark 1.56 (On Proposition I.8.1) *The last sentence in the above proof requires some additional details.*

Random variables: *Proposition I.8.1 result states that $\sigma(\mathcal{A}'(\mathcal{A}'_j)) = \mathcal{B}(\mathbb{R}^n)$, where here, $\mathcal{A}'(\mathcal{A}'_j) \equiv \left\{\prod_{j=1}^n (a_j, b_j]\right\}$ is the semi-algebra (Corollary I.7.3) of right semi-closed rectangles formed with component sets from the semi-algebras $\mathcal{A}'_j \equiv \{(a_j, b_j]\}$, of right semi-closed intervals. If $\mathcal{A}(\mathcal{A}'_j)$ is the associated algebra (Exercise I.6.10) of all finite disjoint unions of $\mathcal{A}'(\mathcal{A}'_j)$-sets, then $\mathcal{A}(\mathcal{A}'_j) \subset \mathcal{A}$ of the above proof, since $\mathcal{A}'_j \subset \mathcal{B}(\mathbb{R})$. Thus:*

$$\mathcal{B}(\mathbb{R}^n) = \sigma(\mathcal{A}'(\mathcal{A}'_j)) \subset \sigma(\mathcal{A}), \tag{1}$$

so $\sigma(\mathcal{A})$ at least contains $\mathcal{B}(\mathbb{R}^n)$, which is enough for the above proof.

But note that $\mathcal{A} \subset \mathcal{B}(\mathbb{R}^n)$, so also $\sigma(\mathcal{A}) \subset \mathcal{B}(\mathbb{R}^n)$, and thus:

$$\sigma(\mathcal{A}) = \mathcal{B}(\mathbb{R}^n).$$

Random vectors: *This generalization requires a generalization of the statement and proof of Proposition I.8.1.*

Now $\mathcal{A}'(\mathcal{A}'_j) \equiv \left\{\prod_{j=1}^n \left[\prod_{k=1}^{n_j} (a_k^{(j)}, b_k^{(j)}]\right]\right\}$ is the semi-algebra of product sets with components in $\mathcal{A}'_j \equiv \left\{\prod_{k=1}^{n_j} (a_k^{(j)}, b_k^{(j)}]\right\}$. These component collections are semi-algebras by

Corollary I.7.3, and then $\mathcal{A}'(\mathcal{A}'_j)$ *is a semi-algebra by the same result. Now* $\mathcal{A}'_j \subset \mathcal{B}(\mathbb{R}^{n_j})$ *obtains* $\mathcal{A}'(\mathcal{A}'_j) \subset \mathcal{A}'[\mathcal{B}(\mathbb{R}^{n_j})] \subset \mathcal{B}\left(\prod_{j=1}^n \mathbb{R}^{n_j}\right)$ *and thus:*

$$\sigma(\mathcal{A}'(\mathcal{A}'_j)) \subset \mathcal{B}\left(\prod\nolimits_{j=1}^n \mathbb{R}^{n_j}\right). \tag{2}$$

Conversely, $\mathcal{B}\left(\prod_{j=1}^n \mathbb{R}^{n_j}\right) \subset \sigma(\mathcal{A}'(\mathcal{A}'_j))$ *since the sigma algebra on the right contains the open product sets:*

$$\left\{\prod\nolimits_{j=1}^n \left[\prod\nolimits_{k=1}^{n_j} (a_k^{(j)}, b_k^{(j)})\right]\right\} = \bigcap\nolimits_{r>0} \prod\nolimits_{j=1}^n \left[\prod\nolimits_{k=1}^{n_j} (a_k^{(j)}, b_k^{(j)} + r]\right],$$

defined as an intersection over r rational. Recalling Remark 1.54, the collection on the left is equivalent to open product sets $\prod_{i=1}^N (a_i, b_i)$ *in* $\mathbb{R}^N$ *which generate the open balls and hence all open sets by definition. Hence* $\sigma(\mathcal{A}'(\mathcal{A}'_j))$ *contains the open sets in* $\prod_{j=1}^n \mathbb{R}^{n_j}$*, and then by Definition 1.2:*

$$\mathcal{B}\left(\prod\nolimits_{j=1}^n \mathbb{R}^{n_j}\right) \subset \sigma(\mathcal{A}'(\mathcal{A}'_j)). \tag{3}$$

Thus $\sigma(\mathcal{A}'(\mathcal{A}'_j)) = \mathcal{B}\left(\prod_{j=1}^n \mathbb{R}^{n_j}\right)$ *by (2) and (3), and since* $\mathcal{A}'(\mathcal{A}'_j) \subset \mathcal{A} \subset \mathcal{B}\left(\prod_{j=1}^n \mathbb{R}^{n_j}\right)$*, it follows that* $\sigma(\mathcal{A}) = \mathcal{B}\left(\prod_{j=1}^n \mathbb{R}^{n_j}\right)$.

With Proposition 1.54, we are now ready for the first result on the joint density function for independent random variables or vectors, but note that this is not an existence result. It states that when all density functions exist, they are related in a predictable way.

Corollary 1.57 (Density functions for independent random variables) *Let* $\{X_j\}_{j=1}^n : \mathcal{S} \longrightarrow \mathbb{R}$ *be random variables on* $(\mathcal{S}, \mathcal{E}, \mu)$ *and* $X \equiv (X_1, X_2, ..., X_n)$ *the associated random vector, with associated density functions* $\{f(x_j)\}_{j=1}^n$, $f(x)$ *defined on* $\mathbb{R}$ *and* $\mathbb{R}^n$*, respectively.*

Then $\{X_j\}_{j=1}^n$ *are independent random variables if and only if:*

$$f(x_1, x_2, ..., x_n) = \prod\nolimits_{j=1}^n f_j(x_j), \ m\text{-a.e.} \tag{1.51}$$

More generally, let $\{X_j\}_{j=1}^n$ *be independent random vectors with ranges in* $\{\mathbb{R}^{n_j}\}_{j=1}^n$*, and* $X \equiv (X_1, X_2, ..., X_n)$ *the associated random vector with range in* $\mathbb{R}^N$ *and* $N = \sum_{j=1}^n n_j$. *If* $\{f(x_j)\}_{j=1}^n$, $f(x)$ *are the associated density functions defined on* $\{\mathbb{R}^{n_j}\}_{j=1}^n$ *and* $\mathbb{R}^N$*, then* $\{X_j\}_{j=1}^n$ *are independent random vectors if and only if (1.51) is satisfied.*

Proof. *In either context, Proposition 1.34 obtains that* $\{f(x_j)\}_{j=1}^n$, $f(x)$ *are the density functions for the probability measures of Proposition 1.30. To simplify notation, we prove the case for random variables and leave the verification of the details of the result for random vectors as an exercise.*

If $\{X_j\}_{j=1}^n$ *are independent random variables, let* $A \equiv \prod_{j=1}^n A_j \in \mathcal{B}(\mathbb{R}^n)$ *with* $A_j \in \mathcal{B}(\mathbb{R})$ *for all j. Then by (1.50) and Definition 1.9,* $\lambda_X(A)$ *is given by:*

$$\int_A f(x)dx = \prod\nolimits_{j=1}^n \int_{A_j} f_j(x_j)dx_j = \int_A \prod\nolimits_{j=1}^n f_j(x_j)dx. \tag{1}$$

These are defined as Lebesgue integrals, and the last equality follows from nonnegativity of densities and Tonelli's theorem of Proposition V.5.24.

The collection of such rectangles is a semi-algebra $\mathcal{A}'$, and (1) *states that for all $A \in \mathcal{A}'$:*

$$\lambda_X(A) = \lambda'_X(A), \tag{2}$$

where λ'_X is the Borel measure (Proposition V.4.3) induced by $\prod_{j=1}^n f_j(x_j)$.

The identify in (2) *extends by additivity to the algebra $\mathcal{A}$ of all finite unions of $\mathcal{A}'$-sets, and the uniqueness theorem of Proposition I.6.14 obtains that this identity holds on $\sigma(\mathcal{A})$, the smallest sigma algebra that contains $\mathcal{A}$. Finally, Proposition I.8.1 and Remark 1.56 prove that $\sigma(\mathcal{A}) = \mathcal{B}(\mathbb{R}^n)$, and thus for all $A \in \mathcal{B}(\mathbb{R}^n)$:*

$$\int_A \left[f(x) - \prod_{j=1}^n f_j(x_j)\right] dx = 0. \tag{3}$$

The result in (1.51) is now item 8 of Proposition V.3.42.

Conversely, (1.51) assures (2) *for all $A \in \mathcal{B}(\mathbb{R}^n)$, and then* (1) *by item 3 of Proposition V.3.42. But by* (1), *for all $A \in \mathcal{A}'$:*

$$\lambda_X(A) = \prod_{j=1}^n \int_{A_j} f_j(x_j)dx_j = \prod_{j=1}^n \lambda_{X_j}(A_j).$$

This identity extends to $\mathcal{A}$ as above, and then to $\mathcal{B}(\mathbb{R}^n)$ by Proposition I.6.14, and so for all $A \in \mathcal{B}(\mathbb{R}^n)$:

$$\lambda_X(A) = \left(\prod_{j=1}^n \lambda_{X_j}\right)(A), \tag{4}$$

where $\prod_{j=1}^n \lambda_{X_j}$ is the product measure of Chapter I.7.

By (4) *and Proposition 1.55, $\{X_j\}_{j=1}^n$ are independent random variables, and the proof is complete.* ■

The next result addresses the existence of density functions for independent random variables.

Proposition 1.58 (Existence of density functions: Independent RVs) *Let $\{X_j\}_{j=1}^n : \mathcal{S} \longrightarrow \mathbb{R}$ be independent random variables on $(\mathcal{S}, \mathcal{E}, \mu)$, $X \equiv (X_1, X_2, ..., X_n)$ the associated random vector, and F, $\{F_j\}_{j=1}^n$ the associated distribution functions defined on $\mathbb{R}^n$ and $\mathbb{R}$.*

Then F has a density function if and only if all F_j have density functions, and then:

$$f(x_1, x_2, ..., x_n) = \prod_{j=1}^n f_j(x_j), \ m\text{-a.e.} \tag{1.52}$$

Proof. *Assume that all F_j have density functions:*

$$F_j(x_j) \equiv \int_{-\infty}^{x_j} f_j(y_j)dy_j,$$

expressed as Lebesgue integrals. Then from (1.48):

$$\begin{aligned} F(x_1, x_2, ..., x_n) &= \prod_{j=1}^n \int_{-\infty}^{x_j} f_j(y_j)dy_j \\ &= \int_{-\infty}^{x_n} \cdots \int_{-\infty}^{x_1} \left[\prod_{j=1}^n f_j(y_j)\right] dy_1...dy_n \qquad (1) \\ &= \int_{A_{(x_1,x_2,...,x_m)}} \prod_{j=1}^n f_j dm, \end{aligned}$$

where $A_{(x_1,x_2,...,x_m)} \equiv \prod_{j=1}^n (-\infty, x_j]$.

The middle step in (1) is notational, while the last step follows from Tonelli's theorem of Proposition V.5.24, noting that $\prod_{j=1}^{n} f_j$ *is nonnegative and Lebesgue measurable. Since the iterated integrals in (1) are finite, Tonelli's theorem then assures the m-integrability of* $\prod_{j=1}^{n} f_j$. *Thus F has a density function* $\prod_{j=1}^{n} f_j(x_j)$, *and (1.52) follows since density functions are unique m-a.e. by Proposition 1.10.*

Conversely, if f is the density function for F, we can by Remark 1.14 assume that f is strictly nonnegative by using the equivalent density $\tilde{f} \equiv \max(0, f(x))$. *Then Tonelli's theorem of Proposition V.5.24 obtains:*

$$\begin{aligned} F(x_1, x_2, ..., x_n) &\equiv \int_{A_{(x_1, x_2, ..., x_m)}} f dm \\ &= \int_{-\infty}^{x_n} \cdots \int_{-\infty}^{x_1} f(y_1, y_2, ..., y_n) dy_1 ... dy_n. \end{aligned}$$

Letting all $x_k, \rightarrow \infty$ *for* $k \neq j$ *obtains by (1.34) and Fubini's theorem:*

$$\begin{aligned} F_j(x_j) &= \int_{-\infty}^{\infty} \cdots \int_{-\infty}^{x_j} \cdots \int_{-\infty}^{\infty} f(y_1, y_2, ..., y_n) dy_1 ... dy_n \\ &= \int_{-\infty}^{x_j} \left[\int_{-\infty}^{\infty} \cdots \int_{-\infty}^{\infty} f(y_1, y_2, ..., y_n) dy_1 ... \widehat{dy_j} ... dy_n \right] dy_j, \end{aligned}$$

where $\widehat{dy_j}$ *denotes that this integration variable is omitted.*

To see that:

$$f_j(y_j) \equiv \int_{-\infty}^{\infty} \cdots \int_{-\infty}^{\infty} f(y_1, y_2, ..., y_n) dy_1 ... \widehat{dy_j} ... dy_n,$$

is a density function for $F_j(x_j)$, *first note that* f_j *is measurable and integrable by Tonelli's theorem, with integral 1 since* $F_j(x_j) \rightarrow 1$ *as* $x_j \rightarrow \infty$. *Further,* $f \geq 0$ *assures that* $f_j \geq 0$ *by item 4 of Proposition III.2.49.*

Since $F(x) = \prod_{j=1}^{n} F_j(x_j)$ *from (1.48), another application of Fubini's theorem obtains:*

$$\begin{aligned} \int_{-\infty}^{x_n} \cdots \int_{-\infty}^{x_1} f(y_1, y_2, ..., y_n) dy_1 ... dy_n &= \prod_{j=1}^{n} \int_{-\infty}^{x_j} f_j(y_j) dy_j \\ &= \int_{-\infty}^{x_n} \cdots \int_{-\infty}^{x_1} \prod_{j=1}^{n} f_j(y_j) dy_1 ... dy_n, \end{aligned}$$

and so both $f(y_1, y_2, ..., y_n)$ *and* $\prod_{j=1}^{n} f_j(y_j)$ *are density functions associated with F. By Proposition 1.34, both are then density functions for the probability measure* λ_X, *and so (1.52) follows from Proposition 1.10.* ■

Corollary 1.59 (Marginal density functions: Independent RVs) *Let* $\{X_j\}_{j=1}^{n} : \mathcal{S} \longrightarrow \mathbb{R}$ *be independent random variables on* $(\mathcal{S}, \mathcal{E}, \mu)$, $X \equiv (X_1, X_2, ..., X_n)$ *the associated random vector, and* $F(x_1, x_2, ..., x_n)$ *the associated distribution function of X. If* $I = \{i_1, ..., i_m\} \subset \{1, 2, ..., n\}$, *then the marginal distribution function* F_I *has a density function if and only if all* F_{i_k} *have density functions, and then:*

$$f_I(x_{i_1}, x_{i_2}, ..., x_{i_m}) = \prod_{k=1}^{m} f_{i_k}(x_{i_k}), \text{ } m\text{-a.e.} \tag{1.53}$$

Proof. *By (1.36),* $F_I(x_{i_1}, x_{i_2}, ..., x_{i_m})$ *is the joint distribution function of* $\{X_{i_k}\}_{k=1}^{m}$, *and since independence of* $\{X_j\}_{j=1}^{n}$ *assures independence of* $\{X_{i_k}\}_{k=1}^{m}$ *by Definition 1.53, this result is an application of Proposition 1.58.* ■

1.5 Conditional Density Functions

Conditional distribution functions were introduced in Definition II.3.39, and reflect the notion of a conditional probability measure from Definition II.1.31. We begin with the latter. See Section 4.3.1 for an important generalization of this idea.

Definition 1.60 (Conditional probability measure) *Let $(\mathcal{S}, \mathcal{E}, \mu)$ be a* ***probability space*** *and $B \in \mathcal{E}$ with $\mu(B) > 0$. The* ***conditional probability measure*** *$\mu(\cdot|B)$ is defined on $A \in \mathcal{E}$ by:*

$$\mu(A|B) \equiv \frac{\mu\left(A \bigcap B\right)}{\mu(B)}. \tag{1.54}$$

The expression $\mu(A|B)$ is called the ***conditional probability of*** *A, and if additional emphasis on B is needed, the* ***probability of*** *A,* ***conditional on*** *B.*

Exercise II.1.32 verified that $\mu(\cdot|B)$ is indeed a probability measure on $\mathcal{E}$ when $\mu(B) > 0$, and thus $(\mathcal{S}, \mathcal{E}, \mu(\cdot|B))$ is a probability space. In addition:

$$\mathcal{E}_B \equiv \left\{A \bigcap B \middle| A \in \mathcal{E}\right\},$$

is a sigma algebra by Exercise II.1.29, and so $(\mathcal{S}, \mathcal{E}_B, \mu(\cdot|B))$ is again a probability space.

Turning next to conditional distribution functions, note that the requirement of this definition, that $\mu\left[X_J^{-1}(B)\right] > 0$, can by (1.21) be stated as $\lambda_{X_J}(B) > 0$, using the probability measure induced by X_J.

Definition 1.61 (Conditional distribution function) *Let $X = (X_1, X_2, ..., X_n)$ be a random vector $X : \mathcal{S} \longrightarrow \mathbb{R}^n$ defined on $(\mathcal{S}, \mathcal{E}, \mu)$, $J \equiv \{j_1, ..., j_m\} \subset \{1, 2, ..., n\}$, and $X_J \equiv (X_{j_1}, X_{j_2}, ..., X_{j_m})$ the associated marginal random vector. Given a Borel set $B \in \mathcal{B}(\mathbb{R}^m)$ with $\mu\left[X_J^{-1}(B)\right] > 0$, the* ***conditional distribution function*** *of X given $X_J \in B$, denoted:*

$$F(x|X_J \in B) \equiv F(x_1, x_2, ..., x_n|X_J \in B),$$

is defined by the ***conditional probability measure:***

$$F(x|X_J \in B) \equiv \mu\left[X^{-1}\left(\prod\nolimits_{i=1}^{n}(-\infty, x_i]\right)\middle| \; X_J^{-1}(B)\right],$$

and so by Definition 1.60:

$$F(x|X_J \in B) = \frac{\mu\left[X^{-1}\left(\prod_{i=1}^{n}(-\infty, x_i]\right) \bigcap X_J^{-1}(B)\right]}{\mu\left[X_J^{-1}(B)\right]}. \tag{1.55}$$

This distribution function is sometimes denoted $F_{J|B}(x)$.

The function $F_{J|B}(x)$ is indeed a distribution function by Proposition II.3.41, and thus induces a probability measure $\lambda_{X,J|B}$ on $\mathcal{B}(\mathbb{R}^n)$.

Proposition 1.62 (Conditional probability measure $\lambda_{X,J|B}$) *Let $X = (X_1, X_2, ..., X_n)$ be a random vector $X : \mathcal{S} \longrightarrow \mathbb{R}^n$ defined on $(\mathcal{S}, \mathcal{E}, \mu)$ with induced probability measure λ_X defined on $\mathcal{B}(\mathbb{R}^n)$. For $J \equiv \{j_1, ..., j_m\} \subset \{1, 2, ..., n\}$, let $X_J \equiv (X_{j_1}, X_{j_2}, ..., X_{j_m})$ have induced probability measure λ_{X_J} defined on $\mathcal{B}(\mathbb{R}^m)$.*

If $B \in \mathcal{B}(\mathbb{R}^m)$ with $\mu\left[X_J^{-1}(B)\right] > 0$, the probability measure $\lambda_{X,J|B}$ induced by $F_{J|B}(x)$ is defined on $A \in \mathcal{B}(\mathbb{R}^n)$ by:

$$\lambda_{X,J|B}(A) = \frac{\lambda_X\left(A \bigcap \pi_J^{-1}(B)\right)}{\lambda_{X_J}(B)}, \tag{1.56}$$

where $\pi_J : \mathbb{R}^n \to \mathbb{R}^m$ is the projection mapping $\pi_J : (x_1, x_2, ..., x_n) = (x_{j_1}, x_{j_2}, ..., x_{j_m})$ of (1.37).

Proof. *Let λ' denote the expression on the right in (1.56). Then λ' is well-defined on $\mathcal{B}(\mathbb{R}^n)$ since $\pi_J^{-1}(\mathcal{B}(\mathbb{R}^m)) \subset \mathcal{B}(\mathbb{R}^n)$ by Exercise 1.48, and as noted above:*

$$\lambda_{X_J}(B) \equiv \mu\left[X_J^{-1}(B)\right] > 0. \tag{1}$$

Further, $X_J = \pi_J X$ obtains $X_J^{-1} = X^{-1}\pi_J^{-1}$, so λ' can be rewritten by (1.21):

$$\begin{aligned} \lambda'(A) &= \frac{\mu\left[X^{-1}\left[A \bigcap \pi_J^{-1}(B)\right]\right]}{\mu\left[X_J^{-1}(B)\right]} \\ &= \frac{\mu\left[X^{-1}A \bigcap X_J^{-1}(B)\right]}{\mu\left[X_J^{-1}(B)\right]}. \end{aligned}$$

It follows that $\lambda'(A) \geq 0$ for all such A, and it is an exercise to check countable additivity and thus verify that λ' is a measure on $\mathcal{B}(\mathbb{R}^n)$.

To prove that λ' so defined in (1.56) is the probability measure $\lambda_{X,J|B}$ induced by $F_{J|B}(x)$, recall that by (1.18):

$$\lambda_{X,J|B}\left[\prod\nolimits_{i=1}^{n}(-\infty, x_i]\right] = F_{J|B}(x). \tag{2}$$

Turning to the definition of $F_{J|B}(x)$ in (1.55), since $X_J = \pi_J X$:

$$\begin{aligned} F_{J|B}(x) &\equiv \frac{\mu\left[X^{-1}\left(\prod_{i=1}^{n}(-\infty, x_i]\right) \bigcap X^{-1}(\pi_J^{-1}(B))\right]}{\mu\left[X_J^{-1}(B)\right]} \\ &= \frac{\mu\left[X^{-1}\left(\prod_{i=1}^{n}(-\infty, x_i] \bigcap \pi_J^{-1}(B)\right)\right]}{\lambda_{X_J}(B)} \\ &\equiv \lambda'\left(\prod\nolimits_{i=1}^{n}(-\infty, x_i]\right). \end{aligned} \tag{3}$$

Thus by (1) and (2), $F_{J|B}(x)$ is a distribution function for both λ' and $\lambda_{X,J|B}$, and so by the uniqueness result of Proposition 1.27, $\lambda' = \lambda_{X,J|B}$ on $\mathcal{B}(\mathbb{R}^n)$. ■

Remark 1.63 (On $\lambda_{X,J|B}$) *By (1.39) of Proposition 1.47, the conditional probability measure $\lambda_{X,J|B}$ can also be expressed in terms of λ_X:*

$$\lambda_{X,J|B}(A) = \frac{\lambda_X\left(A \bigcap \pi_J^{-1}(B)\right)}{\lambda_X\left(\pi_J^{-1}(B)\right)}.$$

Turning to density functions, we have the following.

Proposition 1.64 (Existence of density functions: Conditional DFs) *Let $X = (X_1, X_2, ..., X_n)$ be a random vector $X : \mathcal{S} \longrightarrow \mathbb{R}^n$ defined on $(\mathcal{S}, \mathcal{E}, \mu)$. Assume that the joint distribution function $F(x_1, x_2, ..., x_n)$ has a density function $f(x_1, x_2, ..., x_n)$, and that $B \in \mathcal{B}(\mathbb{R}^m)$ with $\mu\left[X_J^{-1}(B)\right] > 0$ for $J \equiv \{j_1, ..., j_m\} \subset \{1, 2, ..., n\}$ and $X_J \equiv (X_{j_1}, X_{j_2}, ..., X_{j_m})$.*

Then $F_{J|B}(x_1, x_2, ..., x_n)$ has a density function $f_{J|B}(x_1, x_2, ..., x_n) \equiv f(x|x_J \in B)$ given by:

$$f(x|x_J \in B) = \frac{f(x)\chi_{\pi_J^{-1}(B)}(x)}{\lambda_{X_J}(B)}, \quad m\text{-a.e.}, \tag{1.57}$$

where $\chi_{\pi_J^{-1}(B)}(x_1, x_2, ..., x_n) = 1$ if $(x_{j_1}, x_{j_2}, ..., x_{j_m}) \in B$ and is 0 otherwise, and $\pi_J : \mathbb{R}^n \to \mathbb{R}^m$ is the projection mapping of (1.37).

This can be equivalently expressed:

$$f(x|x_J \in B) = \frac{f(x)\chi_{\pi_J^{-1}(B)}(x)}{\int_B f_J dm}, \quad m\text{-a.e.}$$

where $f_J(x_{j_1}, x_{j_2}, ..., x_{j_m})$ denotes the marginal density function defined on $\mathbb{R}^m$.

Proof. *Denote by $g(x)$ the expression on the right in (1.57). Note that $g(x)$ is well-defined on $\mathbb{R}^n$ since by (1.21), $\lambda_{X_J}(B) \equiv \mu\left[X_J^{-1}(B)\right] > 0$. Further, $g(x)$ is nonnegative and measurable by these properties for $f(x)$ and the Borel measurability of $\chi_{\pi_J^{-1}(B)}(x)$, recalling that $\pi_J^{-1}(\mathcal{B}(\mathbb{R}^m)) \subset \mathcal{B}(\mathbb{R}^n)$ by Exercise 1.48.*

Now $\chi_{\pi_J^{-1}(B)}(x) \leq 1$ obtains $g(x) \leq cf(x)$ for $c = 1/\lambda_{X_J}(B)$, and thus $g(x)$ is Lebesgue integrable. Integrating $g(x)$ over $A \equiv A_{(x_1,x_2,...,x_n)} = \prod_{i=1}^n(-\infty, x_i]$ yields by Definition III.2.9:

$$\begin{aligned}\int_A g(x)dm &= \frac{1}{\lambda_{X_J}(B)}\int_A f(x)\chi_{\pi_J^{-1}(B)}(x)dm \\ &= \frac{1}{\lambda_{X_J}(B)}\int_{A\bigcap\pi_J^{-1}(B)} f(x)dm. \end{aligned} \tag{1}$$

Since $f(x)$ is the density for λ_X, (1.23) and then (1.56) obtain:

$$\begin{aligned}\int_A g(x)dm &= \frac{\lambda_X\left(A\bigcap\pi_J^{-1}(B)\right)}{\lambda_{X_J}(B)} \\ &= \lambda_{X,J|B}(A_{(x_1,x_2,...,x_n)}).\end{aligned}$$

Now $\lambda_{X,J|B}(A_{(x_1,x_2,...,x_n)}) = F_{J|B}(x_1, x_2, ..., x_n)$ by (1.18), and so for all $(x_1, x_2, ..., x_n)$:

$$F_{J|B}(x_1, x_2, ..., x_n) = \int_{A_{(x_1,x_2,...,x_n)}} g(x)dm.$$

Thus $g(x)$ is a density function of $F_{J|B}(x_1, x_2, ..., x_n)$ by definition, and then by Proposition 1.10, $g(x) = f_{J|B}(x_1, x_2, ..., x_n)$, m-a.e.

The last expression follows because $f_J(x_{j_1}, x_{j_2}, ..., x_{j_m})$ is also the density function of the marginal probability measure λ_{X_J} by Proposition 1.34. ■

Corollary 1.65 (Existence of density functions: Conditional probability measures) *Assume that the joint distribution function $F(x_1, x_2, ..., x_n)$ has a density function $f(x_1, x_2, ..., x_n)$, and that $X_J \equiv (X_{j_1}, X_{j_2}, ..., X_{j_m})$ and $B \in \mathcal{B}(\mathbb{R}^m)$ with $\mu\left[X_J^{-1}(B)\right] \neq 0$.*

Then the conditional probability measure $\lambda_{X,J|B}(A)$ of (1.56) is equivalently defined on $A \in \mathcal{B}(\mathbb{R}^n)$ by:

$$\lambda_{X,J|B}(A) = \frac{1}{\lambda_{X_J}(B)}\int_{A\bigcap\pi_J^{-1}(B)} f(x)dm. \tag{1.58}$$

Proof. *This follows from (1) of the above proof, since $f(x|x_J \in B)$ is a density function for $\lambda_{X,J|B}$.* ■

Exercise 1.66 (Discrete density function on $\mathbb{R}^n$) *Let $\{y_j\}_{j=1}^N \subset \mathbb{R}^n$ and $\{c_j\}_{j=1}^N \subset \mathbb{R}$ be nonnegative with $\sum_{j=1}^N c_j = 1$, where if $N = \infty$ we assume that $\{y_j\}_{j=1}^N$ has no accumulation points. On $\mathbb{R}^n$, define the distribution function:*

$$F(x) = \sum\nolimits_{y_j \leq x} c_j,$$

where $y_j \leq x$ is shorthand for $y_{jk} \leq x_k$ for $1 \leq k \leq n$.

Let X be a random vector defined on $(\mathcal{S}, \mathcal{E}, \mu)$ with range $\{y_j\}_{j=1}^N$ and with $c_j \equiv \mu\left[X^{-1}(y_j)\right]$.

General hint: For the following, work on derivations for $n = 2$ and $m = 1$, then $n = 3$ and $m = 1$, 2 say, before attempting general n, m.

1. *Confirm that $F(x)$ is the distribution function of X on $\mathbb{R}^n$. Hint: The n-increasing condition is proved by confirming that if $A = \prod_{i=1}^n (a_i, b_i]$, then:*

$$\sum\nolimits_x sgn(x) F(x) = \sum\nolimits_{y_j \in A} c_j.$$

For continuity from above, recall that $\{y_j\}_{j=1}^N$ has no accumulation points.

2. *Given $J \equiv \{j_1, ..., j_m\} \subset \{1, 2, ..., n\}$ and $X_J \equiv (X_{j_1}, X_{j_2}, ..., X_{j_m})$, explicitly define the marginal density function $f_J(x_{j_1}, x_{j_2}, ..., x_{j_m})$. Hint: Note that f_J is defined on the set $\{\pi_J y_j\}_{j=1}^N \subset \mathbb{R}^m$, and that these elements need not be distinct.*

3. *What does $\mu\left[X_J^{-1}(B)\right] > 0$ mean for $B \in \mathcal{B}(\mathbb{R}^m)$? Hint: $X_J = \pi_J X$ so $X_J^{-1} = X^{-1}\pi_J^{-1}$. See also item 4.*

4. *Derive the formula for $f_{J|B}(x_1, x_2, ..., x_n) \equiv f(x|x_J \in B)$ assuming that $\mu\left[X_J^{-1}(B)\right] > 0$:*

$$f(x|x_J \in B) = \frac{f(x)\chi_{\pi_J^{-1}(B)}(x)}{\sum_B f_J(x_{j_1}, x_{j_2}, ..., x_{j_m})}, \tag{1.59}$$

where the summation is over all $(x_{j_1}, x_{j_2}, ..., x_{j_m}) \in B$.

5. *Derive the analogous formula for $\lambda_{F_{J|B}}[A]$ with $A \in \mathcal{B}(\mathbb{R}^n)$, assuming that $\mu\left[X_J^{-1}(B)\right] > 0$:*

$$\lambda_{F_{J|B}}[A] = \sum\nolimits_A f_{J|B}(x). \tag{1.60}$$

We close this section with a couple of examples.

Example 1.67 *1. If $f(x, y)$ is a discrete density function, $B = \{y_0\}$ for some y_0 with $f_2(y_0) > 0$, where $f_2 \equiv f_J$ is the marginal density with $J = \{2\}$, then from (1.59):*

$$f\left((x, y)|(x, y)_J \in B\right) = \frac{f(x, y_0)}{f_2(y_0)},$$

a familiar formula from elementary probability theory.

Note that $f\left((x, y)|(x, y)_J \in B\right)$ is independent of y, and depends only on x. While the $\chi_{\pi_J^{-1}(B)}(x)$ term is omitted for notational simplicity, the above formula obtains 0 unless (x, y_0) is in the domain of $f(x, y)$, meaning $f(x, y_0) > 0$.

A similar formula is possible when $f(x, y)$ is a continuously differentiable density function, even though $\mu\left[(X, Y)_J^{-1}(B)\right] = 0$. The derivation in Example II.3.42 begins with $B^+ \equiv [y_0, y_0 + \Delta y]$, and then investigates the marginal distribution function $f\left((x, y)|(x, y)_J \in B^+\right)$ as $\Delta y \to 0$.

2. *Let F be a distribution function for a random variable X defined on $(\mathcal{S},\mathcal{E},\mu)$, and $B=(x_0,\infty)$. To have $\mu\left[X^{-1}(B)\right]>0$ implies by (1.21) that $\lambda_X[B]>0$. Since:*

$$\lambda_X\left[\widetilde{B}\right]\equiv\lambda_X\left[(-\infty,x_0]\right]=F(x_0),$$

$\mu\left[X^{-1}(B)\right]>0$ is equivalent to $F(x_0)<1$.

Then by (1.55):

$$\begin{aligned} F(x|X\in B) &\equiv \frac{\mu\left[X^{-1}((-\infty,x])\bigcap X^{-1}((x_0,\infty))\right]}{\mu\left[X^{-1}((x_0,\infty))\right]} \\ &= \begin{cases} 0, & x<x_0, \\ \frac{\lambda_X[(x_0,x]]}{1-F(x_0)}, & x\geq x_0. \end{cases} \end{aligned}$$

Thus by (1.16):

$$F(x|X\in B)=\begin{cases} 0, & x<x_0, \\ \frac{F(x)-F(x_0)}{1-F(x_0)}, & x\geq x_0. \end{cases}$$

If F has a density function f, then by (1.57):

$$f(x|x\in B)=\frac{f(x)\chi_B(x)}{1-F(x_0)},$$

and equivalently:

$$f(x|x\in B)=\begin{cases} 0, & x<x_0, \\ \frac{f(x)}{1-F(x_0)}, & x\geq x_0. \end{cases}$$

2

Transformations of Random Vectors

In Chapter IV.2, the distribution and density functions of monotonically transformed random variables were investigated, as were various results on sums and ratios of such variables, sometimes with omitted details. Results on joint distributions were also informally introduced, since some of the details of these analyses required the tools of Book V.

Transformations of random vectors are more formally investigated in this chapter, starting with a useful technical result known as Cavalieri's principle. Though formulated centuries earlier, it is now a corollary of Fubini's theorem of Book V, and provides an elegant tool for generalizing the Book II studies on the distribution functions of sums of random vectors, as well as the associated density functions when they exist.

These results then provide for an important generalization of the Book V result on the integrability of the convolution of integrable functions.

The final sections of this chapter derive the distribution and density functions for a ratio of independent random variables, and then address distribution and density function results on general measurable transformations of random vectors.

2.1 Cavalieri's Principle

Let X, Y be **independent random vectors** defined on a probability space $(\mathcal{S}, \mathcal{E}, \mu)$:

$$X : \mathcal{S} \to \mathbb{R}^m, \qquad Y : \mathcal{S} \to \mathbb{R}^n.$$

Thus by Definition 1.53, if $A \in \mathcal{B}(\mathbb{R}^m)$ and $B \in \mathcal{B}(\mathbb{R}^n)$:

$$\mu\left[X^{-1}(A) \bigcap Y^{-1}(B)\right] = \mu\left[X^{-1}(A)\right] \mu\left[Y^{-1}(B)\right].$$

Let $F_X(x)$ and $F_Y(y)$ be the associated joint distribution functions, and λ_X and λ_Y the induced Borel probability measures on $\mathcal{B}(\mathbb{R}^m)$ and $\mathcal{B}(\mathbb{R}^n)$, respectively, given in Proposition 1.27. Thus $(\mathbb{R}^m, \mathcal{B}(\mathbb{R}^m), \lambda_X)$ and $(\mathbb{R}^n, \mathcal{B}(\mathbb{R}^n), \lambda_Y)$ are probability spaces.

By independence, the random vector $(X, Y) : \mathcal{S} \to \mathbb{R}^m \times \mathbb{R}^n$ has joint distribution function given by Proposition II.3.53:

$$F(x, y) = F_X(x) F_Y(y).$$

By Proposition 1.55, the probability measure $\lambda_{(X,Y)}$ induced by F and defined on $\mathcal{B}(\mathbb{R}^m \times \mathbb{R}^n)$ is given by:

$$\lambda_{(X,Y)} = \lambda_X \times \lambda_Y, \tag{2.1}$$

the product measure of Proposition I.7.20 for the product space defined by $(\mathbb{R}^m, \mathcal{B}(\mathbb{R}^m), \lambda_X)$ and $(\mathbb{R}^n, \mathcal{B}(\mathbb{R}^n), \lambda_Y)$.

The next exercise provides a more constructive derivation of this result. Recall that we can identify $\mathbb{R}^m \times \mathbb{R}^n$ with $\mathbb{R}^{m+n}$, and $\mathcal{B}(\mathbb{R}^m \times \mathbb{R}^n)$ with $\mathcal{B}(\mathbb{R}^{m+n})$, by Remark 1.54.

DOI: 10.1201/9781003275770-2

Exercise 2.1 *Given a joint distribution function* $F(x,y) = F_X(x)F_Y(y)$, *and the right semi-closed rectangle* $A = \prod_{i=1}^{m+n}(a_i, b_i]$, *define* μ_F *as in (1.17):*

$$\mu_F\left(\prod_{i=1}^{m+n}(a_i, b_i]\right) \equiv \sum_{(x,y)} sgn(x,y)F(x,y).$$

Here the summation is over all vertexes (x,y) *of* A, *where* x *denotes the first* m *components and* y *the last* n *components, and* $sgn(x,y)$ *equals* -1 *if the total number of* a_i *components in* (x,y) *is odd, and equals* $+1$ *otherwise.*

Show that $sgn(x,y) = sgn(x)sgn(y)$, *and then prove that:*

$$\begin{aligned}\mu_F\left(\prod_{i=1}^{m+n}(a_i, b_i]\right) &= \mu_{F_X}\left(\prod_{i=1}^{m}(a_i, b_i]\right)\mu_{F_Y}\left(\prod_{i=m+1}^{m+n}(a_i, b_i]\right) \\ &\equiv \sum_x sgn(x)F_X(x)\sum_y sgn(y)F_Y(y),\end{aligned}$$

where the first sum is over the vertexes of $A_X \equiv \prod_{i=1}^{m}(a_i, b_i]$, *the second over the vertexes of* $A_Y \equiv \prod_{i=m+1}^{m+n}(a_i, b_i]$.

Thus by (1.22):

$$\lambda_{(X,Y)}\left(\prod_{i=1}^{m+n}(a_i, b_i]\right) = \lambda_X\left(\prod_{i=1}^{m}(a_i, b_i]\right)\lambda_Y\left(\prod_{i=m+1}^{m+n}(a_i, b_i]\right). \tag{2.2}$$

Hint: Note for any fixed x *that* $\sum_{(x,y)}$ *is a summation over all the vertexes of* A_Y *and thus* $\sum_{(x,y)} = \sum_x \sum_y$.

The next result can be understood as a probabilistic application of the Fubini-Tonelli theorems of Book V, but it also has an intuitive application in what is known as **Cavalieri's principle,** and hence we refer to it by that name.

This principle is named for **Bonaventura Cavalieri** (1598–1647), who, quite remarkably, developed these ideas before the formality of an integral calculus was developed. This principle is easy to state:

Cavalieri's Principle: *Given two 3-dimensional objects labeled* A *and* A', *if the areas of every horizontal cross-section of these objects agree, then the volumes agree.*

Here, cross-sections are defined as the intersections of these objects with a complete set of parallel planes. To formalize this result, recall Definition V.5.9.

Definition 2.2 (Cross-sections of $A \in \mathcal{B}(\mathbb{R}^m \times \mathbb{R}^n)$**)** *Given* $A \in \mathcal{B}(\mathbb{R}^m \times \mathbb{R}^n) = \mathcal{B}(\mathbb{R}^{m+n})$, *the* ***x-cross-section*** $A_x \subset \mathbb{R}^n$ *for* $x \in \mathbb{R}^m$, *and the* ***y-cross-section*** $A_y \subset \mathbb{R}^m$ *for* $y \in \mathbb{R}^n$, *are defined:*

$$A_x = \{y|(x,y) \in A\}, \quad A_y = \{x|(x,y) \in A\}.$$

It is by no means apparent what measurability properties such cross-sections possess. But we have the following.

Proposition 2.3 ($A_x \in \mathcal{B}(\mathbb{R}^n)$, $A_y \in \mathcal{B}(\mathbb{R}^m)$) *Given* $A \in \mathcal{B}(\mathbb{R}^m \times \mathbb{R}^n)$, *then* $A_x \in \mathcal{B}(\mathbb{R}^n)$ *for all* $x \in \mathbb{R}^m$, *and* $A_y \in \mathcal{B}(\mathbb{R}^m)$ *for all* $y \in \mathbb{R}^n$.

Proof. *Let* $\mathcal{A}'$ *denote the semi-algebra (Proposition I.7.2) on* $\mathbb{R}^m \times \mathbb{R}^n$ *of measurable rectangles* $E \times F$ *with* $E \in \mathcal{B}(\mathbb{R}^m)$ *and* $F \in \mathcal{B}(\mathbb{R}^n)$, *and* $\mathcal{A}$ *the associated algebra of all finite disjoint unions of* $\mathcal{A}'$ *sets by Exercise I.6.10. We leave as an exercise the details of the proof that the smallest sigma algebra* $\sigma'(\mathcal{A})$ *that contains* $\mathcal{A}$ *is* $\mathcal{B}(\mathbb{R}^m \times \mathbb{R}^n)$. *This requires another generalization of Proposition I.8.1 as seen in Remark 1.56. First,* $\sigma'(\mathcal{A})$ *contains the open sets of* $\mathbb{R}^m \times \mathbb{R}^n$ *and so by definition* $\mathcal{B}(\mathbb{R}^m \times \mathbb{R}^n) \subset \sigma'(\mathcal{A})$. *The final step requires an* *application of Corollary I.7.23, noting that* $\sigma'(\mathcal{A}) \subset \sigma(\mathbb{R}^m \times \mathbb{R}^n)$, *the latter sigma algebra denoting the complete sigma algebra of Proposition I.7.20 and denoted* $\sigma(X)$.

Then by Proposition V.5.15, given $A \in \mathcal{B}(\mathbb{R}^m \times \mathbb{R}^n)$ *with* $m(A) < \infty$, $A_x \in \mathcal{B}(\mathbb{R}^n)$ *for all* $x \in \mathbb{R}^m$, *and* $A_y \in \mathcal{B}(\mathbb{R}^m)$ *for all* $y \in \mathbb{R}^n$. *For general* A, *let* $A_M \equiv A \bigcap \{|(x,y)| \leq M\}$, *where* $|(x,y)|$ *denotes the standard norm of (2.14) on* $\mathbb{R}^m \times \mathbb{R}^n$. *Then given* $x \in \mathbb{R}^m$, $(A_M)_x \in \mathcal{B}(\mathbb{R}^n)$ *for all* M, *so:*

$$A_x = \bigcup_M (A_M)_x \in \mathcal{B}(\mathbb{R}^n).$$

Similarly, $A_y \in \mathcal{B}(\mathbb{R}^m)$ *for all* $y \in \mathbb{R}^n$, *and the proof is complete.* ■

If $W : \mathcal{S} \to \mathbb{R}^k$ is a random vector defined on a probability space $(\mathcal{S}, \mathcal{E}, \mu)$ and $A \in \mathcal{B}(\mathbb{R}^k)$, recall the notational shortcut for probability statements:

$$\Pr[W \in A] \equiv \mu\left[W^{-1}(A)\right] \equiv \lambda_W(A), \tag{2.3}$$

where the last identity is (1.21).

Proposition 2.4 (Cavalieri's Principle) *Let* X, Y *be independent random vectors defined on a probability space* $(\mathcal{S}, \mathcal{E}, \mu)$:

$$X : \mathcal{S} \to \mathbb{R}^m, \qquad Y : \mathcal{S} \to \mathbb{R}^n,$$

with induced Borel measures λ_X *and* λ_Y *defined on* $\mathcal{B}(\mathbb{R}^m)$ *and* $\mathcal{B}(\mathbb{R}^n)$, *respectively.*

Then for all $A \in \mathcal{B}(\mathbb{R}^{m+n})$:

$$\Pr[(X,Y) \in A] = \int_{\mathbb{R}^j} \lambda_Y(A_x) d\lambda_X = \int_{\mathbb{R}^k} \lambda_X(A_y) d\lambda_Y. \tag{2.4}$$

Further, for $A \in \mathcal{B}(\mathbb{R}^{m+n})$:

$$\begin{aligned} \Pr[X \in B,\ (X,Y) \in A] &= \textstyle\int_B \lambda_Y(A_x) d\lambda_X, \quad \textit{for } B \in \mathcal{B}(\mathbb{R}^m), \\ \Pr[Y \in C,\ (X,Y) \in A] &= \textstyle\int_C \lambda_X(A_y) d\lambda_Y, \quad \textit{for } C \in \mathcal{B}(\mathbb{R}^n). \end{aligned} \tag{2.5}$$

Proof. *If* $A \in \mathcal{B}(\mathbb{R}^{m+n})$, *then by (2.3):*

$$\Pr[(X,Y) \in A] \equiv \mu\left[(X,Y)^{-1}(A)\right] \equiv \lambda_{(X,Y)}[A].$$

By independence of X *and* Y, $\lambda_{(X,Y)} = \lambda_X \times \lambda_Y$ *by (1.50). Recalling Definitions V.3.2 and V.3.7 on the integral of a simple functions:*

$$\lambda_{(X,Y)}[A] \equiv \int_A d\lambda_{(X,Y)} = \int_{\mathbb{R}^{m+n}} \chi_A(x,y) d(\lambda_X \times \lambda_Y), \tag{1}$$

where χ_A *denotes the characteristic function of* A, *defined to equal* 1 *on* A *and* 0 *otherwise.*

Since $\mathcal{B}(\mathbb{R}^{m+n}) = \sigma(\mathcal{A})$ *by Remark 1.54, where* $\mathcal{A}$ *is the algebra generated by the semi-algebra of measurable rectangles* $E \times F$ *defined by* $(\mathbb{R}^m, \mathcal{B}(\mathbb{R}^m), \lambda_X)$ *and* $(\mathbb{R}^n, \mathcal{B}(\mathbb{R}^n), \lambda_Y)$, *Fubini's theorem of Proposition V.5.20 applies and obtains from* (1):

$$\begin{aligned} \lambda_{(X,Y)}[A] &= \int_{\mathbb{R}^k} \left[\int_{\mathbb{R}^j} \chi_A(x,y) d\lambda_X(x)\right] d\lambda_Y(y) \\ &= \int_{\mathbb{R}^j} \left[\int_{\mathbb{R}^k} \chi_A(x,y) d\lambda_Y(y)\right] d\lambda_X(x). \end{aligned} \tag{2}$$

By Proposition 2.3, we obtain as in the justification for (1)*:*

$$\lambda_X(A_y) = \int_{\mathbb{R}^j} \chi_A(x,y)d\lambda_X(x), \qquad \lambda_Y(A_x) = \int_{\mathbb{R}^k} \chi_A(x,y)d\lambda_Y(y),$$

and this with (2) *proves (2.4).*

Focusing on the first statement in (2.5):

$$\begin{aligned}\Pr[X \in B,\ (X,Y) \in A] &\equiv \mu\left[(X,Y)^{-1}(A)\bigcap(X,Y)^{-1}(B \times \mathbb{R}^n)\right] \\ &\equiv \lambda_{(X,Y)}\left[A\bigcap(B \times \mathbb{R}^n)\right].\end{aligned}$$

Then since:

$$\chi_{A\cap(B\times\mathbb{R}^n)}(x,y) = \chi_B(x)\chi_A(x,y),$$

it follows as in (1) *then* (2)*:*

$$\begin{aligned}\Pr[X \in B,\ (X,Y) \in A] &= \int_{\mathbb{R}^{m+n}} \chi_B(x)\chi_A(x,y)d(\lambda_X \times \lambda_Y) \\ &= \int_{\mathbb{R}^m} \chi_B(x)\lambda_Y(A_x)d\lambda_X \\ &= \int_B \lambda_Y(A_x)d\lambda_X.\end{aligned}$$

The second statement in (2.5) follows as an exercise, and the proof is complete. ■

Example 2.5 (Cavalieri's principle) *We illustrate two geometric examples as in introduction to Cavalieri's principle, and then one from probability theory.*

1. *The formula for the area of a triangle is $\frac{1}{2}bh$, where b denotes the length of the base, and h the vertical height measured from the line that contains the base. This area is the same independent of the location of the vertex opposite the base, as long as h is fixed.*

   Similarly, the volume of a circular cone is $\frac{1}{3}\pi r^2 h$, where r is the radius of the base, and h the vertical height measured from the plane that contains the base. Again the volume is independent of the location of the top vertex, as long as h is fixed.

   These are both applications of Cavalieri's principle since one observes that the areas of the cross-sections are independent of the location of the vertex.

2. *For a probability theory application, let X and Y denote independent random variables with X uniformly distributed on $[0,1]$, and Y a standard normal variable. Define:*

$$A = \{Y \leq X\}.$$

   Given $x \in [0,1]$, $A_x = \{Y \leq x\}$ and $\lambda_Y(A_x) = \Phi(x)$ where Φ is the distribution function for the standard normal. Thus by (2.4), since $d\lambda_X = dx$:

$$\lambda_{(X,Y)}[A] = \int_0^1 \Phi(x)dx \approx 0.684\,37.$$

   Alternatively, $A_y = \{y \leq X\}$:

$$\lambda_X(A_y) = \begin{cases} 1, & y \leq 0, \\ 1-y, & 0 \leq y \leq 1, \\ 0, & 1 \leq y, \end{cases}$$

and so:

$$\lambda_{(X,Y)}[A] = \int_{-\infty}^{1} \lambda_X(A_y) d\lambda_Y.$$

Now the Borel measure λ_Y is defined relative to the standard normal density function by:

$$\lambda_Y[(-\infty, x]] = \frac{1}{\sqrt{2\pi}} \int_{-\infty}^{x} e^{-y^2/2} dy,$$

and thus by Proposition V.4.8:

$$\begin{aligned}\lambda_{(X,Y)}[A] &= \frac{1}{\sqrt{2\pi}} \int_{-\infty}^{1} \mu(A_y) e^{-y^2/2} dy \\ &= \frac{1}{\sqrt{2\pi}} \int_{-\infty}^{0} e^{-y^2/2} dy + \frac{1}{\sqrt{2\pi}} \int_{0}^{1} (1-y) e^{-y^2/2} dy \\ &\approx 0.684\,37.\end{aligned}$$

2.2 Sums of Independent Random Vectors

Section IV.2.2 introduced various results for the distribution and density functions for sums of independent random variables in special cases which largely required only the integration theories from Book III. This introduction was of necessity somewhat informal in places, but identified the then forthcoming mathematics of Book V that a formal development required.

With the mathematical tools of Book V now in hand, we are able to develop the general results, including the extension to random vectors.

2.2.1 Distribution Functions

Let X and Y be independent random vectors on a probability space $(\mathcal{S}, \mathcal{E}, \mu)$:

$$X, Y : \mathcal{S} \to \mathbb{R}^n,$$

with respective joint distribution functions F_X and F_Y and induced probability measures λ_X and λ_Y defined on $\mathcal{B}(\mathbb{R}^n)$.

As a combined random vector $(X,Y) : \mathcal{S} \to \mathbb{R}^n \times \mathbb{R}^n$ with joint distribution function F, the Borel measure $\lambda_{(X,Y)}$ on $\mathcal{B}(\mathbb{R}^n \times \mathbb{R}^n)$ and induced by F is $\lambda_X \times \lambda_Y$ by independence and (2.1). Since $\mathcal{B}(\mathbb{R}^n \times \mathbb{R}^n) = \mathcal{B}(\mathbb{R}^{2n})$ by Remark 1.54, this implies that for $A \in \mathcal{B}(\mathbb{R}^{2n})$:

$$\mu\left[(X,Y)^{-1}[A]\right] \equiv \lambda_{(X,Y)}[A] = (\lambda_X \times \lambda_Y)[A]. \tag{2.6}$$

Define the random vector $Z : \mathcal{S} \to \mathbb{R}^n$ by $Z \equiv X + Y$. Equivalently, $Z = f(X,Y)$ with $f(x,y) = x + y$:

$$(\mathcal{S}, \mathcal{E}, \mu) \to_{(X,Y)} (\mathbb{R}^{2n}, \mathcal{B}(\mathbb{R}^{2n}), m) \to_f (\mathbb{R}^n, \mathcal{B}(\mathbb{R}^n), m).$$

Given a Borel set $B \in \mathcal{B}(\mathbb{R}^n)$, we will with the aid of (2.4) determine:

$$\Pr[Z \in B] \equiv \mu\left[Z^{-1}[B]\right],$$

or in terms of the component mappings:

$$\Pr[Z \in B] = \mu\left[(X,Y)^{-1} f^{-1}[B]\right] \equiv \lambda_{(X,Y)}\left[f^{-1}[B]\right].$$

Since $f(x,y) = x + y$ is continuous and thus Borel measurable by Proposition I.3.13, $f^{-1}(B) \in \mathcal{B}(\mathbb{R}^{2n})$ for all $B \in \mathcal{B}(\mathbb{R}^n)$. Then by (2.6):

$$\Pr[Z \in B] = (\lambda_X \times \lambda_Y)\left[f^{-1}[B]\right]. \tag{2.7}$$

Defining $A \equiv f^{-1}(B)$:

$$A = \{(x,y)|x + y \in B\},$$

the cross-sections of A are:

$$A_x = \{y|y \in B - x\}, \qquad A_y = \{x|x \in B - y\}.$$

Here $B - x$ and $B - y$ are defined as **translates** of the Borel set B by x and y, respectively. For example:

$$B - x \equiv \{b - x|b \in B\}.$$

Exercise 2.6 (Translates of Borel sets) *Confirm that the translate of an open set is an open set (Hint: Proposition I.2.12), and from this prove that the translate of $B \in \mathcal{B}(\mathbb{R}^n)$ is a Borel set (Hint: Check sigma algebra operations).*

Definition 2.7 (Convolution of measures) *Define $f : \mathbb{R}^n \times \mathbb{R}^n \to \mathbb{R}^n$ by $f(x,y) = x+y$. The set function $\lambda_X * \lambda_Y$ defined on $B \in \mathcal{B}(\mathbb{R}^n)$ by:*

$$(\lambda_X * \lambda_Y)[B] \equiv (\lambda_X \times \lambda_Y)[f^{-1}(B)], \tag{2.8}$$

is called the ***convolution of the Borel measures λ_X and λ_Y****.*

In the terminology and notation of Definition 1.28, $\lambda_X * \lambda_Y$ is **the measure on $\mathcal{B}(\mathbb{R}^n)$ induced by the measurable transformation** $f : (\mathbb{R}^{2n}, \mathcal{B}(\mathbb{R}^{2n}), m) \to (\mathbb{R}^n, \mathcal{B}(\mathbb{R}^n), m)$:

$$\lambda_X * \lambda_Y \equiv (\lambda_X \times \lambda_Y)_f,$$

and $\lambda_X * \lambda_Y$ is indeed a measure by Proposition V.4.13.

Exercise 2.8 ($\lambda_X * \lambda_Y$ is a probability measure) *Show directly that $\lambda_X * \lambda_Y$ is a probability measure on $\mathcal{B}(\mathbb{R}^n)$. In other words, show that $\lambda_X * \lambda_Y$ is countable additive on $\mathcal{B}(\mathbb{R}^n)$, $\lambda_X * \lambda_Y(A) \geq 0$ for all $A \in \mathcal{B}(\mathbb{R}^n)$, and $\lambda_X * \lambda_Y(\mathbb{R}^n) = 1$.*

If $A \equiv f^{-1}(B)$, then by (2.8):

$$(\lambda_X * \lambda_Y)[B] \equiv (\lambda_X \times \lambda_Y)[A].$$

Recalling the A cross-sections above, it follows from (2.4) that:

$$(\lambda_X * \lambda_Y)[B] = \int_{\mathbb{R}^n} \lambda_X(B - y)d\lambda_Y = \int_{\mathbb{R}^n} \lambda_Y(B - x)d\lambda_X, \tag{2.9}$$

where the latter integral is sometimes denoted $\lambda_Y * \lambda_X$. Thus:

$$\lambda_X * \lambda_Y = \lambda_Y * \lambda_X. \tag{2.10}$$

Exercise 2.9 (Convolution is associative) *Prove that convolution is associative:*

$$(\lambda_X * \lambda_Y) * \lambda_Z = \lambda_X * (\lambda_Y * \lambda_Z). \tag{2.11}$$

Hint: Fubini's theorem of Proposition V.5.20 generalizes to triple integrals as did Proposition V.5.16 generalize to Proposition V.5.18. Assume this to prove (2.11), then prove this Fubini generalization.

The next result derives the distribution function of $Z = X + Y$, expressed in terms of the induced probability measures λ_X and λ_Y. By (1.22), these can be equivalently stated in terms of the Borel measures λ_{F_X} and λ_{F_Y} induced by the respective distribution functions in Proposition 1.27.

Proposition 2.10 (Distribution function of $Z = X + Y$) *Let X and Y be independent random vectors on a probability space $(\mathcal{S}, \mathcal{E}, \mu)$, $X, Y : \mathcal{S} \to \mathbb{R}^n$, with respective joint distribution functions F_X and F_Y defined on $\mathbb{R}^n$, and induced probability measures λ_X and λ_Y defined on $\mathcal{B}(\mathbb{R}^n)$.*

Then the joint distribution function of $Z \equiv X + Y$ is given by the Lebesgue-Stieltjes integral:

$$F_Z(z) = \int_{\mathbb{R}^n} F_X(z-y)d\lambda_Y(y) = \int_{\mathbb{R}^n} F_Y(z-x)d\lambda_X(x). \tag{2.12}$$

If F_Y has an associated joint density function f_Y, then F_Z in (2.12) is given by the Lebesgue integral:

$$F_Z(z) = \int_{\mathbb{R}^n} F_X(z-y)f_Y(y)dy, \tag{2.13}$$

with an analogous result if F_X has a joint density function f_X.

Proof. *Let $z = (z_1, ..., z_n)$ and $B \equiv \prod_{i=1}^n(-\infty, z_i]$. Then from (2.7) and (2.8):*

$$F_Z(z) = (\lambda_X \times \lambda_Y)\left[f^{-1}[B]\right] \equiv (\lambda_X * \lambda_Y)[B]. \tag{1}$$

Since:

$$B - y = \prod_{i=1}^n(-\infty, z_i - y_i],$$

(1.21) obtains:

$$\lambda_X(B-y) \equiv \mu\left[X^{-1}(B-y)\right] = F_X(z-y). \tag{2}$$

Then by (1), *(2.9) and* (2)*:*

$$F_Z(z) = \int_{\mathbb{R}^n} F_X(z-y)d\lambda_Y,$$

proving the first expression in (2.12). The second expression is derived analogously and left as an exercise.

If Y has a joint density function f_Y, then for all $A \in \mathcal{B}(\mathbb{R}^n)$, $\lambda_Y(A)$ is defined in (1.2) as a Lebesgue integral:

$$\lambda_Y(A) = \int_A f_Y(t)dt.$$

So by Proposition V.4.8:

$$\int_{\mathbb{R}^n} F_X(z-y)d\lambda_Y = \int_{\mathbb{R}^n} F_X(z-y)f_Y(y)dy,$$

which is (2.13). ■

2.2.2 Density Functions

The goal of this section is to derive the density function of $F_Z(z)$ of Proposition 2.10 when both $F_X(x)$ and $F_Y(y)$ have density functions. The existence of one of $f_X(x)$ and $f_Y(y)$ allows $F_Z(z)$ to be expressed as a Lebesgue integral in (2.13), while the existence of the other will be seen to ensure the existence of $f_Z(z)$.

For this proof, we will use Fubini's theorem of Proposition V.5.16, which requires a number of assumptions. Lebesgue product spaces $(\mathbb{R}^n, \mathcal{M}_L(\mathbb{R}^n), m)$ are complete by Proposition I.7.20 as discussed in Section I.7.6. Further, Lebesgue spaces are σ-finite (Exercise 1.3), by noting that the collection of rectangles defined by $A_M = \{(x_1, x_2, ..., x_n) | -M \leq x_j \leq M$ for all $j\}$ union to $\mathbb{R}^n$, and each A_M has Lebesgue measure $(2M)^n$.

Thus to justify the use of Fubini's theorem, we must prove that if both $f(x)$ and $g(y)$ are Lebesgue integrable functions on $(\mathbb{R}^n, \mathcal{M}_L(\mathbb{R}^n), m)$, then $f(x-y)g(y)$ is a Lebesgue integrable function on the Lebesgue product space $(\mathbb{R}^{2n}, \mathcal{M}_L(\mathbb{R}^{2n}), m)$ constructed with two copies of $(\mathbb{R}^n, \mathcal{M}_L(\mathbb{R}^n), m)$. This was proved for $n = 1$ in Propositions V.6.15 and V.6.16, so the first task for this section is to "up-dimension" these results. Then in addition to obtaining the density function $f_Z(z)$, this work will also result in the up-dimensioning of Proposition V.6.17, on the integrability of the convolution $f * g(x)$ defined in (2.16).

Given Lebesgue integrable $f(x)$ and $g(y)$ on $(\mathbb{R}^n, \mathcal{M}_L(\mathbb{R}^n), m)$, the first question is that of Lebesgue measurability of $h(x,y) \equiv f(x-y)g(y)$:

$$h : (\mathbb{R}^{2n}, \mathcal{M}_L(\mathbb{R}^{2n}), m) \to (\mathbb{R}, \mathcal{B}(\mathbb{R}), m),$$

for which it is enough to establish measurability of the component functions by Proposition V.2.13. While the measurability of $g(y)$ is readily established, perhaps surprisingly, the Lebesgue measurability of $f(x-y)$ on $\mathbb{R}^{2n}$ is somewhat challenging to prove.

One reference that explicitly addresses this question is Hewitt and Stromberg (1965), whose approach was followed in the Book V development and will be repeated here. Given Lebesgue measurable $f : \mathbb{R}^n \to \mathbb{R}$, their idea is to identify a property of $\varphi : \mathbb{R}^{2n} \to \mathbb{R}^n$ which assures the Lebesgue measurability of $f(\varphi(x,y))$, and then to prove that $\varphi(x,y) \equiv x - y$ has this property.

To appreciate the subtlety of this question on measurability of composite functions, the reader may want to recall Remark I.3.31 and Proposition I.3.33. While developed in the context of functions on $\mathbb{R}$, it is an exercise to extend these results to functions on $\mathbb{R}^n$.

Proposition 2.11 (Lebesgue measurability of $f(\varphi) : \mathbb{R}^{2n} \to \mathbb{R}^n$) *Let $\varphi : \mathbb{R}^{2n} \to \mathbb{R}^n$ be Borel measurable and with the additional property that if $N \in \mathcal{M}_L(\mathbb{R}^n)$ with $m(N) = 0$, then $\varphi^{-1}(N) \in \mathcal{M}_L(\mathbb{R}^{2n})$ and $m\left[\varphi^{-1}(N)\right] = 0$.*

Then $f(\varphi) : \mathbb{R}^{2n} \to \mathbb{R}$ is Lebesgue measurable for every Lebesgue measurable function $f : \mathbb{R}^n \to \mathbb{R}$.

Proof. *Given a set $A \in \mathcal{M}_L(\mathbb{R}^n)$, let $f(x) \equiv \chi_A(x)$, the characteristic function of A and defined to equal 1 when $x \in A$ and equal 0 otherwise. Corollary I.7.23 proves that given such A, there exists disjoint $G \in \mathcal{B}(\mathbb{R}^n)$ and $N \in \mathcal{M}_L(\mathbb{R}^n)$ so that $A = G \bigcup N$ and $m(N) = 0$.*

Then $\chi_A(\varphi(x,y)) \equiv \chi_{\varphi^{-1}(A)}(x,y)$, and since $\varphi^{-1}(A) = \varphi^{-1}(G) \bigcup \varphi^{-1}(N)$ as a disjoint union:

$$\chi_{\varphi^{-1}(A)}(x,y) = \chi_{\varphi^{-1}(G)}(x,y) + \chi_{\varphi^{-1}(N)}(x,y).$$

Since $\varphi^{-1}(G) \in \mathcal{B}(\mathbb{R}^{2n})$ by Borel measurability of φ, and $\varphi^{-1}(N) \in \mathcal{M}_L(\mathbb{R}^{2n})$ by hypothesis, $\chi_A(\varphi)$ is Lebesgue measurable for all $A \in \mathcal{M}_L(\mathbb{R}^n)$.

By Proposition V.2.13, $f(\varphi)$ is Lebesgue measurable for all simple functions $f = \sum_{j=1}^n a_j \chi_{A_j}$, since then $f(\varphi) = \sum_{j=1}^n a_j \chi_{A_j}(\varphi)$.

Given a nonnegative Lebesgue measurable function $f(x)$, Proposition V.2.28 assures the existence of an increasing sequence of simple functions $\psi_k(x)$ with $\psi_k(x) \to f(x)$ for all x. Thus $\psi_k(\varphi)(x,y) \to f(\varphi)(x,y)$ for all (x,y), and measurability of $f(\varphi)$ follows from Proposition V.2.19.

This result then extends to a general Lebesgue measurable function $f(x)$ by the decomposition $f(x) = f^+(x) - f^-(x)$ of Definition V.3.37, since these components are nonnegative by definition, and Lebesgue measurable by Proposition V.2.19. ■

The next result proves that the function $\varphi(x-y) = x - y$ is an example of a function that satisfies the requirements of Proposition 2.11. It's proof requires some of the results of Chapter I.8, and thus we first require a discussion on the Lebesgue measure space $(\mathbb{R}^n, \mathcal{M}_L(\mathbb{R}^n), m)$.

Remark 2.12 (On $(\mathbb{R}^n, \mathcal{M}_L(\mathbb{R}^n), m)$) *The Lebesgue measure space $(\mathbb{R}^n, \mathcal{M}_L(\mathbb{R}^n), m)$ can be constructed as a product space of n-copies of $(\mathbb{R}, \mathcal{M}_L(\mathbb{R}), m)$ as discussed in Section I.7.6, and also a general n-dimensional Borel space of Chapter I.8 using $F(x) = \prod_{j=1}^n x_j$. Recall that such $F(x)$ is an n-increasing and continuous from above function by Exercise 1.22.*

Chapter I.7: *$\mathcal{A}' \equiv \left\{\prod_{j=1}^n (a_j, b_j]\right\}$ is a semi-algebra of sets by Corollary I.7.3. The set function μ_0 of I.(7.3):*

$$\mu_0 \left[\prod_{j=1}^n (a_j, b_j]\right] \equiv \prod_{j=1}^n m\left[(a_j, b_j]\right], \tag{1}$$

is a pre-measure on $\mathcal{A}'$ by Proposition V.3.34, and thus by finite additivity extends to a measure on the algebra $\mathcal{A}$ of all finite disjoint unions of $\mathcal{A}'$-sets.

This measure then extends by Proposition I.7.20 to a measure $\mu_{\mathbb{R}^n}$ on a complete sigma algebra $\sigma(\mathbb{R}^n)$, where $\mathcal{A}' \subset \sigma(\mathbb{R}^n)$, and thus by Proposition I.8.1:

$$\sigma(\mathcal{A}') = \sigma(\mathcal{A}) = \mathcal{B}(\mathbb{R}^n) \subset \sigma(\mathbb{R}^n),$$

where $\sigma(\mathcal{A}')$, respectively $\sigma(\mathcal{A})$, is the smallest sigma algebra that contains $\mathcal{A}'$, respectively $\mathcal{A}$.

Chapter I.8: *On the subset $\mathcal{A}'_B = \left\{\prod_{j=1}^n (a_j, b_j] \mid -\infty < a_i \leq b_i < \infty\right\}$ of bounded right semi-closed rectangles, the set function μ'_0 is defined in I.(8.14) by:*

$$\mu'_0 \left[\prod_{j=1}^n (a_j, b_j]\right] = \sum_x sgn(x) F(x), \tag{2}$$

where each $x = (x_1, ..., x_n)$ in this summation is one of the 2^n vertices of this rectangle, so $x_i = a_i$ or $x_i = b_i$, and $sgn(x)$ is defined as -1 if the number of a_i components of x is odd, and $+1$ otherwise.

This set function extends by Proposition I.8.15 to a measure μ_F on a complete sigma algebra $\mathcal{M}_F(\mathbb{R}^n)$, with $\mathcal{A}'_B \subset \mathcal{M}_F(\mathbb{R}^n)$, and thus by Proposition I.8.1:

$$\sigma(\mathcal{A}'_B) = \sigma(\mathcal{A}) = \mathcal{B}(\mathbb{R}^n) \subset \mathcal{M}_F(\mathbb{R}^n).$$

Now by (1), (2) and Exercise 1.22, if $A = \prod_{j=1}^n (a_j, b_j] \in \mathcal{A}'_B$, then:

$$\mu_{\mathbb{R}^n}(A) = \mu_F(A) = \prod_{j=1}^n (b_j - a_j),$$

and this extends to $\mathcal{A}$ by countable additivity. Thus by the uniqueness result Proposition I.6.14,

$$\mu_{\mathbb{R}^n}(A) = \mu_F(A), \tag{3}$$

for all $A \in \sigma(\mathcal{A}) = \mathcal{B}(\mathbb{R}^n)$.

Finally, by Corollary I.7.23, any set in $\sigma(\mathbb{R}^n)$ can be bounded by sets in $\mathcal{B}(\mathbb{R}^n)$ within $\mu_{\mathbb{R}^n}$-measure 0, and by Proposition I.8.18, any set in $\mathcal{M}_F(\mathbb{R}^n)$ can be bounded by sets in $\mathcal{B}(\mathbb{R}^n)$ within μ_F-measure 0.

Thus if $B \in \mathcal{M}_F(\mathbb{R}^n)$ there exist $A, C \in \mathcal{B}(\mathbb{R}^n)$ with $A \subset B \subset C$ and $\mu_F(A) = \mu_F(C)$, so $\mu_F(B)$ equals this common value. However, by completeness, it follows that $B \in \sigma(\mathbb{R}^n)$, and since $\mu_{\mathbb{R}^n} = \mu_F$ on $\mathcal{B}(\mathbb{R}^n)$ this obtains that $\mu_{\mathbb{R}^n}(B)$ also equals this common value. Thus $\mathcal{M}_F(\mathbb{R}^n) \subset \sigma(\mathbb{R}^n)$ and $\mu_{\mathbb{R}^n} = \mu_F$ on $\mathcal{M}_F(\mathbb{R}^n)$ sets. The reverse inclusion is identical.

This obtains that $\sigma(\mathbb{R}^n) = \mathcal{M}_F(\mathbb{R}^n)$, which we denote by $\mathcal{M}_L(\mathbb{R}^n)$, and $\mu_{\mathbb{R}^n} = \mu_F$, which we denote by m.

Exercise 2.13 (m is translation invariant on $\mathcal{M}_L(\mathbb{R}^n)$) *Recall item 5 of Proposition I.2.28 that proved that Lebesgue outer measure m^* is translation invariant on the power set $\mathcal{P}(\mathbb{R})$ of all subsets of $\mathbb{R}$. Since $m \equiv m^*$ on $\mathcal{M}_L(\mathbb{R})$, and $\mathcal{M}_L(\mathbb{R})$ is closed under translations by Proposition I.2.35, meaning $A \in \mathcal{M}_L(\mathbb{R})$ if and only if $A + x \in \mathcal{M}_L(\mathbb{R})$ for any x, it follows that m is translation invariant on $\mathcal{M}_L(\mathbb{R})$.*

Prove that m is translation invariant on $\mathcal{M}_L(\mathbb{R}^n)$. Hint: Check that m has this property on $\mathcal{A}'$ and then $\mathcal{A}$ using (1) of the above proof, and then also on the sets of $\mathcal{A}_{\delta\sigma}$ and $\mathcal{A}_{\sigma\delta}$ of Proposition I.8.18. Complete the proof with item 3 of that result.

Proposition 2.14 ($\varphi(x-y) = x - y$) *The function $\varphi : \mathbb{R}^n \times \mathbb{R}^n \to \mathbb{R}^n$, defined by $\varphi(x-y) = x - y$, satisfies the requirements of Proposition 2.11.*

Proof. *First, $\varphi : \mathbb{R}^{2n} \to \mathbb{R}^n$ is Borel measurable by continuity and Proposition V.2.12.*

If $N \in \mathcal{M}_L(\mathbb{R}^n)$ with $m(N) = 0$, then $\varphi^{-1}(N) = \bigcup_{m=1}^{\infty} P_m$ with:

$$P_m \equiv \{(x,y) | x - y \in N \text{ and } |y| \leq m\},$$

where $|y|$ denotes the Euclidean or standard norm in $\mathbb{R}^n$:

$$|y| \equiv \left(\sum\nolimits_{j=1}^{n} y_j^2\right)^{1/2}. \tag{2.14}$$

If it can be proved that $P_m \in \mathcal{M}_L(\mathbb{R}^{2n})$ and $m[P_m] = 0$ for all m, then $m\left[\varphi^{-1}(N)\right] = 0$ will follow from countable subadditivity of Lebesgue measure.

To this end, m is outer regular by Proposition I.8.20, so there exists open sets $\{G'_k\}_{k=1}^{\infty} \subset \mathbb{R}^n$ with $N \subset G'_k$ for all k, and $\inf_k m[G'_k] = m(N) = 0$. Thus given $\epsilon > 0$, we can assume by selection that $m[G'_k] < \epsilon$ for all k. Defining $G_k = \bigcap_{j \leq k} G'_j$, then $\{G_k\}_{k=1}^{\infty}$ are open since open sets are a topology on $\mathbb{R}^n$ by Remark 1.54, are nested with $G_{k+1} \subset G_k$ for all k, $m[G_1] < \epsilon$, and $\lim_{k\to\infty} m[G_k] = m(N) = 0$.

Fixing m, define $\{H_k\}_{k=1}^{\infty} \subset \mathbb{R}^{2n}$ by:

$$H_k = \{(x,y) | x - y \in G_k\} \bigcap \{|y| \leq m\}.$$

As the pre-image of an open set under a continuous function, $\{(x,y) | x-y \in G_k\} = \varphi^{-1}(G_k)$ is open by Proposition V.2.10. Thus $H_k \in \mathcal{B}(\mathbb{R}^{2n})$ as the intersection of an open set $\varphi^{-1}(G_k)$ and a closed set $\{|y| \leq m\}$, and $H_{k+1} \subset H_k$ for all k. In addition,

$$P_m \subset \bigcap\nolimits_{k=1}^{\infty} H_k, \tag{1}$$

since $N \subset G'_k$ for all k obtains $N \subset G_k$ for all k, and this assures $P_m \subset H_k$ for all k by definition.

To prove that $P_m \in \mathcal{M}_L(\mathbb{R}^{2n})$ and $m[P_m] = 0$, we prove that:

$$m\left[\bigcap\nolimits_{k=1}^{\infty} H_k\right] = 0. \tag{2}$$

Then $m[P_m] = 0$ by (1) and monotonicity of Lebesgue measure, while $P_m \in \mathcal{M}_L(\mathbb{R}^{2n})$ will follow from completeness of m.

For (2), since $\{H_k\}_{k=1}^{\infty}$ are nested with $H_{k+1} \subset H_k$, if $m[H_1] < \infty$, Proposition I.2.45 assures by continuity from above of m that:

$$m\left[\bigcap\nolimits_{k=1}^{\infty} H_k\right] = \lim_{k\to\infty} m[H_k]. \tag{3}$$

Thus (2) will follow if:

$$m[H_1] < \infty, \qquad \lim_{k\to\infty} m[H_k] = 0.$$

For $m[H_1]$, recall that $G_1 = G_1'$ is an open set in $\mathbb{R}^n$ with $m[G_1'] < \epsilon$. Since m is translation invariant on $\mathcal{M}_L(\mathbb{R}^n)$ by Exercise 2.13, for any fixed $y \in \mathbb{R}^n$:

$$\begin{aligned} m[\{(x,y)|x - y \in G_1'\}] &= m[\{x \in G_1' + y\}] \\ &= m(G_1') < \epsilon. \end{aligned}$$

By Cavalieri's principle of Proposition 2.4, expressed as a Lebesgue integral:

$$m[H_1] = \int_{|y|\le n} m(G_1')\,dy < \epsilon m\left[\bar{B}_n(0)\right],$$

where $\bar{B}_n(0) = \{z|\,|z| \le n\}$ is the closed ball about 0 of radius n. Thus $m[H_1] < \infty$, and the application of Proposition I.2.45 is justified.

For $\lim_{k\to\infty} m[H_k] = 0$, let χ_{H_k} denote the characteristic function of the set H_k. Then Fubini's theorem of Proposition V.5.16 obtains that as Lebesgue integrals:

$$\begin{aligned} m[H_k] &\equiv \int_{\mathbb{R}^{2n}} \chi_{H_k}(x,y)dm \\ &= \int_{\mathbb{R}^n}\left[\int_{\mathbb{R}^n} \chi_{H_k}(x,y)dx\right]dy \\ &= \int_{|y|\le n}\left[\int_{\mathbb{R}^n} \chi_{G_k}(x-y)dx\right]dy. \end{aligned}$$

By substitution (Proposition V.4.44 with $T : x \to x - y$), for any fixed y:

$$\int_{\mathbb{R}^n} \chi_{G_k}(x-y)dx = \int_{\mathbb{R}^n} \chi_{G_k}(x)dx = m(G_k).$$

Hence, defining $\bar{B}_n(0) \equiv \{|y| \le n\}$:

$$m[H_k] = m(G_k)m\left[\bar{B}_n(0)\right],$$

and $\lim_{k\to\infty} m[H_k] = 0$ since $\lim_{k\to\infty} m[G_k] = m(N) = 0$.

Thus $m[P_m] = 0$ for all m by (3) and (1), which obtains $m\left[\varphi^{-1}(N)\right] = 0$ as was to be proved. ■

With the technical matters of Propositions 2.11 and 2.14 settled, we return to the existence of a density function for $Z = X + Y$. See also Definition 2.17.

Proposition 2.15 (Density function of $Z = X + Y$) *Let X and Y be independent random vectors on a probability space $(\mathcal{S}, \mathcal{E}, \mu)$, and define $Z = Y + Y$.*

If F_X and F_Y have density functions f_X and f_Y, then F_Z in (2.13) has a density function f_Z, expressible as Lebesgue integrals by:

$$\begin{aligned} f_Z(z) &= \int_{\mathbb{R}^n} f_X(z-y)f_Y(y)dy, \ m\text{-a.e.} \\ f_Z(z) &= \int_{\mathbb{R}^n} f_Y(z-x)f_X(x)dx, \ m\text{-a.e.} \end{aligned} \tag{2.15}$$

Proof. *Since X has a density function:*

$$F_X(z-y) \equiv \int_{A_{z-y}} f_X(x)dx = \int_{A_z} f_X(x-y)dx, \tag{1}$$

where $A_x \equiv \prod_{j=1}^{n}(-\infty, x_j]$ for $x = z-y$ or $x = z$, and the second integral follows from Proposition V.4.44 with transformation $T : x \to x - y$.

In detail, for given y define $T : (\mathbb{R}^n, \mathcal{M}_L(\mathbb{R}^n), m) \to (\mathbb{R}^n, \mathcal{M}_L(\mathbb{R}^n), m)$ by $T(x) = x - y$. Then by V.(4.44), for any measurable set A and measurable function g:

$$\int_{T^{-1}A} g(T(x)) \left|\det T'(x)\right| dm = \int_A g(x)dm.$$

Letting $g = f_X$ and $A = A_{z-y}$, (1) *follows since $T^{-1}A = A_z$ and $\det T'(x) = 1$.*

Then by (1) *and (2.13):*

$$F_Z(z) = \int_{\mathbb{R}^n}\int_{A_z} f_X(x-y)f_Y(y)dxdy. \tag{2}$$

For the integrand in (2), *$f_X(x-y)$ is Lebesgue measurable as a function defined on $\mathbb{R}^{2n}$ by Propositions 2.11 and 2.14. In addition, $f_Y(y)$ is Lebesgue measurable on $\mathbb{R}^n$ and hence measurable as a function $\tilde{f}_Y(x,y) = f_Y(y)$ on $\mathbb{R}^{2n}$ since if $A \in \mathcal{B}(\mathbb{R})$:*

$$\tilde{f}_Y^{-1}(A) = \mathbb{R}^n \times f_Y^{-1}(A) \in \mathcal{M}_L(\mathbb{R}^n).$$

Thus the integrand in (2) *is Lebesgue measurable and nonnegative, and Tonelli's theorem of Proposition V.5.23 allows the reversal of the iterated integrals to obtain:*

$$F_Z(z) = \int_{A_z}\left[\int_{\mathbb{R}^n} f_X(x-y)f_Y(y)dy\right]dx. \tag{3}$$

If a density f_Z for F_Z exists, then f_Z is Lebesgue measurable, and:

$$F_Z(z) = \int_{A_z} f_Z(x)dx.$$

Comparing with (3), *the first expression for $f_Z(z)$ in (2.15) will be a density function for F_Z if this function is Lebesgue measurable, and this follows from Tonelli's theorem. By item 4 of Proposition V.3.42, any Lebesgue measurable function that equals this integral expression m-a.e. is also a density for F_Z, and the proof of the first statement is complete.*

The second expression in (2.15) is derived similarly, reversing the roles of f_X and f_Y, and is left as an exercise. ■

Example 2.16 (Sum of independent, multivariate normals) *We will see in the Chapter 6 development of the multivariate normal distribution that if $x \equiv (x_1, ..., x_n)$:*

$$f(x) = \frac{1}{(2\pi)^{n/2}\left[\det C\right]^{1/2}} \exp\left[-\frac{1}{2}(x-\mu)^T C^{-1}(x-\mu)\right] \tag{1}$$

is a density function for a ***multivariate normal distribution*** *of the random vector* $X \equiv (X_1, ..., X_n)$*, with* ***mean vector*** $\mu \equiv (\mu_1, ..., \mu_n)$*, where* $\mu_j = E[X_j]$ *for all* j*, and* $n \times n$ ***covariance matrix*** C *defined by:*

$$C_{ij} = Cov[X_i, X_j] \equiv E\left[(X_i - \mu_i)(X_j - \mu_j)\right].$$

For more on expectations, see Chapter IV.4.

In addition, $(x-\mu)^T$ *denotes the row vector transpose of the column vector* $(x-\mu)$*,* C^{-1} *is the inverse of the covariance matrix, and* $\det C$ *denotes the determinant of this matrix.*

The ***multivariate standard normal distribution*** *is defined with* $E[X_j] = 0$ *for all* j*, and* $C = I$*, the identity matrix, and the above formula reduces to:*

$$f(x) = \frac{1}{(2\pi)^{n/2}} \exp\left[-\frac{1}{2}\sum\nolimits_{j=1}^{n} x_j^2\right]. \tag{2}$$

We now derive the density function of $Z = X + Y$ *for independent* X *and* Y*, when both have the relatively simple density function in* (2).

By (2.15):

$$\begin{aligned}
f_Z(z) &= (2\pi)^{-n} \int_{\mathbb{R}^n} \exp\left[-\frac{1}{2}\left(\sum\nolimits_{j=1}^{n} (x_j - y_j)^2 + \sum\nolimits_{j=1}^{n} y_j^2\right)\right] dy \\
&= (2\pi)^{-n} \exp\left[-\frac{1}{2}\sum\nolimits_{j=1}^{n} x_j^2\right] \int_{\mathbb{R}^n} \exp\left[-\left(\sum\nolimits_{j=1}^{n} y_j^2 - \sum\nolimits_{j=1}^{n} x_j y_j\right)\right] dy \\
&= (2\pi)^{-n} \exp\left[-\frac{1}{4}\sum\nolimits_{j=1}^{n} x_j^2\right] \int_{\mathbb{R}^n} \exp\left[-\sum\nolimits_{j=1}^{n} \left(y_j - \frac{x_j}{2}\right)^2\right] dy.
\end{aligned}$$

Tonelli's theorem of Proposition V.5.23 and a change of variable by Proposition V.4.24 obtains:

$$\begin{aligned}
\int_{\mathbb{R}^n} \exp\left[-\sum\nolimits_{j=1}^{n} \left(y_j - \frac{x_j}{2}\right)^2\right] dy &= \prod\nolimits_{j=1}^{n} \int_{\mathbb{R}} \exp\left[-\left(y_j - \frac{x_j}{2}\right)^2\right] dy_j \\
&= \left(\int_{\mathbb{R}} \exp\left[-y_1^2\right] dy_1\right)^n.
\end{aligned}$$

This integrand is seen IV.(1.65) as the functional part of the density function of a normal variate with $\mu_1 = 0$ *and* $\sigma_1 = 2^{-1/2}$*, which consequently has an integral of* $\sigma_1\sqrt{2\pi} = 2^{-1/2}(2\pi)^{1/2}$*. Taking the nth power and substituting obtains:*

$$f_Z(z) = \frac{1}{(2\pi)^{n/2}\,2^{n/2}} \exp\left[-\frac{1}{2}\sum\nolimits_{j=1}^{n} \frac{x_j^2}{2}\right]. \tag{3}$$

This again is a density function of a multivariate normal. Comparing with (1)*, it follows that* Z *has mean vector* $\mu \equiv 0$*, the sum of the* X *and* Y *mean vectors, and covariance matrix with* $C_{ii} = \sigma_i^2 = 2$ *for all* i *and* $C_{ij} = 0$ *otherwise, the sum of the* X *and* Y *covariance matrices. Further,* $\det C = 2^n$ *as a diagonal matrix.*

This result on μ *and* C *for* Z *generalizes, but the above calculation is nearly impossible with general covariance matrices, and quite messy even with general diagonal covariance matrices. Fortunately, once multivariate moment generating functions and characteristic functions have been investigated in Chapter 5, the derivation of such conclusions will be greatly simplified.*

2.3 A Result on Convolutions

In Book V, the one-dimensional analogues of Propositions 2.11 and 2.14 were used to obtain the result in Proposition V.6.17 on the Lebesgue integrability of the convolution of Lebesgue integrable functions on $\mathbb{R}$. Thus it can be of no surprise that here we use to above results to obtain the analogous result on $\mathbb{R}^n$.

Recalling Definition V.6.13, but now framed on $\mathbb{R}^n$:

Definition 2.17 (Convolution of functions) *Let $f(x)$ and $g(x)$ be measurable functions on the Lebesgue measure space $(\mathbb{R}^n, \mathcal{M}_L(\mathbb{R}^n), m)$. The* **convolution of** f **and** g, *denoted $f * g$, is defined by the Lebesgue integral:*

$$f * g(x) = \int f(x-y)g(y)dy, \tag{2.16}$$

when this integral exists.

Exercise 2.18 ($f * g = g * f$) *Show that for any x for which the integral in (2.16) exists, that $f * g(x) = g * f(x)$. In other words:*

$$\int f(x-y)g(y)dy = \int g(x-y)f(y)dy. \tag{2.17}$$

Hint: Recall the transformation $T : \mathbb{R}^n \to \mathbb{R}^n$ in the proof of Proposition 2.15 and apply Proposition V.4.44.

Exercise 2.19 (Existence of $f * g$) *Prove that if one of f and g is Lebesgue integrable, and the other bounded, then $f * g(x)$ exists. Thus, existence does not require that both f and g be Lebesgue integrable. Hint: The triangle inequality of Proposition V.3.42.*

The integral expression for the distribution function F_Z in (2.13) thus gives the **convolution of F_X and f_Y**:

$$F_Z(z) = F_X * f_Y(z).$$

By Exercise 2.19, this convolution is well-defined on $\mathbb{R}^n$ since f_Y is integrable and F_X is bounded.

Similarly, the expression for the density function f_Z in (2.13) gives the **convolution of f_X and f_Y**:

$$f_Z(z) = f_X * f_Y(z).$$

Proposition 2.15 proves that if f_X and f_Y are density functions of distributions, and thus nonnegative and Lebesgue integrable, then this convolution is well-defined.

The next result generalizes Proposition V.6.17 to $\mathbb{R}^n$, proving that the convolution of Lebesgue integrable functions is well-defined, and indeed, $f * g(x)$ is Lebesgue integrable.

Proposition 2.20 (Lebesgue integrability of $f * g(x)$) *Let $f(x)$ and $g(x)$ be integrable functions on the Lebesgue measure space $(\mathbb{R}^n, \mathcal{M}_L(\mathbb{R}^n), m)$. Then:*

1. *The function $f(x-y)g(y)$ is Lebesgue integrable in y for almost all x. That is:*

$$\int |f(x-y)g(y)|\, dy < \infty, \ m\text{-a.e.}$$

2. *Defining* $f*g(x)$ *as in (2.15), then* $f*g(x)$ *is Lebesgue integrable with:*

$$\int |f*g(x)|\,dx \leq \int |f(x)|\,dx \int |g(y)|\,dy, \tag{2.18}$$

and:

$$\int f*g(x)dx = \int f(x)dx \int g(y)dy. \tag{2.19}$$

Proof. *By Propositions 2.11 and 2.14, if* $f(x)$ *is Lebesgue measurable on* $\mathbb{R}^n$, *then* $f(x-y)$ *is Lebesgue measurable on* $\mathbb{R}^{2n}$. *Similarly, if* $g(y)$ *is Lebesgue measurable on* $\mathbb{R}^n$ *then* $g(y)$ *is Lebesgue measurable on* $\mathbb{R}^{2n}$ *as demonstrated in the proof of Proposition 2.15. Thus* $F(x,y) \equiv f(x-y)g(y)$ *is Lebesgue measurable on* $\mathbb{R}^{2n}$.

To prove integrability of $|F(x,y)|$, *an application of Tonelli's theorem of Proposition V.5.23 obtains:*

$$\int_{\mathbb{R}^{2n}} |f(x-y)g(y)|\,dm = \int_{\mathbb{R}^n}\left(\int_{\mathbb{R}^n} |f(x-y)|\,dx\right)|g(y)|\,dy.$$

By a change of variable as in Exercise 2.18:

$$\int |f(x-y)|\,dx = \int |f(x)|\,dx, \text{ for all } y,$$

and hence:

$$\int_{\mathbb{R}^{2n}} |f(x-y)g(y)|\,dm = \int_{\mathbb{R}^n} |f(x)|\,dx \int_{\mathbb{R}^n} |g(y)|\,dy. \tag{1}$$

By (1), *Lebesgue integrability of* $f(x)$ *and* $g(x)$ *on* $\mathbb{R}^n$ *assures Lebesgue integrability of* $|f(x-y)g(y)|$ *on* $\mathbb{R}^{2n}$, *and thus integrability of* $f(x-y)g(y)$ *on* $\mathbb{R}^{2n}$ *by Definition V.3.38. Fubini's theorem of Proposition V.5.16 then obtains that* $f(x-y)g(y)$ *is Lebesgue integrable in* y *for almost all* x, *which is item 1.*

By the triangle inequality of Proposition V.3.42, and a change of variables as above:

$$\begin{aligned}\int |f*g(x)|\,dx &\leq \int\int |f(x-y)g(y)|\,dydx \\ &= \int |f(x)|\,dx \int |g(y)|\,dy.\end{aligned}$$

This proves the Lebesgue integrability of $|f*g(x)|$ *and (2.18).*

Applying Fubini's theorem to integrable $f(x-y)g(y)$, *and another change of variables, obtains:*

$$\begin{aligned}\int f*g(x)dx &= \int\int f(x-y)g(y)dydx \\ &= \int\int f(x-y)dxg(y)dy \\ &= \int f(x)dx \int g(y)dy,\end{aligned}$$

which is (2.19). ■

Recalling the Banach space development of Chapter V.9, and in particular the $L_p(X)$-spaces of Definition V.9.16, we have the following restatement of the above result.

Proposition 2.21 ($L_1(\mathbb{R}^n)$ is closed under convolution) *If* $f,g \in L_1(\mathbb{R}^n)$, *then* $f*g \in L_1(\mathbb{R}^n)$ *and:*

$$\|f*g\|_1 \leq \|f\|_1 \|g\|_1. \tag{2.20}$$

Proof. *This is a restatement of (2.18) in the notation of the* $L_1(\mathbb{R}^n)$*-norm of Definition V.9.16 with* $L_1(\mathbb{R}^n)$ *defined on* $(\mathbb{R}^n, \mathcal{M}_L(\mathbb{R}^n), m)$. ■

2.4 Ratios of Independent Random Variables

As was the case for sums of independent random variables, Section IV.2.3 introduced results for the distribution and density functions for ratios of independent random variables using the integration theories from Book III. With the mathematical tools of Book V now in hand, we are able to develop more general results.

Let X and Y be independent random variables on a probability space $(\mathcal{S}, \mathcal{E}, \mu)$:

$$X, Y : \mathcal{S} \to \mathbb{R},$$

with respective distribution functions F_X and F_Y and induced probability measures λ_X and λ_Y defined on $\mathcal{B}(\mathbb{R})$. If taken as a random vector $(X, Y) : \mathcal{S} \to \mathbb{R}^2$ with joint distribution function F, the Borel measure on $\mathcal{B}(\mathbb{R} \times \mathbb{R})$ and induced by F is given by $\lambda_X \times \lambda_Y$ due to independence and (2.1). Since $\mathcal{B}(\mathbb{R} \times \mathbb{R}) = \mathcal{B}(\mathbb{R}^2)$ by Remark 1.54, this implies as in (2.6) that for $A \in \mathcal{B}(\mathbb{R}^2)$:

$$\mu\left[(X, Y)^{-1}[A]\right] \equiv (\lambda_X \times \lambda_Y)[A].$$

Define $Z : \mathcal{S} \to \mathbb{R}$ by $Z \equiv X/Y$, and assume that $\mu\left[Y^{-1}[0]\right] \equiv \lambda_Y[0] = 0$. Then Z is Borel measurable on $\mathcal{S} - \{Y^{-1}[0]\}$ by item 4 of Proposition V.2.13. Defining:

$$\bar{Z} = \begin{cases} Z, & s \in \mathcal{S} - \{Y^{-1}[0]\}, \\ 0, & s \in \{Y^{-1}[0]\}, \end{cases}$$

$\bar{Z}$ is Borel measurable on $\mathcal{S}$, $\bar{Z} = Z$ μ-a.e., and thus $\bar{Z}$ and Z have the same distribution function.

To simplify notation, assume that Z is Borel measurable on $\mathcal{S}$. Formally, $Z = g(X, Y)$ with $g(x, y) = x/y$:

$$(\mathcal{S}, \mathcal{E}, \mu) \to_{(X,Y)} \left(\mathbb{R}^2, \mathcal{B}(\mathbb{R}^2), m^2\right) \to_g (\mathbb{R}, \mathcal{B}(\mathbb{R}), m).$$

Given a Borel set $B \in \mathcal{B}(\mathbb{R})$, we will with the aid of (2.4) determine:

$$\Pr[Z \in B] \equiv \mu\left[Z^{-1}[B]\right],$$

or in terms of the component mappings with $Z = g(X, Y)$:

$$\Pr[Z \in B] = \mu\left[(X, Y)^{-1} g^{-1}[B]\right].$$

To justify that this expression is well-defined, note that if $B \in \mathcal{B}(\mathbb{R})$ then $g^{-1}(B) \in \mathcal{B}(\mathbb{R}^2)$. This follows since $g(x, y) = x/y$ is a continuous for $y \neq 0$ and hence Borel measurable function for $y \neq 0$ by Proposition I.3.13, and for any such B, $g^{-1}(B) \subset \mathbb{R}^2 - \{y = 0\}$. Thus as stated above:

$$\Pr[Z \in B] = (\lambda_X \times \lambda_Y)\left[g^{-1}[B]\right]. \tag{2.21}$$

Let $A \equiv g^{-1}(B)$:

$$A = \{(x, y) | x/y \in B\}.$$

Then since $A \subset \mathbb{R}^2 - \{y = 0\}$, $\Pr[Z \in B]$ in (2.21) can be expressed:

$$(\lambda_X \times \lambda_Y)[A] = (\lambda_X \times \lambda_Y)\left[A^+\right] + (\lambda_X \times \lambda_Y)\left[A^-\right], \tag{2.22}$$

where A^+, respectively A^-, denotes A restricted to $y > 0$, respectively, $y < 0$.

The y-cross-sections of $A^{\pm}$ are defined:

$$A_y^+ = \{x|x \in B \cdot y,\ y > 0\}, \qquad A_y^- = \{x|x \in B \cdot y,\ y < 0\},$$

where $B \cdot y$ is defined as the dilation/contraction of the Borel set B by y:

$$B \cdot y \equiv \{by|b \in B\}.$$

Exercise 2.22 **($B \cdot y \in \mathcal{B}(\mathbb{R})$ for $B \in \mathcal{B}(\mathbb{R})$)** *Prove that $B \cdot y \in \mathcal{B}(\mathbb{R})$ if $B \in \mathcal{B}(\mathbb{R})$. Hint: Start with the semi-algebra $\mathcal{A}' = \{(a, b]\}$, and then progress as in Exercise 2.13.*

Proposition 2.23 (Distribution function of $Z = X/Y$) *Let X and Y be independent random variables on a probability space $(\mathcal{S}, \mathcal{E}, \mu)$, $X, Y : \mathcal{S} \to \mathbb{R}$, with $\mu\left[Y^{-1}[0]\right] \equiv \lambda_Y[0] = 0$, and with respective distribution functions F_X and F_Y and induced probability measures λ_X and λ_Y defined on $\mathcal{B}(\mathbb{R})$.*

Then the distribution function of $Z \equiv X/Y$ is given as a Lebesgue-Stieltjes integral:

$$F_Z(z) = \int_{-\infty}^{0} \left[1 - F_X(yz^-)\right] d\lambda_Y(y) + \int_0^{\infty} F_X(yz) d\lambda_Y(y), \tag{2.23}$$

where $F_X(yz^-)$ denotes the left limit of F_X at yz.

If F_Y has a density function $f_Y(y)$, then expressed as a Lebesgue integral:

$$F_Z(z) = \int_{-\infty}^{0} \left[1 - F_X(yz^-)\right] f_Y(y) dy + \int_0^{\infty} F_X(yz) f_Y(y) dy. \tag{2.24}$$

Proof. *For $B \in \mathcal{B}(\mathbb{R})$, let $A \equiv g^{-1}(B)$. Then by (2.21), (2.22), and then (2.4):*

$$\begin{aligned} \Pr[Z \in B] &= (\lambda_X \times \lambda_Y)\left[A^+\right] + (\lambda_X \times \lambda_Y)\left[A^-\right] \\ &= \int_{-\infty}^{0} \lambda_X(A_y^+) d\lambda_Y(y) + \int_0^{\infty} \lambda_X(A_y^-) d\lambda_Y(y). \end{aligned}$$

These integrals are well-defined since $\lambda_Y[0] = 0$.

Taking $B = (-\infty, z]$:

$$A_y^+ = (-\infty, zy], \qquad A_y^- = [zy, \infty),$$

and thus $\lambda_X\left(A_y^-\right) = \lambda_X[zy, \infty) \equiv 1 - F_X(yz^-)$, and $\lambda_X(A_y^+) \equiv F_X(yz)$. This now obtains (2.23).

If F_Y has a density function, meaning that as a Lebesgue integral:

$$F_Y(y) = \int_{-\infty}^{y} f_Y(t) dt,$$

then (2.24) follows by Proposition V.4.8. ∎

Turning next to the existence of a density function for $Z = X/Y$, we again assume that both F_X and F_Y have density functions.

Proposition 2.24 (Density function of $Z = X/Y$) *Let X and Y be independent random variables on a probability space $(\mathcal{S}, \mathcal{E}, \mu)$ with $\mu\left[Y^{-1}[0]\right] = 0$.*

If both F_X and F_Y have density functions f_X and f_Y, then the distribution function F_Z of $Z \equiv X/Y$ in (2.24) has a density function expressible as a Lebesgue integral:

$$f_Z(z) = \int_{-\infty}^{\infty} |y| f_X(yz) f_Y(y) dy, \ m\text{-a.e.} \tag{2.25}$$

Proof. *If X has a density function f_X, then F_X is continuous by Proposition III.3.35, noting that for fixed $a < x$:*

$$F_X(x) = \int_{-\infty}^{x} f_X(t)dt = F(a) + \int_{a}^{x} f_X(t)dt.$$

Thus $F_X(yz^-) = F_X(yz)$, so by (2.24):

$$F_Z(z) = \int_{-\infty}^{0} \left[\int_{yz}^{\infty} f_X(x)dx\right] f_Y(y)dy + \int_{0}^{\infty} \left[\int_{-\infty}^{yz} f_X(x)dx\right] f_Y(y)dy. \tag{1}$$

For fixed y, define a transformation T on $\mathbb{R}$ by $T : x \to yx$, so $|T'(x)| = |y|$. By the Lebesgue change of variable formula of Proposition V.4.44, and $A = [yz, \infty)$ with $y < 0$:

$$\int_{yz}^{\infty} f_X(x)dx \equiv \int_{A} f_X(x)dx = \int_{T^{-1}A} f_X(yx)\,|y|\,dx = \int_{-\infty}^{z} |y|\, f_X(yx)dx.$$

Similarly with $A = (-\infty, yz]$ and $y > 0$:

$$\int_{-\infty}^{yz} f_X(x)dx = \int_{A} f_X(x)dx = \int_{T^{-1}A} f_X(yx)\,|y|\,dx = \int_{-\infty}^{z} |y|\, f_X(yx)dx.$$

Combining into (1)*:*

$$\begin{aligned} F_Z(z) &= \int_{-\infty}^{0}\int_{-\infty}^{z} |y|\, f_X(yx)f_Y(y)dxdy + \int_{0}^{\infty}\int_{-\infty}^{z} |y|\, f_X(yx)f_Y(y)dxdy \\ &= \int_{-\infty}^{\infty}\int_{-\infty}^{z} |y|\, f_X(yx)f_Y(y)dxdy. \end{aligned}$$

As this integrand is nonnegative and Lebesgue measurable, we can reverse the iterated integrals by Tonelli's theorem of Proposition V.5.23:

$$F_Z(z) = \int_{-\infty}^{z}\int_{-\infty}^{\infty} |y|\, f_X(yx)f_Y(y)dydx. \tag{2}$$

Tonelli's result also obtains that:

$$f_Z(x) = \int_{-\infty}^{\infty} |y|\, f_X(yx)f_Y(y)dy$$

is Lebesgue measurable, and so by (2)*, $f_Z(x)$ is a density function for F_Z.*

By item 3 of Proposition V.3.42, any Lebesgue measurable function that equals this integral expression m-a.e. is also a density for F_Z, and the proof is complete. ■

Example 2.25 (Ratio of standard normals is standard Cauchy) *Assume that X and Y are independent standard normal variables on a probability space $(\mathcal{S}, \mathcal{E}, \mu)$ with common density $\phi(x)$ given in (6.2):*

$$\phi(x) = \frac{1}{\sqrt{2\pi}} \exp\left(-x^2/2\right),$$

and define $Z = X/Y$.

Since $\mu\left[Y^{-1}[0]\right] = \lambda_Y[0] = 0$, *we can apply (2.25), and simplify by symmetry:*

$$\begin{aligned} f_Z(z) &= \frac{1}{2\pi}\int_{-\infty}^{\infty} |y| \exp\left(-\left(1+z^2\right)y^2/2\right)dy \\ &= \frac{1}{\pi}\int_0^{\infty} y \exp\left(-\left(1+z^2\right)y^2/2\right)dy. \end{aligned}$$

This is a Lebesgue integral, but because the integrand is continuous and the integral exists as a Riemann integral, Proposition III.2.56 assures that these integral values agree.

This Riemann integral can then be evaluated using a simple substitution of $u = \left(1+z^2\right)y^2/2$ *(recall Section III.1.1.6), and so* $du = \left(1+z^2\right)ydy$ *and:*

$$\begin{aligned} f_Z(z) &= \frac{1}{\pi\left(1+z^2\right)}\int_0^{\infty} \exp(-u)\,du \\ &= \frac{1}{\pi\left(1+z^2\right)}. \end{aligned}$$

This is the ***standard Cauchy density function*** *in IV.(1.69), named for* ***Augustin Louis Cauchy*** *(1789–1857).*

Thus the ratio of independent standard normal variates is standard Cauchy.

2.5 Densities of Transformed Random Vectors

Section IV.2.1 investigated the distribution function and density function for the random variable $Y \equiv g(X)$, where $X : \mathcal{S} \to \mathbb{R}$ is a random variable defined on $(\mathcal{S}, \mathcal{E}, \mu)$ with associated distribution function $F(x)$, and $g : \mathbb{R} \to \mathbb{R}$ is Borel measurable. In Proposition IV.2.1, $F_Y(y)$ was derived for monotonically increasing or monotonically decreasing g. When $F(x)$ is continuously differentiable with density function $f(x)$, the density function for Y was derived for continuously differentiable g with $g'(x) \neq 0$:

$$f_Y(y) = f_X\left(g^{-1}(y)\right)\left|\frac{dg^{-1}(y)}{dy}\right|. \tag{2.26}$$

In this section, we generalize these results to a random vector $X : \mathcal{S} \to \mathbb{R}^n$ and one-to-one transformation $g : \mathbb{R}^n \to \mathbb{R}^n$, using the integration theory of Book V. The first result addresses existence of $f_Y(y)$ generally, where the ultimate formula reflects the generally unknown Radon-Nikodým derivative. With more assumptions on g, the second result provides an explicit formula for $f_Y(y)$.

For these results, recall Definition 1.11 on the notion that ν is **absolutely continuous with respect to** μ, denoted $\nu \ll \mu$, and the related Radon-Nikodým theorem of Proposition 1.12. For the first result, note that we require a **Borel measurable version of the density** $f(x)$. This is always possible since by Proposition 1.13, a density exists if and only if $\lambda_X \ll m$ on $\mathcal{B}(\mathbb{R}^n)$, and in this case, a Borel measurable $f(x)$ is provided by the Radon-Nikodým theorem.

The reason for this technical restriction on $f(x)$ is the need to identify the measurability of the composite function $f(g^{-1})$, and recalling the discussion before Proposition I.3.33, Lebesgue measurable functions do not compose predictably.

Proposition 2.26 (Density function for a transformed RV) *Let $X \equiv (X_1, X_2, ..., X_n)$ be a random vector defined on $(\mathcal{S}, \mathcal{E}, \mu)$ with a joint distribution function $F(x_1, x_2, ..., x_n)$ and an associated Borel measurable density function $f(x_1, x_2, ..., x_n)$. Let $g : (\mathbb{R}^n, \mathcal{B}(\mathbb{R}^n), m) \rightarrow (\mathbb{R}^n, \mathcal{B}(\mathbb{R}^n))$ be Borel measurable, one-to-one, and let m_g be the induced measure defined on $A \in \mathcal{B}(\mathbb{R}^n)$ as in (2.8):*

$$m_g(A) \equiv m(g^{-1}(A)),$$

where m is Lebesgue measure on $\mathbb{R}^n$.

If $m_g \ll m$ and g^{-1} is Borel measurable, then $Y = (Y_1, Y_2, ..., Y_n) \equiv g(X)$ has a density function given by:

$$f_Y(y) = f\left(g^{-1}(y)\right)\rho(y), \quad m\text{-a.e.}, \tag{2.27}$$

where $\rho(y) \equiv \frac{\partial m_g}{\partial m}$ is the Radon-Nikodým derivative of m_g with respect to m.

Proof. *Borel measurability of g assures that $g^{-1}(A) \in \mathcal{B}(\mathbb{R}^n)$ for all $A \in \mathcal{B}(\mathbb{R}^n)$. Hence $Y \equiv g(X)$ is a random vector on $(\mathcal{S}, \mathcal{E}, \mu)$ since if $A \in \mathcal{B}(\mathbb{R}^n)$:*

$$Y^{-1}(A) = X^{-1}(g^{-1}(A)) \in \mathcal{E}.$$

If λ_Y is the probability measure of (1.21) induced by Y, then for $A \in \mathcal{B}(\mathbb{R}^n)$:

$$\lambda_Y(A) \equiv \mu\left[Y^{-1}(A)\right] = \mu\left[X^{-1}(g^{-1}A)\right] \equiv \lambda_X(g^{-1}A).$$

Thus by (1.8) and (1.22), with $A_y \equiv \prod_{j=1}^n(-\infty, y_j]$:

$$F_Y(y_1, x_2, ..., y_n) = \lambda_Y(A_y) = \lambda_X(g^{-1}A_y).$$

Since a density $f(x)$ of X is a density of λ_X by Proposition 1.34:

$$F_Y(y_1, x_2, ..., y_n) = \int_{g^{-1}A_y} f(x)dm. \tag{1}$$

As g is assumed one-to-one:

$$f(x) = f(g^{-1}[g(x)]),$$

so by Proposition V.4.18, (1) can be expressed:

$$F_Y(y_1, x_2, ..., y_n) = \int_{A_y} f(g^{-1}y)dm_g. \tag{2}$$

If $m_g \ll m$, the Radon–Nikodym theorem of Proposition 1.12 obtains a Borel measurable function $\rho : \mathbb{R}^n \rightarrow \mathbb{R}$, also denoted $\rho \equiv \frac{\partial m_g}{\partial m}$, so that for all $A \in \mathcal{B}(\mathbb{R}^n)$:

$$m_g(A) = \int_A \rho(y)dm.$$

Proposition V.4.8 then provides the restatement of (2):

$$F_Y(y_1, x_2, ..., y_n) = \int_{A_y} f(g^{-1}y)\rho(y)dm. \tag{3}$$

The integrand in (3) is nonnegative, and Borel measurable by the measurability assumptions on f, g, and ρ, recalling Proposition I.3.33 on compositions (generalized to $\mathbb{R}^n$ as an exercise) and Proposition V.2.13. Since the identity in (3) is true for all $(y_1, x_2, ..., y_n)$, $f_Y(y) = f\left(g^{-1}(y)\right)\rho(y)$ is a density function for Y, as is by item 3 of Proposition III.2.49, any function equal to $f_Y(y)$, m-a.e. ■

In order to obtain a more explicit formula for $f_Y(y)$, we will need to make additional assumptions on the transformation $g : \mathbb{R}^n \to \mathbb{R}^n$ that will allow us to identify $\rho(y) \equiv \frac{\partial m_g}{\partial m}$, the Radon-Nikodým derivative of m_g with respect to m. Specifically, we will assume that $g : \mathbb{R}^n \to \mathbb{R}^n$ is a continuously differentiable, one-to-one transformation with $\det(g'(x)) \neq 0$, where $\det(g'(x))$ denotes the **Jacobian determinant**.

As background, letting $x = (x_1, x_2, ..., x_n)$:

$$g(x) \equiv (g_1(x), g_2(x), \cdots, g_n(x)),$$

where $g_i : \mathbb{R}^n \to \mathbb{R}$ for all i. We will say that g is continuous, differentiable, continuously differentiable, etc., if the component functions $\{g_i\}_{i=1}^n$ have such properties.

When this transformation has differentiable component functions, the **Jacobian matrix** associated with g, denoted $g'(x)$, is defined:

$$g'(x) \equiv \begin{pmatrix} \frac{\partial g_1}{\partial x_1} & \frac{\partial g_1}{\partial x_2} & \cdots & \frac{\partial g_1}{\partial x_n} \\ \frac{\partial g_2}{\partial x_1} & \frac{\partial g_2}{\partial x_2} & \cdots & \frac{\partial g_2}{\partial x_n} \\ \vdots & \vdots & \vdots & \vdots \\ \frac{\partial g_n}{\partial x_1} & \frac{\partial g_n}{\partial x_2} & \cdots & \frac{\partial g_n}{\partial x_n} \end{pmatrix}. \tag{2.28}$$

The **Jacobian determinant** associated with g, denoted $\det(g'(x))$, is defined as the determinant of $g'(x)$. This matrix and its determinant are named for **Carl Gustav Jacob Jacobi** (1804–1851), an early developer of determinants and their applications in analysis.

Notation 2.27 (On $g'(x)$) *The Jacobian matrix is denoted in many ways:*

$$g'(x) \equiv \frac{\partial(g_1, g_2, ..., g_n)}{\partial(x_1, x_2, ..., x_n)} = \left(\frac{\partial(g_1, g_2, ..., g_n)}{\partial(g_1, g_2, ..., g_n)}\right) = \left(\frac{\partial g}{\partial x}\right).$$

Proposition 2.28 (Density function of a smoothly transformed RV) *Let $X \equiv (X_1, X_2, ..., X_n)$ be a random vector defined on $(\mathcal{S}, \mathcal{E}, \mu)$ with a joint distribution function $F(x_1, x_2, ..., x_n)$ and an associated Borel measurable density function $f(x_1, x_2, ..., x_n)$.*

If $g : \mathbb{R}^n \to \mathbb{R}^n$ is continuously differentiable, one-to-one, and with $\det(\frac{\partial g^{-1}}{\partial y}) \neq 0$ for all y, then $Y \equiv g(X)$ has a density function given by:

$$f_Y(y) = f\left(g^{-1}(y)\right) \left|\det\left(\frac{\partial g^{-1}}{\partial y}\right)\right|, \quad m\text{-a.e.}, \tag{2.29}$$

where $\frac{\partial g^{-1}}{\partial y}$ denotes the Jacobian matrix of g^{-1} as defined in (2.28).

Proof. *Since g is continuously differentiable and one-to-one, $T \equiv g^{-1}$ is continuously differentiable by the inverse function theorem of Proposition V.4.38. It then follows from Proposition V.4.46 that for all $A \in \mathcal{B}(\mathbb{R}^n)$:*

$$m\left[(T(A)\right] = \int_A |\det(T'(y))|\, dm.$$

Recalling the induced measure $m_g(A) \equiv m(g^{-1}(A))$, this can be rewritten:

$$m_g(A) = \int_A \left|\det\left(\frac{\partial g^{-1}}{\partial y}\right)\right| dm.$$

Now $\left|\det\left(\frac{\partial g^{-1}}{\partial y}\right)\right|$ *is nonnegative and continuous, and thus Borel measurable by Proposition V.2.12, and so* $\left|\det\left(\frac{\partial g^{-1}}{\partial y}\right)\right|$ *is a density function for the measure* m_g*. Stated another way:*

$$\left|\det\left(\frac{\partial g^{-1}}{\partial y}\right)\right| = \frac{\partial m_g}{\partial m},$$

the Radon-Nikodým derivative of m_g *with respect to* m*.*

Thus (2.29) follows from (2.27). ■

Gamma and beta distribution functions were discussed in Section IV.1.4.2. As an application of the above result we generalize somewhat Proposition IV.2.30, which proved that if X and Y are independent gamma, then:

$$Z = \frac{X}{X+Y}$$

is a beta variate.

Proposition 2.29 (Transformation of gamma variates) *Let* (X_1, X_2) *be a random vector of independent gamma random variables defined on* $(\mathcal{S}, \mathcal{E}, \mu)$ *with parameters* α_1, λ *and* α_2, λ*, respectively. Define the random vector:*

$$(Y_1, Y_2) \equiv g(X_1, X_2) = \left(\frac{X_1}{X_1 + X_2}, X_1 + X_2\right).$$

Then Y_1 *and* Y_2 *are independent random variables, where* Y_1 *is beta with parameters* α_1 *and* α_2*, and* Y_2 *is gamma with parameters* $\alpha_1 + \alpha_2, \lambda$*.*

Proof. *By IV.(1.55) and Proposition 1.58, for* $x_1 > 0$*,* $x_2 > 0$*:*

$$f(x_1, x_2) = \frac{\lambda^{\alpha_1+\alpha_2} x_1^{\alpha_1-1} x_2^{\alpha_2-1} e^{-\lambda(x_1+x_2)}}{\Gamma(\alpha_1)\Gamma(\alpha_2)},$$

where the gamma function $\Gamma(\alpha)$ *is defined for* $\alpha > 0$*:*

$$\Gamma(\alpha) = \int_0^\infty x^{\alpha-1} e^{-x} dx. \tag{2.30}$$

The inverse transformation is defined by:

$$g^{-1}(y_1, y_2) = (y_1 y_2, (1 - y_1) y_2),$$

and so:

$$\det\left(\frac{\partial g^{-1}}{\partial y}\right) = \det\begin{pmatrix} y_2 & y_1 \\ -y_2 & 1 - y_1 \end{pmatrix} = y_2.$$

Now:

$$g\{(x_1, x_2)\,|x_1 > 0, x_2 > 0\} = \{(y_1, y_2)\,|y_1 > 0, 0 < y_2 < 1\},$$

and so from (2.29):

$$\begin{aligned} f_Y(y_1, y_2) &= \frac{\lambda^{\alpha_1+\alpha_2} (y_1 y_2)^{\alpha_1-1} ((1-y_1) y_2)^{\alpha_2-1} y_2 e^{-\lambda y_2}}{\Gamma(\alpha_1)\Gamma(\alpha_2)} y_2 \\ &= \frac{\Gamma(\alpha_1+\alpha_2)}{\Gamma(\alpha_1)\Gamma(\alpha_2)} y_1^{\alpha_1-1} (1-y_1)^{\alpha_2-1} \times \frac{\lambda^{\alpha_1+\alpha_2} y_2^{\alpha_1+\alpha_2-1} e^{-\lambda y_2}}{\Gamma(\alpha_1+\alpha_2)}. \end{aligned}$$

Thus $f_Y(y_1, y_2)$ *is a product of density functions, a beta density with parameters* α_1 *and* α_2 *on* $y_1 > 0$ *by IV.(1.61) and IV.(1.63), and a gamma density with parameters* $\alpha_1 + \alpha_2, \lambda$ *on* $y_2 \in (0, 1)$*. As a product of density functions, independence follows from Proposition 1.58.* ■

3

Weak Convergence of Probability Measures

With the integration theory of Book V, this chapter derives deeper results on the weak convergence of random vectors, extending the studies and applications of Books II and IV.

Starting with a briefer version of the portmanteau theorem on $\mathbb{R}$ applicable to random variables, a general version of this result is derived on $\mathbb{R}^m$. This theorem provides a host of useful, equivalent formulations for the weak convergence of a sequence of probability measures, or equivalently, the convergence in distribution of a sequence of random vectors.

Turning next to applications, the chapter generalizes Book II results from $\mathbb{R}$ to $\mathbb{R}^m$ for the mapping theorem, Mann-Wald theorem, Slutsky's theorem, the Delta method, and Prokhorov's theorem, and then introduces Scheffé's theorem and part one of a very useful result called the Cramér-Wold device. This last result is completed in Section 5.6.1.

Remark 3.1 **(λ_F vs. λ_X)** *Recall that from Proposition 1.30 that on $\mathcal{B}(\mathbb{R}^n)$:*

$$\lambda_F = \lambda_X,$$

where λ_X is the probability measure induced by a random vector X as defined in (1.21), and λ_F is the probability measure induced by the joint distribution function $F(x)$ of X of Proposition 1.27.

In this chapter, we will alternate between notations depending on the context. Given a discussion related to X, λ_X is the more natural designation, while for any discussion related to F, we will use λ_F.

3.1 Portmanteau Theorem on $\mathbb{R}$

In this section we present two important implications of the weak convergence of a sequence of measures on $\mathbb{R}$. Recall Definition II.8.2:

Definition 3.2 (Weak convergence: Distributions and measures on $\mathbb{R}$) *A sequence of* ***distribution functions*** *$\{F_n(x)\}_{n=1}^{\infty}$ on $\mathbb{R}$ is said to* ***converge weakly*** *to a distribution function $F(x)$, denoted:*

$$F_n \Rightarrow F,$$

if $F_n(x) \to F(x)$ for every continuity point of $F(x)$.

A sequence of ***probability measures*** *$\{\lambda_n\}_{n=1}^{\infty}$ on $\mathbb{R}$ is said to* ***converge weakly*** *to a probability measure λ, denoted:*

$$\lambda_n \Rightarrow \lambda,$$

if $\lambda_n((-\infty, x])$ converges to $\lambda((-\infty, x])$ for all x for which $\lambda(\{x\}) = 0$.

DOI: 10.1201/9781003275770-3

A sequence of random variables $\{X_n\}_{n=1}^{\infty}$ ***converges in distribution*** *or* ***converges in law*** *to a random variable* X*, denoted:*

$$X_n \to_d X, \quad or, \quad X_n \Rightarrow X,$$

if $F_n \Rightarrow F$ *for the associated distribution functions.*

Exercise II.8.4 proved that these definitions are equivalent for probability measures and distribution functions induced from one another by (1.18) and Proposition 1.27. In particular, x is a continuity point of a distribution function F if and only if $\lambda_F(\{x\}) = 0$ for the induced measure of Proposition 1.27. Conversely, if λ is a probability measure with $\lambda(\{x\}) = 0$, then x is a continuity point of the induced distribution function F_λ by (1.16), letting $a \to b$.

Corollary II.8.8 proved that if $\lambda_n \Rightarrow \lambda$ and $\lambda(\{a, b\}) = 0$ for $a < b$, where $\{a, b\}$ is a set of two points, then $\lambda_n(I) \to \lambda(I)$ for any interval $I = \langle a, b \rangle$, whether open, closed or semi-closed. If for example $\lambda = m$, Lebesgue measure on $[0, 1]$, this obtains from $\lambda_n \Rightarrow m$ that $\lambda_n(I) \to m(I)$ for all intervals $I \subset [0, 1]$ since $m(\{x\}) = 0$ for all x.

From this result, it was natural to wonder if $\lambda_n \Rightarrow \lambda$ implies that $\lambda_n(A) \to \lambda(A)$ for all Borel measurable sets. Indeed, is this even true in this special case with $\lambda = m$? But Example II.8.10 proved that such weak convergence does not imply this, there defining A as the set of rational numbers in $[0, 1]$, and $\lambda = m$.

As promised in Remark II.8.9, it will be proved in Proposition 3.6 that $\lambda_n \Rightarrow \lambda$ implies that $\lambda_n(A) \to \lambda(A)$ for any measurable set A if the "boundary" of A has λ-measure 0. When $A = \langle a, b \rangle$ above, this boundary is the set $\{a, b\}$, so if $\lambda(\{a, b\}) = 0$, this general result reduces to Corollary II.8.8. A measurable set A with this property is called a **λ-continuity set**.

Proposition 3.6 will also document an important result for the convergence of integrals defined by these measures. In fact, it is not uncommon to see weak convergence of measures **defined** in terms of this property. Since these will be proved to be equivalent definitions, it is a matter of taste which is introduced first as the definition. However, the approach taken in Book II has the advantage that the basic properties of weak convergence can be introduced and applied before an integration theory is developed.

Remark 3.3 (On the Portmanteau theorem) *The result below is often called the* ***Portmanteau theorem,*** *or perhaps surprisingly, the* **portmanteau theorem.** *This result is not named for a mathematician and thus is often not capitalized. It was first proved in 1940 by* ***Aleksandr Aleksandrov*** *(1912–1999), and in some references is identified with his name.*

The portmanteau label derives from an altogether different reference.

The word "portmanteau" is used in linguistics, and sometimes qualified as a "portmanteau word," to describe a word that is formed by packing two other words together. Examples include "smog," which derives from smoke and fog, as well as "guesstimate" from guess and estimate, "brunch" from breakfast and lunch, and there are literally hundreds of examples. This use was coined by ***Lewis Carroll,*** *the pen name of* ***Charles Lutwidge Dodgson*** *(1832–1898), in* ***Through the Looking-Glass****. A portmanteau is also a large trunk or suitcase that opens into two equal halves.*

Returning to Aleksandrov's theorem, it is commonly referred to by this name because it combines together many seemingly different ways to think about weak convergence of measures, and proves them to be equivalent.

We begin with a definition that formalizes the above discussion on boundaries of sets. See (3.4) for an alternative formulation of boundary, and Section III.4.3.2 for additional results and exercises on set closures.

Definition 3.4 (Boundary of a set ∂A; λ-continuity set) *Given a set $A \subset \mathbb{R}^m$, the **boundary of** A, denoted ∂A, is the set of points that are simultaneously the limit of a sequence of points in A, and the limit of a sequence of points in $\widetilde{A} = A^c \equiv \mathbb{R}^m - A$, the complement of A. Equivalently:*

$$\partial A \equiv \overline{A} \bigcap \overline{A^c}, \tag{3.1}$$

where $\overline{A}$, respectively $\overline{A^c}$, denote the "closures" of these sets, defined as the collection of limit points of A, respectively, A^c. For example:

$$\overline{A} = \{x |\text{ there exist } \{x_n\} \subset A \text{ with } x_n \to x\}. \tag{3.2}$$

*Given a measure λ, a Borel measurable set A is said to be a **λ-continuity set** if $\lambda(\partial A) = 0$.*

Example 3.5 *For any bounded interval, $I = \langle a, b \rangle$, whether open, closed, or semi-closed, it is apparent that $\partial A = \{a, b\}$ since one can construct sequences $\{a \pm 1/n\}_{n=1}^{\infty}$ and $\{b \pm 1/n\}_{n=1}^{\infty}$ as required by this definition. Alternatively, $\overline{A} = [a, b]$, $\overline{A^c} = (-\infty, a] \bigcup [b, \infty)$, and so again $\partial A = \{a, b\}$ by (3.1). Hence if $\lambda(\{a, b\}) = 0$, then I is a **λ-continuity set**.*

*If $A = [0, 1] \cap \mathbb{Q}$, the rationals in $[0, 1]$, then $\partial A = [0, 1]$ since any point in this interval is the limit point of rational numbers or irrational numbers. Alternatively, $\overline{A} = [0, 1]$ while $\overline{A^c} = \mathbb{R}$. If $\lambda \equiv m$, Lebesgue measure, then A is not an **λ-continuity set**.*

We now state a simplified version of the portmanteau theorem as a warm-up. There are in fact other characterizations beyond those identified below. See the portmanteau theorem for $\mathbb{R}^m$, which for $m = 1$ generalizes the result below.

Proposition 3.6 (Portmanteau theorem on $\mathbb{R}$) *Let $\{\lambda_n\}_{n=1}^{\infty}$, λ be probability measures on $\mathbb{R}$. Then the following are equivalent:*

1. *$\lambda_n \Rightarrow \lambda$.*
2. *$\int g(x) d\lambda_n \to \int g(x) d\lambda$ for every bounded, continuous $g : \mathbb{R} \to \mathbb{R}$.*
3. *$\lambda_n(A) \to \lambda(A)$ for every **λ-continuity** set A.*

Proof.

a. *1⇒2: Assume $\lambda_n \Rightarrow \lambda$ and let F, $\{F_n\}_{n=1}^{\infty}$ be the distribution functions of (1.18) induced by these measures. By Skorokhod's representation theorem of Proposition II.8.30, there exist random variables X, $\{X_n\}_{n=1}^{\infty}$ defined on the probability space $((0, 1), \mathcal{B}((0, 1)), m)$ with these distribution functions, and for all $s \in (0, 1)$:*

$$X_n(s) \to X(s). \tag{1}$$

Since X_n has distribution function F_n, the distribution function induced by the probability measure λ_n, it follows from Proposition 1.30 that λ_n is the measure on $\mathcal{B}(\mathbb{R})$ induced by X_n in the sense of (1.21):

$$\lambda_n(A) \equiv m(X_n^{-1}(A)), \quad \text{all } A \in \mathcal{B}(\mathbb{R}). \tag{2}$$

Similarly, λ is the measure on $\mathcal{B}(\mathbb{R})$ induced by X.

The random variables X, $\{X_n\}_{n=1}^{\infty}$ are by definition, measurable transformations:

$$X_n : ((0, 1), \mathcal{B}((0, 1)), m) \to (\mathbb{R}, \mathcal{B}(\mathbb{R}), \lambda_n),$$
$$X : ((0, 1), \mathcal{B}((0, 1)), m) \to (\mathbb{R}, \mathcal{B}(\mathbb{R}), \lambda).$$

And by (2), $\lambda_n = m_{X_n}$*, the measure on* $(\mathbb{R}, \mathcal{B}(\mathbb{R}))$ *induced by* X_n*, while* $\lambda = m_X$*, the measure on* $(\mathbb{R}, \mathcal{B}(\mathbb{R}))$ *induced by* X*.*

If $g(x)$ *is a bounded, continuous function on* $\mathbb{R}$*, the change of variable formula of Proposition V.4.18 obtains:*

$$\begin{aligned} \int g(x)d\lambda_n &= \int_{(0,1)} g(X_n(s))dm, \\ \int g(x)d\lambda &= \int_{(0,1)} g(X(s))dm. \end{aligned} \tag{3}$$

By continuity of g *and (1),* $g(X_n(s)) \to g(X(s))$ *for all* $s \in (0,1)$*. Then since* g *is bounded, the bounded convergence theorem of Proposition III.2.22 yields:*

$$\int_{(0,1)} g(X_n(s))dm \to \int_{(0,1)} g(X(s))dm,$$

and the result in item 2 follows from (3).

b. $2 \Rightarrow 1$*: Given* y *and* $\epsilon > 0$*, define:*

$$g_\epsilon^+(x) = \begin{cases} 1, & x \leq y, \\ (y + \epsilon - x)/\epsilon, & y \leq x \leq y + \epsilon, \\ 0, & x \geq y + \epsilon. \end{cases}$$

Then if F*,* $\{F_n\}_{n=1}^\infty$ *are the distribution functions associated with* λ*,* $\{\lambda_n\}_{n=1}^\infty$*, it follows by definition that*

$$F_n(y) = \lambda_n\left[(-\infty, y]\right] \leq \int g_\epsilon^+(x)d\lambda_n \leq F(y+\epsilon) = \lambda_n\left[(-\infty, y+\epsilon]\right].$$

Now $\int g_\epsilon^+(x)d\lambda_n \to \int g_\epsilon^+(x)d\lambda$ *by item 2, so* $\limsup_n F_n(y) \leq F(y+\epsilon)$ *for all* $\epsilon > 0$*, and by right continuity of* F*:*

$$\limsup F_n(y) \leq F(y). \tag{4}$$

Similarly, define:

$$g_\epsilon^-(x) = \begin{cases} 1, & x \leq y - \epsilon, \\ (y - x)/\epsilon, & y - \epsilon \leq x \leq y, \\ 0, & x \geq y. \end{cases}$$

Then $F(y-\epsilon) \leq \int g_\epsilon(x)d\lambda \leq F_n(y)$*, so* $F(y-\epsilon) \leq \liminf_n F_n(y)$ *for all* $\epsilon > 0$*, which implies that*

$$F(y^-) \leq \liminf F_n(y). \tag{5}$$

Combining (4) and (5):

$$F(y^-) \leq \liminf F_n(y) \leq \limsup F_n(y) \leq F(y).$$

If y *is a continuity point of* F*, then* $F(y^-) = F(y)$ *and thus* $F_n(y) \to F(y)$ *by Corollary I.3.46. By Exercise II.8.4, this is equivalent to* $\lambda_n \Rightarrow \lambda$*.*

c. $1 \Rightarrow 3$*: Given a* λ*-continuity set* A*, let* $g(x) = \chi_A(x)$*, the characteristic function of* A*, and defined to equal* 1 *on* A *and* 0 *otherwise. Then by (3.1), the set* D_g *of discontinuities of* g *satisfies* $D_g = \partial A$*, and so* $\lambda(\partial A) = 0$ *implies that* $\lambda(D_g) = 0$*. In other words,* $g(X_n) \to g(X)$ m*-a.e for the random variables of part a.*

By the change of variable formula and the bounded convergence theorem noted in part a:

$$\lambda_n(A) = \int_{(0,1)} g(x)d\lambda_n = \int_{(0,1)} g(X_n(s))dm \to \int_{(0,1)} g(X(s))dm = \lambda(A),$$

which is item 3.

d. $3 \Rightarrow 1$*: Let* $A = (-\infty, x]$*. Then* $\partial A = \{x\}$ *and by item* 3*,* $\lambda_n(A) \to \lambda(A)$ *for all* x *with* $\lambda(\{x\}) = 0$*. Thus* $\lambda_n \Rightarrow \lambda$ *by definition.*

■

Remark 3.7 (Applications) *A number of important results relating to weak convergence have already been developed in Chapters 5, 8, and 9 of Book II, and Chapters 4, 6, and 7 of Book IV. With the results of the portmanteau theorem, some of these earlier results could have been derived with different tools. This will be seen in a later section when some earlier results are generalized from distributions on* $\mathbb{R}$ *to distributions on* $\mathbb{R}^m$*.*

One application of item 2 of the portmanteau theorem was referenced in Proposition V.7.35, in the proof of the continuity theorem for the Fourier-Stieltjes transform.

3.2 Portmanteau Theorem on $\mathbb{R}^m$

The primary goal of this section is to generalize the portmanteau theorem of Proposition 3.6 to measures on $\mathbb{R}^m$, adding a number of equivalent criteria. In the next section we investigate some applications. Other applications will be found throughout this book.

We begin with a generalization of Definition 3.2.

Definition 3.8 (Weak convergence: Distributions and measures on $\mathbb{R}^m$) *A sequence of distribution functions* $\{F_n(x)\}_{n=1}^{\infty}$ *on* $\mathbb{R}^m$ *will be said to* ***converge weakly*** *to a distribution function* $F(x)$*, denoted:*

$$F_n \Rightarrow F,$$

if $F_n(x) \to F(x)$ *for every continuity point of* $F(x)$*.*

A sequence of probability measures $\{\lambda_n\}_{n=1}^{\infty}$ *on* $\mathbb{R}^m$ *will be said to* ***converge weakly*** *to a probability measure* λ*, denoted:*

$$\lambda_n \Rightarrow \lambda,$$

if $F_n \Rightarrow F$ *for the induced distribution functions of (1.18).*

A sequence of random vectors $\{X_n\}_{n=1}^{\infty}$ ***converges in distribution*** *or* ***converges in law*** *to a random vector* X*, denoted:*

$$X_n \to_d X, \quad \textit{or}, \quad X_n \Rightarrow X,$$

if $F_n \Rightarrow F$ *for the associated distribution functions.*

For Proposition 3.17, the notion of a **λ-continuity set** A in $\mathbb{R}^m$ is then given as in Definition 3.4. But we will require an alternative formulation for the **boundary** ∂A of a set $A \subset \mathbb{R}^m$.

By Definition 3.4, the boundary of A is the collection of points that are limit points both of sequences in A, and sequences in $A^c \equiv \mathbb{R}^m - A$, the complement of A. Since $\overline{A}$, the closure of A, is by definition in (3.2) the collection of limit points of sequences in A, it follows by definition that $\partial A \subset \overline{A}$.

Exercise 3.9 ($\overline{A}$ is closed) *Prove that for any* $A \subset \mathbb{R}^m$*, that* $\overline{A}$ *is a closed set, where* $\overline{A}$ *denotes the collection of limit points of* A *and defined as in (3.2). Hint: By Definition V.1.4, prove that the complement* $\left(\overline{A}\right)^c$ *is open. See Lemma 3.10.*

To characterize ∂A now requires that we remove from $\overline{A}$ all points of A which are not limit points of the complement A^c. The **interior** of A, denoted $\mathring{A}$, is defined to be this set:

$$\mathring{A} = \{x \in A | x \text{ is not a limit point of } A^c\}.$$

Note that $\mathring{A} \subset A$ by definition, and $\mathring{A}$ is an **open** set.

Lemma 3.10 ($\mathring{A}$ is an open set) *$\mathring{A}$ is an open set.*
Proof. *By Definition V.1.4, we must prove that if $x \in \mathring{A}$, there exists $r > 0$ so that:*

$$B_r(x) \subset \mathring{A},$$

where with $|\cdot|$ the standard norm on $\mathbb{R}^m$ defined in (2.14):

$$B_r(x) \equiv \{y |\, |x - y| < r\}. \tag{3.3}$$

This follows since if $x \in \mathring{A}$, to not be a limit point of A^c implies the existence of $r > 0$ so that $B_r(x) \bigcap A^c = \emptyset$, and thus $B_r(x) \subset A$. In fact $B_r(x) \subset \mathring{A}$ since no point in $B_r(x)$ can be a limit point of A^c.

To see this, note that if $z \in B_r(x)$, then $|x - z| < s < r$, and this implies by the triangle inequality that $B_{r-s}(z) \subset B_r(x) \subset A$, so z is not a limit point of A^c. ■

The above-noted reformulation of ∂A now follows.

Lemma 3.11 ($\partial A \equiv \overline{A} - \mathring{A}$.) *Given $A \subset \mathbb{R}^m$:*

$$\mathring{A} \equiv \left(\overline{A^c}\right)^c,$$

and:

$$\partial A = \overline{A} - \mathring{A}. \tag{3.4}$$

Proof. *Since $\overline{A^c}$ is the collection of limit points of A^c, it follows by definition that:*

$$\mathring{A} \equiv \mathbb{R}^m - \overline{A^c} = \left(\overline{A^c}\right)^c.$$

Then from (3.1):

$$\partial A \equiv \overline{A} \bigcap \overline{A^c} = \overline{A} - \left(\overline{A^c}\right)^c.$$

■

For the statement of Proposition 3.17, recall the notion of **Lipschitz continuity,** named for **Rudolf Lipschitz** (1832–1903).

Definition 3.12 (Lipschitz continuity) *A function $f : \mathbb{R}^m \to \mathbb{R}$ is said to be **Lipschitz continuous** if there exists $L > 0$ so that for all $x, y \in \mathbb{R}^m$:*

$$|f(x) - f(y)| \leq L\,|x - y|, \tag{3.5}$$

where $|x - y|$ denotes the standard norm in $\mathbb{R}^m$ of (2.14).

Lipschitz continuous functions are certainly continuous, since continuity only requires:

$$\lim_{y \to x} f(y) = f(x),$$

while (3.5) actually specifies a bound on the rate of convergence. On the other hand, the continuous function $f(x) = \sqrt{x}$ on $[0, \infty)$ is not Lipschitz continuous at $x = 0$, and so Lipschitz continuity is indeed more restrictive than continuity. It is also the case that continuously differentiable functions are Lipschitz continuous, but again not conversely, as $f(x) = |x|$ is Lipschitz continuous and not continuously differentiable.

The following example is needed for the next proof.

Example 3.13 ($d(x,F)$ and $\max(1-kd(x,F),0)$ are Lipschitz continuous) *Let $F \subset \mathbb{R}^m$ be a closed set, meaning by Definition V.1.4 that $F^c \equiv \mathbb{R}^m - F$ is open. Define $d(x,F)$, the* ***distance to*** F, *by:*

$$d(x,F) = \inf\{|x-y| \mid y \in F\}. \tag{3.6}$$

By Exercise 3.14, for any x there is a $y \in F$ so that $d(x,F) = |x-y|$, but in general such y is not unique.

To show that $d(x,F)$ is Lipschitz continuous with $L = 1$, let x_1, x_2 be given. Then by the triangle inequality, for all $y \in F$:

$$d(x_1,F) \leq |x_1 - y| \leq |x_1 - x_2| + |x_2 - y|,$$

and taking an infimum:

$$d(x_1,F) \leq |x_1 - x_2| + d(x_2,F).$$

Interchanging variates obtains:

$$|d(x_1,F) - d(x_2,F)| \leq |x_1 - x_2|,$$

which is (3.5) with $L = 1$.

Similarly, $g_k(x) \equiv \max(1-kd(x,F),0)$ is Lipschitz continuous with $L = k$. If x_1, x_2 are given with both $d(x_i,F) \geq 1/k$, then both $g_k(x_i) = 0$ and the conclusion follows. If both $d(x_i,F) < 1/k$ then:

$$g_k(x_2) - g_k(x_1) = k\left[d(x_1,F) - d(x_2,F)\right],$$

and the result follows from Lipschitz continuity of $d(x,F)$. Finally, if $d(x_1,F) \geq 1/k$ and $d(x_2,F) < 1/k$ then:

$$0 \leq g_k(x_2) - g_k(x_1) = 1 - kd(x_2,F) \leq k\left[d(x_1,F) - d(x_2,F)\right].$$

Since nonnegative, absolute values change nothing and the result again follows from Lipschitz continuity of $d(x,F)$.

Exercise 3.14 (On $d(x,F)$) *If $F \subset \mathbb{R}^m$ is a closed set, prove that for any x there is a $y \in F$ so that $d(x,F) = |x-y|$. Show by example that such y need not be unique. Hint: Prove that F^c open assures that if $\{x_n\}_{n=1}^{\infty} \subset F$ and $x_n \to x$ then $x \in F$.*

Remark 3.15 ($\mathbb{R} - \bigcup_{j=1}^{\infty} r_j$ is always dense in $\mathbb{R}$) *There is a technical detail in the proof of part a below which we address here, related to the notion of a dense set in $\mathbb{R}$.*

Introduced in Section III.4.3.2, recall that a set D is dense in $\mathbb{R}$ if given any x, $(x-\epsilon, x+\epsilon)\bigcap D \neq \emptyset$ for all $\epsilon > 0$. If $x \in D$, then this criterion is satisfied by definition, so the essence of being dense is that elements of D are arbitrarily close to elements outside D.

Given a countable collection of reals $\{r_j\}_{j=1}^{\infty}$, it is apparent that $\mathbb{R} - \bigcup_{j=1}^{\infty} r_j$ is an uncountable set, but we want to conclude that this set is ***dense*** *in $\mathbb{R}$.*

Assume otherwise, that there exists $x \in \mathbb{R}$ so that for some $\epsilon > 0$:

$$\left(\mathbb{R} - \bigcup\nolimits_{j=1}^{\infty} r\right)\bigcap(x-\epsilon, x+\epsilon) = \emptyset. \tag{1}$$

Letting B^c denote the complement of a set B :

$$\mathbb{R} - \bigcap\nolimits_{j=1}^{\infty} r_j \equiv \mathbb{R}\bigcap\left(\bigcup\nolimits_{j=1}^{\infty} r_j\right)^c,$$

and thus (1) implies that $(x-\epsilon, x+\epsilon)\bigcap\left(\bigcup_{j=1}^{\infty} r_j\right)^c = \emptyset$. This obtains $(x-\epsilon, x+\epsilon) \subset \bigcup_{j=1}^{\infty} r_j$, a contradiction.

For the proof below, the collection $\{r_j\}_{j=1}^{\infty}$ is the union of m countable sets, one identified for each component.

Exercise 3.16 (Baire's category theorem) *Develop another proof that* $\mathbb{R}-\bigcup_{j=1}^{\infty} r_j$ *is a dense set in* $\mathbb{R}$ *using Baire's category theorem of Proposition III.4.51. Hint: Let* $A_j \equiv \mathbb{R}-r_j$, *and note that* A_j *is open and dense in* $\mathbb{R}$.

We are now ready for the statement and proof of the portmanteau theorem on $\mathbb{R}^m$. Letting $m=1$, this result provides additional criteria for Proposition 3.6.

For background on the proof, the reader is referred to Section I.3.4.2 for limits inferior and superior of numerical and function sequences, and Section II.2.1 for limits inferior and superior of sequences of sets.

Proposition 3.17 (Portmanteau theorem on $\mathbb{R}^m$**)** *Let* $\{\lambda_n\}_{n=1}^{\infty}$, λ *be probability measures on* $\mathbb{R}^m$. *Then the following are equivalent:*

1. $\lambda_n \Rightarrow \lambda$.
2. $\lambda(G) \leq \liminf \lambda_n(G)$ *for all open sets* $G \subset \mathbb{R}^m$.
3. $\limsup \lambda_n(F) \leq \lambda(F)$ *for all closed sets* $F \subset \mathbb{R}^m$.
4. $\lim \lambda_n(A) = \lambda(A)$ *for all* λ*-continuity sets* A.
5. $\int g(x)d\lambda_n \rightarrow \int g(x)d\lambda$, *for every bounded, continuous* $g: \mathbb{R}^m \rightarrow \mathbb{R}$.
6. $\int g(x)d\lambda_n \rightarrow \int g(x)d\lambda$, *for every bounded, Lipschitz continuous* $g: \mathbb{R}^m \rightarrow \mathbb{R}$.

Proof. *We first show that items 1–4 are equivalent, then complete the proof with* $4 \Rightarrow 5 \Rightarrow 6$ *and* $6 \Rightarrow 3$.

a. $1 \Rightarrow 2$: *For any component index* k, *the hyperplane defined by* $\{x \in \mathbb{R}^m | x_k = r\}$ *has nonzero* λ*-measure for at most countably many reals* $\{r_{k_j}\}_{j=1}^{\infty}$, *since otherwise,* λ *could not be a finite measure. Letting:*

$$D = \mathbb{R} - \bigcup_{k=1}^{m} \bigcup_{j=1}^{\infty} r_{k_j},$$

D is a dense and uncountable set $D \subset \mathbb{R}$ by Remark 3.15, and $\lambda[\{x \in \mathbb{R}^m | x_k = r\}] = 0$ *for any k and all $r \in D$.*

Define a class of rectangles: $\mathcal{R} = \{\prod_{i=1}^{m}(a_i, b_i] |\ a_i, b_i \in D\}$. *We claim that every vertex of a rectangle in $\mathcal{R}$ is a continuity point of the distribution function F induced by λ.*

If $y = (y_1, ..., y_n)$ *denotes such a vertex, then since by construction the bounding hyperplanes* $\{x \in \mathbb{R}^m | x_k = y_k\}$ *have* λ*-measure 0, it follows from (1.18):*

$$F(y) \equiv \lambda\left[\prod_{i=1}^{m}(-\infty, y_i]\right] = \lambda\left[\prod_{i=1}^{m}(-\infty, y_i)\right]. \tag{1}$$

We claim that this assures that F is continuous from below at such y, meaning that if $y^{(k)} < y$ *and* $y^{(k)} \rightarrow y$, *where both statements are to be interpreted componentwise, then* $F(y^{(k)}) \rightarrow F(y)$.

To this end, first note:

$$\bigcup_{k=1}^{\infty}\left[\prod_{i=1}^{m}(-\infty, y_i^{(k)}]\right] = \prod_{i=1}^{m}(-\infty, y_i).$$

The union on the left can be made into a nested union by replacing $y^{(k)}$ *with* $\max\{y^{(j)} | j \leq k\}$, *and thus continuity from below of* λ *by Proposition I.2.45 assures that* $F(y^{(k)}) \rightarrow F(y)$.

As any distribution function F is continuous from above by Proposition 1.23 and increasing in each variable by definition, we can now prove that F is in fact continuous at each such vertex y as follows.

If $z^{(j)} \to y$ as $j \to \infty$, define $z^{(j)\pm}$ componentwise by $z_i^{(j)\pm} = y_i \pm \max \left| z_k^{(j)} - y_k \right|$. Then $z^{(j)^-} \leq z^{(j)} \leq z^{(j)^+}$ for all j, interpreted componentwise as above, and thus by definition of F it follows that:

$$F\left(z^{(j)^-}\right) \leq F\left(z^{(j)}\right) \leq F\left(z^{(j)^+}\right). \tag{2}$$

Now $z^{(j)^\pm} \to y$ as $j \to \infty$, with $z^{(j)^-}$ converging from below and $z^{(j)^+}$ from above, and thus it follows from (2) and continuity from above and below that $F\left(z^{(j)}\right) \to F(y)$. That is, F is continuous at all vertexes of rectangles in $\mathcal{R}$.

Since $\lambda_n \Rightarrow \lambda$ by assumption of item 1, Definition 3.8 obtains that $F_n(y) \to F(y)$ for any such vertex. Then by Proposition 1.27 and (1.17), it then follows that $\lambda_n(A) \to \lambda(A)$ if $A \in \mathcal{R}$, or for any A that is a finite intersection of $\mathcal{R}$-sets, since such intersections obtain $\mathcal{R}$-sets. Then by the inclusion-exclusion formula of Proposition I.8.8, $\lambda_n(B) \to \lambda(B)$ for any finite union of elements of $\mathcal{R}$.

Now if $G \subset \mathbb{R}^m$ is open, then G is a countable union of open balls defined in (3.3) by Proposition I.2.12, and by virtually the same proof that is left as an exercise, G is a countable union of open rectangles. Each such open rectangle is a countable union of $\mathcal{R}$-sets since D is dense in $\mathbb{R}^m$. Hence $G = \bigcup_k A_k$, a countable union of $\mathcal{R}$-sets, and for any M:

$$\begin{aligned} \lambda\left[\bigcup_{m \leq M} A_m\right] &= \lim_{n\to\infty} \lambda_n \left[\bigcup_{m \leq M} A_m\right] \\ &= \liminf_n \lambda_n \left[\bigcup_{m \leq M} A_m\right] \leq \liminf_n \lambda_n(G). \end{aligned}$$

The second equality is Definition II.2.1, that when a limit exists, this limit equals the limit inferior. By continuity from below of λ, $\lambda\left[\bigcup_{m \leq M} A_m\right] \to \lambda(G)$ as $M \to \infty$, proving item 2.

b. $2 \Leftrightarrow 3$: *Given closed F, let $G = \mathbb{R}^m - F$, an open set by Definition V.1.4. Then by Proposition II.2.5 and item 2:*

$$\limsup \lambda_n (\mathbb{R}^m - F) \geq \liminf \lambda_n (\mathbb{R}^m - F) \geq \lambda (\mathbb{R}^m - F). \tag{3}$$

By finite additivity:

$$\lambda_n (\mathbb{R}^m - F) + \lambda_n (F) = \lambda_n (\mathbb{R}^m) \equiv 1,$$

and similarly for λ, and so from (3):

$$\liminf \left[1 - \lambda_n (F)\right] \geq 1 - \lambda (F).$$

By Definition I.3.42 this is equivalent to:

$$\liminf \left[-\lambda_n (F)\right] \geq -\lambda (F),$$

and from I.(3.13) this obtains $\limsup_n \lambda_n (F) \leq \lambda (F)$, which is item 3.

Given open $G \subset \mathbb{R}^m$, define $F = \mathbb{R}^m - G$, which is closed by Definition I.2.10. It is left as an exercise to proceed as above to prove that item 3 implies item 2.

c. $2,3 \Rightarrow 4$: *For any set* A, *since* $\mathring{A}$ *is an open set by Lemma 3.10,* $\overline{A}$ *is closed by Exercise 3.9, and* $\mathring{A} \subset A \subset \overline{A}$ *by definition, it follows by items 2 and 3, the monotonicity of measures, and II.(2.5):*

$$\begin{aligned} \lambda\left(\mathring{A}\right) &\leq \liminf_n \lambda_n\left(\mathring{A}\right) \leq \liminf_n \lambda_n(A) \\ &\leq \limsup_n \lambda_n(A) \leq \limsup_n \lambda_n\left(\overline{A}\right) \leq \lambda\left(\overline{A}\right). \end{aligned} \tag{4}$$

If A *is a* λ*-continuity set then* $\lambda[\partial A] = 0$ *by definition, and so by (3.4) and finite additivity,* $\lambda\left(\mathring{A}\right) = \lambda\left(\overline{A}\right)$. *Thus by* (4) *and Corollary I.3.46:*

$$\lambda\left(\mathring{A}\right) = \lim \lambda_n(A) = \lambda\left(\overline{A}\right),$$

and $\lim \lambda_n(A) = \lambda(A)$ *follows from monotonicity of measures and* $\mathring{A} \subset A \subset \overline{A}$.

d. $4 \Rightarrow 1$: *Let* $F(y)$ *and* $F_n(y)$ *denote the distribution functions induced by* λ *and* λ_n *as in (1.18), for example,* $F(y) = \lambda[\prod_{i=1}^m(-\infty, y_i]] \equiv \lambda[A_y]$. *If* x *is a continuity point of* F, *then* $\lim_{y\to x^-} F(y) = F(x)$, *where this notation implies that* $y_j < x_j$ *for all* j. *Then by continuity from below of* λ *(Proposition I.2.45):*

$$\lim_{y\to x^-} F(y) = \lambda\left[\prod_{i=1}^m(-\infty, x_i)\right] \equiv \lambda\left[\mathring{A}_x\right].$$

Thus if x *is a continuity point of* F, *then* $F(x) = \lambda\left[\mathring{A}_x\right]$ *and hence* $\lambda[\partial A_x] = 0$ *by (3.4). This implies that* $A_x \equiv \prod_{i=1}^j(-\infty, x_i]$ *is a* λ*-continuity set and hence by item 4,* $\lambda_n[A_x] \to \lambda[A_x]$. *In other words,* $F_n(x) \to F(x)$ *for all continuity points of* F *and thus* $\lambda_n \Rightarrow \lambda$ *by definition.*

e. $4 \Rightarrow 5$: *Given measurable* $g: \mathbb{R}^m \to \mathbb{R}$, $\{g^{-1}(y)\}_{y\in\mathbb{R}}$ *are disjoint so it follows as in part a that* $\lambda\left[g^{-1}(y)\right] = 0$ *for all but at most countably many* y, *and the collection of* y *with* $\lambda\left[g^{-1}(y)\right] = 0$ *is dense in* $\mathbb{R}$. *Thus if* g *is continuous and bounded with* $|g| \leq C$, *then for any* $\epsilon > 0$ *we can choose increasing* $\{y_k\}_{k=0}^N$ *with* $y_0 < -C$, $y_N > C$, *and for all* k, $\lambda\left[g^{-1}(y_k)\right] = 0$ *and* $y_{k+1} - y_k < \epsilon$.

Define $A_k \subset \mathbb{R}^m$ *by* $A_k = \{x | y_{k-1} < g(x) \leq y_k\}$. *Since* g *is continuous and hence sequentially continuous, the closure* $\overline{A}_k \subset \{x | y_{k-1} \leq g(x) \leq y_k\}$. *The same argument with* A_k^c, *the complement of* A_k, *obtains* $\overline{A_k^c} \subset \{x | g(x) \leq y_{k-1}$ *or* $g(x) \geq y_k\}$, *and thus* $\{x | y_{k-1} < g(x) < y_k\} \subset \mathring{A}_k$ *since* $\mathring{A}_k \equiv \left(\overline{A_k^c}\right)^c$ *by Lemma 3.11. By (3.4):*

$$\partial A_k = \overline{A}_k - \mathring{A}_k \subset \{x | g(x) = y_{k-1} \text{ or } g(x) = y_k\},$$

and by the selection of y_k *it follows that* $\lambda[\partial A_k] = 0$ *for all* k. *Thus each* A_k *is a* λ*-continuity set, so by item 4, as* $n \to \infty$:

$$\sum_{k=1}^N y_k \lambda_n[A_k] \to \sum_{k=1}^N y_k \lambda[A_k].$$

Given the above ϵ, *there exists* $M(\epsilon)$ *so that for* $n \geq M(\epsilon)$:

$$\left|\sum_{k=1}^N y_k \lambda_n[A_k] - \sum_{k=1}^N y_k \lambda[A_k]\right| < \epsilon. \tag{5}$$

By construction of the A_k*-sets, for either* $\nu = \lambda_n$ *or* $\nu = \lambda$, *recalling that these are probability measures:*

$$\left|\int g(x)d\nu - \sum_{k=1}^N y_k \nu[A_k]\right| \leq \sum_{k=1}^N \max|g - y_k| \nu[A_k] < \epsilon,$$

where $\max|g - y_k|$ *is defined on* A_k.

Thus by (5) *and the triangle inequality, for* $n \geq M(\epsilon)$*:*

$$\left|\int g(x)d\lambda_n - \int g(x)d\lambda\right| \leq 3\epsilon,$$

which proves item 5.

f. $5 \Rightarrow 6$*: This is true by definition.*

g. $6 \Rightarrow 3$*: Given closed* $F \subset \mathbb{R}^m$*, define the function* $d(x, F)$ *as in (3.6) and let* $g_k(x) = \max(1 - kd(x, F), 0)$ *as in Example 3.13. Then* $g_k(x)$ *is bounded and Lipschitz continuous, monotonically decreasing in* k *for each* x*, and since* F *is closed,* $g_k(x) \to \chi_F(x)$ *pointwise as* $k \to \infty$*. Here* $\chi_F(x)$ *is the characteristic function of* F*, defined to be* 1 *for* $x \in F$ *and* 0 *otherwise. This limit is true by definition if* $x \in F$*, while if* $x \in F^c$*, which is open, there exists* $r > 0$ *so that* $B_r(x) \subset F^c$*, and so* $d(x, F) \geq r$ *and* $g_k(x) = 0$ *for* $k > 1/r$*.*

Hence for any k*, recalling the integral of a simple function in Definition V.3.7:*

$$\lambda_n(F) \equiv \int \chi_F(x)d\lambda_n \leq \int g_k(x)d\lambda_n.$$

Taking a limit superior and applying item 6 obtains for all k*:*

$$\limsup_n \lambda_n(F) \leq \lim_{n\to\infty} \int g_k(x)d\lambda_n = \int g_k(x)d\lambda. \tag{6}$$

But by the bounded convergence theorem of Proposition V.3.50:

$$\lim_{k\to\infty} \int g_k(x)d\lambda \to \lambda(F),$$

and this with (6) *obtains item 3.*

■

Example 3.18 (Unbounded functions) *That items 5 and 6 cannot in general be extended beyond bounded functions is not difficult to illustrate.*

Define λ_n *on* $\mathbb{R}$ *by* $\lambda_n[\{n\}] = 1/n$ *and* $\lambda_n[\{0\}] = 1 - 1/n$*, and* λ *on* $\mathbb{R}$ *by* $\lambda[\{0\}] = 1$*. Then* $\lambda_n \Rightarrow \lambda$*, but with* $g(x) \equiv x$*:*

$$1 = \int g(x)d\lambda_n \nrightarrow \int g(x)d\lambda = 0.$$

In this example, λ_n *also converges to* λ ***strongly*** *and* ***in total variation****. See* ***Scheffé's theorem*** *of Proposition 3.45.*

3.3 Applications

In this section we generalize several of the results originally developed in Book II from distributions on $\mathbb{R}$ to distributions on $\mathbb{R}^m$:

- Mapping theorem

- Mann-Wald theorem
- Slutsky's theorem
- The Delta method
- Prokhorov's theorem

In addition we introduce Scheffé's theorem and part one of a result called the Cramér-Wold device. This latter result was called into service in Proposition II.9.52 for the development of multivariate extreme value theory. Part two of the Cramér-Wold device is of necessity deferred to Chapter 5, as an application of characteristic functions.

3.3.1 The Mapping Theorem

With the aid of the portmanteau theorem, Proposition 3.20 proves a significant generalization of the mapping theorem on $\mathbb{R}$ of Proposition II.8.35. This result addresses weak convergence of measures induced by transformations, for which the Book II proof required **Skorokhod's representation theorem.** The following section then provides an application of this mapping theorem to various modes of convergence of transformed random vectors, and summarized in the Mann-Wald theorem.

For the current investigation, let $h : \mathbb{R}^j \to \mathbb{R}^k$ be a Borel measurable transformation by Definition 1.4, which is also called $\mathcal{B}(\mathbb{R}^j)/\mathcal{B}(\mathbb{R}^k)$-measurable.

Let $D_h \subset \mathbb{R}^j$ denote the collection of discontinuities of h. The next exercise proves that $D_h \in \mathcal{B}(\mathbb{R}^j)$, and thus is a measurable set.

Exercise 3.19 **($D_h \in \mathcal{B}(\mathbb{R}^j)$)** *Prove that for any function, meaning measurable or not, that D_h is Borel measurable. Hint: Define the set $A(\epsilon, \delta)$ for arbitrary $\epsilon, \delta > 0$ by:*

$$A(\epsilon, \delta) \equiv \{x | \text{there exists } y, z \in B_\delta(x) \text{ and } |h(y) - h(z)| > \epsilon\},$$

where the open ball $B_\delta(x)$ is defined as in (3.3).

First prove that $A(\epsilon, \delta)$ is open. That is, if $x \in A(\epsilon, \delta)$ then $B_{\delta'}(x) \subset A(\epsilon, \delta)$ for some $\delta' > 0$. Hint: Consider $\delta' = \delta - \max\{|x - y|, |x - z|\}$.

Then prove that:

$$D_h = \bigcup\nolimits_\epsilon \bigcap\nolimits_\delta A(\epsilon, \delta),$$

where these set functions are over all positive rational ϵ, δ.

Thus not only is D_h Borel measurable, but it is a $\mathcal{G}_{\delta\sigma}$-set of Notation I.2.16, which means it is a countable union (the "σ") of sets, each of which is a countable intersection (the "δ") of open (the "$\mathcal{G}$") sets.

Generalizing the mapping theorem on $\mathbb{R}$ of Proposition II.8.35, we now show that if $\lambda(D_h) = 0$, then weak convergence $\lambda_n \Rightarrow \lambda$ on $(\mathbb{R}^j, \mathcal{B}(\mathbb{R}^j))$ is preserved in the induced measures of (1.21) defined on $(\mathbb{R}^k, \mathcal{B}(\mathbb{R}^k))$. This result is called the **continuous mapping theorem** when h is assumed to be continuous. In this case, $D_h = \emptyset$ and need not even be mentioned.

More generally, it is sometimes called the continuous mapping theorem where term "continuous" implies "convergence preserving." Recall that continuous functions $h : \mathbb{R}^j \to \mathbb{R}^k$ have this property by Exercise 1.52, that if $x_n \to x$ in $\mathbb{R}^j$, then $h(x_n) \to h(x)$ in $\mathbb{R}^k$. The general theorem states that something less than continuity of h is sufficient to preserve weak convergence of measures. See also the Mann-Wald theorem.

Proposition 3.20 (Mapping theorem on $\mathbb{R}^j/\mathbb{R}^k$) *Let $\{\lambda_n\}_{n=1}^{\infty}$, λ be probability measures on $(\mathbb{R}^j, \mathcal{B}(\mathbb{R}^j))$ with $\lambda_n \Rightarrow \lambda$, and $h : (\mathbb{R}^j, \mathcal{B}(\mathbb{R}^j)) \to (\mathbb{R}^k, \mathcal{B}(\mathbb{R}^k))$ a Borel measurable transformation with $\lambda(D_h) = 0$.*

Then on $(\mathbb{R}^k, \mathcal{B}(\mathbb{R}^k))$:

$$(\lambda_n)_h \Rightarrow \lambda_h,$$

where λ_h and $(\lambda_n)_h$ are the measures induced by h and defined on $A \in \mathcal{B}(\mathbb{R}^k)$ by (1.21).

Proof. *For closed $F \in \mathcal{B}(\mathbb{R}^k)$, $h^{-1}(F) \in \mathcal{B}(\mathbb{R}^j)$ by Borel measurability, and $\overline{h^{-1}(F)} \in \mathcal{B}(\mathbb{R}^j)$ since $\mathcal{B}(\mathbb{R}^j)$ contains all open and thus also all closed sets. If $x \in \overline{h^{-1}(F)}$, then by definition of closure there exists $\{x_m\}_{m=1}^{\infty} \subset h^{-1}(F)$ with $x_m \to x$. If x is a continuity point of h, then $h(x_m) \to h(x)$ by Exercise 1.52, and since F is closed it follows that $h(x) \in F$ and thus $x \in h^{-1}(F)$. Hence, $\overline{h^{-1}(F)} \subset h^{-1}(F) \bigcup D_h$.*

By monotonicity and subadditivity of measures and item 3 of the Proposition 3.17:

$$\limsup_n \lambda_n \left[h^{-1}(F)\right] \leq \limsup_n \lambda_n \left[\overline{h^{-1}(F)}\right] \leq \lambda \left[\overline{h^{-1}(F)}\right] \leq \lambda \left[h^{-1}(F)\right] + \lambda \left[D_h\right].$$

If $\lambda[D_h] = 0$, this obtains by (1.21):

$$\limsup_n (\lambda_n)_h (F) \leq \lambda_h(F).$$

As this is true for all closed $F \in \mathcal{B}(\mathbb{R}^k)$, item 3 of Proposition 3.17 proves that $(\lambda_n)_h \Rightarrow \lambda_h$. ■

Corollary 3.21 (Continuous mapping theorem on $\mathbb{R}^j/\mathbb{R}^k$) *If $\{\lambda_n\}_{n=1}^{\infty}$, λ are probability measures on $(\mathbb{R}^j, \mathcal{B}(\mathbb{R}^j))$ with $\lambda_n \Rightarrow \lambda$, and $h : (\mathbb{R}^j, \mathcal{B}(\mathbb{R}^j)) \to (\mathbb{R}^k, \mathcal{B}(\mathbb{R}^k))$ is a continuous transformation, then $(\lambda_n)_h \Rightarrow \lambda_h$.*

Proof. *Here $\mu(D_h) = 0$ follows from $D_h = \emptyset$.* ■

3.3.2 Mann-Wald Theorem

In this section, we derive a generalization of the Mann-Wald theorem on $\mathbb{R}$ of Proposition II.8.37. This result provides an application of the mapping theorem to various modes of convergence of transformed random vectors. For its statement, we generalize some definitions from Chapter II.5 from random variables to random vectors.

In the following, recall the abbreviated set notation, for example in item 1:

$$\{|X_n - X| \geq \epsilon\} \equiv \{s \in \mathcal{S} | \, |X_n(s) - X(s)| \geq \epsilon\}.$$

Definition 3.22 (Random vector modes of convergence) *Given a probability space $(\mathcal{S}, \mathcal{E}, \mu)$ and random vectors X, $\{X_n\}_{n=1}^{\infty} : \mathcal{S} \to \mathbb{R}^j$:*

1. *X_n **converges to X in probability,** denoted:*

$$X_n \to_P X,$$

if for every $\epsilon > 0$:

$$\lim_{n \to \infty} \mu(\{|X_n - X| \geq \epsilon\}) = 0, \tag{3.7}$$

where $|X_n - X|$ is the standard norm on $\mathbb{R}^j$ defined in (2.14).

2. *X_n **converges to X with probability** 1, or, X_n **converges to X almost surely,** denoted:*

$$X_n \to_1 X, \text{ or, } X_n \to_{a.s.} X,$$

if:

$$\mu(\{\lim_{n\to\infty} X_n = X\}) = 1, \tag{3.8}$$

where $\lim_{n\to\infty} X_n = X$ *is defined in terms of the standard norm on* $\mathbb{R}^j$.

3. X_n ***converges in distribution to*** X***,*** *or,* ***converges in law,*** *denoted:*

$$X_n \to_d X, \text{ or, } X_n \Rightarrow X,$$

if $F_n \Rightarrow F$*, meaning by Definition 3.8 that* $F_n(x) \to F(x)$ *for every continuity point of* F.

Equivalently, $X_n \to_d X$ *if* $\lambda_n \Rightarrow \lambda$ *for the associated probability measures defined in (1.21).*

Remark 3.23 (Other norms) *For the definition of* $X_n \to_P X$*, it is sometimes convenient to use an alternative norm on* $\mathbb{R}^j$.

Recalling the discussion on Banach spaces in Chapter V.9, the standard norm is also called the l_2*-norm, while the general* l_p*-norm is defined for* $1 \leq p < \infty$ *by:*

$$\|x\|_p \equiv \left[\sum\nolimits_{i=1}^{j} |x_i|^p\right]^{1/p}, \tag{3.9}$$

and for $p = \infty$ *by:*

$$\|x\|_\infty \equiv \max\{|x_i|\}. \tag{3.10}$$

Any one of these norms can be used in the definition of convergence in probability since they are ***equivalent norms.*** *That is, given* p *and* p'*, there exists positive* $c_{p,p'}$ *and* $C_{p,p'}$ *so that for all* x*:*

$$c_{p,p'}\|x\|_{p'} \leq \|x\|_p \leq C_{p,p'}\|x\|_{p'}. \tag{3.11}$$

Thus if (3.7) is true for $p = 2$ *it is true for any* p*, and conversely.*

For more on norms, see Section V.9.2, and for norms and norm equivalence, see Remark V.4.43 or Chapter 3 of ***Reitano*** *(2010).*

The following result is called the **Mann-Wald theorem,** named for **Henry Mann** (1905–2000) and **Abraham Wald** (1902–1950), who published it in 1943. Given $X_n \to_\bullet X$ for some mode of convergence of random vectors defined on $(\mathcal{S}, \mathcal{E}, \mu)$ with range in $\mathbb{R}^j$, it addresses conditions on a measurable transformation $h : \mathbb{R}^j \to \mathbb{R}^k$ which assure that $h(X_n) \to_\bullet h(X)$ as random vectors with range in $\mathbb{R}^k$.

When the mode of convergence is $X_n \to_d X$, or convergence in distribution, this equivalent to $\lambda_{X_n} \Rightarrow \lambda_X$ for the associated probability measures. The Mann-Wald conclusion of $h(X_n) \to_d h(X)$ is then seen to be equivalent to weak convergence of the induced measures $(\lambda_{X_n})_h \Rightarrow (\lambda_X)_h$, and thus this part of the result is sometimes presented as a version of the mapping theorem.

The proof of the Mann-Wald theorem requires a few results from Book II on relationships between these modes of convergence. Generalizing Proposition II.5.21:

Proposition 3.24 **(**$X_n \to_1 X \Rightarrow X_n \to_P X \Rightarrow X_n \to_d X$**)** *Let* $\{X_n\}_{n=1}^\infty$, X *be random vectors on* $(\mathcal{S}, \mathcal{E}, \mu)$ *with range in* $\mathbb{R}^j$.

1. *If* X_n *converges to* X *with probability* 1*, then* X_n *converges to* X *in probability:*

$$X_n \to_1 X \;\Rightarrow\; X_n \to_P X.$$

2. If X_n converges to X in probability, then X_n converges to X in distribution:

$$X_n \to_P X \ \Rightarrow \ X_n \to_d X.$$

Proof. *If $X_n \to_1 X$, then by Definition 3.22:*

$$\mu\left[A\bigcup B\right] = 0,$$

where $A \equiv \{\lim_{n\to\infty} |X_n - X| > 0\}$ and $B \equiv \{\lim_{n\to\infty} |X_n - X|$ does not exist$\}$, with $|X_n - X|$ defined as in (2.14).

Given $\epsilon > 0$, define $A_n(\epsilon) \in \mathcal{E}$ by:

$$A_n(\epsilon) \equiv \{|X_n - X| \geq \epsilon\}.$$

By Definition II.2.1:

$$\limsup_n A_n(\epsilon) \equiv \bigcap\nolimits_{n=1}^{\infty} \bigcup\nolimits_{k=n}^{\infty} A_k(\epsilon), \tag{3.12}$$

and II.(2.5) obtains that for any $\epsilon > 0$:

$$\mu\left(\limsup_n A_n(\epsilon)\right) \geq \limsup_n \mu(A_n(\epsilon)). \tag{1}$$

Assume that $X_n \nrightarrow_P X$, and thus there exists $\epsilon > 0$ such that $\lim\limits_{n\to\infty} \mu(A_n(\epsilon)) > 0$. Then $\limsup_n \mu(A_n(\epsilon)) > 0$ and from (1):

$$\mu\left(\limsup_n A_n(\epsilon)\right) > 0.$$

By (3.12), $\limsup_n A_n(\epsilon)$ is the measurable set on which $|X_n - X| \geq \epsilon$ for infinitely many n, and so $\limsup_n A_n(\epsilon) \subset A\bigcup B$ defined above. This is a contradiction, and so $X_n \to_P X$.

For item 2, convergence $X_n \to_P X$ means that for every $\epsilon > 0$:

$$\lim_{n\to\infty} \mu(\{|X_n - X| \geq \epsilon\}) = 0. \tag{2}$$

For any $x \in \mathbb{R}^j$ and $\epsilon > 0$, let $\bar{\epsilon} \equiv (\epsilon, ..., \epsilon) \in \mathbb{R}^j$. As events in $\mathcal{S}$, where we define $\mathbb{R}^j$-inequalities componentwise:

$$\begin{aligned} \{X_n \leq x\} &\subset \{X \leq x + \bar{\epsilon}\} \bigcup \{|X_n - X| \geq \epsilon\}, \\ \{X \leq x - \bar{\epsilon}\} &\subset \{X_n \leq x\} \bigcup \{|X_n - X| \geq \epsilon\}. \end{aligned}$$

These inclusions are left as an exercise, and justified by considering complements, that $A \subset B$ if and only if $\widetilde{B} \subset \widetilde{A}$, and applying de Morgan's laws of Exercise I.2.2.

By monotonicity and subadditivity of μ, this implies that:

$$F(x - \bar{\epsilon}) - \mu(\{|X_n - X| \geq \epsilon\}) \leq F_n(x) \leq F(x + \bar{\epsilon}) + \mu(\{|X_n - X| \geq \epsilon\}). \tag{3}$$

By Corollary I.3.46, (2) implies that both the limit inferior and limit superior of $\mu(\{|X_n - X| \geq \epsilon\})$ equal 0. Taking limits inferior and superior in (3) obtains that for any $\epsilon > 0$:

$$F(x - \bar{\epsilon}) \leq \liminf F_n(x) \leq \limsup F_n(x) \leq F(x + \bar{\epsilon}).$$

If x is a continuity point of $F(x)$ then both $F(x - \bar{\epsilon})$ and $F(x + \bar{\epsilon})$ converge to $F(x)$ as $\epsilon \to 0$, and thus $\lim F_n(x) = F(x)$ and $X_n \to_d X$. ■

Next, we have the generalization of Proposition II.5.25:

Proposition 3.25 **($X_n \to_P X \Rightarrow X_{n_m} \to_{a.e} X$)** *Let $\{X_n\}_{n=1}^{\infty}$, X be random vectors on $(\mathcal{S}, \mathcal{E}, \mu)$ with range in $\mathbb{R}^j$.*

If X_n converges to X in probability, then there exists a subsequence $\{X_{n_m}\}_{m=1}^{\infty}$ that converges to X with probability 1:

$$X_n \to_P X \Rightarrow X_{n_m} \to_1 X.$$

Proof. *If $X_n \to_P X$, then for any $\epsilon > 0$, $\mu\{|X_n - X| \geq \epsilon\} \to 0$ as $n \to \infty$. Letting $\epsilon = 2^{-m}$, choose n_m so that for $n \geq n_m$:*

$$\mu\{|X_n - X| \geq 2^{-m}\} \leq 2^{-m}.$$

We can assume that $\{n_m\}_{m=1}^{\infty}$ is an increasing sequence by using $n'_m \equiv \max_{k \leq m}\{n_k\}$.

Define:

$$A_m = \{|X_{n_m} - X| \geq 2^{-m}\},$$

and note that $\sum_{m=1}^{\infty} \mu(A_m) < \infty$. The Borel-Cantelli lemma of Proposition II.2.6 then obtains that $\mu(\limsup A_m) = 0$. Equivalently, with $B^c \equiv \widetilde{B}$ the complement of B:

$$\mu\left[(\limsup A_m)^c\right] = 1. \tag{1}$$

Applying de Morgan's laws to (3.12):

$$\begin{aligned} (\limsup A_m)^c &= \bigcup\nolimits_{m=1}^{\infty} \bigcap\nolimits_{k \geq m}^{\infty} \widetilde{A}_k \\ &= \bigcup\nolimits_{m=1}^{\infty} \bigcap\nolimits_{k \geq m}^{\infty} \{|X_{n_k} - X| < 2^{-k}\}. \end{aligned}$$

If $s \in (\limsup A_m)^c$, then $s \in \bigcap\nolimits_{k \geq m}^{\infty} \{|X_{n_k} - X| < 2^{-k}\}$ for some m. Thus $X_{n_k}(s) \to X(s)$ on $(\limsup A_m)^c$, and the proof is complete by (1). ■

With the preliminary work done, we are now ready for the Mann-Wald theorem. The proof of this result is very similar to the random variable counterpart of Proposition II.8.37, with the exception that item 3 will have a simpler proof based on the mapping theorem of Proposition 3.20.

For this result, recall that for any function $h : \mathbb{R}^j \to \mathbb{R}^k$, that $D_h \in \mathcal{B}(\mathbb{R}^j)$ by Exercise 3.19 and thus $\mu\{X^{-1}(D_h)\}$ is well-defined. Also by (1.21), the condition $\mu\{X^{-1}(D_h)\} = 0$ below is identical to that in the mapping theorem of Proposition 3.20, that $\lambda(D_h) = 0$, when $\lambda = \lambda_X$ is the probability measure on $\mathbb{R}^j$ induced by X.

Proposition 3.26 (Mann-Wald theorem) *Let $\{X_n\}_{n=1}^{\infty}$, X be random vectors defined on $(\mathcal{S}, \mathcal{E}, \mu)$ with range in $\mathbb{R}^j$, and h a Borel measurable function $h : \mathbb{R}^j \to \mathbb{R}^k$ with discontinuity set D_h with $\mu\{X^{-1}(D_h)\} = 0$.*

1. *If $X_n \to_{a.e.} X$, then $h(X_n) \to_{a.e.} h(X)$.*
2. *If $X_n \to_P X$, then $h(X_n) \to_P h(X)$.*
3. *If $X_n \to_d X$, then $h(X_n) \to_d h(X)$.*

In particular, if h is a continuous function, then $X_n \rightarrow_{\bullet} X$ implies that $h(X_n) \rightarrow_{\bullet} h(X)$ for all three modes of convergence.

Proof. *1. If $X_n \rightarrow_{a.e.} X$, then there is an exceptional set $E \in \mathcal{E}$ with $\mu(E) = 0$ and $X_n(s) \rightarrow X(s)$ for $s \in \widetilde{E}$. Letting $D = \{X^{-1}(D_h)\}$, it follows that for $s \in (D \bigcup E)^c$, $X(s)$ is a continuity point of h and $X_n(s) \rightarrow X(s)$.*

Thus $h(X_n(s)) \rightarrow h(X(s))$ for $s \in (D \bigcup E)^c$ by Exercise 1.52, and since $\lambda(D \bigcup E) = 0$, this obtains $h(X_n) \rightarrow_{a.e.} h(X)$.

2. By contradiction, assume that $X_n \rightarrow_P X$ and $h(X_n) \nrightarrow_P h(X)$. Then given $\delta > 0$ and $\epsilon > 0$, there is a subsequence $\{X_{n_m}\}_{m=1}^{\infty}$ so that for all m:

$$\mu\{|h(X_{n_m}) - h(X)| > \epsilon\} > \delta. \tag{1}$$

Since $\mu\{X^{-1}(D_h)\} = 0$, it can be assumed that the sets $\{|h(X_{n_m}) - h(X)| > \epsilon\}$ include only continuity points of h.

As $X_{n_m} \rightarrow_P X$ by definition, Proposition 3.25 obtains a subsequence of $\{X_{n_m}\}_{m=1}^{\infty}$ so that $X_{n_{m_k}} \rightarrow_1 X$, and thus by part 1, it follows that $h\left(X_{n_{m_k}}\right) \rightarrow_1 h(X)$. This new subsequence retains the property that $h\left(X_{n_{m_k}}\right) \nrightarrow_P h(X)$ by (1), and this contradicts Proposition 3.24, that convergence with probability 1 assures convergence in probability.

3. By Definition 3.22, $X_n \rightarrow_d X$ means that $\lambda_{X_n} \Rightarrow \lambda_X$ for the associated probability measures on $(\mathbb{R}^j, \mathcal{B}(\mathbb{R}^j))$. Since $\lambda_X(D_h) \equiv \mu\left(X^{-1}(D_h)\right) = 0$, Proposition 3.20 states that $(\lambda_{X_n})_h \Rightarrow (\lambda_X)_h$. But if $A \in \mathcal{B}(\mathbb{R}^k)$:

$$(\lambda_{X_n})_h[A] \equiv \lambda_{X_n}\left[h^{-1}(A)\right] \equiv \mu\left[X_n^{-1}\left(h^{-1}(A)\right)\right] = \mu\left[\left(h(X_n)\right)^{-1}(A)\right] = \lambda_{h(X_n)}(A),$$

and similarly, $(\lambda_X)_h = \lambda_{h(X)}$.

Thus the mapping theorem conclusion that $(\lambda_{X_n})_h \Rightarrow (\lambda_X)_h$ obtains $\lambda_{h(X_n)} \Rightarrow \lambda_{h(X)}$, and so $h(X_n) \rightarrow_d h(X)$ by Definition 3.22. ■

Exercise 3.27 (Marginal distributions) *Prove that if $F_n \Rightarrow F$, then each of the $2^j - 2$ proper marginal distribution functions of Definition 1.45 of the sequence $\{F_n\}_{n=1}^{\infty}$ converges weakly to the respective marginal distribution of F. Hint: (1.38) implies a continuous mapping h.*

3.3.3 Cramér-Wold Theorem: Part 1

In the Remark II.9.51 discussion on multivariate extreme value theory, the mapping theorem of Proposition 3.20 was referenced as the key ingredient for the result introduced below. There we were given weak convergence of joint distribution functions $F_n \Rightarrow F$, and required the result that the marginal distribution functions of $\{F_n\}_{n=1}^{\infty}$ also converged weakly to the respective marginal distribution functions of F. We now demonstrate that this result is a corollary to the next result, which we call Part 1 of the Cramér-Wold theorem.

A quite remarkable result, the **Cramér-Wold device** or the **Cramér-Wold theorem,** is named for **Harald Cramér** (1893–1985) and **Herman Wold** (1908–1992). It states that $X_n \rightarrow_d X$ for random vectors defined on $(\mathcal{S}, \mathcal{E}, \mu)$ with range in $\mathbb{R}^j$, if and only if $t \cdot X_n \rightarrow_d t \cdot X$ for all $t \in \mathbb{R}^j$.

Recall that the **dot product** or **inner product** of j-vectors $x = (x_1, ..., x_j)$ and $y = (y_1, ..., y_j)$ is defined:

$$x \cdot y \equiv \sum\nolimits_{k=1}^{j} x_k y_k. \tag{3.13}$$

What we call Part 1 of this result is the easier part, that $X_n \rightarrow_d X$ for random vectors implies that $t \cdot X_n \rightarrow_d t \cdot X$ as random variables for all t.

To introduce this result, given $t \in \mathbb{R}^j$ and $\alpha \in \mathbb{R}$, the set $H_t(\alpha)$ defined as:

$$H_t(\alpha) \equiv \{x \in \mathbb{R}^j : t \cdot x \leq \alpha\}$$

is a half-space in $\mathbb{R}^j$ with bounding hyperplane $\{x \in \mathbb{R}^j : t \cdot x = \alpha\}$. The distribution function $F_{X'}(\alpha)$ associated with $X' \equiv t \cdot X$ is then given by:

$$F_{X'}(\alpha) = \lambda_X \left[H_t(\alpha)\right],$$

where λ_X is the probability measure on $\mathbb{R}^j$ induced by the random variable X of (1.21).

Thus Part 1 states that:

$$\text{If } X_n \rightarrow_d X \text{ then } X'_n \rightarrow_d X',$$

where $X'_n = t \cdot X_n$ and $X' = t \cdot X$. By Definition 3.22, this can be equivalently stated in terms of distribution functions:

$$\text{If } F_n \Rightarrow F \text{ then } F_{X'_n} \Rightarrow F_{X'},$$

or in terms of the associated probability measures:

$$\text{If } \lambda_{X_n} \Rightarrow \lambda_X \text{ then } \lambda_{X'_n} \Rightarrow \lambda_{X'}.$$

Part 2 of this result states that if $X'_n \rightarrow_d X'$ for all t, meaning that all of the one-dimensional distributions defined relative to half-spaces converge weakly for all t, this assures the weak convergence of the random vectors $X_n \Rightarrow X$. Again this can be stated multiple ways as above. The proof of this will require the tools of characteristic functions, and in particular, a **continuity theorem on** $\mathbb{R}^j$ extending Proposition V.7.35 on $\mathbb{R}$. This result is thus deferred to Chapter 5.

Proposition 3.28 (Cramér-Wold theorem: Part 1) *Let $\{X_n\}_{n=1}^{\infty}$, X, be random vectors defined on $(\mathcal{S}, \mathcal{E}, \mu)$ with range in $\mathbb{R}^j$, and assume that $X_n \rightarrow_d X$.*

Then for any $t \in \mathbb{R}^j$:

$$t \cdot X_n \rightarrow_d t \cdot X. \tag{3.14}$$

Proof. *Recall that $X_n \rightarrow_d X$ is equivalent to $\lambda_{X_n} \Rightarrow \lambda_X$ for the associated probability measures. Given $t \in \mathbb{R}^j$, define $h : \mathbb{R}^j \rightarrow \mathbb{R}$ by $h(x) = t \cdot x$. Then h is continuous, so $D_h = \emptyset$ and $\lambda(D_h) = 0$, and h is Borel measurable by Proposition V.2.12. Thus $(\lambda_{X_n})_h \Rightarrow (\lambda_X)_h$ by Proposition 3.20.*

If $A \in \mathcal{B}(\mathbb{R})$, then by (1.21):

$$(\lambda_X)_h [A] \equiv \lambda_x \left[h^{-1}(A)\right] \equiv \mu \left[X^{-1}(h^{-1}(A)\right] = \mu \left[[h(X)]^{-1}(A)\right] = \lambda_{h(X)} [A],$$

and similarly $(\lambda_{X_n})_h = \lambda_{h(X_n)}$.

Thus $(\lambda_{X_n})_h \Rightarrow (\lambda_X)_h$ can be restated as $\lambda_{h(X_n)} \Rightarrow \lambda_{h(X)}$, and then $t \cdot X_n \rightarrow_d t \cdot X$ by Definition 3.22. ■

We now turn to the result on weak convergence of marginal distribution functions needed in the proof of Proposition II.9.52 on multivariate extreme value theory.

Corollary 3.29 (Cramér-Wold theorem: Part 1) *Let $\{X_n\}_{n=1}^{\infty}$, X, be random vectors defined on $(\mathcal{S}, \mathcal{E}, \mu)$ with range in $\mathbb{R}^j$, with associated distribution functions $\{F_n\}_{n=1}^{\infty}$, F.*

If $X_n \rightarrow_d X$, then for all k, $X_{n,k} \rightarrow_d X_k$ for the kth component random variates $X_{n,k}$ and X_k, where $X_n = (X_{n,1}, ..., X_{n,j})$ and $X = (X_1, ..., X_j)$.

Equivalently, if $F_n \Rightarrow F$, then $F_{n,k} \Rightarrow F_k$ for the marginal distribution functions $\{F_{n,k}\}_{n=1}^{\infty}$, F_k of the kth component variates $X_{n,k}$ and X_k.

Proof. *Define $t \in \mathbb{R}^j$ by $t_k = 1$ and $t_i = 0$ otherwise. Then $t \cdot X_n = X_{n,k}$ and $t \cdot X = X_k$, and this is a restatement of (3.14).* ■

3.3.4 Slutsky's Theorem

Slutsky's theorem, named for **Evgeny "Eugen" Slutsky** (1880–1948), was introduced in Proposition II.5.29, and addressed the following question.

If $X_n \rightarrow_d X$ and $Y_n \rightarrow_\bullet Y$ where "$\rightarrow_\bullet$" denotes convergence in some manner, does $X_n + Y_n \rightarrow_d X + Y$, or $X_nY_n \rightarrow_d XY$, etc.?

The Book II version of Slutsky's theorem and associated Exercise II.5.30 provided affirmative results:

If $X_n \rightarrow_d X$, $Y_n \rightarrow_P a$ and $Z_n \rightarrow_P b$, then:

1. $X_n + Y_n \rightarrow_d X + a$,
2. $X_nY_n \rightarrow_d aX$,
3. $X_n/Y_n \rightarrow_d X/a$ if $a \neq 0$,
4. $X_nY_n + Z_n \rightarrow_d aX + b$.

Example II.5.31 then demonstrated that perhaps surprisingly, if $X_n \rightarrow_d X$ and $Y_n \rightarrow_d Y$, then it need not be the case that $X_n + Y_n \rightarrow_d X + Y$, nor that $X_nY_n \rightarrow_d XY$.

The earlier result can now be extended with the aid of the portmanteau theorem and the mapping theorem.

Proposition 3.30 (Slutsky's theorem) *Let $\{X_n\}_{n=1}^{\infty}$, $\{Y_n\}_{n=1}^{\infty}$ be sequences of random vectors defined on $(\mathcal{S}, \mathcal{E}, \mu)$ with $X_n : \mathcal{S} \rightarrow \mathbb{R}^j$ and $Y_n : \mathcal{S} \rightarrow \mathbb{R}^k$. Assume $X_n \rightarrow_d X$ and $Y_n \rightarrow_P a$, where X is a random vector $X : \mathcal{S} \rightarrow \mathbb{R}^j$ and $a \in \mathbb{R}^k$ is a constant.*

If $h : \mathbb{R}^{j+k} \rightarrow \mathbb{R}$ is continuous, then:

$$h(X_n, Y_n) \rightarrow_d h(X, a). \tag{3.15}$$

Proof. *By item 3 of the Mann-Wald theorem of Proposition 3.26, it is enough to prove that $(X_n, Y_n) \rightarrow_d (X, a)$.*

To this end, let λ_n denote the probability measure on $\mathbb{R}^{j+k}$ induced by (X_n, Y_n), and λ analogously defined for (X, a). Thus for $A \in \mathcal{B}(\mathbb{R}^{j+k})$:

$$\lambda_n(A) = \mu\left[(X_n, Y_n)^{-1}(A)\right], \qquad \lambda(A) = \mu\left[(X, a)^{-1}(A)\right]. \tag{1}$$

Note that λ is well-defined with $a \in \mathbb{R}^k$ constant since:

$$(X, a)^{-1}(A) = \{s | (X(s), a) \in A\} = X^{-1}(A_a),$$

where $A_a = \{x | (x, a) \in A\} \subset \mathbb{R}^j$ is the a-cross-section of A of Definition 2.2, and Borel measurable by Proposition 2.3. Thus $X^{-1}(A_a) \in \mathcal{E}$ and $\lambda(A)$ is well-defined.

By definition of $Y_n \rightarrow_P a$, given $\epsilon > 0$, $\delta > 0$, there exists N so that $\lambda_n[|Y_n - a| \geq \delta] \leq \epsilon$ for $n \geq N$. Let $A_{\epsilon,\delta} \subset \mathbb{R}^{j+k}$ be defined by $A_{\epsilon,\delta} \equiv \{(x, y) | \, |y - a| < \delta\}$, and denote the complement set $A^c_{\epsilon,\delta} \equiv \{(x, y) | \, |y - a| \geq \delta\}$. For notational convenience we denote the respective characteristic functions of these sets as $\chi_A(x, y)$ and $\chi_{A^c}(x, y)$.

To prove that $(X_n, Y_n) \rightarrow_d (X, a)$, we equivalently prove that $\lambda_n \Rightarrow \lambda$ using item 6 of the portmanteau theorem.

Let $g : \mathbb{R}^{j+k} \rightarrow \mathbb{R}$ be bounded by $K > 0$ and Lipschitz continuous with constant L. Then by the triangle inequality:

$$\left|\int g(x,y)d\lambda_n - \int g(x,y)d\lambda\right| \leq \left|\int [g(x,y) - g(x,a)]\,\chi_A(x,y)d\lambda_n\right| \tag{2.a}$$

$$+ \left|\int [g(x,y) - g(x,a)]\,\chi_{A^c}(x,y)d\lambda_n\right| \tag{2.b}$$

$$+\left|\int g(x,a)d\lambda_n - \int g(x,a)d\lambda\right| \tag{2.c}$$

$$+\left|\int [g(x,a) - g(x,y)]\, d\lambda\right|. \tag{2.d}$$

Taking these four bounding integrals in turn:

2.a. *By the Lipschitz condition and definition of* χ_A *we obtain for* $n \geq N$*:*

$$\left|\int [g(x,y) - g(x,a)]\, \chi_A(x,y)d\lambda_n\right| \leq L\int |y-a|\, \chi_A(x,y)d\lambda_n \leq L\delta, \tag{3}$$

since λ_n *is a probability measure.*

2.b. *Since g is bounded by K, and then by the definition of* χ_{A^c}*, obtains for* $n \geq N$*:*

$$\left|\int [g(x,y) - g(x,a)]\, \chi_{A^c}(x,y)d\lambda_n\right| \leq 2K\int \chi_{A^c}(x,y)d\lambda_n \leq 2K\epsilon, \tag{4}$$

since:

$$\lambda_n(A^c) \leq \lambda_n(Y^{-1}A^c) = \lambda_n\left[|Y_n - a| \geq \delta\right] \leq \epsilon.$$

2.c. *Let* ν_n *and* ν *be the probability measures on* $\mathbb{R}^j$ *induced by* X_n *and* X*. If* $F_n(x,y)$ *is the distribution function associated with* λ_n *and* $G_n(x)$ *is the distribution function associated with* ν_n*, then by Proposition II.3.36,* $G_n(x)$ *is the marginal distribution of* $F_n(x,y)$*. Using Fubini's theorem of Proposition V.5.20:*

$$\int_{\mathbb{R}^{j+k}} g(x,a)d\lambda_n = \int_{\mathbb{R}^j}\int_{\mathbb{R}^k} g(x,a)d\lambda_n = \int_{\mathbb{R}^j} g(x,a)d\nu_n.$$

Deriving the same identity for the $d\lambda$*-integral:*

$$\left|\int_{\mathbb{R}^{j+k}} g(x,a)d\lambda_n - \int_{\mathbb{R}^{j+k}} g(x,a)d\lambda\right| = \left|\int_{\mathbb{R}^j} g(x,a)d\nu_n - \int_{\mathbb{R}^j} g(x,a)d\nu\right|.$$

Now $X_n \to_d X$ *by assumption, which by definition means that* $\nu_n \Rightarrow \nu$*. Since* $f(x) \equiv g(x,a)$ *is Lipschitz continuous on* $\mathbb{R}^j$*, part* 6 *of the portmanteau theorem assures that this expression converges to zero as* $n \to \infty$*. That is, for* ϵ *as above and* $n \geq N'$*:*

$$\left|\int g(x,a)d\lambda_n - \int g(x,a)d\lambda\right| \leq \epsilon. \tag{5}$$

2.d. *Using the Lipschitz continuity of g:*

$$\left|\int [g(x,a) - g(x,y)]\, d\lambda\right| \leq L\int |y-a|\, \chi_A(x,y)d\lambda,$$

since $\lambda(A^c_{\epsilon,\delta}) = 0$ *by definition of* λ *in* (1)*. Thus:*

$$\left|\int [g(x,a) - g(x,y)]\, d\lambda\right| \leq L\delta. \tag{6}$$

Combining (3)–(6) *into* (2) *obtains for* $n \geq \max(N, N')$*:*

$$\left|\int g(x,y)d\lambda_n - \int g(x,y)d\lambda\right| \leq 2L\delta + (2K+1)\epsilon.$$

As ϵ, δ are arbitrary, this proves that for all bounded, Lipschitz continuous functions $g : \mathbb{R}^{j+k} \rightarrow \mathbb{R}$:

$$\int g(x,y)d\lambda_n \rightarrow \int g(x,y)d\lambda.$$

Thus $(X_n, Y_n) \rightarrow_d (X, a)$ by item 6 of the portmanteau theorem of Proposition 3.17, and the proof is complete ■

Remark 3.31 (Generalization) *The assumption that h is continuous can be further generalized to $h : \mathbb{R}^{j+k} \rightarrow \mathbb{R}$ Borel measurable and:*

$$\mu(D_h) \equiv \lambda\left[(X,a)^{-1}(D_h)\right] = 0.$$

The Mann-Wald theorem remains applicable in the first step, and the proof that $(X_n, Y_n) \rightarrow_d (X, a)$ does not change since this proof was independent of h.

3.3.5 The Delta Method

The **Delta method,** also written as the **Δ-method,** was introduced in Proposition II.8.40 in the context of transformations of **normalized** sequences of random variables. In this case, the result is:

Proposition 3.32 (Proposition II.8.40: The Δ-method) *Let $\{X_n\}_{n=1}^{\infty}$, X be random variables defined on a probability space $(\mathcal{S}, \mathcal{E}, \mu)$, and assume that there exists a positive sequence $\{c_n\}_{n=1}^{\infty} \subset \mathbb{R}$ with $c_n \rightarrow \infty$ as $n \rightarrow \infty$, and a constant x_0, so that*

$$c_n\left(X_n - x_0\right) \rightarrow_d X.$$

If g is a function that is differentiable at x_0, then:

$$c_n\left[g(X_n) - g(x_0)\right] \rightarrow_d g'(x_0)X. \tag{3.16}$$

A common application of this result is to the case where X has a standard normal distribution, and thus the delta method provides additional asymptotic results in cases where the **central limit theorem** applies. One version of this theorem was seen in Proposition IV.6.13, and others will be derived in Chapter 7.

Example 3.33 (The Δ-method) *If $\{Y_j\}_{j=1}^{\infty}$ are independent (μ, σ^2), meaning variates with mean μ and variance σ^2, let:*

$$X_n = \frac{1}{n}\sum\nolimits_{j=1}^{n} Y_j.$$

If the Y-variate has a moment generating function, then by Proposition IV.6.13:

$$\frac{\sqrt{n}}{\sigma}\left[X_n - \mu\right] \rightarrow_d Z,$$

with $Z \sim N(0,1)$, a standard normal variate. The assumption on existence of a moment generating function will be eliminated with Proposition 7.4.

Thus if g is differentiable at μ:

$$\frac{\sqrt{n}}{\sigma}\left[g\left(X_n\right) - g\left(\mu\right)\right] \rightarrow_d g'(\mu)Z.$$

In this section we address two generalizations of this Book II result:

1. The first generalization addresses the case where $g'(x_0) = 0$.

   The above result then states that $c_n\left[g(X_n) - g(x_0)\right] \to_d 0$, a constant, and by Proposition II.5.27 this implies that $c_n\left[g(X_n) - g(x_0)\right] \to_P 0$. The answer to this question will only use the tools of Book II.

2. The second generalization is to determine the multivariate analog of the above result, so that now $\{X_n\}_{n=1}^{\infty}$, X are random vectors. This proof will require the tools of this chapter.

For the first question, recall that if a function of a single variable $g(x)$ is m-times differentiable at x_0, then the *mth*-**order Taylor polynomial of** $g(x)$ **about** x_0, $T_m(x)$, is given by:

$$T_m(x) = \sum\nolimits_{i=1}^{m} \frac{g^{(i)}(x_0)}{i!}(x - x_0)^i, \tag{3.17}$$

where $g^{(i)}(x_0)$ denotes the ith derivative of $g(x)$ evaluated at x_0, and $g^{(0)}(x_0) \equiv g(x_0)$ to simplify notation. This polynomial, and the associated **Taylor series** $T_\infty(x)$, when it converges to $g(x)$, is named for **Brook Taylor** (1685–1731).

An important property of $T_m(x)$ is that:

$$g(x) = T_m(x) + R_m(x),$$

with **remainder term** $R_m(x)$ given by:

$$R_m(x) = h_m(x)(x - x_0)^m,$$

where $h_m(x) \to 0$ as $x \to x_0$.

In the **little-*o*** notation of Definition V.7.13:

$$g(x) - T_m(x) = o\left[(x - x_0)^m\right],$$

which means that as $x \to x_0$:

$$\frac{|g(x) - T_m(x)|}{(x - x_0)^m} \to 0.$$

Assuming a little more of the function $g(x)$, this remainder term can be more explicitly characterized.

If $g^{(m+1)}(x)$ is continuous in an open interval $I \equiv (x_0 - a, x_0 + a)$ about x_0, the **Lagrange form of the remainder,** named for **Joseph-Louis Lagrange** (1736–1813), states that for any $x \in I$:

$$g(x) = T_m(x) + \frac{g^{(m+1)}(x')}{(m+1)!}(x - x_0)^{m+1}, \tag{3.18}$$

where x' is "between" x and x_0. That is, either $x < x' < x_0$ or $x_0 < x' < x$, and this is often expressed as $x' = tx + (1 - t)x_0$ for $0 < t < 1$.

Proposition 3.34 (The general Δ-method) *Let* $\{X_n\}_{n=1}^{\infty}$, X *be random variables defined on a probability space* $(\mathcal{S}, \mathcal{E}, \mu)$, *and assume that there exists a positive sequence* $\{c_n\}_{n=1}^{\infty} \subset \mathbb{R}$ *with* $c_n \to \infty$ *as* $n \to \infty$, *and a constant* x_0, *so that:*

$$c_n\left(X_n - x_0\right) \to_d X.$$

If g is $(m+1)$-times differentiable at x_0 with $g^{(m+1)}(x)$ continuous, and:

$$g'(x_0) = \cdots = g^{(m)}(x_0) = 0,$$

then:

$$c_n^{m+1}\left[g(X_n) - g(x_0)\right] \to_d \frac{g^{(m+1)}(x_0)}{(m+1)!}X^{m+1}. \tag{3.19}$$

Proof. *Since $g^{(j)}(x_0) = 0$ for $1 \leq j \leq m$, $T_m(x) = g(x_0)$ and (3.18) obtains:*

$$c_n^{m+1}\left[g(X_n) - g(x_0)\right] = \frac{g^{(m+1)}(X_n')}{(m+1)!}c_n^{m+1}(X_n - x_0)^{m+1}, \tag{1}$$

where $X_n' = tX_n + (1-t)x_0$ for $0 < t < 1$. Here $t = t(s)$ is defined pointwise for $s \in \mathcal{S}$.

Now $c_n(X_n - x_0) \to_d X$ and $c_n \to \infty$ obtain that $X_n \to_P x_0$ by Proposition II.8.39. Since $X_n' - x_0 = t(X_n - x_0)$ for $0 < t < 1$, it follows that $|X_n' - x_0| < |X_n - x_0|$ and hence:

$$\mu\left(\{|X_n' - x_0| \geq \epsilon\}\right) \leq \mu\left(\{|X_n - x_0| \geq \epsilon\}\right),$$

and hence $X_n \to_P x_0$ assures that $X_n' \to_P x_0$. By the Mann-Wald theorem of Proposition 3.26 with continuous $h(x) \equiv g^{(m+1)}(x)/(m+1)!$, this obtains:

$$\frac{g^{(m+1)}(X_n')}{(m+1)!} \to_P \frac{g^{(m+1)}(x_0)}{(m+1)!}.$$

Similarly, $c_n(X_n - x_0) \to_d X$ with continuous $h(x) = x^{m+1}$ obtains that:

$$c_n^{m+1}(X_n - x_0)^{m+1} \to_d X^{m+1}.$$

Applying Slutsky's theorem of Proposition II.5.29 to the product in (1) yields (3.19). ■

Example 3.35 (Powers of the standard normal) *Extending Example 3.33, let $\{Y_j\}_{j=1}^{\infty}$ be independent (μ, σ^2) with a moment generating function, and define X_n as above. Letting $g(x) = (x-\mu)^{m+1}$, then $g(\mu) = g'(\mu) = \ldots = g^{(m)}(\mu) = 0$.*

Since:

$$\frac{\sqrt{n}}{\sigma}\left[X_n - \mu\right] \to_d Z$$

by the central limit theorem of Proposition IV.6.13, it follows from (3.19) that:

$$\left(\frac{\sqrt{n}}{\sigma}\right)^{m+1}(X_n - \mu)^{m+1} \to_d Z^{m+1}.$$

In particular when $m = 1$:

$$\frac{(X_n - \mu)^2}{\sigma^2/n} \to_d Z^2.$$

Recalling Example IV.2.4, Z^2 is a chi-squared variate with 1-degree of freedom, denoted $\chi^2_{1\ d.f.}$.

For the multivariate version of the delta method we again begin with Taylor series expansions. We will not need all of what follows but provide the extra details for completeness.

Multivariate Taylor series have nearly identical properties to the one-variable series because they can be derived from this model. Given $g : \mathbb{R}^j \to \mathbb{R}$, x_0, $x \in \mathbb{R}^j$, and parameter $t \in \mathbb{R}$, define:

$$\tilde{g}(t) \equiv g(x_0 + t(x - x_0)).$$

We can then apply (3.17) to $\tilde{g}(t)$ with $t = 1$ and $t_0 = 0$ to obtain $\tilde{g}(1) = g(x)$ as a multivariate expansion in terms of derivatives $\tilde{g}^{(m)}(0)$, which are derivatives of $g(x)$ evaluated at x_0.

Each derivative of $\tilde{g}(t)$ of order j will then equal a summation of partial derivatives of that order. For example:

$$\begin{aligned}\tilde{g}'(0) &= \sum_{i=1}^{j} \frac{\partial g(x_0)}{\partial x_i}(x_i - x_{0,i}),\\ \tilde{g}''(0) &= \sum_{j=1}^{j} \sum_{i=1}^{j} \frac{\partial^2 g(x_0)}{\partial x_i \partial x_j}(x_i - x_{0,i})(x_j - x_{0,j}).\end{aligned}$$

There will be duplications in these summations, and it becomes clear a more efficient approach and notation would be helpful.

For the approach, the chain rule applied above can be expressed:

$$\frac{d}{dt} = \sum_{i=1}^{j} (x_i - x_{0,i})\frac{\partial}{\partial x_i},$$

where we evaluate the $\tilde{g}$-derivatives at $t = 0$ and the g-partial derivatives at $x = x_0$. Thus by the multinomial theorem in (9.71):

$$\begin{aligned}\frac{d^m}{dt^m} &= \left(\sum_{i=1}^{j} (x_i - x_{0,i})\frac{\partial}{\partial x_i}\right)^m\\ &= \sum_{\alpha_1,\alpha_2,..\alpha_j} \frac{m!}{\alpha_1!\alpha_2!...\alpha_j!}(x_1 - x_{0,1})^{\alpha_1}...(x_j - x_{0,j})^{\alpha_j}\frac{\partial^{\alpha_1}}{\partial x_1^{\alpha_1}}...\frac{\partial^{\alpha_j}}{\partial x_j^{\alpha_j}}, \qquad (1)\end{aligned}$$

where the sum is over all nonnegative integer j-tuples $(\alpha_1, \alpha_2, ..\alpha_j)$ with $\sum_{i=1}^{j} \alpha_i = m$.

Definition 3.36 (Multi-index notation) *Let* $\alpha \equiv (\alpha_1, ..., \alpha_j)$ *denote a multi-index of nonnegative integers,* $x = (x_1, ..., x_j)$*, and define:*

$$|\alpha| = \sum_{i=1}^{j} \alpha_i, \qquad \alpha! = \prod_{i=1}^{j} \alpha_i!, \qquad x^{\alpha} = \prod_{i=1}^{j} x_i^{\alpha_i}.$$

The partial derivative operator ∂^{α} *has order* $|\alpha| = m$ *and is defined by:*

$$\partial^{\alpha} g \equiv \frac{\partial^{|\alpha|} g}{\partial x_1^{\alpha_1}...\partial x_j^{\alpha_j}}.$$

Then (1) can be simplified to:

$$\frac{d^m}{dt^m} = \sum_{|\alpha|=m} \frac{m!}{\alpha!}(x - x_0)^{\alpha}\partial^{\alpha},$$

If the function $g(x)$ is m-times differentiable at x_0, meaning $\partial^{\alpha} g$ exists for all α with $|\alpha| \leq m$, then the mth-**order Taylor polynomial of** $g(x)$ **about** x_0, $T_m(x)$, is derived from (3.17) with $\tilde{g}(t)$ as noted above:

$$T_m(x) = \sum_{|\alpha| \leq m} \frac{\partial^{\alpha} g(x_0)}{\alpha!}(x - x_0)^{\alpha}, \qquad (3.20)$$

where $\partial^{\alpha} g(x_0)$ denotes the derivative $\partial^{\alpha} g$ evaluated at x_0, and $\partial^0 g(x_0) \equiv g(x_0)$ as in the one-dimensional notation.

It then follows that:

$$g(x) = T_m(x) + R_m(x),$$

with **remainder term** $R_m(x)$ given by:

$$R_m(x) = \sum\nolimits_{|\alpha|=m} h_\alpha(x)(x - x_0)^\alpha,$$

where $h_\alpha(x) \to 0$ for all α as $x \to x_0$.

If $\partial^\alpha g(x)$ is continuous in an open ball $B \equiv \{x| \, |x - x_0| < a\}$ about x_0 for all α with $|\alpha| = m + 1$, the **Lagrange form of the remainder** states that for any $x \in B$:

$$g(x) = T_m(x) + \sum\nolimits_{|\alpha|=m+1} \frac{\partial^\alpha g(x')}{\alpha!}(x - x_0)^\alpha, \tag{3.21}$$

where x' is "between" x and x_0, meaning $x' = tx + (1-t)x_0$ for $0 < t < 1$.

For the next proposition on the multivariate delta method, we also require the notion of **tightness** of a collection of probability measures on $\mathbb{R}^j$, generalizing the one variable notion in Definition II.8.16.

Definition 3.37 (Tight sequence) *A sequence of probability measures $\{\lambda_n\}_{n=1}^\infty$ on $\mathbb{R}^j$ is said to be **tight** if for any $\epsilon > 0$, there is a bounded rectangle $A = \prod_{i=1}^j (a_i, b_i]$, so that $\lambda_n(A) > 1 - \epsilon$ for all n.*

Exercise 3.38 ($\lambda_n \Rightarrow \lambda$ implies tightness) *Generalize Proposition II.8.18 to show that if $\{\lambda_n\}_{n=1}^\infty$ is a sequence of probability measures on $\mathbb{R}^j$ with $\lambda_n \Rightarrow \lambda$ for a probability measure λ, then $\{\lambda_n\}_{n=1}^\infty$ is tight. Hint: Recalling that measures are continuous from below (Proposition I.2.45), first prove that given $A = \prod_{i=1}^j (a_i, b_i]$ and defining $A_M \equiv \prod_{i=1}^j (a_i - M, b_i + M]$, that $\lambda[A_M] \to 1$ as $M \to \infty$ for any probability measure λ. Now complete the steps of the Book II proof.*

Exercise 3.39 (If $c_n(X_n - x_0) \to_d X$, then $X_n \to_P x_0$) *Generalize Proposition II.8.39 to show that if $\{X_n\}_{n=1}^\infty$, X are random vectors defined on a probability space $(\mathcal{S}, \mathcal{E}, \mu)$ with range in $\mathbb{R}^j$, and there exists a positive sequence $\{c_n\}_{n=1}^\infty \subset \mathbb{R}$ with $c_n \to \infty$ as $n \to \infty$, and a constant vector $x_0 \in \mathbb{R}^j$ so that $c_n(X_n - x_0) \to_d X$, then $X_n \to_P x_0$. Hint: Recalling the notation of Remark 3.23 on norms, $X_n \to_P x_0$ means that for any $\epsilon > 0$:*

$$\lim_{n\to\infty} \mu(\{\|X_n - x_0\|_2 \geq \epsilon\}) = 0,$$

where $\|\cdot\|_2$ as defined in (3.9) with $p = 2$ and therefore is the standard norm of (2.14). Then by (3.11), there exists positive $c_{2,\infty}$ and $C_{2,\infty}$ so that

$$c_{2,\infty} \|X_n - x_0\|_\infty \leq \|X_n - x_0\|_2 \leq C_{2,\infty} \|X_n - x_0\|_\infty,$$

with $\|\cdot\|_\infty$ as defined in (3.10), and thus:

$$\{\|X_n - x_0\|_2 \geq \epsilon\} \subset \{\|X_n - x_0\|_\infty \geq \epsilon / C_{2,\infty}\}.$$

Now note that the set on the right is a rectangle centered on x_0 and use the tightness result of Exercise 3.38.

The multivariate delta method is presented next. We leave it as an exercise for the interested reader to generalize this as in Proposition 3.34 to the case where $\partial^\alpha g(x_0) = 0$ for all $|\alpha| \leq m$.

Proposition 3.40 (The Multivariate Δ-Method) *Let $\{X_n\}_{n=1}^{\infty}$, X be random vectors defined on a probability space $(\mathcal{S}, \mathcal{E}, \mu)$ with range in $\mathbb{R}^j$, and assume that there exists a positive sequence $\{c_n\}_{n=1}^{\infty} \subset \mathbb{R}$ with $c_n \to \infty$ as $n \to \infty$, and a constant vector $x_0 \in \mathbb{R}^j$, so that*

$$c_n (X_n - x_0) \to_d X.$$

If $g : \mathbb{R}^j \to \mathbb{R}$ is a continuously differentiable function, then:

$$c_n [g(X_n) - g(x_0)] \to_d \nabla g(x_0) \cdot X, \tag{3.22}$$

where:

$$\nabla g(x_0) \equiv \left(\frac{\partial g(x_0)}{\partial x_1}, ..., \frac{\partial g(x_0)}{\partial x_j} \right), \tag{3.23}$$

is the ***gradient of*** g ***at*** x_0, *and $x \cdot y \equiv \sum_{i=1}^{j} x_i y_i$ is the* ***dot product*** *of x and y.*

In other words:

$$c_n [g(X_n) - g(x_0)] \to_d \sum_{i=1}^{j} \frac{\partial g(x_0)}{\partial x_i} X_i.$$

Proof. *Letting $m = 0$ in (3.21) obtains:*

$$c_n [g(X_n) - g(x_0)] = \nabla g(X_n') \cdot c_n (X_n - x_0), \tag{1}$$

where $X_n' = tX_n + (1-t)x_0$ for $0 < t < 1$, and as in the proof of Proposition 3.34, $t = t(s)$ for $s \in \mathcal{S}$.

Now $c_n (X_n - x_0) \to_d X$ and $c_n \to \infty$ assure that $X_n \to_P x_0$ by Exercise 3.39, and this again obtains that $X_n' \to_P x_0$ as in the proof of Proposition 3.34. In detail, since $X_n' - x_0 = t (X_n - x_0)$ for $0 < t < 1$, it follows that $|X_n' - x_0| < |X_n - x_0|$ in the standard norm of (2.14), and hence:

$$\mu (\{|X_n' - x_0| \geq \epsilon\}) \leq \mu (\{|X_n - x_0| \geq \epsilon\}).$$

Thus by the Mann-Wald theorem of Proposition 3.26 with continuous $h(x) \equiv \nabla g(x)$:

$$\nabla g(X_n') \to_P \nabla g(x_0). \tag{2}$$

Since $c_n (X_n - x_0) \to_d X$ by assumption, (3.22) follows from Slutsky's theorem of Proposition 3.30 by defining continuous $k(y, z) \equiv y \cdot z$ on $\mathbb{R}^{2j}$. Then by (3.15):

$$k(\nabla g(X_n'), c_n (X_n - x_0)) \to_d k(\nabla g(x_0), X),$$

which obtains (3.22) by (1). ■

3.3.6 Scheffé's Theorem

Scheffé's theorem is named for **Henry Scheffé** (1907–1977), who investigated the implication of almost everywhere convergence of probability density functions:

$$f_n \to f, \ m\text{-a.e.},$$

where m denotes Lebesgue measure on $\mathbb{R}^j$. Perhaps not surprisingly, item 1 of Scheffé's theorem states that $\lambda_n \Rightarrow \lambda$ in this case, for the associated probability measures as defined in (1.23). But in fact, almost everywhere convergence of density functions assures something stronger than weak convergence of the associated measures, as will be seen in items 2 and 3 of Scheffé's theorem.

The proof of this result does not require the tools of the portmanteau theorem, only the integration theory of Book V. And with this theory, it is also possible to prove Scheffé's result for density functions defined more generally relative to Borel measures.

Generalizing somewhat the terminology of Definition 1.9:

Definition 3.41 (Probability density w.r.t. ν) *Given a Borel measure space $(\mathbb{R}^j, \mathcal{B}(\mathbb{R}^j), \nu)$, a Borel measurable function $f : \mathbb{R}^j \to \mathbb{R}$ is a **probability density function with respect to** ν if $f \geq 0$ ν-a.e. and $\int f(x)d\nu = 1$.*

By Proposition V.4.5, given such a density function f, the associated set function λ defined on $A \in \mathcal{B}(\mathbb{R}^j)$ by:

$$\lambda(A) \equiv \int_A f(x)d\nu, \tag{3.24}$$

is in fact a probability measure on $\mathcal{B}(\mathbb{R}^j)$. Thus every probability density function with respect to ν on $\mathbb{R}^j$ induces a probability measure λ on $\mathbb{R}^j$. Then by Definition 1.9, $f(x)$ is a density of λ with respect to ν.

So there is nothing new in Definition 3.41. The point of this terminology is to allow reference to such density functions in contexts where we are not interested in the measure λ induced by $f(x)$.

Example 3.42 (All $g \in L_1(\mathbb{R}^j, \mathcal{B}(\mathbb{R}^j), \nu)$ induce probability densities) *Recalling Definition V.9.16, $L_1(\mathbb{R}^j, \mathcal{B}(\mathbb{R}^j), \nu)$ denotes the Banach space (Definition V.9.9) of functions $g : \mathbb{R}^j \to \mathbb{R}$ with ν-integrable $|g|$ and norm defined by:*

$$\|g\|_1 \equiv \int |g|\, d\nu < \infty.$$

Each $g \in L_1(\mathbb{R}^j, \mathcal{B}(\mathbb{R}^j), \nu)$ is in fact an equivalence class (Remark V.9.20) of functions, with any two functions in a given class equal ν-a.e., and thus by item 4 of Proposition V.3.29, have equal norms.

Every $g \in L_1(\mathbb{R}^j, \mathcal{B}(\mathbb{R}^j), \nu)$ induces a density function f on $(\mathbb{R}^j, \mathcal{B}(\mathbb{R}^j), \nu)$ defined by $f = |g| / \|g\|_1$, and thus an associated Borel measure λ defined in (3.24).

Example 3.43 (Densities w.r.t. ν and distribution functions) *Given a probability density f with respect to ν, let λ be the probability measure defined as in (3.24). Then with $A = A_{(x_1,\ldots,x_j)} \equiv \prod_{i=1}^{j}(-\infty, x_i]$:*

$$F(x_1, ..., x_j) \equiv \lambda\left(A_{(x_1,\ldots,x_j)}\right),$$

*is a **distribution function** on $\mathbb{R}^j$ by Proposition I.8.10. But note that f is **not necessarily a density function associated with** F as given in Definition 1.33.*

However, in the special case where ν is absolutely continuous with respect to Lebesgue measure m, so $\nu \ll m$ as in Definition 1.11, then a density function associated with F can be identified in terms of f.

When $\nu \ll m$, the Radon-Nikodým theorem of Proposition 1.12 identifies a Borel measurable function $g_\nu(x) \equiv \frac{d\nu}{dm}$ so that for all $A \in \mathcal{B}(\mathbb{R}^j)$:

$$\nu(A) = \int_A g_\nu(x)dm.$$

Then by Proposition V.4.8, for all $A \in \mathcal{B}(\mathbb{R}^j)$*:*

$$\lambda(A) \equiv \int_A f(x)d\nu = \int_A f(x)g_\nu(x)dm. \tag{3.25}$$

Thus in this special case of $\nu \ll m$, $f(x)g_\nu(x)$ *is then a density function associated with the distribution function* F *as in Definition 1.33.*

In the general case, F *need not have a density function in this sense.*

For example, let $\{y_i\}_{i=1}^N \subset \mathbb{R}^j$, *where if* $N = \infty$ *we assume that this collection has no cluster points, and let* $\{c_i\}_{i=1}^N \subset \mathbb{R}^+$ *with* $\sum_{i=1}^N c_i = 1$. *For* $A \in \mathcal{B}(\mathbb{R}^j)$, *define:*

$$\nu(A) = \sum\nolimits_{y_i \in A} c_i.$$

It is an exercise to check that ν *is indeed a measure, and thus* $(\mathbb{R}^j, \mathcal{B}(\mathbb{R}^j), \nu)$ *is a Borel measure space.*

If f *is a probability density with respect to* ν, *an exercise in Definition V.3.11 for the value of this integral obtains:*

$$\int f(x)d\nu = \sum\nolimits_{i=1}^N c_i f(y_i) = 1.$$

The measure λ *induced by* f *in (3.24) is then defined:*

$$\lambda(A) = \sum\nolimits_{y_i \in A} c_i f(y_i),$$

and the induced distribution function F *is defined as above on* $x \equiv (x_1, ..., x_j)$*:*

$$F(x_1, ..., x_j) = \sum\nolimits_{y_i \leq x} c_i f(y_i). \tag{1}$$

It is not the case that $\nu \ll m$ *since for all* i, $m(y_i) = 0$ *yet* $\nu(y_i) = c_i > 0$. *In the notation of Definition V.8.3,* $\nu \perp m$, *meaning that these measures are* ***mutually singular****. In this case,* F *is a* ***saltus distribution function*** *(Definition IV.1.15), and it is an exercise to prove that* F *has no density function in the sense of Definition 1.33. Hint: If* μ_F *is the probability measure induced by* F *of Proposition 1.27, and* F *has a density function* g, *then by Proposition 1.34, for all* $A \in \mathcal{B}(\mathbb{R}^j)$*:*

$$\mu_F(A) = \int_A g dm.$$

Given a collection $\{A_n\}_{n=1}^\infty \subset \mathcal{B}(\mathbb{R}^j)$ *that only contain one* y_i *and with* $m(A_n) \to 0$, *use Definition V.3.7 and Lebesgue's dominated convergence theorem of Proposition V.3.45 to show that* $\int_{A_n} g dm \to 0$, *while* $\mu_F(A_n) = c_i$ *for all* i.

For **Scheffé's theorem**, while item 1 is a familiar statement on weak convergence of measures, items 2 and 3 are stronger statements, defined next.

Definition 3.44 (Strong convergence; Convergence in total variation) *Let* $\{\lambda_n\}_{n=1}^\infty$, λ *be probability measures on* $\mathcal{B}(\mathbb{R}^j)$. *Then:*

1. $\{\lambda_n\}_{n=1}^{\infty}$ ***converges strongly to*** λ *if for all* $A \in \mathcal{B}(\mathbb{R}^j)$*:*

$$\lambda_n(A) \to \lambda(A).$$

2. $\{\lambda_n\}_{n=1}^{\infty}$ ***converges to*** λ ***in total variation*** *if:*

$$\sup_{A \in \mathcal{B}(\mathbb{R}^j)} |\lambda_n(A) - \lambda(A)| \to 0.$$

Note that item 1 is called strong convergence because by item 4 of the portmanteau theorem, weak convergence only assures that $\lambda_n(A) \to \lambda(A)$ for all λ-continuity sets A. By Definition 3.4, this means that weak convergence only assures that $\lambda_n(A) \to \lambda(A)$ for $A \in \mathcal{B}(\mathbb{R}^j)$ with $\lambda(\partial A) = 0$.

Proposition 3.45 (Scheffé's Theorem) *Let* $\{f_n\}_{n=1}^{\infty}$*,* f *be density functions with respect to* ν *on* $(\mathbb{R}^j, \mathcal{B}(\mathbb{R}^j), \nu)$ *with* $f_n \to f$*,* ν*-a.e. Then:*

$$\int |f_n - f| \, d\nu \to 0. \tag{3.26}$$

Further, if $\{\lambda_n\}_{n=1}^{\infty}$*,* λ*, denote the associated Borel measures given in (3.24), then:*

1. $\lambda_n \Rightarrow \lambda$.

2. $\lambda_n(A) \to \lambda(A)$ ***for all*** $A \in \mathcal{B}(\mathbb{R}^j)$.

3. $\sup_{A \in \mathcal{B}(\mathbb{R}^j)} |\lambda_n(A) - \lambda(A)| \to 0.$

Proof. *Certainly item* $3 \Rightarrow 2$ *by definition, while item* $2 \Rightarrow 1$ *by item 4 of the portmanteau theorem of Proposition 3.17. So left to prove is (3.26), and that this result obtains item 3.*

To prove (3.26), first note that:

$$|f_n - f| = f_n - f + 2\max(f - f_n, 0).$$

Since f_n *and* f *are density functions with respect to* ν*, this obtains:*

$$\int |f_n - f| \, d\nu = 2\int \max(f - f_n, 0) \, d\nu.$$

Now $\max(f - f_n, 0) \to 0$ *pointwise* ν*-a.e., and* $\max(f - f_n, 0) \leq f$*,* ν*-a.e. Thus (3.26) follows from Lebesgue's dominated convergence theorem of Corollary V.3.47.*

To prove that (3.26) implies item 3, we apply (3.24), and then the triangle inequality of Proposition V.3.42:

$$\begin{aligned}
\sup_{A \in \mathcal{B}(\mathbb{R}^j)} |\lambda_n(A) - \lambda(A)| &\equiv \sup_{A \in \mathcal{B}(\mathbb{R}^j)} \left| \int_A (f_n - f) \, d\nu \right| \\
&\leq \sup_{A \in \mathcal{B}(\mathbb{R}^j)} \int_A |f_n - f| \, d\nu \\
&\leq \int |f_n - f| \, d\nu,
\end{aligned}$$

and the proof is complete by (3.26). ■

With a little additional work, we can characterize the supremum in item 3 in terms of the ν-integral of $|f_n - f|$.

Corollary 3.46 (On $\sup_{A\in\mathcal{B}(\mathbb{R}^j)} |\lambda_n(A) - \lambda(A)|$) *With the assumptions above:*

$$\sup_{A\in\mathcal{B}(\mathbb{R}^j)} |\lambda_n(A) - \lambda(A)| = \frac{1}{2}\int |f_n - f|\, d\nu. \tag{3.27}$$

Proof. *Let $g_n \equiv f_n - f$, and recalling Definition V.3.37 define $g_n^+ \equiv \max(g_n, 0)$ and $g_n^- \equiv \max(-g_n, 0)$. Then $g_n = g_n^+ - g_n^-$, and since $\int g_n d\nu = 0$ it follows that:*

$$\int g_n^+ d\nu = \int g_n^- d\nu.$$

Now $\{f_n \geq f\} = \{f_n - f \geq 0\} \in \mathcal{B}(\mathbb{R}^j)$, since $f_n - f$ is Borel measurable, and hence:

$$\begin{aligned}
\sup_{A\in\mathcal{B}(\mathbb{R}^j)} (\lambda_n(A) - \lambda(A)) &= \sup_{A\in\mathcal{B}(\mathbb{R}^j)} \int_A g_n d\nu \\
&= \int_{\{f_n \geq f\}} g_n d\nu \\
&= \int g_n^+ d\nu.
\end{aligned}$$

Similarly,

$$\sup_{A\in\mathcal{B}(\mathbb{R}^j)} (\lambda(A) - \lambda_n(A)) = \int_{\{f \geq f_n\}} (-g_n) d\nu = \int g_n^- d\nu.$$

Since both supremums are nonnegative, these equal the respective absolute values and addition obtains:

$$2\sup_{A\in\mathcal{B}(\mathbb{R}^j)} |\mu_n(A) - \mu(A)| = \int g_n^+ d\nu + \int g_n^- d\nu.$$

Then $|g_n| = g_n^+ + g_n^-$, and hence:

$$2\sup_{A\in\mathcal{B}(\mathbb{R}^j)} |\mu_n(A) - \mu(A)| = \int |f_n - f|\, d\nu.$$

■

Remark 3.47 (On Scheffé and Riesz) *Scheffé published his results in 1947, but it turns out that the density function convergence conclusion in (3.26) is a special case of a 1928 result by **Frigyes Riesz (1880–1956)** and sometimes known as **Riesz's lemma.** This is the same Riesz whose work was prominent in Chapter V.9 on Banach spaces and particularly $L_p(X, \sigma(X), \lambda)$.*

Riesz's lemma: *If $\{f_n\}_{n=1}^{\infty}$, $f \in L_p(X, \sigma(X), \lambda)$ for $p \geq 1$ with $f_n \to f$, λ-a.e. and $\|f_n\|_p \to \|f\|_p$, then $\|f_n - f\|_p \to 0$.*

As density functions are elements of $L_1(X, \sigma(X), \lambda)$ and all have $\|f\|_1 = 1$, Riesz's lemma states that if $f_n \to f$, λ-a.e., then $\|f_n - f\|_1 \to 0$, which is (3.26).

Example 3.48 *Two applications of Scheffé's theorem follow. The first result was also derived in Proposition IV.6.4 using Slutsky's theorem, and the second was derived in Proposition IV.6.5 using moment generating functions, and in Proposition II.1.11 using a calculus-based analysis.*

1. ***Student T ⇒ Standard Normal:*** *Recall the density function of the* ***Student T distribution*** *or* ***Student's T distribution*** *with $\nu > 0$ degrees of freedom introduced in item 3 of Example IV.2.28:*

$$f_T(t) = \frac{\Gamma\left((\nu+1)/2\right)}{\sqrt{\pi\nu}\Gamma\left(\nu/2\right)}\left(1+\frac{t^2}{\nu}\right)^{-(\nu+1)/2}, \tag{3.28}$$

where the ***gamma function*** *$\Gamma(\alpha)$ is defined in (2.30). It is named for* ***William Sealy Gosset*** *(1876–1937) who published under the pen name of* ***Student.***

Now as $n \to \infty$:

$$\left(1+\frac{a}{n}\right)^{-n} \to e^{-a}, \tag{1}$$

since $n\ln(1+\frac{a}{n}) \to a$, and this is proved by the formula for the derivative of $f(x) = \ln(1+ax)$ at $x = 0$. Letting $\nu \to \infty$ obtains:

$$\begin{aligned}\left(1+\frac{t^2}{\nu}\right)^{-(\nu+1)/2} &= \left[\left(1+\frac{t^2}{\nu}\right)^{-\nu}\right]^{(\nu+1)/2\nu} \\ &\to e^{-t^2/2}.\end{aligned}$$

For the coefficient of $f_T(t)$, we require the gamma function version of ***Stirling's formula*** *of IV.(4.105). Stirling's formula, also known as* ***Stirling's approximation,*** *is named for* ***James Stirling*** *(1692–1770) and states that as $z \to \infty$:*

$$\Gamma\left(z+1\right) \approx \sqrt{2\pi}z^{z+1/2}e^{-z}. \tag{3.29}$$

Since $\Gamma(n+1) = n!$ for $n \in \mathbb{N}$ by IV.(1.58), it can be observed that this formula is identical to the Book IV version when $z = n$.

After making substitutions and doing a bit of algebra, taking limits, including the use of (1), we obtain:

$$\frac{\Gamma\left((\nu+1)/2\right)}{\sqrt{\pi\nu}\Gamma\left(\nu/2\right)} \to \frac{1}{\sqrt{2\pi}}. \tag{2}$$

We leave the details as an exercise.

Thus the Student T density converges pointwise to the standard normal density in (6.2) as $\nu \to \infty$. By Scheffé's theorem, the associated probability measures also converge in the various modes identified in Proposition 3.45.

2. ***Poisson limit theorem:*** *The density function of the* ***binomial distribution*** *with parameters $0 < p < 1$ and $n \in \mathbb{N}$ is given in IV(1.40):*

$$f_{B_n}(j) = \binom{n}{j}p^j(1-p)^{n-j}, \quad j = 0, 1, .., n. \tag{3.30}$$

The density function of the ***Poisson distribution,*** *named for* ***Siméon-Denis Poisson*** *(1781–1840), has parameter $\lambda > 0$ and is given in IV.(1.47):*

$$f_P(j) = e^{-\lambda}\lambda^j/j!, \quad j = 0, 1, 2, ..., \tag{3.31}$$

Fix λ and consider the binomial density with parameters $p \equiv \lambda/n$ and n:

$$\begin{aligned}f_{B_n}(j) &= \binom{n}{j}(\lambda/n)^j(1-\lambda/n)^{n-j} \\ &= \frac{\lambda^j}{j!}\left(1-\frac{\lambda}{n}\right)^n\left(1-\frac{\lambda}{n}\right)^{-j}\prod_{k=0}^{j-1}\left(1+\frac{k}{n}\right).\end{aligned}$$

Equivalent to (1) *above,* $\left(1-\frac{\lambda}{n}\right)^n \to e^{-\lambda}$ *as* $n\to\infty$*, while the third and fourth terms converge to* 1.

Thus the binomial density with $p \equiv \lambda/n$ *and* n *converges pointwise to the Poisson density in (6.2) as* $n\to\infty$*. By Scheffé's theorem, the associated probability measures also converge in the various modes identified in Proposition 3.45. To apply this result, it is an exercise to identify a Borel measure space on which these densities are defined.*

3.3.7 Prokhorov's theorem

The final investigation for this chapter is into the generalization of **Prokhorov's theorem,** named for **Yuri Vasilyevich Prokhorov** (1929–2013). Stated in Proposition II.8.24 in the context of distribution functions, this result was proved using a summary of the various results developed in Section II.8.2.

For its general statement, we recall the notion of **tightness** of a collection of measures introduced in Definition 3.37. The conclusion of Exercise 3.38 was that if $\{\lambda_n\}_{n=1}^{\infty}$, λ are probability measures on $\mathbb{R}^j$ and $\lambda_n \Rightarrow \lambda$, then $\{\lambda_n\}_{n=1}^{\infty}$ are tight. The next result provides a converse.

Proposition 3.49 (Prokhorov's theorem) *Given a tight sequence of probability measures* $\{\lambda_n\}_{n=1}^{\infty}$ *on* $\mathbb{R}^j$*, there exists a subsequence* $\{\lambda_{n_k}\}_{k=1}^{\infty}$*, and a probability measure* λ*, so that* $\lambda_{n_k} \Rightarrow \lambda$.

Proof. *Let* $\{F_n\}_{n=1}^{\infty}$ *denote the associated sequence of distribution functions, so* $F_n(x) \equiv \lambda_n\left[\prod_{i=1}^{j}(-\infty, x_i]\right]$. *The first step is in essence the* j*-dimensional version of* ***Helly's selection theorem*** *of Proposition II.8.14, named for* ***Eduard Helly*** *(1884–1943). The goal of this step is to identify a subsequence* $\{F_{n_k}\}_{k=1}^{\infty}$ *and candidate increasing function* F *that will obtain* λ *once* F *proves to be a distribution function. The second step will be to prove that due to tightness, the identified function* F *is indeed a distribution function, and that* $F_{n_k}(x) \to F(x)$ *for all continuity points of* F*. The conclusion that* $\lambda_{n_k} \Rightarrow \lambda$ *will then follow by definition.*

To simplify notation, let $x \leq y$ *or* $x < y$ *be interpreted as component-wise statements, so that* $x_i \leq y_i$ *or* $x_i < y_i$ *for all* i*, respectively.*

Let $\{r_i\}_{i=1}^{\infty} \subset \mathbb{R}^j$ *denote an enumeration of all points with rational components, a countable collection. Since* $\{F_n(r_1)\}_{n=1}^{\infty}$ *is a sequence bounded by* 1*, there is an accumulation point which we denote as* $A(r_1)$*, and a subsequence* $\{n_{1,k}\}_{k=1}^{\infty}$ *so that* $F_{n_{1,k}}(r_1) \to A(r_1)$*. Next, since* $\{F_{n_{1,k}}(r_2)\}_{k=1}^{\infty}$ *is a bounded sequence, there is an accumulation point* $A(r_2)$ *and a subsequence* $\{n_{2,k}\}_{k=1}^{\infty} \subset \{n_{1,k}\}_{k=1}^{\infty}$ *so that* $F_{n_{2,k}}(r_2) \to A(r_2)$*. Continuing in this way, define sequences* $\{n_{i+1,k}\}_{k=1}^{\infty} \subset \{n_{i,k}\}_{k=1}^{\infty}$ *with* $F_{n_{i+1,k}}(r_{i+1}) \to A(r_{i+1})$.

Now define $n_k \equiv n_{k,k}$*. By construction,* $F_{n_k}(r_i) \to A(r_i)$ *for all points in* $\mathbb{R}^j$ *with rational components, and* $0 \leq A(r_i) \leq 1$ *for all* r_i *by construction. Also, given* r_m, r_n *with* $r_m \leq r_n$*, if* $k > \max(n,m)$ *then* $F_{n_k}(r_m) \leq F_{n_k}(r_n)$ *by definition since each* F_{n_k} *is a distribution function. Hence if* $r_m \leq r_n$*:*

$$A(r_m) = \lim_k F_{n_k}(r_m) \leq \lim_k F_{n_k}(r_n) = A(r_n). \tag{1}$$

Now define:

$$F(x) = \inf_{r>x} A(r). \tag{2}$$

Then $0 \leq F(x) \leq 1$ *for all* x*, and* $F(x)$ *is an increasing function because* $A(r_i)$ *is increasing by* (1)*. Thus if* $x \leq y$*, then* $F(x) \leq F(y)$.

To prove that F *is a distribution function we first prove that* F *is* ***continuous from above*** *in the sense of (1.13), and satisfies the* n***-increasing condition*** *of (1.14), and then address limits.*

1. ***For continuity from above,*** *let $x \in \mathbb{R}^j$ and $\epsilon > 0$ be given. By definition of the infimum in (2), there is an $r_i > x$ so that $A(r_i) < F(x) + \epsilon$. Also, for any y with $x \leq y < r_i$, it follows as above that:*

$$F(x) \leq F(y) \leq A(r_i) < F(x) + \epsilon.$$

If $y \to x+$, again meaning componentwise, then:

$$F(x) \leq \lim_{y \to x+} F(y) < F(x) + \epsilon.$$

As $\epsilon > 0$ is arbitrary, F is continuous from above.

2. ***For the n-increasing condition,*** *let $B = \prod_{i=1}^{j}(a_i, b_i]$ be a bounded rectangle and $\epsilon > 0$ be given. By continuity from above, there exists $c = (\delta, \delta, ..., \delta)$ so that if x denotes any one of the 2^j vertices of B, then $|A(r) - F(x)| < \epsilon/2^j$ for rational r with $x < r < x + c$. Choosing rational points q and r with $a < q < a + c$ and $b < r < b + c$, define $B' = \prod_{i=1}^{j}(q_i, r_i]$, and denote by s any one of the 2^j rational vertices of B'. Since $F_{n_k}(r) \to A(r)$ for all rational points in $\mathbb{R}^j$, and each F_{n_k} satisfies (1.14):*

$$\sum\nolimits_s sgn(s)A(s) = \lim_{k\to\infty} \sum\nolimits_s sgn(s)F_{n_k}(s) \geq 0.$$

Pairing the respective vertices of B and B', let s_x denote the rational vertex of B' associated with B-vertex x, then:

$$\left|\sum\nolimits_s sgn(s)A(s) - \sum\nolimits_x sgn(x)F(x)\right| \leq \sum\nolimits_x |A(s_x) - F(x)| < \epsilon,$$

since this is a summation of 2^j terms and by above, $|A(r) - F(x)| < \epsilon/2^j$ for rational r with $x < r < x + c$. Combining estimates it follows that $\sum_x sgn(x)F(x) \geq 0$ since ϵ was arbitrary.

Because F is continuous from above and satisfies the n-increasing condition, there exists a Borel measure λ associated with F by Proposition I.8.15. We next prove that λ is a ***probability measure,*** *and for this the tightness of $\{\lambda_n\}_{n=1}^{\infty}$ is essential.*

By the definition of tightness, for any $\epsilon > 0$ there exists T so that with apparent notation, the rectangle $(-T, T]^j \subset \mathbb{R}^j$ satisfies $\lambda_n\left[(-T, T]^j\right] > 1 - \epsilon$ for all n. Choose $x \in \mathbb{R}^j$ with $x_i > T$ for all i. Then for all n, $F_n(r) > 1 - \epsilon$ for any rational r with $r > x$ since:

$$F_n(r) \equiv \lambda_n\left[\prod\nolimits_{i=1}^{j}(-\infty, r_i]\right] \geq \lambda_n\left[(-T, T]^j\right].$$

But then $A(r) = \lim_{n_k\to\infty} F_{n_k}(r) \geq 1 - \epsilon$ for all $r > x$, and thus by (2):

$$F(x) \geq 1 - \epsilon, \tag{3}$$

for x with all $x_i > T$. Thus, $F(x) \to 1$ as $x \to \infty$.

On the other hand, choose $x \in \mathbb{R}^j$ with $x_i < -T$ for some i. Then $F_n(r) < \epsilon$ for all n for any rational r with $r < x$ since:

$$F_n(r) \equiv \lambda_n\left[\prod\nolimits_{i=1}^{j}(-\infty, r_i]\right] \leq \lambda_n\left[\mathbb{R}^j - (-T, T]^j\right] < \epsilon.$$

As before, this implies that $A(r) \leq \epsilon$ for all $r < x$, and thus by (2), $F(x) \leq \epsilon$ for x with $x_i < -T$ for all i. Since ϵ is arbitrary, this proves that $F(x) \to 0$ as $x \to -\infty$, and thus F is a distribution function.

The final step is to prove that $\lambda_{n_k} \Rightarrow \lambda$*, which is to prove that* $F_{n_k}(x) \to F(x)$ *for all continuity points of* F*. Given such a continuity point* x *and* $\epsilon > 0$*, choose rational* $r > x$ *so that* $A(r) < F(x) + \epsilon$ *as above. By continuity at* x*, there is* $y < x$ *so that* $F(x) - \epsilon < F(y)$*, and choosing rational* q *with* $y < q < x$ *it follows that* $F(y) \leq A(q)$*. Then since* $A(q) \leq A(r)$ *follows from* $q < x < r$ *as above:*

$$F(x) - \epsilon < A(q) \leq A(r) < F(x) + \epsilon.$$

By construction $F_{n_k}(r_i) \to A(r_i)$ *for all rational* r_i*, and this obtains:*

$$F(x) - \epsilon \leq \lim_{k\to\infty} F_{n_k}(q) \leq \lim_{k\to\infty} F_{n_k}(r) < F(x) + \epsilon.$$

But $F_{n_k}(q) \leq F_{n_k}(x) \leq F_{n_k}(r)$ *for all* n_k*, and this implies:*

$$F(x) - \epsilon \leq \liminf_k F_{n_k}(x) \leq \limsup_k F_{n_k}(x) \leq F(x) + \epsilon. \tag{5}$$

As $\epsilon > 0$ *is arbitrary,* (5) *yields that* $\liminf_k F_{n_k}(x) = \limsup_k F_{n_k}(x)$*, and then by Corollary I.3.46, we conclude that* $\lim_k F_{n_k}(x) = F(x)$*.* ■

Remark 3.50 (On $\lambda_{n_k} \Rightarrow \lambda$) *A natural question is: If probability measures* $\{\lambda_n\}_{n=1}^{\infty}$ *on* $\mathbb{R}^j$ *are tight, and there is a subsequence* $\{\lambda_{n_k}\}_{k=1}^{\infty}$ *with* $\lambda_{n_k} \Rightarrow \lambda$*, is it true that* $\lambda_n \Rightarrow \lambda$*?*

Perhaps the simplest answer follows from an exercise. If $\{\lambda_n\}_{n=1}^{\infty}$ *and* $\{\lambda'_n\}_{n=1}^{\infty}$ *are tight sequences of probability measures on* $\mathbb{R}^j$*, prove that* $\{\lambda''_n\}_{n=1}^{\infty}$ *is also tight where* $\lambda''_{2n} = \lambda_n$ *and* $\lambda''_{2n+1} = \lambda'_n$*. The same is true for any assignment of these sequences which preserve their orders.*

Prokhorov's theorem now assures existence of weakly convergence subsequences, but this result cannot assert more. If $\lambda_n \Rightarrow \lambda$ *and* $\lambda'_n \Rightarrow \lambda'$ *with* $\lambda \neq \lambda'$*, then* $\{\lambda''_n\}_{n=1}^{\infty}$ *has at least two subsequential limits by construction. And this example generalizes to have at least* N *subsequential limits for any* N*.*

But what if all convergence subsequences converge weakly to the same λ*?*

Corollary 3.51 (Prokhorov's theorem) *Given a tight sequence of probability measures* $\{\lambda_n\}_{n=1}^{\infty}$ *on* $\mathbb{R}^j$*, if every subsequence* $\{\lambda_{n_k}\}_{k=1}^{\infty}$ *that converges weakly, converges to a given probability measure* λ*, then* $\lambda_n \Rightarrow \lambda$*.*

Proof. *Assume that for the given* λ *that* $\lambda_n \not\Rightarrow \lambda$*, and thus there is a continuity point* x *of the associated distribution function* F *so that* $F_n(x) \not\to F(x)$*, where* F_n *is the distribution function associated with* λ_n*.*

Then for any $\epsilon > 0$*, there is a subsequence* $\{n_k\}_{k=1}^{\infty}$ *so that* $|F_{n_k}(x) - F(x)| \geq \epsilon$ *for all* k*. But* $\{\lambda_{n_k}\}_{k=1}^{\infty}$ *are tight, and thus by Proposition 3.49, there is a subsequence* $\{\lambda_{n_{k_j}}\}_{j=1}^{\infty}$ *which converges weakly, and hence by hypothesis must converge to* λ*. But this is a contradiction, since for the* F*-continuity point* x*,* $\left|F_{n_{k_j}}(x) - F(x)\right| \geq \epsilon$ *for all* j*. Hence* $\lambda_n \Rightarrow \lambda$*.* ■

Remark 3.52 (On $\lambda_n \Rightarrow \lambda$) *Corollary 3.51 states that if every subsequence* $\{\lambda_{n_k}\}_{k=1}^{\infty}$ *that is weakly convergent satisfies* $\lambda_{n_k} \Rightarrow \lambda$ *for fixed* λ*, then* $\lambda_n \Rightarrow \lambda$*. This perhaps seems too strong a conclusion when the assumptions are apparently silent on subsequences* $\{\lambda_{n_k}\}_{k=1}^{\infty}$ *that are not weakly convergent.*

First, tightness is key here. Otherwise, we can modify the exercise in Remark 3.50, to blend tight $\{\lambda_n\}_{n=1}^{\infty}$ *with* $\lambda_n \Rightarrow \lambda$ *and an arbitrary sequence* $\{\lambda'_n\}_{n=1}^{\infty}$ *that is not tight and has no tight subsequences. For example on* $\mathbb{R}$*, define* $\lambda'_n = m$*, as Lebesgue measure defined on* $[n, n+1]$ *and* 0 *elsewhere. It then follows that* $\{\lambda''_n\}_{n=1}^{\infty}$ *of Remark 3.50 is not tight, yet has many weakly convergent subsequence* $\{\lambda''_{n_k}\}_{k=1}^{\infty}$ *which all converge to* λ*.*

In detail, if $\{\lambda''_{n_k}\}_{k=1}^{\infty}$ *contains more than finitely many measures from* $\{\lambda'_n\}_{n=1}^{\infty}$*, then* $\{\lambda''_{n_k}\}_{k=1}^{\infty}$ *cannot be tight and thus by Exercise 3.38 cannot converge. On the other hand, if* $\{\lambda''_{n_k}\}_{k=1}^{\infty}$ *contains finitely many measures from* $\{\lambda'_n\}_{n=1}^{\infty}$ *and any subsequence of* $\{\lambda_n\}_{n=1}^{\infty}$*, then* $\{\lambda''_{n_k}\}_{k=1}^{\infty}$ *is tight and* $\lambda''_{n_k} \Rightarrow \lambda$*.*

Thus other than tightness, $\{\lambda''_n\}_{n=1}^{\infty}$ *satisfies the assumption of Corollary 3.51, that all weakly convergence subsequences converge to* λ*. But* $\{\lambda''_n\}_{n=1}^{\infty}$ *cannot be weakly convergent, since again by Exercise 3.38, such convergence assures tightness.*

Thus tightness of $\{\lambda_n\}_{n=1}^{\infty}$ *is essential for the conclusion of this corollary. From this it follows that* ***all subsequences*** $\{\lambda_{n_k}\}_{k=1}^{\infty} \subset \{\lambda_n\}_{n=1}^{\infty}$ *with* $n_k \to \infty$ *satisfy* $\lambda_{n_k} \Rightarrow \lambda$*. In other words, the assumptions of Corollary 3.51 are sufficient to assure that there can be no subsequence of* $\{\lambda_n\}_{n=1}^{\infty}$ *that is not weakly convergent to* λ*.*

4

Expectations of Random Variables 2

This chapter completes and extends the introductory studies of Book IV on expectations of random variables. With the integration theory of Book V, expectations can now be formally defined in terms of integrals on the underlying probability space. Using various transformations and change of variable results, this definition is seen to have various guises, including the familiar formulas from continuous and discrete probability theory.

Moments and the moment generating functions follow as examples of expectations, and various properties are developed, including an investigation into weak convergence of distribution functions and the implications for convergence of moments and moment generating functions.

The final sections study the general notions of conditional probabilities and conditional expectations. After introducing the ideas from the perspective of elementary probability theory, these notions are formally defined, and then proved to exist with the aid of the Radon-Nikodým theorem. Various foundational properties of conditional expectations are then derived, as is Jensen's inequality and $L_p(\mathcal{S})$-space properties. The final section develops various results related to integration to the limit.

4.1 Expectations and Moments

We begin by formalizing the definition given in Section IV.4.1.2. In short, expectations can be defined for any λ-integrable random variable.

Definition 4.1 (Expectation) *Let $X : \mathcal{S} \to \mathbb{R}$ be a random variable defined on a probability space $(\mathcal{S}, \mathcal{E}, \lambda)$. The* ***expectation of*** *X, denoted $E[X]$, is defined by:*

$$E[X] = \int_{\mathcal{S}} X(s)d\lambda, \tag{4.1}$$

when this integral exists, meaning by Definition V.3.38:

$$\int_{\mathcal{S}} |X(s)|\, d\lambda < \infty. \tag{4.2}$$

If $g : \mathbb{R} \to \mathbb{R}$ is a Borel measurable function, then $Y = g(X)$ is a random variable on $(\mathcal{S}, \mathcal{E}, \lambda)$ and the ***expectation of*** *$g(X)$, denoted $E[g(X)]$, is analogously defined by:*

$$E[g(X)] = \int_{\mathcal{S}} g(X(s))d\lambda, \tag{4.3}$$

when this integral exists:

$$\int_{\mathcal{S}} |g(X(s))|\, d\lambda < \infty. \tag{4.4}$$

DOI: 10.1201/9781003275770-4

If $X : \mathcal{S} \rightarrow \mathbb{R}^j$ is a random vector, $E[X]$ is defined componentwise:

$$E[X] \equiv (E[X_1], ..., E[X_j]) . \tag{4.5}$$

If $g : \mathbb{R}^j \rightarrow \mathbb{R}^k$ is a Borel measurable transformation, then $E[g(X)]$ is defined by (4.1) if $k = 1$, and by (4.5) if $k \geq 2$.

While such integrals on $(\mathcal{S}, \mathcal{E}, \lambda)$ are well-defined by the integration theory of Chapter V.3, in order to evaluate $E[g(X)]$ for given g, X, and $(\mathcal{S}, \mathcal{E}, \lambda)$, it is always necessary to transform this λ-integral to an integral on $\mathbb{R}$ which can then be evaluated directly. Such transformations will utilize the change of variables results of Chapter V.4.

This first step is a transformation from the λ-integral to a Lebesgue-Stieltjes integral with respect to λ_X, the probability measure on $\mathbb{R}$ induced by X of (1.21). In the special case of absolutely continuous λ_X, meaning $\lambda_X \ll m$ in the sense of Definition 1.11 with m Lebesgue measure, this Lebesgue-Stieltjes integral can then be transformed into a Lebesgue integral. By Proposition V.8.18, $\lambda_X \ll m$ if and only if $F(x)$, the distribution function of X, is absolutely continuous in the sense of Definition 1.40.

For notational simplicity, we investigate transformations of $E[g(X)]$, since then $g(x) = x$ provides the results for $E[X]$.

1. **Transformation to a Lebesgue-Stieltjes Integral on $\mathbb{R}$:**

   The random variable $X : \mathcal{S} \rightarrow \mathbb{R}$ is a **measurable transformation** between measure spaces $(\mathcal{S}, \mathcal{E}, \lambda)$ and $(\mathbb{R}, \mathcal{B}(\mathbb{R}), m)$, and induces a probability measure λ_X on $\mathbb{R}$ by (1.21). Specifically, for $A \in \mathcal{B}(\mathbb{R})$:

   $$\lambda_X(A) \equiv \lambda \left[X^{-1}(A)\right] .$$

   Applying the change of variables result of Proposition V.4.18, where the transformation X here is denoted there by T:

   $$E[g(X)] \equiv \int_{\mathcal{S}} g(X(s))d\lambda = \int_{\mathbb{R}} g(x)d\lambda_X .$$

   The integral on the right is a **Lebesgue-Stieltjes integral.**

   If $A = (-\infty, x]$ then $\lambda_X(A) = F(x)$ where F is the distribution function of X, and so as noted in (1.22), $\lambda_X = \lambda_F$ on $\mathcal{B}(\mathbb{R})$, where λ_F denotes the Borel measure on $\mathbb{R}$ of Proposition 1.27 induced by F. That is, $d\lambda_X = d\mu_F$, and this expectation can also be expressed as:

   $$E[g(X)] = \int_{\mathbb{R}} g(x)d\lambda_F. \tag{4.6}$$

   For **continuous** g, Proposition V.3.63 obtains that $E[g(X)]$ can also be expressed as a **Riemann-Stieltjes integral** of Book III:

   $$E[g(X)] = \int_{\mathbb{R}} g(x)dF. \tag{4.7}$$

   This is the approach seen in Definition IV.4.1 noted above.

Example 4.2 (Discrete distribution) *Consider the special case where X is defined on $(\mathcal{S}, \mathcal{E}, \lambda)$ and has a discrete range, meaning $X(\mathcal{S}) = \{x_i\}_{i=1}^N$. We assume for simplicity that if $N = \infty$, that this collection has no accumulation points and is indexed in increasing order. If this collection is unbounded positively and negatively, it can be partitioned into two monotonic sequences and the following logic applied to each separately.*

Denoting $\lambda_X[x_i] \equiv \lambda\left[X^{-1}(x_i)\right] = f(x_i)$, it follows that $\sum_{i=1}^{N} f(x_i) = 1$ and the distribution function of X can be expressed:

$$F(x) = \sum\nolimits_{x_i \leq x} f(x_i),$$

recalling that by (1.5) and (1.21):

$$F(x) \equiv \lambda\left[X^{-1}((-\infty, x])\right] \equiv \lambda_X[(-\infty, x]].$$

*Such $F(x)$ is often called a **discrete distribution** function.*

It then follows similarly that for all i:

$$\begin{aligned} \lambda_X[(x_i, x_{i+1})] = 0, \quad \lambda_X[x_i] = f(x_i), \\ \lambda_X[(-\infty, x_1)] = 0, \quad \lambda_X[(x_N, \infty)] = 0, \end{aligned} \tag{1}$$

noting that $(x_N, \infty) = \emptyset$ if $N = \infty$.

*The **standard expectations formula** for a discrete distribution is usually defined:*

$$E[g(X)] = \sum\nolimits_{i=1}^{\infty} g(x_i) f(x_i), \tag{4.8}$$

provided that as in (4.4):

$$\sum\nolimits_{i=1}^{\infty} |g(x_i)| f(x_i) < \infty.$$

We claim that the formula for $E[g(X)]$ in (4.8) follows from (4.6).

To this end, recalling Definitions V.3.11 and V.3.38:

$$\int_{\mathbb{R}} g(x) d\lambda_F \equiv \sup_{\varphi \leq g^+} \int_{\mathbb{R}} \varphi(x) d\lambda_F - \sup_{\psi \leq g^-} \int_{\mathbb{R}} \psi(x) d\lambda_F.$$

Here φ and ψ are simple functions, and g^+ and g^- the positive and negative parts of g (Definition V.3.37), each of which is a nonnegative function. To evaluate this integral, we assume for simplicity that g is nonnegative, since otherwise, the following can be applied to each of g^+ and g^- separately.

For $i \geq 1$, let $A_i \equiv \{x_i\}$ and $A_i' \equiv (x_i, x_{i+1})$, and also define $A_0' = (-\infty, x_1)$ and $A_N' = (x_N, \infty)$, noting that $A_N' = \emptyset$ if $N = \infty$. As a disjoint union:

$$\mathbb{R} = \bigcup\nolimits_{i=1}^{N} A_i \bigcup\nolimits_{i=1}^{N-1} A_i' \bigcup A_0' \bigcup A_N'. \tag{2}$$

Given an arbitrary simple function, $\varphi_n(x) = \sum_{j=1}^{n} b_j \chi_{B_j}(x)$ with disjoint $\{B_j\}_{j=1}^{n}$ and $\varphi_n \leq g$, it follows from Definition V.3.2:

$$\int_{\mathbb{R}} \varphi_n(x) d\lambda_F = \sum\nolimits_{j=1}^{n} b_j \lambda_F(B_j).$$

By (2), each B_j-set can be expressed as a disjoint union:

$$B_j = \bigcup\nolimits_k \left(B_j \bigcap C_k\right),$$

where $\{C_k\}$ are all the sets in (2). Then by countable additivity and (1), $\lambda_F(B_j) = \sum_{x_i \in B_j} f(x_i)$, recalling that $\lambda_F = \lambda_X$ as noted in (1.22). Further, $\varphi_n \leq g$ implies that each $b_j \leq \inf_{x \in B_j} g(x)$, and so:

$$\int_{\mathbb{R}} \varphi_n(x) d\lambda_F = \sum\nolimits_{j=1}^{n} \inf_{x \in B_j} g(x) \sum\nolimits_{x_i \in B_j} f(x_i). \tag{3}$$

Taking a supremum of the integral in (3) over all such $\varphi \leq g$, this supremum is obtained by letting $n \to \infty$ and reducing the B_j-sets to contain only one x_i.

Alternatively, define $\{\varphi_n(x)\}$ so that $\varphi_n \leq g$ for all n, and $\varphi_n \to g$ pointwise. See Proposition V.2.28 for example. Then since $g(x)$ is assumed integrable, Lebesgue's dominated convergence theorem of Proposition V.3.45 obtains (4.8):

$$\int_{\mathbb{R}} g(x)d\lambda_F = \lim_{n\to\infty} \int_{\mathbb{R}} \varphi_n(x)d\lambda_F = \sum\nolimits_{i=1}^{\infty} g(x_i)f(x_i).$$

Exercise 4.3 (Continuous $g(x)$ and Riemann-Stieltjes) *If $g(x)$ is continuous, derive (4.8) from the Riemann-Stieltjes integral representation in (4.7). Hint: This integral exists over every interval $[a,b]$ by item 1 of Proposition III.4.17, and then exists by assumption as $a \to -\infty$ and $b \to \infty$. By Proposition III.4.12, you can evaluate this integral over any interval $[a,b]$ as a limit of Riemann-Stieltjes sums of your choosing. Alternatively, this evaluation can be done with Proposition III.4.28 over $[a,b]$, and then take a limit.*

2. **Transformation to a Lebesgue Integral on $\mathbb{R}$**

   In the special case where λ_X is absolutely continuous with respect to Lebesgue measure m, denoted $\lambda_X \ll m$ in Definition 1.11, the Lebesgue-Stieltjes integral in (4.6) can be transformed into a Lebesgue integral.

   By Proposition V.8.18, $\lambda_X \ll m$ if and only if $F(x)$, the distribution function of X, is absolutely continuous in the sense of Definition 1.40. Thus there are two ways to derive this transformation.

   (a) **When $F(x)$ is absolutely continuous**

   By Proposition III.3.59, $F'(x)$ exists m-a.e., meaning almost everywhere relative to Lebesgue measure, and is Lebesgue integrable. Proposition III.3.62 then yields that if $f(x)$ is Lebesgue measurable and $f(x) = F'(x)$ m-a.e., then letting $a \to \infty$:

$$F(x) = F(a) + \int_a^x f(y)dm \to \int_{-\infty}^x f(y)dm,$$

   recalling that $F(a) \to 0$ as $a \to -\infty$. Such f is a density function associated with F by Definition 1.33.

   Since $\lambda_X = \lambda_F$ by (1.22), it follows that from Proposition 1.34 that λ_F can also be defined in terms of this density, so for all $A \in \mathcal{B}(\mathbb{R})$:

$$\lambda_F(A) = \int_A f(y)dm. \tag{4}$$

   (b) **When $\lambda_X \ll m$**

   By the Radon-Nikodým theorem of Proposition 1.12, if $\lambda_F \ll m$, then there exists a Borel measurable function $f(x)$, which is unique m-a.e, so that $\lambda_F(A)$ is given as in (4) for all $A \in \mathcal{B}(\mathbb{R})$.

   Thus by either approach we obtain (4), that the measure λ_F has a density function relative to Lebesgue measure. Then by Proposition V.4.8, $E[g(X)]$ in (4.6) can be expressed as a **Lebesgue integral:**

$$E[g(X)] \equiv \int_{\mathbb{R}} g(x)d\lambda_F = (\mathcal{L})\int_{\mathbb{R}} g(x)f(x)dx. \tag{4.9}$$

3. **Transformation to a Riemann Integral on $\mathbb{R}$**

   In the additionally special case where both $g(x)$ and $f(x) = F'(x)$ are continuous, then $E[g(X)]$ in (4.7) can be expressed as a **Riemann integral.** This result follows from Proposition III.4.28, integrating over any interval $[a, b]$, and then the limits as $a \to -\infty$ and $b \to \infty$ exist by assumption, obtaining:

$$E[g(X)] \equiv \int_{\mathbb{R}} g(x)dF = (\mathcal{R}) \int_{\mathbb{R}} g(x)f(x)dx. \tag{4.10}$$

4.1.1 Expectations of Independent RV Products

In this section we investigate $E[XY]$ for independent random variables X and Y defined on $(\mathcal{S}, \mathcal{E}, \lambda)$, but first we recharacterize Definition 1.53.

By Definition II.3.47, random variables X and Y defined on $(\mathcal{S}, \mathcal{E}, \lambda)$ are said to be **independent** if the associated sigma algebras of Definition II.3.43, $\sigma(X)$ and $\sigma(Y)$, are **independent sigma algebras** by Definition II.1.15. While this reference to the associated sigma algebras can be suppressed as in Definition 1.53, these sigma algebras play a more prominent role for the next result.

To formalize:

Definition 4.4 (The sigma algebra $\sigma(X)$) *If $X : \mathcal{S} \longrightarrow \mathbb{R}$, is a random variable on $(\mathcal{S}, \mathcal{E}, \lambda)$,* ***the sigma algebra generated by*** *X, denoted $\sigma(X)$, is the smallest sigma algebra with respect to which X is measurable. If $X : \mathcal{S} \longrightarrow \mathbb{R}^n$ is a random vector on $(\mathcal{S}, \mathcal{E}, \lambda)$,* ***the sigma algebra generated by*** *X, denoted $\sigma(X)$, is defined analogously.*

Remark 4.5 (Characterization of $\sigma(X)$) *It was derived in Exercise II.3.44 that $\sigma(X)$ is given by:*

$$\sigma(X) = X^{-1}\left(\mathcal{B}(\mathbb{R}^n)\right), \tag{4.11}$$

where $n = 1$ for the random variable definition, and general n in the case of a random vector.

Thus $B \in \sigma(X)$ if and only if there exists $A \in \mathcal{B}(\mathbb{R})$, respectively $\mathcal{B}(\mathbb{R}^n)$, so that $B = X^{-1}(A)$. Thus Definition 1.53 can be equivalently stated:

Definition 4.6 (Independent random variables/vectors) *If $\{X_j\}_{j=1}^n : \mathcal{S} \longrightarrow \mathbb{R}$ are random variables on $(\mathcal{S}, \mathcal{E}, \lambda)$, we say that $\{X_j\}_{j=1}^n$ are* ***independent random variables*** *if given $\{B_j\}_{j=1}^n$ with $B_j \in \sigma(X_j)$:*

$$\lambda\left(\bigcap\nolimits_{j=1}^n B_j\right) = \prod\nolimits_{j=1}^n \lambda(B_j). \tag{4.12}$$

An infinite collection of random variables is said to be independent if every finite subcollection is independent. Independent random vectors are defined analogously.

The main result of this section follows. Note that we must assume that XY is integrable, as this does not follow from the integrability of X and Y. The reader can derive an example, or see Example IV.4.53. See also Remark 4.8.

Proposition 4.7 ($E[XY]$, independent X, Y) *Let X and Y be independent, integrable random variables on a probability space $(\mathcal{S}, \mathcal{E}, \lambda)$, with XY integrable. Then:*

$$E[XY] = E[X]E[Y]. \tag{4.13}$$

By induction, (4.13) is valid for all finite products of independent, integrable variates $\{X_j\}_{j=1}^n$ *with* $\prod_{j=1}^n X_j$ *integrable:*

$$E\left[\prod_{j=1}^n X_j\right] = \prod_{j=1}^n E[X_j].$$

Proof. *If* $A \in \sigma(X)$ *and* $B \in \sigma(Y)$, *then by Definitions V.3.2 and V.3.7, (4.12) can be restated:*

$$\lambda\left(A\bigcap B\right) = \int \chi_{A\bigcap B} d\lambda = \int \chi_A d\lambda \int \chi_B d\lambda, \tag{1}$$

where χ_C *is the characteristic function of* C, *defined to be* 1 *on* C *and* 0 *otherwise.*

Given disjoint $\{A_i\}_{i=1}^n \subset \sigma(X)$ *and* $\{B_j\}_{j=1}^m \subset \sigma(Y)$, *and* $\{\{a_i\}_{i=1}^n, \{b_j\}_{j=1}^m\} \subset \mathbb{R}$, *define simple functions* $\varphi = \sum_{i=1}^n a_i\chi_{A_i}$ *and* $\psi = \sum_{j=1}^m b_j\chi_{B_j}$. *Then since* $\chi_{A_i}\chi_{B_j} = \chi_{A_i\bigcap B_j}$, *linearity of the integral and* (1) *obtain:*

$$\int \varphi\psi d\lambda = \sum_{i=1}^n \sum_{j=1}^m a_i b_j \int \chi_{A_i\bigcap B_j} d\lambda = \int \varphi d\lambda \int \psi d\lambda. \tag{2}$$

If X *is an integrable random variable on* $(\mathcal{S}, \mathcal{E}, \lambda)$, *express* $X = X^+ - X^-$ *as in Definition V.3.37, with* $X^{\pm}$ *nonnegative and integrable. By Proposition V.2.28, there exist increasing sequences of simple functions* $\{\varphi_n^{\pm}\}_{n=1}^{\infty}$ *so that* $\varphi_n^{\pm} \to X^{\pm}$ *pointwise. By the construction of that result, the associated* A_i*-sets are sets in* $\sigma(X)$.

Further, $\varphi_n \equiv \varphi_n^+ - \varphi_n^-$ *satisfies* $\varphi_n \to X$ *pointwise, and:*

$$|\varphi_n| \equiv \varphi_n^+ + \varphi_n^- \leq X^+ + X^- \equiv |X|.$$

Thus since X *is integrable, Lebesgue's dominated convergence theorem of Proposition V.3.45 obtains:*

$$\int \varphi_n d\lambda \to \int X d\lambda. \tag{3}$$

Analogously, there exists a simple function sequence with $\psi_n \to Y$ *pointwise with* $|\psi_n| \leq |Y|$, *where the associated* B_j*-sets are sets in* $\sigma(Y)$, *and with* (3) *satisfied with a change of notation.*

Now consider the simple function sequence $\{\varphi_n\psi_n\}_{n=1}^{\infty}$. *By construction,* $\varphi_n\psi_n \to XY$ *pointwise and* $|\varphi_n\psi_n| \leq |XY|$, *so by integrability of* XY, *dominated convergence applies with:*

$$\int \varphi_n\psi_n d\lambda \to \int XY d\lambda.$$

Applying (2) *to the integral on the left, then* (3) *to both terms obtains:*

$$\int XY d\lambda = \int X d\lambda \int Y d\lambda,$$

which is (4.13). ■

Remark 4.8 (On integrability of XY**)** *As noted above and illustrated in Example IV.4.53, integrability of* X *and* Y *on* $(\mathcal{S}, \mathcal{E}, \lambda)$ *does not assure integrability of* XY.

But if $|X|^p$ *and* $|Y|^q$ *are integrable random variables for* $1 \leq p, q \leq \infty$ *with* $\frac{1}{p} + \frac{1}{q} = 1$, *and where* $\frac{1}{\infty} \equiv 0$ *and* $|Y|^{\infty} \equiv \sup|Y|$, *then* XY *is indeed an integrable random.*

By ***Hölder's inequality*** *of Proposition IV.4.54, and derived by* **Otto Hölder** *(1859–1937):*

$$E[|XY|] \leq E[|X|^p]^{1/p} E[|Y|^q]^{1/q}, \tag{4.14}$$

where $E[|X|^p]^{1/p} \equiv \left(E[|X|^p]\right)^{1/p}$ *for notational simplicity. When* $q = \infty$, $E[|Y|^q]^{1/q}$ *is taken as* $\sup|Y|$, *recalling Exercise V.9.19 that* $E[|Y|^q]^{1/q} \to \sup|Y|$ *as* $q \to \infty$.

Note that this result does not require independence of X *and* Y.

Two corollaries follow. The first is relatively transparent. The second may initially surprise, and states that for any $B \in \sigma(Y)$ that:

$$E[X] = \frac{1}{\lambda(B)} \int_B X d\lambda.$$

Since Definition 4.1 can be expressed as:

$$E[X] = \frac{1}{\lambda(\mathcal{S})} \int_{\mathcal{S}} X d\lambda,$$

this result states that the average value of X over any independent set B is the average of X over $\mathcal{S}$, or, the mean of X.

Corollary 4.9 ($E[XY]$, independent X, Y) *Let X and Y be independent, integrable random variables on a probability space $(\mathcal{S}, \mathcal{E}, \lambda)$, with XY integrable. Then:*

$$E[|XY|] = E[|X|]E[|Y|]. \tag{4.15}$$

Proof. *Since $g(x) = |x|$ is continuous and thus Borel measurable by Proposition I.3.11, if X and Y are independent, so too are $|X|$ and $|Y|$ by Proposition II.3.56. Thus (4.15) follows from (4.13).* ■

Corollary 4.10 ($\int_B X d\lambda$, $B \in \sigma(Y)$, independent X, Y) *Let X and Y be independent random variables on a probability space $(\mathcal{S}, \mathcal{E}, \lambda)$, with X integrable and $B \in \sigma(Y)$. Then:*

$$\frac{1}{\lambda(B)} \int_B X d\lambda = E[X]. \tag{4.16}$$

Proof. *By Definition V.3.7:*

$$\int_B X d\lambda = \int \chi_B X d\lambda.$$

With $\sigma(\chi_B)$ denoting the sigma algebra of Definition 4.4 generated by the random variable χ_B:

$$\sigma(\chi_B) = \{\emptyset, B, \mathcal{S}\} \subset \sigma(Y).$$

Thus χ_B and X are independent random variables on $\mathcal{S}$, so (4.16) follows from (4.13), recalling that $\int \chi_B d\lambda = \lambda(B)$ by Definition V.3.2. ■

4.1.2 Moments and the MGF

Three types of moments are commonly defined and were introduced in Section IV.4.2.1, while the moment generating function was introduced in Section IV.4.2.2. We summarize these here within the above framework. See Definition 5.6 for multivariate versions of these notions.

For these definitions, we apply the various transformations discussed above.

1. **Moments About the Origin**

Sometimes referred to as the **raw moments** or simply the **moments,** these are the expectations defined relative to the function $g(X) = X^n$.

Definition 4.11 (Moments; mean) *Let $X : \mathcal{S} \to \mathbb{R}$ be a random variable defined on a probability space $(\mathcal{S}, \mathcal{E}, \lambda)$ with distribution function $F(x)$.*

The nth **moment,** *denoted μ'_n, is defined as $E[X^n]$:*

$$\mu'_n \equiv \int_{\mathcal{S}} X^n(s) d\lambda, \tag{4.17}$$

when (4.4) is satisfied, and is undefined otherwise.

When $n = 1$, $\mu'_1 \equiv E[X]$ is called the **mean of the distribution** *F, or the* **mean of** *X, and denoted by μ:*

$$\mu \equiv \mu'_1. \tag{4.18}$$

When $F(x)$ is continuously differentiable with a continuous density function $f(x)$, moments can be defined as Riemann integrals:

$$\begin{aligned} \mu'_n &\equiv (\mathcal{R}) \int_{-\infty}^{\infty} x^n f(x) dx, \\ \mu &\equiv (\mathcal{R}) \int_{-\infty}^{\infty} x f(x) dx. \end{aligned} \tag{4.19}$$

These representations are valid as Lebesgue integrals when $F(x)$ is absolutely continuous and $f(x) = F'(x)$, m-a.e.

When $F(x)$ is a discrete distribution function with discontinuities on $\{x_i\}_{i=\infty}^{\infty}$, where these points have no accumulation points:

$$\begin{aligned} \mu'_n &\equiv \sum_{i=-\infty}^{\infty} x_i^n f(x_i), \\ \mu &\equiv \sum_{i=-\infty}^{\infty} x_i f(x_i). \end{aligned} \tag{4.20}$$

2. Central Moments

The **central moments** are defined with $g(X) = (X - \mu)^n$, where μ denotes the mean of the distribution.

Definition 4.12 (Central moments; variance) *Let $X : \mathcal{S} \to \mathbb{R}$ be a random variable defined on a probability space $(\mathcal{S}, \mathcal{E}, \lambda)$ with distribution function $F(x)$.*

If $\mu = E[X]$ exists, the nth **central moment,** *denoted μ_n, is defined as $E[(X - \mu)^n]$:*

$$\mu_n \equiv \int_{\mathcal{S}} (X(s) - \mu)^n d\lambda, \tag{4.21}$$

when (4.4) is satisfied, and undefined otherwise.

When $n = 2$, $\mu_2 \equiv E\left[(X - \mu)^2\right]$ is called the **variance of the distribution** *F, or the* **variance of** *X, and denoted by σ^2:*

$$\sigma^2 \equiv \mu_2, \tag{4.22}$$

and the positive square root is called the **standard deviation of the distribution** *F or* **of** *X:*

$$\sigma \equiv \sqrt{\mu_2}. \tag{4.23}$$

When $F(x)$ is continuously differentiable with a continuous density function $f(x)$:

$$\begin{aligned} \mu_n &\equiv (\mathcal{R}) \int_{-\infty}^{\infty} (x - \mu)^n f(x) dx, \\ \sigma^2 &\equiv (\mathcal{R}) \int_{-\infty}^{\infty} (x - \mu)^2 f(x) dx. \end{aligned} \tag{4.24}$$

These representations are valid as Lebesgue integrals when $F(x)$ is absolutely continuous and $f(x) = F'(x)$, m-a.e.

When $F(x)$ is a discrete distribution function with discontinuities on $\{x_i\}_{i=\infty}^{\infty}$, where these points have no accumulation points:

$$\begin{aligned} \mu_n &\equiv \sum\nolimits_{i=-\infty}^{\infty}(x_i - \mu)^n f(x_i), \\ \sigma^2 &\equiv \sum\nolimits_{i=-\infty}^{\infty}(x_i - \mu)^2 f(x_i). \end{aligned} \tag{4.25}$$

Remark 4.13 **($\sigma^2 = \mu_2' - \mu^2$)** *The reader is undoubtedly familiar the common restatement of the variance formula:*

$$\sigma^2 = \mu_2' - \mu^2, \tag{4.26}$$

which is derived with a little algebra and properties of integrals.

3. Absolute Moments

There are both **absolute moments** and **absolute central moments** defined respectively in terms of $g(X) = |X|^n$ and $g(X) = |X - \mu|^n$. Of course, the absolute value is redundant when n is an even integer. By definition, these moments exist whenever the associated moments and central moments exist due to the constraint in (4.4).

There is no standard notation for these moments, but $\mu'_{|n|}$ and $\mu_{|n|}$ seem self-explanatory and are used in these books.

4. Moment Generating Function

In contrast to the above moment definitions which produce numerical values, the moment generating function is defined as an expectation of a parametrized exponential function. Specifically, $g(X)$ is defined by $g(X) = e^{tX}$. Thus the moment generating function is an expectation parametrized by t, and is thus truly a function of t.

Note that $M_X(t)$ always exists for $t = 0$, and as may be recalled from studies in Book IV, all useful results with this function are obtained when $M_X(t)$ exists for $t \in (-a, a)$ for $a > 0$. See Section 5.1 for multivariate moment generating functions.

Since exponential functions are nonnegative, there is no explicit mention of the constraint in (4.4).

Definition 4.14 **(Moment generating function)** *Let $X : \mathcal{S} \to \mathbb{R}$ be a random variable defined on a probability space $(\mathcal{S}, \mathcal{E}, \lambda)$ with distribution function $F(x)$.*

*The **moment generating function of** X, denoted $M_X(t)$, is defined by:*

$$M_X(t) \equiv \int_{\mathcal{S}} e^{tX(s)} d\lambda, \tag{4.27}$$

for all t for which the integral is finite.

When $F(x)$ is continuously differentiable with a continuous density function $f(x)$:

$$M_X(t) \equiv (\mathcal{R}) \int_{-\infty}^{\infty} e^{tx} f(x)dx. \tag{4.28}$$

This representation is valid as Lebesgue integral when $F(x)$ is absolutely continuous and $f(x) = F'(x)$, m-a.e.

When $F(x)$ is a discrete distribution function with discontinuities on $\{x_i\}_{i=\infty}^{\infty}$, where these points have no accumulation points:

$$M_X(t) \equiv \sum\nolimits_{i=-\infty}^{\infty} e^{tx_i} f(x_i). \tag{4.29}$$

4.1.3 Properties of Moments

In this section, we derive a number of the identities seen in Section IV.4.2. These identities were originally derived under the simplifying yet restrictive assumption of the existence of density functions, while here we require only the properties of integrals developed in Book V.

We start with two exercises that utilize the linearity and other properties of the integral from Proposition V.3.42.

Exercise 4.15 ($\mu'_n \Rightarrow \mu'_m$ **for** $m < n$) *Let X be a random variable defined on a probability space $(\mathcal{S}, \mathcal{E}, \lambda)$. Prove that if μ'_n exists, then μ'_m exists for $m < n$. The same statement then holds for μ_n by the next exercise. Hint: You only need to verify (4.2) for existence. Split the integral into $|X| \leq 1$ and $|X| > 1$.*

Exercise 4.16 ($\mu_n \Leftrightarrow \mu'_n$) *Let X be a random variable defined on a probability space $(\mathcal{S}, \mathcal{E}, \lambda)$. Prove that for any n, μ_n exists if and only if μ'_n exists, and:*

$$\mu_n = \sum\nolimits_{j=0}^{n} (-1)^{n-j} \binom{n}{j} \mu'_j \mu^{n-j}, \qquad \mu'_n = \sum\nolimits_{j=0}^{n} \binom{n}{j} \mu_j \mu^{n-j}. \tag{4.30}$$

*Hint: Recall the **binomial theorem**:*

$$(a+b)^n = \sum\nolimits_{j=0}^{n} \binom{n}{j} a^j b^{n-j}. \tag{4.31}$$

For the next results, recall the **multinomial theorem** in IV(4.40), and derived in Exercise IV.4.22:

$$\left(\sum\nolimits_{i=1}^{n} a_i\right)^m = \sum\nolimits_{m_1, m_2, ..m_n} \frac{m!}{m_1! m_2! ... m_n!} a_1^{m_1} a_2^{m_2} ... a_n^{m_n}, \tag{4.32}$$

where this summation is over all distinct n-tuples $(m_1, m_2, ..m_n)$ with $m_j \geq 0$ and $\sum_{j=1}^{n} m_j = m$.

Proposition 4.17 ($E\left[\sum_{i=1}^{n} X_i\right]$; $E\left[\left(\sum_{i=1}^{n} X_i\right)^m\right]$, $m > 1$, **independent** $\{X_i\}_{i=1}^{n}$) *Let $\{X_i\}_{i=1}^{n}$ be random variables on a probability space $(\mathcal{S}, \mathcal{E}, \lambda)$, and $X \equiv \sum_{i=1}^{n} X_i$.*

If $\mu_i \equiv E[X_i]$ exists for all i, then $\mu \equiv E[X]$ exists with:

$$\mu = \sum\nolimits_{i=1}^{n} \mu_i. \tag{4.33}$$

For $m > 1$, if $\{X_i\}_{i=1}^{n}$ are independent and $\mu'_{m_i} \equiv E[X_i^{m_i}]$ exists for all i and $m_i \leq m$, then $\mu'_m \equiv E[X^m]$ exists with:

$$\mu'_m = \sum\nolimits_{m_1, m_2, ..m_n} \frac{m!}{m_1! m_2! ... m_n!} \mu'_{m_1} ... \mu'_{m_n}, \tag{4.34}$$

where the summation is over all distinct n-tuples $(m_1, m_2, ..., m_n)$ with $m_j \geq 0$ and $\sum_{j=1}^{n} m_j = m$, noting that $\mu_{0_i} = 1$ for all i.

Proof. *The identity in (4.33) is linearity of the integral of Proposition V.3.42.*

By the multinomial theorem and linearity of the integral:

$$E[X^m] = \sum\nolimits_{m_1, m_2, ..m_n} \frac{m!}{m_1! m_2! ... m_n!} E\left[X_1^{m_1} X_2^{m_2} ... X_n^{m_n}\right].$$

If $\{X_i\}_{i=1}^{n}$ are independent, then so too are $\{X_i^{m_i}\}_{i=1}^{n}$ for any $\{m_i\}_{i=1}^{n}$ by Proposition II.3.56, since $g_i(x) = x^{m_i}$ is continuous and thus Borel measurable for $m_i \geq 0$.

Then by Proposition 4.7:

$$E\left[X_1^{m_1}X_2^{m_2}...X_n^{m_n}\right] = E\left[X_1^{m_1}\right]...E\left[X_n^{m_n}\right],$$

and the result follows. ■

This result then provides a comparable expression for central moments of sums of independent of random variables.

Proposition 4.18 ($E\left[(X-\mu)^m\right]$, $X = \sum_{i=1}^n X_i$, **independent** $\{X_i\}_{i=1}^n$) *Let $\{X_i\}_{i=1}^n$ be independent random variables on a probability space $(\mathcal{S}, \mathcal{E}, \lambda)$, and $X \equiv \sum_{i=1}^n X_i$.*

If $\mu_{m_i} \equiv E\left[(X_i - \mu_i)^{m_i}\right]$ exists for all i and $m_i \leq m$, then $\mu_m \equiv E\left[(X-\mu)^m\right]$ exists where $\mu \equiv E[X]$, and:

$$\mu_m = \sum\nolimits_{m_1, m_2, ..m_n} \frac{m!}{m_1! m_2! ... m_n!} \mu_{m_1} ... \mu_{m_n}. \tag{4.35}$$

The summation is over all distinct n-tuples $(m_1, m_2, ..., m_n)$ with $m_j \geq 0$ and $\sum_{j=1}^n m_j = m$, noting that $\mu_{0_i} = 1$ and $\mu_{1_i} = 0$ for all i.

Thus for $m = 2$, with $\mu_2 \equiv \sigma^2$ and $\mu_{2_i} \equiv \sigma_i^2$:

$$\sigma^2 = \sum\nolimits_{i=1}^n \sigma_i^2. \tag{4.36}$$

If $\{X_i\}_{i=1}^n$ are not independent:

$$\sigma^2 = \sum\nolimits_{i=1}^n \sum\nolimits_{j=1}^n \sigma_{ij} = \sum\nolimits_{i=1}^n \sigma_i^2 + 2\sum\nolimits_{j<i} \sigma_{ij}, \tag{4.37}$$

with $\sigma_{ii} \equiv \sigma_i^2$, and $\sigma_{ij} = E\left[(X_i - \mu_i)(X_j - \mu_j)\right]$.

Proof. *By the multinomial theorem:*

$$\begin{aligned} (X-\mu)^m &= \left[\sum\nolimits_{i=1}^n (X_i - \mu_i)\right]^m \\ &= \sum\nolimits_{m_1, m_2, ..m_n} \frac{m!}{m_1! m_2! ... m_n!} (X_1 - \mu_1)^{m_1} ... (X_n - \mu_n)^{m_n}. \end{aligned}$$

Letting $Y_i \equiv X_i - \mu_i$, independence of $\{X_i\}_{i=1}^n$ assures independence of $\{Y_i^{m_i}\}_{i=1}^n$ for any $\{m_i\}_{i=1}^n$ by Proposition II.3.56, and thus (4.35) follows from Proposition 4.7.

Since $\mu_{0_i} = 1$ and $\mu_{1_i} = 0$ for all i, (4.36) restates (4.35) with $m = 2$.

Without independence, apply the binomial theorem to obtain:

$$(X-\mu)^2 = \sum\nolimits_{i=1}^n \sum\nolimits_{j=1}^n (X_i - \mu_i)(X_j - \mu_j).$$

The first formula in (4.37) is derived by taking an expectation and applying (4.33), the second by reorganizing this double sum. ■

Remark 4.19 (Covariance; correlation) *Given random variables X_i, X_j on a probability space $(\mathcal{S}, \mathcal{E}, \lambda)$, the* ***covariance of*** X_i ***and*** X_j*, denoted $cov(X_i, X_j)$ or σ_{ij} as above, is defined:*

$$cov(X_i, X_j) \equiv E[(X_i - \mu_i)(X_j - \mu_j)] = E[X_i X_j] - \mu_i \mu_j. \tag{4.38}$$

The second expression follows from the first by linearity of the integral.

The ***correlation between*** X_i ***and*** X_j*, denoted $corr(X_i, X_j)$ and often ρ_{X_i, X_j} or ρ_{ij}, is defined by:*

$$corr(X_i, X_j) \equiv \frac{cov(X_i, X_j)}{\sigma_i \sigma_j}, \tag{4.39}$$

where σ_i, σ_j are the standard deviations of these variates. By linearity of the integral, correlation can also be defined as:

$$\rho_{ij} = E\left[\left(\frac{X_i - \mu_i}{\sigma_i}\right)\left(\frac{X_j - \mu_j}{\sigma_j}\right)\right].$$

Then (4.37) is expressed:

$$\sigma^2 = \sum\nolimits_{i=1}^{n} \sigma_i^2 + 2\sum\nolimits_{j<i} \rho_{ij}\sigma_i\sigma_j.$$

The final result is for the moment generating function.

Proposition 4.20 **($M_X(t)$, $X = \sum_{i=1}^n X_i$, independent $\{X_i\}_{i=1}^n$)** *Let $\{X_i\}_{i=1}^n$ be independent random variables on a probability space $(\mathcal{S}, \mathcal{E}, \lambda)$, and $X \equiv \sum_{i=1}^n X_i$.*

If $M_{X_i}(t)$ exists on $(-a_i, a_i)$ with $a_i > 0$ for all i, then $M_X(t)$ exists on $(-a, a)$ with $a \equiv \min\{a_i\}$, and on this interval:

$$M_X(t) = \prod\nolimits_{i=1}^{n} M_{X_i}(t). \tag{4.40}$$

Proof. *For any t, independence of $\{X_i\}_{i=1}^n$ assures independence of $\{e^{tX_i}\}_{i=1}^n$ by Proposition II.3.56 since $g(x) = e^{tx}$ is continuous and thus Borel measurable. For $t \in (-a, a)$, $\{e^{tX_i}\}_{i=1}^n$ are integrable by assumption, and then (4.40) follows from Proposition 4.7 since $e^{tX} = \prod_{i=1}^n e^{tX_i}$.* ■

4.2 Weak Convergence and Moment Limits

Section IV.4.5 investigated various results connecting weak convergence of distribution functions and limits of the associated moments and moment generating functions. Reflecting the uniqueness of moments results of the previous Section IV.4.4, it was seen that under certain conditions, weak convergence of distributions assured the convergence of moments and moment generating functions. Conversely, again under certain conditions, convergence of moments or moment generating functions assured the weak convergence of the underlying distribution functions.

These results are sometimes categorized under the heading, **method of moments.**

In this section we return to this investigation, to answer the following question:

If $\{X_n\}_{n=1}^\infty$, $X : \mathcal{S} \to \mathbb{R}$ are random variables defined on a probability space $(\mathcal{S}, \mathcal{E}, \lambda)$ with associated distribution functions $\{F_n(x)\}_{n=1}^\infty$ and $F(x)$, and F_n converges weakly to F, $F_n \Rightarrow F$, does:

$$E[g(X_n)] \to E[g(X)]$$

for certain measurable functions g, or *equivalently in the notation of (4.6), does:*

$$\int_{\mathbb{R}} g(x)d\lambda_{F_n} \to \int_{\mathbb{R}} g(x)d\lambda_F?$$

When g is a continuous and bounded real value function, the portmanteau theorem of Proposition 3.6 assures this result. For unbounded functions of interest, we present two results related to the convergence of moments for which $g(x) = x^m$.

For this investigation, we recall the definition of **uniform integrability** of a random variable sequence. Originally introduced in Definition IV.4.66 in this context, it reappeared in Definition V.3.53 for a result on **integration to the limit.** Here we restate this definition for a sequence of random variables, the current context of interest.

Definition 4.21 (Uniformly integrable random variables) *A sequence of random variables $\{X_n\}_{n=1}^{\infty}$ defined on a probability space $(\mathcal{S},\mathcal{E},\lambda)$ is said to be **uniformly integrable (U.I.)** if:*

$$\lim_{N\to\infty}\sup_n \int_{|X_n|\geq N} |X_n(s)|\,d\lambda = 0. \tag{4.41}$$

Remark 4.22 (On U.I. $\{X_n\}_{n=1}^{\infty}$) *While Definition 4.21 defines uniform integrability in terms of integrals of random variables on the underlying probability space $(\mathcal{S},\mathcal{E},\lambda)$, with the tools of Book V we note that this criterion can be restated in other ways.*

*By the change of variables of Proposition V.4.18, (4.41) can be expressed in terms of a **Lebesgue-Stieltjes integral** on $\mathbb{R}$:*

$$\lim_{N\to\infty}\sup_n \int_{|x|\geq N} |x|\,d\lambda_{X_n} = 0, \tag{4.42}$$

where λ_{X_n} is the Borel measure induced by X_n of (1.21).

*By (1.22), $\lambda_{X_n} = \lambda_{F_n}$, the Borel measure induced the distribution function F_n of X_n. Thus by Proposition V.3.63, since $F_n(x)\to 0$ as $x\to -\infty$, the Lebesgue-Stieltjes criterion in (4.42) can be restated in terms of a **Riemann-Stieltjes integral** on $\mathbb{R}$:*

$$\lim_{N\to\infty}\sup_n \int_{|x|\geq N} |x|\,dF_n = 0. \tag{4.43}$$

The above representations apply in situations where random variables $\{X_n\}_{n=1}^{\infty}$ and associated distribution functions $\{F_n\}_{n=1}^{\infty}$ are given, but the probability space $(\mathcal{S},\mathcal{E},\lambda)$ is not explicitly identified.

That said, Proposition 1.20 asserts that given X and associated F, a random variable Y and probability space $(\mathcal{S},\mathcal{E},\lambda)$ always exist, where $F_Y = F$. And indeed, this probability space is independent of the given random variable, so we can assume that all equivalent variates $\{Y_n\}_{n=1}^{\infty}$ are defined on this same space.

Proposition 4.23 (When $F_n \Rightarrow F$ implies $E\left[X_n^M\right] \to E\left[X^M\right]$) *Let $\{X_n\}_{n=1}^{\infty}$, $X : \mathcal{S}\to\mathbb{R}$ be random variables defined on a probability space $(\mathcal{S},\mathcal{E},\lambda)$ with associated distribution functions $\{F_n(x)\}_{n=1}^{\infty}$ and $F(x)$.*

If $F_n \Rightarrow F$ and $\{X_n^M\}_{n=1}^{\infty}$ are uniformly integrable for some integer $M\geq 1$, then $E\left[|X|^M\right] < \infty$ and:

$$E\left[X_n^M\right] \to E\left[X^M\right]. \tag{4.44}$$

Proof. *By Skorokhod's representation theorem of Proposition II.8.30, there exist random variables $\{Y_n\}_{n=1}^{\infty}$ and Y on the Lebesgue measure space $((0,1),\mathcal{B}(0,1),m)$ with the given distribution functions $\{F_n\}_{n=1}^{\infty}$ and F, and for which $Y_n \to Y$ for all $t\in(0,1)$. By Exercise 4.24:*

$$\int_{\mathcal{S}} |X_n(s)|^M\,d\lambda = \int_{(0,1)} |Y_n(t)|^M\,dm, \tag{1}$$

and thus $\{Y_n^M\}_{n=1}^{\infty}$ are uniformly integrable.

Choosing N large enough so that the supremum of the associated Lebesgue integrals in (4.41) is less than δ, it follows that for all n:

$$\begin{aligned}\int_{(0,1)} |Y_n(t)|^M\,dm &= \int_{|Y_n|<N} |Y_n(t)|^M\,dm + \int_{|Y_n|\geq N} |Y_n(t)|^M\,dm \\ &\leq N^M + \delta.\end{aligned}$$

Since $Y_n \to Y$ for all t, Fatou's lemma of Proposition III.2.34 obtains:

$$\int_{(0,1)} |Y(t)|^M dm \leq \liminf_n \int_{(0,1)} |Y_n(t)|^M dm \leq N^M + \delta.$$

Since X and Y have the same distribution function F, (1) *also applies with these variates, so $E\left[|X|^M\right] \leq N^M + \delta$ and $E\left[X^M\right]$ exists.*

To prove (4.44), the triangle inequality of Proposition III.2.49 obtains for any N:

$$\begin{aligned}\left|\int_{(0,1)} Y_n^M dm - \int_{(0,1)} Y^M dm\right| &\leq \left|\int_{|Y_n|\leq N} Y_n^M dm - \int_{|Y|\leq N} Y^M dm\right| \\ &\quad + \sup_n \int_{|Y_n|\geq N} \left|Y_n^M\right| dm + \int_{|Y|\geq N} \left|Y^M\right| dm \\ &= I + II + III.\end{aligned}$$

Since $Y_n \to Y$ pointwise, the bounded convergence theorem of Proposition III.2.22 obtains that $I \to 0$ as $n \to \infty$ for any N. By uniform integrability of $\{Y_n^m\}_{n=1}^\infty$, for any $\epsilon > 0$ there exists N_1 so that $II < \epsilon$ for $n \geq N_1$. Similarly, $E\left[|Y|^m\right] < \infty$ obtains for any $\epsilon > 0$ there exists N_2 so that $III < \epsilon$ for $n \geq N_2$.

Letting $N = \max\{N_1, N_2\}$ yields as $n \to \infty$:

$$\limsup \left|\int_{(0,1)} Y_n^M dm - \int_{(0,1)} Y^M dm\right| < 2\epsilon.$$

As ϵ is arbitrary, and the lim inf *of this expression satisfies the same bound by I.(3.14), it follows from Corollary I.3.46 that:*

$$\lim_{n\to\infty} \left|\int_{(0,1)} Y_n^M dm - \int_{(0,1)} Y^M dm\right| = 0. \tag{2}$$

The result in (4.44) follows by restating (2) *in terms of λ-integrals as in* (1). ■

Exercise 4.24 *Derive* (1) *in the above proof. Hint: Use the change of variables results of Section 4.1.*

The next proposition leads to the same conclusion but is often easier to apply in practice. It replaces the requirement on uniform integrability of $\{X_n^M\}_{n=1}^\infty$ with $\sup_n E\left[|X_n|^{M+\epsilon}\right] < \infty$ for any $\epsilon > 0$.

Proposition 4.25 (When $F_n \Rightarrow F$ implies $E\left[X_n^M\right] \to E\left[X^M\right]$) *Let $\{X_n\}_{n=1}^\infty, X : \mathcal{S} \to \mathbb{R}$ be random variables defined on a probability space $(\mathcal{S}, \mathcal{E}, \lambda)$ with associated distribution functions $\{F_n(x)\}_{n=1}^\infty$ and $F(x)$.*

If $F_n \Rightarrow F$ and for some integer $M \geq 1$ and $\epsilon > 0$:

$$\sup_n E\left[|X_n|^{M+\epsilon}\right] < \infty,$$

then $E\left[|X|^M\right] < \infty$ and (4.44) holds.

Proof. *Similar to the proof of Chebyshev's inequality of Proposition IV.4.34:*

$$\int_{\mathcal{S}} |X_n(s)|^{M+\epsilon} d\lambda \geq \int_{|X_n|\geq N} |X_n(s)|^{M+\epsilon} d\lambda \geq N^\epsilon \int_{|X_n|\geq N} |X_n(s)|^M d\lambda.$$

Thus as $N \to \infty$:

$$\sup_n \int_{|X_n| \geq N} |X_n(s)|^M \, d\lambda \leq \frac{1}{N^\epsilon} \sup_n E\left[|X_n|^{M+\epsilon}\right] \to 0.$$

Hence $\{X_n^M\}_{n=1}^\infty$ is uniformly integrable, and Proposition 4.23 applies. ■

Example 4.26 (On uniform integrability) *For the above conclusion of convergence of moments, it is not difficult to exemplify why an assumption is needed on uniform integrability.*

With $\delta > 0$, define distribution functions $\{F_n(x)\}_{n=1}^\infty$, $F(x)$:

$$F_n(x) = \begin{cases} 0, & x < 0, \\ 1 - 1/n, & 0 \leq x < n^{1+\delta}, \\ 1, & n^{1+\delta} \leq x, \end{cases} \qquad F(x) = \begin{cases} 0, & x < 0, \\ 1, & 0 \leq x. \end{cases}$$

Then $F_n \Rightarrow F$, but $E[X_n] \nrightarrow E[X]$. Indeed, $E[X_n] = n^\delta$ is unbounded while $E[X] = 0$.

Not surprisingly, $\{X_n\}_{n=1}^\infty$ is not uniformly integrable:

$$\sup_n \int_{|X_n| \geq N} |X_n(s)| \, d\lambda \geq N^\delta,$$

nor is the requirement of the second proposition satisfied, that $\sup_n E\left[|X_n|^{1+\epsilon}\right] < \infty$ for some $\epsilon > 0$.

4.3 Conditional Expectations

Given a random variable X defined on a probability space $(\mathcal{S}, \mathcal{E}, \mu)$, and a sigma (sub)algebra $\mathcal{F} \subset \mathcal{E}$, the goal of this section is to define $E[X|\mathcal{F}]$, or in words, **the conditional expectation of X given $\mathcal{F}$**. As will be seen, contrary to the name and the notation, the conditional expectation is not a number, but is actually a function with specified measurability and integrability properties vis-à-vis X and $\mathcal{F}$.

While this initially appears to be an entirely different concept than that seen in the Book II developments of conditional probabilities in Section II.1.5 and conditional distributions in Section II.3.3.2, these prior notions will be seen to be special cases of the current more general development.

We begin with a generalization of the earlier work on conditional probability measures, then turn to conditional expectations and their properties.

Applications of this development will occur frequently in Book VII and later.

4.3.1 Conditional Probability Measures

Given a probability space $(\mathcal{S}, \mathcal{E}, \mu)$ and fixed $B \in \mathcal{E}$ with $\mu(B) > 0$, the **conditional probability measure** $\mu(\cdot|B)$ was defined on $A \in \mathcal{E}$ in (1.54) of Definition 1.60:

$$\mu(A|B) \equiv \frac{\mu(A \bigcap B)}{\mu(B)}.$$

For any set $B \in \mathcal{E}$ with $\mu(B) > 0$, $\mu(\cdot|B)$ is indeed a probability measure on $(\mathcal{S}, \mathcal{E})$ by Exercise II.1.32, and it is natural to wonder if this definition can be extended to all $B \in \mathcal{E}$.

Put another way by now fixing A, this definition provides $\mu(A|\cdot)$, the conditional measure of A relative to every set in the collection $\{B \in \mathcal{E}|\mu(B) > 0\}$, and one might wonder if this definition can be extended to all $B \in \mathcal{E}$.

It turns out that this latter extension is possible, and while we will not make much explicit use of this result in this book, its development introduces machinery that will be fundamental in the study of conditional expectations.

To this end, let $A \in \mathcal{E}$ be given and assume that $B \in \mathcal{F}$ where $\mathcal{F}$ is a sigma subalgebra, $\mathcal{F} \subset \mathcal{E}$. While this inclusion allows for the case $\mathcal{F} = \mathcal{E}$, it gives an important generalization that will be of use below. The idea of defining $\mu(A|B)$ relative to every set $B \in \mathcal{F}$ is a small step from defining $\mu(A|s)$ as a function, or random variable, for $s \in \mathcal{S}$:

$$\mu(A|s) : \mathcal{S} \to \mathbb{R}.$$

The following definition identifies the properties that we want this function to have. The remark following provides some intuition.

Notation 4.27 (Conditional probability "function") *It should be noted that $\mu(A|\mathcal{F})$ is often called the* ***conditional probability of** A **given** $\mathcal{F}$*, and the fact this notation reflects a function is usually suppressed. To the experienced reader, this causes no confusion. But to those new to these ideas it can be confounding, as the notation is reminiscent of conditional probability, which is a numerical value.*

So at the risk of annoying the more experienced readers, we will explicitly refer to $\mu(A|\mathcal{F})$ as a function in this section, so those new to this material will get comfortable with this notational convention.

We will, however, largely suppress this temporary "function" label for the study of conditional expectations, $E[X|\mathcal{F}]$, which again are defined as functions as noted in the introduction to this section. Hopefully, all readers will be more comfortable with this convention by that time.

Definition 4.28 (Conditional probability function $\mu(A|\mathcal{F})$) *Given a probability space $(\mathcal{S}, \mathcal{E}, \mu)$, $A \in \mathcal{E}$, and a sigma subalgebra $\mathcal{F} \subset \mathcal{E}$, the* ***conditional probability function*** *$\mu(A|\mathcal{F})$, read "μ* ***of** A **given** $\mathcal{F}$*" and sometimes denoted $Pr[A|\mathcal{F}]$, is defined as any function $f_A(s)$ with the following properties:*

1. *$f_A(s) \equiv \mu(A|\mathcal{F})$ is an $\mathcal{F}$-measurable function on $\mathcal{S}$, so for all Borel sets $H \in \mathcal{B}(\mathbb{R})$:*

$$f_A^{-1}(H) \in \mathcal{F}.$$

2. *$f_A(s) \equiv \mu(A|\mathcal{F})$ is μ-integrable, and for all $B \in \mathcal{F}$:*

$$\int_B \mu(A|\mathcal{F})d\mu = \mu\left(A \bigcap B\right). \tag{4.45}$$

In particular:

$$\int_{\mathcal{S}} \mu(A|\mathcal{F})d\mu = \mu(A). \tag{4.46}$$

Remark 4.29 (On $\mu(A|\mathcal{F})$) *Admittedly, this definition appears quite abstract compared with the natural notion of $\mu(A|B)$ defined above. So we attempt to reveal some of the ideas reflected.*

a. ***The big idea:*** *This definition states that the conditional probability function is a measurable function $f_A(s)$ relative to $(\mathcal{S}, \mathcal{F}, \mu)$, noting the sigma algebra $\mathcal{F}$ here. By taking various expectations of this function, we recover the μ-measures of various sets.*

For example, (4.46) states that:

$$E\left[f_A(s)\right] = \mu(A).$$

Recalling Definition V.3.7 on $\int_B f d\mu$, *(4.45) states that for all* $B \in \mathcal{F}$ *and* $\chi_B(s)$ *the characteristic function of* B:

$$E\left[\chi_B(s) f_A(s)\right] = \mu\left(A \bigcap B\right).$$

Looked at this way, it seems that this is not such a big idea since it is easy to reproduce the μ*-measures of such sets with integration. Indeed, why not simply define* $f_A(s) \equiv \chi_A(s)$, *the characteristic function of* A? *This function is* μ*-integrable since* $A \in \mathcal{E}$, *and satisfies (4.45) and (4.46) as integrals of simple functions (Section V.3.1).*

The problem is that $\chi_A(s)$ *is not in general* $\mathcal{F}$*-measurable. Specifically,*

$$\chi_A^{-1}\left[\mathcal{B}(\mathbb{R})\right] = \{\emptyset, A, \widetilde{A}, \mathcal{S}\},$$

so while $\chi_A(s)$ *is always* $\mathcal{E}$*-measurable, it is* $\mathcal{F}$*-measurable if and only if* $A \in \mathcal{F}$. *For general* $A \in \mathcal{E} - \mathcal{F}$, $\chi_A(s)$ *has the right integrability properties, but the wrong measurability.*

Hence, the big idea in the above definition is not the integration requirements of (4.45) and (4.46), it is $\mathcal{F}$*-measurability! Put another way,* $\mu(A|\mathcal{F})$ *is basically* $\chi_A(s)$, *but changed just enough to be* $\mathcal{F}$*-measurable, but not so much as to change the value of its integrals over* $\mathcal{F}$*-sets.*

With that said, it should not be obvious that such a function exists.

b. ***On Existence:*** *Based on the above discussion, we arrive at the first existence result, that when* $\mathcal{F} = \mathcal{E}$:

$$\mu(A|\mathcal{E}) = \chi_A(s), \ \mu\text{-a.e.} \tag{4.47}$$

First, $A \in \mathcal{E}$ *assures* $\mathcal{E}$*-measurability, while satisfaction of (4.45) and (4.46) was addressed in part a.*

If $f_A(s)$ *is another conditional probability function, then by (4.45):*

$$\int_B \left[f_A(s) - \chi_A(s)\right] d\mu = 0,$$

for all $B \in \mathcal{E}$. *Item 8 of Proposition V.3.42 now obtains (4.47).*

c. ***On Book II:*** *The connection between the new definition of conditional probability* $f_A(s) \equiv \mu(A|\mathcal{F})$ *as a function on* $\mathcal{S}$, *and the previous Book II notion of a set function* $\mu(A|B)$ *definable for* B *with* $\mu(B) > 0$, *follows from item 2 of the definition. Specifically, for* $B \in \mathcal{F}$ *with* $\mu(B) > 0$, *(4.45) and (1.54) produce:*

$$\frac{1}{\mu(B)} \int_B \mu(A|\mathcal{F}) d\mu = \mu(A|B).$$

In other words, $\mu(A|\mathcal{F})$ *is an* $\mathcal{F}$*-measurable function, such that for every* $B \in \mathcal{F}$ *with* $\mu(B) > 0$, *its **average value** over* B *equals the probability of* A *conditional on* B.

d. ***On "any" function:*** *This definition characterizes* $\mu(A|\mathcal{F})$ *as **any** function* $f_A(s)$ *with the identified properties. Thus,* $\mu(A|\mathcal{F})$ *will not in general be unique since if* $g(s) = f_A(s)$, μ*-a.e., (4.45) and (4.46) are satisfied with* $g(x)$ *by item 3 of Proposition V.3.42. But for*

$g(s)$ to be another conditional probability function, it must also satisfy one additional criterion, and that is $\mathcal{F}$-measurability.

It is an exercise to check using measurability and (4.45), that if $f_A(s)$ exists, then $f_A(s) \geq 0$, μ-a.e. Hint: Let $B = \{f_A(s) < 0\}$.

Example 4.30 ($\mathcal{F} = \sigma(\{B_i\}_{i=1}^{\infty})$, disjoint partition $\{B_i\}_{i=1}^{\infty}$) *Let $\{B_i\}_{i=1}^{\infty} \subset \mathcal{E}$ form a disjoint partition of $\mathcal{S}$, so that $B_i \bigcap B_j = \emptyset$ for $i \neq j$ and $\bigcup_{i=1}^{\infty} B_i = \mathcal{S}$, and assume $\mu(B_i) > 0$ for all i. Let $\mathcal{F} = \sigma(\{B_i\}_{i=1}^{\infty})$, the smallest sigma algebra that contains these sets. Then $\mathcal{F} \subset \mathcal{E}$, and a general set $B \in \mathcal{F}$ is a finite or countable union of the B_i-sets.*

In this case $\mu(A|B_i)$ is well-defined for all $A \in \mathcal{E}$ by (1.54). For fixed $A \in \mathcal{E}$, define a function $f_A(s)$ on $\mathcal{S}$ by:

$$f_A(s) = \mu(A|B_i) \text{ for } s \in B_i.$$

Thus $f_A(s)$ is definable as a countable version of a simple function:

$$f_A(s) = \sum\nolimits_{i=1}^{\infty} \mu(A|B_i)\chi_{B_i}(s). \tag{1}$$

To see that $f_A(s) \equiv \mu(A|\mathcal{F})$, first note that $f_A(s)$ is an $\mathcal{F}$-measurable function on $\mathcal{S}$ since for any Borel set $H \in \mathcal{B}(\mathbb{R})$, $f_A^{-1}(H)$ is empty or a finite or countable union of B_i-sets. Also, $f_A(s)$ is integrable since $f_A(s) \leq 1$ and μ is a probability measure. Thus by Corollary V.3.49 applied to the disjoint union $\mathcal{S} = \bigcup_{i=1}^{\infty} B_i$, and then countable additivity of μ:

$$\begin{aligned}
\int_{\mathcal{S}} \mu(A|\mathcal{F})d\mu &= \sum\nolimits_{i=1}^{\infty} \mu(A|B_i) \int_{B_i} \chi_{B_i}(s)d\mu \\
&= \sum\nolimits_{i=1}^{\infty} \mu(A|B_i)\mu(B_i) \\
&\equiv \sum\nolimits_{i=1}^{\infty} \mu(A \bigcap B_i) \\
&= \mu(A).
\end{aligned}$$

Further, if $B = \bigcup_{k=1}^{N} B_{j_k} \in \mathcal{F}$ for $N \leq \infty$, then by the same calculations:

$$\int_B \mu(A|\mathcal{F})d\mu = \mu(A \bigcap B).$$

Hence, for $\mathcal{F}$ given as a sigma algebra generated by a disjoint partition of sets of positive measure, the elementary definition of $\mu(A|\mathcal{F})$ in (1) provides an $\mathcal{F}$-measurable function with the requisite integration properties. Then as in item b of Remark 4.29, if $g_A(s)$ is any $\mathcal{F}$-measurable function on $\mathcal{S}$ that satisfies (4.45), then:

$$g_A(s) = f_A(s) \ \mu\text{-a.e.}$$

Item b of Remark 4.9 and Example 4.30 notwithstanding, given $A \in \mathcal{E}$ and a general sigma subalgebra $\mathcal{F} \subset \mathcal{E}$, it is not at all obvious that an $\mathcal{F}$-measurable function exists that possesses the requisite properties of Definition 4.28. Indeed, to prove existence requires the power of the **Radon-Nikodým theorem** of Proposition 1.12, which then also assures uniqueness μ-a.e.

Proposition 4.31 (Existence of $\mu(A|\mathcal{F})$, unique μ-a.e.) *Let a probability space $(\mathcal{S}, \mathcal{E}, \mu)$, a set $A \in \mathcal{E}$, and a sigma subalgebra $\mathcal{F} \subset \mathcal{E}$ be given.*

Then the conditional probability function $\mu(A|\mathcal{F})$ exists, and is unique μ-a.e.

Proof. *Consider the probability space $(\mathcal{S}, \mathcal{F}, \mu)$, noting the use of the sigma algebra $\mathcal{F}$ here. Given $A \in \mathcal{E}$, define a set function ν_A on $B \in \mathcal{F}$ by:*

$$\nu_A(B) = \mu\left(A \bigcap B\right).$$

Since $A\bigcap B \in \mathcal{E}$ *for all such* B, $\nu_A(B)$ *is well-defined and it is in fact a measure on* $\mathcal{F}$ *by Exercise II.1.29.*

It is also the case that $\mu|_{\mathcal{F}}$, *the restriction of* μ *to* $\mathcal{F}$ *and defined simply on* $B \in \mathcal{F}$ *by:*

$$\mu|_{\mathcal{F}}(B) = \mu(B)$$

is another measure on $\mathcal{F}$. *Further, if* $\mu|_{\mathcal{F}}(B) = 0$ *then* $\nu_A(B) = \mu\left(A\bigcap B\right) = 0$. *Thus by Definition 1.11,* ν_A *is absolutely continuous with respect to* $\mu|_{\mathcal{F}}$:

$$\nu_A \ll \mu|_{\mathcal{F}}.$$

The Radon-Nikodým theorem then asserts that there exists a nonnegative $\mathcal{F}$*-measurable function* $f_A(s)$ *so that for all* $B \in \mathcal{F}$:

$$\int_B f_A(s)d\mu = \nu_A(B) \equiv \mu\left(A\bigcap B\right). \tag{1}$$

The function f_A *is necessarily* μ*-integrable since* $\mathcal{S} \in \mathcal{F}$, *and so by* (1),

$$\int_{\mathcal{S}} f_A(s)d\mu = \nu_A(\mathcal{S}) \equiv \mu(A).$$

Defining $\mu(A|\mathcal{F}) = f_A(s)$ *proves existence.*

The Radon-Nikodým theorem assures that f_A *is unique* μ*-a.e. That is, if* g_A *is an* $\mathcal{F}$*-measurable function so that* (1) *is true for all* $B \in \mathcal{F}$ *with* g_A *in place of* f_A, *then* $g_A = f_A$, μ*-a.e.* ■

Example 4.32 (Extreme sigma subalgebras) *It is interesting to explicitly identify* $f_A(s)$ *in the two extreme cases of sigma subalgebras* $\mathcal{F} \subset \mathcal{E}$:

1. *If* $\mathcal{F} = \{\emptyset, \mathcal{S}\}$, *then* $\mu(A|\mathcal{F}) = \mu(A)$.

2. *If* $\mathcal{F} = \mathcal{E}$, *then* $\mu(A|\mathcal{F}) = \chi_A(s)$.

Item 1 is left as an exercise, while item 2 is (4.47).

4.3.2 Conditional Expectation: An Introduction

In this section, we take an informal approach to conditional expectations to motivate some ideas and the forthcoming development, using more familiar manipulations and results from earlier books. As an informal development, we will not attempt to reconcile all of these notions from elementary probability theory with the framework of Section 4.3.4.

Assume that X, Y are random variables on a probability space $(\mathcal{S}, \mathcal{E}, \mu)$, so $X, Y : \mathcal{S} \to \mathbb{R}$ with a joint density function $f(x,y) \equiv f_{(X,Y)}(x,y)$ and marginal densities $f(x) \equiv f_X(x)$ and $f(y) \equiv f_Y(y)$. For any point y for which the marginal density function $f(y) \neq 0$, the conditional density function $f(x|y) \equiv f_{X|Y}(x|y)$ can be defined by:

$$f(x|y) = \frac{f(x,y)}{f(y)}. \tag{4.48}$$

For discrete densities this follows as a conditional probability statement from (1.59), while a similar formula can be derived in the limit when $f(x,y)$ is continuously differentiable by Example II.3.42.

This density function obtains a distribution function $F(x|y)$ on $\mathbb{R}$ when $f(y) \neq 0$, defined in the continuous case by:

$$F(x|y) = \int_{-\infty}^{x} \frac{f(t,y)}{f(y)} dt, \tag{1}$$

and analogously in the discrete case. Then by Proposition 1.20, there exists a probability space $(\mathcal{S}', \mathcal{E}', \mu')$ and random variable X' with this distribution function.

In such cases, **the conditional expectation of X given y**, denoted $E[X|y]$, is well-defined as the expectation $E[X']$ on $(\mathcal{S}', \mathcal{E}', \mu')$. Circumventing the formal transformations of Section 4.1, this transforms for continuous densities to:

$$E[X|y] \equiv \int_{\mathbb{R}} x f(x|y) dx, \tag{4.49}$$

while in the discrete case:

$$E[X|y] \equiv \sum x_i f(x_i|y). \tag{4.50}$$

Similarly, one can define **the conditional variance of X given y**, $E\left[(X - E[X|y])^2 |y\right]$ and denoted by $Var[X|y]$, as the variance of X'. With a little algebra as in (4.26), $Var[X|y]$ becomes for continuous densities:

$$Var[X|y] = \int_{\mathbb{R}} x^2 f(x|y) dx - (E[X|y])^2, \tag{4.51}$$

with a comparable formula in the discrete case.

Remark 4.33 (On informality) *The above approach, while perhaps reflecting familiar manipulations, suffers from the following shortcoming.*

The conditional expectation $E[X|y]$ has been defined in terms of a random variable X' defined on $(\mathcal{S}', \mathcal{E}', \mu')$, rather than a random variable $Z \equiv X|y$ on $(\mathcal{S}, \mathcal{E}, \mu)$ for which:

$$E[X|y] \equiv \int_{\mathcal{S}} Z d\mu.$$

This random variable would then have the distribution function $F(x|y) = F(z)$, defined as in (1) *above, in terms of the given density functions.*

And in the general case without a density function, it is not apparent how to formalize this calculation in terms of a random variable Z on $(\mathcal{S}, \mathcal{E}, \mu)$.

We will return to this question in the following section.

Continuing with informality, we have the following exercise to derive perhaps familiar identities. We state these results in the integral format, but note that similar results follow for discrete random variables.

See Remark 4.47 for generalizations of items 1 and 2, and Exercise 4.48 for item 3.

Exercise 4.34 ("Total" laws) *Derive the following identities, where $f(x|y)$ is defined in (4.48) with $f(y) \neq 0$ for all y. Hint: Recall Tonelli's theorem of Proposition V.5.23.*

1. **Law of Total Probability**

$$f(x) = \int_{\mathbb{R}} f(x|y) f(y) dy. \tag{4.52}$$

This is the continuous version. The discrete version may be more familiar and consistent with the name, since then $f(x)$, $f(x|y)$, and $f(y)$ are true probabilities:

$$f(x_k) = \sum\nolimits_j f(x_k|y_j) f(y_j).$$

2. **Law of Total Expectation**

$$E[X] = E[E[X|Y]], \tag{4.53}$$

where with $E[X|y]$ as defined in (4.49):

$$E[E[X|Y]] \equiv \int_{\mathbb{R}} E[X|y] f(y)dy.$$

While (4.53) is expressed in the conventional notation, it can also be expressed as:

$$E_X[X] = E_Y[E[X|Y]],$$

since the conditional expectation $E[X|Y]$ is a function of Y.

3. **Law of Total Variance**

$$Var[X] = E[Var[X|Y]] + Var[E[X|Y]], \tag{4.54}$$

where $E[Var[X|Y]]$ is defined analogously to $E[E[X|Y]]$ with $Var[X|y]$ as in (4.51), and $Var[E[X|Y]]$ is defined as in (4.26):

$$Var[E[X|Y]] \equiv \int_{\mathbb{R}} (E[X|y])^2 f(y)dy - (E[E[X|Y]])^2.$$

Again, (4.54) is expressed in the conventional notation, and can also be expressed as:

$$Var_X[X] = E_Y[Var[X|Y]] + Var_Y[E[X|Y]],$$

since $E[X|Y]$ and $Var[X|Y]$ are functions of Y.

Example 4.35 (Estimating default losses) *An insurance company has a medium quality bond portfolio of* 250 *bonds, each with a par value of* \$10 *million, and each with a* ***probability of default*** *over the next year of* $p = 0.003$. *On default, the company's* ***loss given default (LGD)*** *is assumed to be uniformly distributed with losses of* 25–75%. *It is assumed for simplicity that losses on different bonds are independent random variables.*

The goal is to estimate the mean and standard deviation of losses over the next year.

Using the so-called ***individual loss model,*** *the loss random variable* L *can be expressed as:*

$$L = \sum\nolimits_{i=1}^{250} B_i L_i,$$

and $B_i L_i$ *is the loss on the ith bond. This is modeled as the product of a standard binomial* B_i, *which indicates the* ***event of default,*** *and the LGD variable* L_i. *The binomial* B_i *equals* 0 *with probability* 0.997 *and* 1 *with probability* 0.003, *while the LGD distribution function is given as assumed above, with units in millions:*

$$F_L(l) = \begin{cases} 0, & l \leq 2.5, \\ (2l - 5.0)/10, & 2.5 \leq l \leq 7.5, \\ 1, & 7.5 \leq l. \end{cases}$$

Conditioning on $B_i = 0$ *and* $B_i = 1$, *it follows that with par* $P = 10$ *million:*

$$E[B_i L_i | B_i] = 0.5PB_i, \qquad Var[B_i L_i | B_i] = P^2 B_i^2/48,$$

and thus each is apparently a function of B_i. Applying (4.53),

$$E[B_i L_i] = (0.5P)\, E[B_i] = 15000.0.$$

Similarly, with (4.54),

$$\begin{aligned} Var[B_i L_i] &= (P^2/48)\, E[B_i^2] + (0.5P)^2\, Var[B_i] \\ &= (P^2/48)(0.003) + (0.5P)^2 (0.003)(0.997) \\ &= 8.1025 \times 10^{10}. \end{aligned}$$

From the assumed independence of losses $\{B_i L_i\}_{i=1}^{250}$, it follows from Section IV.4.2.4 on moments of sums that:

$$E[L] = 3.75 \times 10^6, \quad Var[L] = 2.0256 \times 10^{13},$$

and a standard deviation of losses of $SD[L] = 4.5007 \times 10^6$.

Using the **aggregate loss model,** *L is expressed as:*

$$L = \sum\nolimits_{i=1}^{N} L_i,$$

where N is the random variable representing the number of defaults, and L_i is the LGD variable defined above on the ith defaulted bond. The distributional model for N is exactly binomial with density:

$$f_N(n) = \binom{250}{n} (0.003)^n (0.997)^{250-n},$$

or approximately Poisson with parameter $\lambda = 250(0.003) = 0.75$. This follows from the Poisson limit theorem of Proposition IV.6.5.

Conditioning on Poisson N obtains:

$$E[L|N] = 0.5PN, \qquad Var[L|N] = P^2 N/48.$$

Completing the calculation with (4.53) and (4.54) produces:

$$E[L] = 3.75 \times 10^6, \qquad Var[L] = 2.0313 \times 10^{13},$$

and a standard deviation of losses of $SD[L] = 4.5070 \times 10^6$.

Exercise 4.36 (Loss models) *Develop general formulas for $E[L]$ and $Var[L]$ for the above models:*

$$\text{IRM:} \qquad L = \sum\nolimits_{i=1}^{n} B_i L_i,$$

$$\text{ARM:} \qquad L = \sum\nolimits_{i=1}^{N} L_i,$$

where B_i is binomial with $q = \Pr[B_i]$, and the LGD variable L_i has mean μ and variance σ^2.

Confirm that as in the example:

$$\mu_{IRM} = \mu_{ARM}, \quad \sigma^2_{IRM} < \sigma^2_{ARM}.$$

4.3.3 Conditional Expectation as a Function

Even in the elementary formulation of (4.49) or (4.50) for random variables X and Y on $(\mathcal{S}, \mathcal{E}, \mu)$ with given density functions, $E[X|y]$ can be imagined to be given by a function $E[X|Y]$ defined on $\mathcal{S}$ by:

$$E[X|Y](s) \equiv E[X|Y(s)].$$

In detail, we could attempt to define $E[X|Y]$ pointwise by:

$$E[X|Y](s) \equiv E[X|y], \text{ if } f(y) \equiv f(Y(s)) > 0.$$

Such a definition has several problems:

1. To state the obvious, we only have a definition when the given random variables have density functions. While most distribution functions in applications have densities, this is limiting in theory.
2. To further state the obvious, we have no definition when $f(y) \equiv f(Y(s)) = 0$, though the extent of this problem depends on the μ-measure of the set $\{s|f(Y(s)) = 0\}$.
3. As defined, it is not clear what measurability properties this function possesses, and so for example, it cannot yet be declared to be a random variable on $\mathcal{S}$, nor a function that can be integrated.
4. A subtle but important point is that the pointwise definition of a function on a measure space can certainly be interpreted as overkill. In such spaces, arbitrary redefinitions of functions on sets of μ-measure zero usually have no influence on the most important properties of interest.

As a second attempt at a general definition of $E[X|Y]$, reflecting item 4, we might instead require only that the μ-integrals of $E[X|Y](s)$ over certain $\mathcal{E}$-sets be defined, and equal to Lebesgue-Stieltjes integrals of $E[X|y]$ over associated Borel sets.

To this end, assume $f(y) > 0$ for all y for simplicity. Recalling Proposition V.4.18, we could require that for all $H \in \mathcal{B}(\mathbb{R})$, that the function $E[X|Y](s)$ satisfy:

$$\int_{Y^{-1}(H)} E[X|Y](s)d\mu = \int_H E[X|y]\,d\mu_Y,$$

where μ_Y is the probability measure of (1.21) induced by Y, or equivalently by Proposition 1.30, induced by the distribution function $F(y)$ of Y. Since $F(y)$ has an associated density function $f(y)$ by assumption, Proposition V.4.8 applies to obtain the requirement that for all $H \in \mathcal{B}(\mathbb{R})$:

$$\int_{Y^{-1}(H)} E[X|Y](s)d\mu = \int_H E[X|y]\,f(y)dy.$$

From this identity, we see that the μ-integral of $E[X|Y](s)$ need only be defined over sets of a sigma subalgebra $\mathcal{F}$ of $\mathcal{E}$, $\mathcal{F} \subset \mathcal{E}$, defined by $\mathcal{F} \equiv Y^{-1}(\mathcal{B}(\mathbb{R}))$, or:

$$\mathcal{F} \equiv \{Y^{-1}(A)|A \in \mathcal{B}(\mathbb{R})\}.$$

This sigma algebra is called the **sigma algebra generated by** Y in Definition 4.4, and denoted by $\sigma(Y)$.

The above requirement can be restated that for all $B \in \mathcal{F} \equiv \sigma(Y)$:

$$\int_B E[X|Y](s)d\mu = \int_{Y(B)} E[X|y]\,f(y)dy. \tag{1}$$

For the integral on the left to make sense, it will be necessary to ensure that $E[X|Y](s)$ is $\mathcal{F}$-measurable, meaning $\sigma(Y)$-measurable. Next, substituting the definition of $E[X|y]$ from (4.49) and (4.48), the μ-integral requirement in (1) can be restated:

$$\int_B E[X|Y](s)d\mu = \int_{\mathbb{R}}\int_{\mathbb{R}} x\chi_{Y(B)}(y)f(x,y)dxdy. \tag{2}$$

Now define the transformation $T : \mathcal{S} \to \mathbb{R}^2$ by $T(s) = (X(s), Y(s))$. With μ_T the measure on $\mathbb{R}^2$ induced by T of Definition 1.28, it follows as an exercise that $f(x, y)$ is the associated density function of μ_T, and so by Proposition V.4.8:

$$\int_{\mathbb{R}}\int_{\mathbb{R}} x\chi_{Y(B)}(y)f(x,y)dxdy = \int_{\mathbb{R}^2} x\chi_{Y(B)}(y)d\mu_T.$$

Then by Proposition V.4.18:

$$\int_{\mathbb{R}^2} x\chi_{Y(B)}(y)d\mu_T = \int_{\mathcal{S}} \chi_B(s)X(s)d\mu \equiv \int_B Xd\mu,$$

where the last step is Definition V.3.7.

Combining steps, the μ-integrals of the yet-to-be-specified, $\mathcal{F}$-measurable function $E[X|Y]$ must equal the respective μ-integrals of X over all $\mathcal{F}$-sets:

$$\int_B E[X|Y](s)d\mu = \int_B Xd\mu, \quad B \in \mathcal{F}. \tag{4.55}$$

This is a promising development, because even though the above manipulations required the strong assumptions that the random variables X, Y had a joint density function and that $f(y) > 0$ for all y, this final conclusion makes sense without these restrictive assumptions.

Thus a good start for the definition of $E[X|Y]$ is that we can find an $\mathcal{F}$-measurable function, where $\mathcal{F} = \sigma(Y)$, so that (4.55) is satisfied for all $B \in \sigma(Y)$.

Does such a function exist?

4.3.4 Existence of Conditional Expectation

In this section, we derive the existence of a $\sigma(Y)$-measurable function $E[X|Y]$ that satisfies the needed integration requirements of (4.55). The derivation will utilize the same approach as for the existence of the **conditional probability function given** $\mathcal{F}$, $\mu(A|\mathcal{F})$. By Definition 4.28, this function had similar measurability requirements, though different integrability criteria.

Reflecting the discussion of the prior section, we begin by defining the **conditional expectation of** X **given** $\mathcal{F}$, denoted $E[X|\mathcal{F}]$, where $X : \mathcal{S} \to \mathbb{R}$ is a random variable and $\mathcal{F} \subset \mathcal{E}$ a sigma subalgebra. When $\mathcal{F} = \sigma(Y)$ for a random variable Y, this is called the **conditional expectation of** X **given** Y and denoted $E[X|Y]$.

Note that here we must assume μ-integrability for X, not just μ-measurability, given (4.56) and (4.57).

Definition 4.37 (Conditional expectation of X **given** $\mathcal{F}$**)** *Given a probability space* $(\mathcal{S}, \mathcal{E}, \mu)$*, a* μ*-integrable random variable* X*, and a sigma subalgebra* $\mathcal{F} \subset \mathcal{E}$*, the* ***conditional expectation of*** X ***given*** $\mathcal{F}$*, denoted* $E[X|\mathcal{F}]$*, is defined as any function* $f_X(s)$ *so that:*

1. $f_X(s) \equiv E[X|\mathcal{F}]$ *is an* $\mathcal{F}$*-measurable function on* $\mathcal{S}$*:*

$$f_X^{-1}(H) \in \mathcal{F}, \quad \text{all } H \in \mathcal{B}(\mathbb{R}).$$

2. $f_X(s) \equiv E[X|\mathcal{F}]$ *is* μ*-integrable, and for all* $B \in \mathcal{F}$*:*

$$\int_B E[X|\mathcal{F}]\,d\mu = \int_B X d\mu, \tag{4.56}$$

and in particular:

$$\int_S E[X|\mathcal{F}]\,d\mu = E[X]. \tag{4.57}$$

If Y *is a random variable on* $(\mathcal{S}, \mathcal{E}, \mu)$*, the* ***conditional expectation of*** X ***given*** Y*, denoted* $E[X|Y]$*, is defined as any function* $f_X(s)$ *that satisfies the above definition with* $\mathcal{F} \equiv \sigma(Y)$*, where* $\sigma(Y) \equiv Y^{-1}[\mathcal{B}(\mathbb{R})]$*. In other words:*

$$E[X|Y] \equiv E[X|\sigma(Y)]. \tag{4.58}$$

Remark 4.38 (The big idea) *This definition states that the conditional expectation* $E[X|\mathcal{F}]$ *is a* ***random variable*** *on* $(\mathcal{S}, \mathcal{F}, \mu)$*, noting the sigma subalgebra* $\mathcal{F}$ *here, with the property that by taking various expectations we derive various expectations of* X*.*

Recalling Definition V.3.7, for any $B \in \mathcal{F}$ *and* $\chi_B(s)$ *the characteristic function of* B*, (4.56) requires:*

$$E[\chi_B(s)E[X|\mathcal{F}]] = E[\chi_B(s)X],$$

while for $B = \mathcal{S}$*, (4.57) requires:*

$$E[E[X|\mathcal{F}]] = E[X].$$

As for the conditional probability (function) $\mu(A|\mathcal{F})$ *of Definition 4.28, it is natural to think that this is not such a big idea, since it is easy to produce such expectations. Indeed, simply define* $E[X|\mathcal{F}] = X$*. This function is* μ*-integrable by assumption, and satisfies (4.56) and (4.57) by definition. Thus if* $\mathcal{F} = \mathcal{E}$*, we obtain the result:*

$$E[X|\mathcal{E}] = X, \ \mu\text{-a.e.} \tag{4.59}$$

First X *satisfies the above definition, and as was derived for (4.47), any other measurable function that satisfies (4.56) must equal* X*,* μ*-a.e.*

But for $\mathcal{F} \subset \mathcal{E}$*,* $f_X(s) = X$ *is not in general* $\mathcal{F}$*-measurable. While* $X^{-1}[\mathcal{B}(\mathbb{R})] \subset \mathcal{E}$*, this need not imply that* $X^{-1}[\mathcal{B}(\mathbb{R})] \subset \mathcal{F}$*, and thus* X *need not be* $\mathcal{F}$*-measurable. Hence, while* X *has the right integration properties, it need not have the right measurability property.*

Thus the big idea in the above definition, as it was for conditional probability, is $\mathcal{F}$*-measurability! In essence,* $E[X|\mathcal{F}]$ *is a version of* X *that is changed just enough to produce the required* $\mathcal{F}$*-measurability, but not so much that the integrals over* $\mathcal{F}$*-sets change.*

In the case of $\mathcal{F} \equiv \sigma(Y)$ *where* $\sigma(Y) = Y^{-1}[\mathcal{B}(\mathbb{R})]$*,* $E[X|Y]$ *is again basically* X*, but is changed just enough to be* $\sigma(Y)$*-measurable.*

Given a sigma subalgebra $\mathcal{F} \subset \mathcal{E}$, while it is not obvious that an $\mathcal{F}$-measurable function with the needed properties exists, the **Radon-Nikodým theorem** of Proposition 1.12 provides a solution, and one that is unique μ-a.e.

Proposition 4.39 (Existence of $E[X|\mathcal{F}]$, unique μ-a.e.) *Given a* μ*-integrable random variable* X *on* $(\mathcal{S}, \mathcal{E}, \mu)$*, and sigma subalgebra* $\mathcal{F} \subset \mathcal{E}$*, the* ***conditional expectation*** $E[X|\mathcal{F}]$ *exists, and is unique* μ*-a.e.*

Proof. *As for conditional probabilities, we again prove the existence of* $E[X|\mathcal{F}]$ *using the Radon-Nikodým theorem, but need an extra step before we can apply it.*

On the probability space $(\mathcal{S}, \mathcal{F}, \mu)$, noting the sigma algebra $\mathcal{F}$ here, define a set function ν_X on $B \in \mathcal{F}$ by:

$$\nu_X(B) = \int_B X d\mu.$$

Since $\mathcal{F} \subset \mathcal{E}$, $\nu_X(B)$ is well-defined. But as such integrals need not be nonnegative, ν_X is by Definition V.8.5, a ***signed*** *measure on $\mathcal{F}$. This follows from integrability of X and Corollary V.3.49.*

Recalling Definition V.3.37, split X into positive and negative parts:

$$X = X^+ - X^-,$$

and define on $B \in \mathcal{F}$:

$$\nu_{X^\pm}(B) = \int_B X^\pm d\mu.$$

Then:

$$\nu_X(B) = \nu_{X^+}(B) - \nu_{X^-}(B),$$

and each of ν_{X^+} and ν_{X^-} is a measure on $\mathcal{F}$ by Proposition V.4.5.

By Definition 1.11, both ν_{X^+} and ν_{X^-} are absolutely continuous with respect to the restriction of μ to $\mathcal{F}$, denoted $\mu|_{\mathcal{F}}$. For example, if $\mu|_{\mathcal{F}}(B) \equiv \mu(B) = 0$, then by item 3 of Proposition V.3.29:

$$\nu_{X^+}(B) = \int_B X^+ d\mu = 0,$$

and hence $\nu_{X^+} \ll \mu|_{\mathcal{F}}$.

The Radon-Nikodým theorem now assures that there exist nonnegative $\mathcal{F}$-measurable functions, $f_{X^+}(s)$ and $f_{X^-}(s)$, so that for all $B \in \mathcal{F}$:

$$\int_B f_{X^\pm}(s) d\mu = \nu_{X^\pm}(B). \tag{1}$$

With $f_X(s) = f_{X^+}(s) - f_{X^-}(s)$, this obtains for all $B \in \mathcal{F}$:

$$\int_B f_X(s) d\mu = \int_B X d\mu. \tag{2}$$

The function f_X is necessarily μ-integrable since $|f_X(s)| = f_{X^+}(s) + f_{X^-}(s)$, and thus by (1) applied to the component functions:

$$\int_S |f_X(s)| \, d\mu = E[|X|] < \infty,$$

since X is integrable.

If g_X is an $\mathcal{F}$-measurable function such that (2) is true with g_X, then (1) is also true with $g_{X^\pm}$. However, the Radon-Nikodým theorem assures that each of $f_{X^+}(s)$ and $f_{X^-}(s)$ is unique μ-a.e., and so $g_{X^\pm} = f_{X^\pm}$ μ-a.e., and then $g_X = f_X$, μ-a.e. ■

The next exercise proves that with the right X, conditional expectation equals conditional probability.

Exercise 4.40 ($E[\chi_A|\mathcal{F}] = \mu(A|\mathcal{F})$, μ-a.e.) *Confirm that if $A \in \mathcal{E}$, the random variable $X \equiv \chi_A(s)$ satisfies:*

$$E[\chi_A|\mathcal{F}] = \mu(A|\mathcal{F}), \ \mu\text{-a.e.} \tag{4.60}$$

That is, the conditional expectation $E[\chi_A|\mathcal{F}]$ of the characteristic function $\chi_A(s)$ equals the conditional probability function $\mu(A|\mathcal{F})$, μ-a.e. Hint: Verify that $E[\chi_A|\mathcal{F}]$ satisfies the conditions of the $\mu(A|\mathcal{F})$ definition, and apply item 8 of Proposition V.3.42.

Example 4.41 (Extreme sigma subalgebras) *It is interesting to explicitly identify* $f_X(s) = E[X|\mathcal{F}]$ *in the two extreme cases of sigma subalgebras,* $\mathcal{F} \subset \mathcal{E}$*:*

1. *If* $\mathcal{F} = \{\emptyset, \mathcal{S}\}$*, then* $E[X|\mathcal{F}] = E[X]$.
2. *If* $\mathcal{F} = \mathcal{E}$*, then* $E[X|\mathcal{F}] = X$.

Item 1 is left as an exercise, while item 2 is (4.59).

Example 4.42 ($\mathcal{F} = \sigma\left(\{B_i\}_{i=1}^N\right)$, **disjoint partition** $\{B_i\}_{i=1}^N$) *Let* X *be a random variable on a probability space* $(\mathcal{S}, \mathcal{E}, \mu)$*,* $X : \mathcal{S} \to \mathbb{R}$*, and let* $\{B_i\}_{i=1}^N \subset \mathcal{S}$ *for* $N \leq \infty$ *be disjoint with* $\bigcup_{i=1}^N B_i = \mathcal{S}$ *and* $\mu(B_i) > 0$ *for all* i*. Define:*

$$\mathcal{F} \equiv \sigma\left(\{B_i\}_{i=1}^N\right),$$

the collection of all unions of such sets.

In order for $E[X|\mathcal{F}]$ *to be* $\mathcal{F}$*-measurable, it must be constant value on each* B_i*. Hence* $E[X|\mathcal{F}](s) = \alpha_i$ *on* B_i*.*

To determine these constants, it follows from (4.56) that:

$$\int_{B_i} E[X|\mathcal{F}]\, d\mu = \int_{B_i} X d\mu.$$

Since $E[X|\mathcal{F}] = \alpha_i$ *on* B_i*, we obtain:*

$$\alpha_i = \frac{1}{\mu(B_i)} \int_{B_i} X d\mu. \tag{1}$$

In other words, $E[X|\mathcal{F}]$ *equals the "average" value of* X *over each* B_i*, where this average is calculated relative to the* μ*-measure.*

In summary:

$$E[X|\mathcal{F}] = \sum\nolimits_{i=1}^N \alpha_i \chi_{B_i},$$

with χ_{B_i} *the characteristic function of* B_i*, and* α_i *defined in* (1).

Exercise 4.43 (Finer and coarser $\mathcal{F}$**)** *Example 4.42 provides a simple basis for understanding how the measurability and integral conditions of* $E[X|\mathcal{F}]$ *vary with* $\mathcal{F}$*.*

Assume that $\mathcal{F}$ *is generated by disjoint* $\{B_i\}_{i=1}^{2M}$ *for finite* M*, with* $\bigcup_{i=1}^{2M} B_i = \mathcal{S}$ *and* $\mu(B_i) > 0$ *for all* i*. Define a* ***coarser sigma algebra*** $\mathcal{F}' \subset \mathcal{F}$ *to be generated by* $\{C_i\}_{i=1}^M$ *with* $C_i = B_{2i-1} \bigcup B_{2i}$*. The same construction works with* $M = \infty$*. Compare measurability and integrability of the functions* $E[X|\mathcal{F}]$ *and* $E[X|\mathcal{F}']$.

This coarsening can be continued, with limiting sigma algebra $\mathcal{F}_0 \equiv \{\emptyset, \mathcal{S}\}$*, recalling item 1 of Example 4.41.*

Similarly, consider ***finer sigma subalgebras*** $\mathcal{F}''$*, where* $\mathcal{F} \subset \mathcal{F}'' \subset \mathcal{E}$*, where each* B_i*-set is split with arbitrary* $\mathcal{E}$*-sets, with limiting sigma algebra* $\mathcal{E}$*, recalling item 2 of Example 4.41.*

4.4 Properties of Conditional Expectations

In future books we will require some facility with manipulations involving conditional expectations. This section summarizes some of the most important properties that will be needed.

To standardize terminology, we state most results in the context of $E[X|\mathcal{F}]$, since that is the context that will appear most often later. If $\mathcal{F} = \sigma(Y)$ for some random variable Y, all these results apply to $E[X|Y]$ without change, by definition. However, these results look quite different in this notation. As an exercise, the reader is encouraged to restate all results in that context.

Remark 4.44 (Proof strategy) *The strategy for many proofs of equalities reflects two steps, which we illustrate with linearity from item 2 of Proposition 4.46:*

$$E[aX + bY|\mathcal{F}] = aE[X|\mathcal{F}] + bE[Y|\mathcal{F}], \ \mu\text{-a.e.}$$

- ***Definitional:*** *Given the definitional properties of $E[X|\mathcal{F}]$ and $E[Y|\mathcal{F}]$, we prove that $aE[X|\mathcal{F}] + bE[Y|\mathcal{F}]$ satisfies the definitional properties for $E[aX + bY|\mathcal{F}]$.*

- ***Uniqueness μ-a.e.:*** *Let $g(s)$ be any other function that satisfies Definition 4.37 for $E[aX + bY|\mathcal{F}]$. The proof that:*

$$g(s) = aE[X|\mathcal{F}] + bE[Y|\mathcal{F}], \ \mu\text{-a.e.}, \tag{1}$$

proceeds as follows.

Proof. *By (4.56), it follows that for all $B \in \mathcal{F}$:*

$$\int_B (g(s) - [aE[X|\mathcal{F}] + bE[Y|\mathcal{F}]])\, d\mu = 0.$$

Thus by item 8 of Proposition V.3.42, (1) *follows.* ■

Below we will largely focus on the Definitional step of the proofs, since the above uniqueness proof will always apply to obtain the μ-a.e. conclusion.

Exercise 4.45 (Making properties concrete) *The reader is encouraged to make the following results more concrete by recasting them in the context of sigma subalgebras such as $\mathcal{F}$ in Example 4.42, generated by disjoint sets $\{B_i\}_{i=1}^N$. For the tower property in item 7, for example, one might choose $\mathcal{F}_2 = \mathcal{F}$, and $\mathcal{F}_1$ generated by the sets $\{C_i\}_{i=1}^M$ with say $C_i = B_{2i} \bigcup B_{2i-1}$ and $N = 2M$.*

4.4.1 Fundamental Properties

The following proposition summarizes many of the foundational results on conditional expectations.

Proposition 4.46 (Foundational properties of $E[X|\mathcal{F}]$) *Given a probability space $(\mathcal{S}, \mathcal{E}, \mu)$, μ-integrable random variables X, Y, and sigma subalgebras $\mathcal{F} \subset \mathcal{E}$, $\mathcal{F}_j \subset \mathcal{E}$, we have the following properties:*

1. ***$\mathcal{F}$-Measurability Property 1:*** *If X is $\mathcal{F}$-measurable, then:*

$$E[X|\mathcal{F}] = X, \ \mu\text{-a.e.} \tag{4.61}$$

In particular, for $a \in \mathbb{R}$:

$$E[a|\mathcal{F}] = a, \ \mu\text{-a.e.} \tag{4.62}$$

2. ***Linearity:*** *For $a, b \in \mathbb{R}$:*

$$E[aX + bY|\mathcal{F}] = aE[X|\mathcal{F}] + bE[Y|\mathcal{F}], \ \mu\text{-a.e.} \tag{4.63}$$

3. ***Monotonicity:*** *If* $X \leq Y$ μ*-a.e., then:*

$$E[X|\mathcal{F}] \leq E[Y|\mathcal{F}], \ \mu\text{-a.e.} \tag{4.64}$$

4. ***Triangle Inequality:***

$$|E[X|\mathcal{F}]| \leq E[|X||\mathcal{F}], \ \mu\text{-a.e.} \tag{4.65}$$

5. ***$\mathcal{F}$-Measurability Property 2:*** *If* Z *is* $\mathcal{F}$*-measurable and* XZ *is* μ*-integrable, then:*

$$E[XZ|\mathcal{F}] = ZE[X|\mathcal{F}], \ \mu\text{-a.e.} \tag{4.66}$$

6. ***Independence:*** *If* X *is independent of* $\mathcal{F}$*, meaning* $\sigma(X)$ *and* $\mathcal{F}$ *are independent sigma algebras, then:*

$$E[X|\mathcal{F}] = E[X], \ \mu\text{-a.e.} \tag{4.67}$$

   Thus if X *and* Y *are independent random variables on* $(\mathcal{S}, \mathcal{E}, \mu)$*, then:*

$$E[X|Y] = E[X], \ \mu\text{-a.e.} \tag{4.68}$$

7. ***Tower Property:*** *If* $\mathcal{F}_1 \subset \mathcal{F}_2 \subset \mathcal{E}$*, then:*

$$\begin{aligned} E[E[X|\mathcal{F}_1]|\mathcal{F}_2] &= E[X|\mathcal{F}_1], \ \mu\text{-a.e.} \\ E[E[X|\mathcal{F}_2]|\mathcal{F}_1] &= E[X|\mathcal{F}_1], \ \mu\text{-a.e.} \end{aligned} \tag{4.69}$$

   In other words, the smallest or "coarsest" sigma subalgebra always wins.

8. ***Law of Total Expectation:***

$$E[E[X|\mathcal{F}]] = E[X]. \tag{4.70}$$

Proof. *We take each statement in turn. For the needed properties of the integrals, see Proposition V.3.42 unless otherwise noted. Also, uniqueness* μ*-a.e. is left as an exercise as noted in Remark 4.44.*

1. *The statement in (4.61) was derived as (4.59) in Remark 4.38, while (4.62) follows since constant functions are measurable with respect to any sigma algebra.*

2. *First,* $aE[X|\mathcal{F}]+bE[Y|\mathcal{F}]$ *is* $\mathcal{F}$*-measurable by Proposition V.2.13 since both conditional expectations are so measurable. If* $B \in \mathcal{F}$*, then by linearity of the integral and Definition 4.37:*

$$\begin{aligned} \int_B (aE[X|\mathcal{F}] + bE[Y|\mathcal{F}])\, d\mu &= a\int_B E[X|\mathcal{F}]\, d\mu + b\int_B E[Y|\mathcal{F}]\, d\mu \\ &= a\int_B X d\mu + b\int_B Y d\mu \\ &= \int_B (aX + bY)\, d\mu. \end{aligned}$$

   Hence $aE[X|\mathcal{F}]+bE[Y|\mathcal{F}]$ *satisfies Definition 4.37 and so equals* $E[aX + bY|\mathcal{F}]$ μ*-a.e. by Remark 4.44.*

3. *For $B \in \mathcal{F}$, monotonicity of the integral obtains:*

$$\int_B E[X|\mathcal{F}]\, d\mu = \int_B X d\mu \leq \int_B Y d\mu = \int_B E[Y|\mathcal{F}]\, d\mu. \tag{1}$$

If $B' \equiv \{s | E[X|\mathcal{F}] - E[Y|\mathcal{F}] \geq \epsilon > 0\}$, then $B' \in \mathcal{F}$ by measurability of $E[X|\mathcal{F}] - E[Y|\mathcal{F}]$ and Proposition V.2.13. Then by (1):

$$\int_{B'} (E[X|\mathcal{F}] - E[Y|\mathcal{F}])\, d\mu \leq 0,$$

while on the other hand:

$$\int_{B'} (E[X|\mathcal{F}] - E[Y|\mathcal{F}])\, d\mu \geq \epsilon\mu[B'].$$

Thus $\mu[B'] = 0$, and hence $E[X|\mathcal{F}] \leq E[Y|\mathcal{F}]$ μ-a.e.

4. *As both $X \leq |X|$ and $-X \leq |X|$, this result follows from items 3 and 2.*

5. *As both Z and $E[X|\mathcal{F}]$ are $\mathcal{F}$-measurable, so too $ZE[X|\mathcal{F}]$, so left to prove is the integrability condition of (4.56), that for $B \in \mathcal{F}$:*

$$\int_B ZE[X|\mathcal{F}]\, d\mu = \int_B ZX d\mu.$$

If Z_n is a simple function, $Z_n = \sum_{i=1}^n a_i \chi_{A_i}(s)$ with $\{A_i\}_{i=1}^n \subset \mathcal{F}$, then by item 2 and properties of the integral:

$$\begin{aligned} \int_B Z_n E[X|\mathcal{F}]\, d\mu &= \sum\nolimits_{i=1}^n a_i \int_{B\bigcap A_i} E[X|\mathcal{F}]\, d\mu \\ &= \sum\nolimits_{i=1}^n a_i \int_{B\bigcap A_i} X d\mu \\ &= \int_B Z_n X d\mu. \end{aligned} \tag{2}$$

For general Z as above, write $Z = Z^+ - Z^-$ as a decomposition into positive and negative parts of Definition V.3.37, and let $\{Z_n^+\}_{n=1}^\infty$ and $\{Z_n^-\}_{n=1}^\infty$ be increasing sequences of simple functions as in Proposition V.2.28, with $Z_n^\pm \leq Z^\pm$ and $Z_n^\pm \to Z^\pm$ pointwise. Similarly splitting $E[X|\mathcal{F}]$ into positive and negative parts and applying Lebesgue's monotone convergence theorem of Proposition V.3.23 on each of the four terms in the product obtains with $Z_n \equiv Z_n^+ - Z_n^-$:

$$\begin{aligned} \int_B Z_n E[X|\mathcal{F}]\, d\mu &\equiv \int_B (Z_n^+ - Z_n^-)\left(E[X|\mathcal{F}]^+ - E[X|\mathcal{F}]^-\right) d\mu \\ &= \int_B Z_n^+ E[X|\mathcal{F}]^+\, d\mu - \int_B Z_n^+ E[X|\mathcal{F}]^-\, d\mu \\ &\quad - \int_B Z_n^- E[X|\mathcal{F}]^+\, d\mu + \int_B Z_n^- E[X|\mathcal{F}]^-\, d\mu \\ &\to \int_B Z^+ E[X|\mathcal{F}]^+\, d\mu - \int_B Z^+ E[X|\mathcal{F}]^-\, d\mu \\ &\quad - \int_B Z^- E[X|\mathcal{F}]^+\, d\mu + \int_B Z^- E[X|\mathcal{F}]^-\, d\mu \\ &= \int_B ZE[X|\mathcal{F}]\, d\mu. \end{aligned} \tag{3}$$

Recalling Remark V.3.24, we cannot yet assert that this final integral in (3) *is finite. Similarly splitting* X *into positive and negative parts.*

$$\int_B Z_n X d\mu \to \int_B ZX d\mu. \tag{4}$$

The result now follows by (2)–(4)*, since integrability of* ZX *also assures that the integral in* (3) *is finite.*

6. *Given* $B \in \mathcal{F}$*, the independence assumption assures that* X *and* χ_B *are independent random variables as in the proof of Corollary 4.10, and by that result:*

$$\int_B X d\mu = E[X]\mu[B] = \int_B E[X] d\mu.$$

Hence, $E[X]$ *satisfies (4.56), and is* $\mathcal{F}$*-measurable since by Definition 4.4,* $\sigma(E[X]) = \{\emptyset, \mathcal{S}\} \subset \mathcal{F}$*.*

This proves (4.67), and (4.68) follows by definition in (4.58).

7. *For the first identity,* $E[X|\mathcal{F}_1]$ *is* $\mathcal{F}_1$*-measurable and hence* $\mathcal{F}_2$*-measurable, since* $\mathcal{F}_1 \subset \mathcal{F}_2$*. Thus from items 1 and 5:*

$$E[E[X|\mathcal{F}_1]|\mathcal{F}_2] = E[X|\mathcal{F}_1]E[1|\mathcal{F}_2] = E[X|\mathcal{F}_1].$$

For the second identity, both functions are $\mathcal{F}_1$*-measurable by definition, and thus only the integrability condition in (4.56) need be checked.*

If $B \in \mathcal{F}_1$*, then from (4.56):*

$$\int_B E[X|\mathcal{F}_1] d\mu = \int_B X d\mu.$$

Since $B \in \mathcal{F}_1 \subset \mathcal{F}_2$*, two applications of (4.56) obtain:*

$$\int_B E[E[X|\mathcal{F}_2]|\mathcal{F}_1] d\mu = \int_B E[X|\mathcal{F}_2] d\mu = \int_B X d\mu,$$

and so $E[E[X|\mathcal{F}_2]|\mathcal{F}_1] = E[X|\mathcal{F}_1]$*.*

8. *As noted in item 1 of Example 4.41, for any integrable random variable* Y*:*

$$E[Y] = E[Y|\mathcal{F}_0], \tag{5}$$

where $\mathcal{F}_0 = \{\emptyset, \mathcal{S}\}$*. Since* $E[X|\mathcal{F}]$ *is integrable by definition,* (5) *and the tower property obtain:*

$$E[E[X|\mathcal{F}]] = E[E[X|\mathcal{F}]|\mathcal{F}_0] = E[X|\mathcal{F}_0] = E[X].$$

■

Remark 4.47 (Laws of total probability and expectation) *The result in (4.70) implies a* ***law of total probability.*** *Recalling (4.60):*

$$E[\chi_A|\mathcal{F}] = \mu(A|\mathcal{F}), \ \mu\text{-a.e.},$$

and then by (4.70):

$$E[E[\chi_A|\mathcal{F}]] = E[\chi_A] \equiv \mu(A).$$

Thus:

$$\mu(A) = E\left[\mu(A|\mathcal{F})\right], \tag{4.71}$$

a result also implied by (4.46) of Definition 4.28.

This result is a generalization of the **law of total probability** *introduced in Proposition II.1.35 in terms of conditional probabilities. There, given a probability space $(\mathcal{S}, \mathcal{E}, \mu)$ and a finite or countable collection of disjoint sets $\{B_i\}_{i=1}^N \subset \mathcal{E}$ with $\mu(B_i) > 0$ and $\bigcup_{i=1}^N B_i = \mathcal{S}$, then for any $A \in \mathcal{E}$, this law obtained:*

$$\mu(A) = \sum\nolimits_{i=1}^{N} \mu(A|B_i)\mu(B_i) \equiv \sum\nolimits_{i=1}^{N} \mu\left(A \bigcap B_i\right). \tag{4.72}$$

To compare with (4.71), if $\mathcal{F}$ is the sigma algebra generated by $\{B_i\}_{i=1}^N$, then $\mu(A|\mathcal{F})$ must be constant on each B_i to be $\mathcal{F}$-measurable. Then by (4.46):

$$\int_{B_i} \mu(A|\mathcal{F})d\mu = \mu\left(A \bigcap B_i\right),$$

and thus $\mu(A|\mathcal{F}) = \mu\left(A \bigcap B_i\right)/\mu\left(B_i\right)$ on B_i. Taking expectations of $\mu(A|\mathcal{F})$ obtains the sum on the right in (4.72), which is $\mu(A)$.

Similarly, the result in (4.70) is equivalent to the **law of total expectation** *in (4.53) with $\mathcal{F} = \sigma(Y)$:*

$$E[X] = E\left[E\left[X|Y\right]\right], \tag{4.73}$$

since then $E\left[X|\mathcal{F}\right] \equiv E\left[X|Y\right]$ by definition in (4.58).

This latter result will be especially useful in later books' developments where for example, a proof that $E\left[X|\mathcal{F}\right] = 0$ for any sigma subalgebra $\mathcal{F} \subset \mathcal{E}$ will be sufficient to obtain that $E\left[X\right] = 0$.

Exercise 4.48 (Law of total variance) *Given a probability space $(\mathcal{S}, \mathcal{E}, \mu)$, μ-integrable random variable X, and sigma subalgebra $\mathcal{F} \subset \mathcal{E}$, if X^2 is μ-integrable, define the* ***conditional variance*** *$V\left[X|\mathcal{F}\right]$ as the conditional expectation of $(X - E\left[X\right])^2$:*

$$V\left[X|\mathcal{F}\right] \equiv E\left[\left(X - E\left[X|\mathcal{F}\right]\right)^2 |\mathcal{F}\right] = E\left[X^2|\mathcal{F}\right] - \left(E\left[X|\mathcal{F}\right]\right)^2. \tag{4.74}$$

When $\mathcal{F} = \sigma(Y)$, this conditional variance is written $V\left[X|Y\right]$.

1. *Justify that $V\left[X|\mathcal{F}\right]$ is well-defined and verify the second equality in (4.74).*
2. *With the variance of X defined:*

   $$V(X) = E\left[\left(X - E\left[X\right]\right)^2\right],$$

   verify the ***law of total variance:***

   $$V(X) = E\left[V\left[X|\mathcal{F}\right]\right] + V\left[E\left[X|\mathcal{F}\right]\right]. \tag{4.75}$$

   Note that this is (4.54) when $\mathcal{F} = \sigma(Y)$. Hint: Verify that $E\left[V\left[X|\mathcal{F}\right]\right]$ and $V\left[E\left[X|\mathcal{F}\right]\right]$ are well-defined, which means that $V\left[X|\mathcal{F}\right]$ is μ-integrable, and $E\left[X|\mathcal{F}\right]$ is square μ-integrable. For the latter verification, look ahead to Jensen's inequality in (4.80).

4.4.2 Conditional Jensen's Inequality

In this section we prove a Jensen's inequality for conditional expectations. Named for **Johan Jensen** (1859–1925), this result was introduced in Proposition IV.4.42, relating $E[\varphi(X)]$ and $\varphi[E(X)]$ for concave and convex functions. While a number of properties of such functions were developed there, for the current result we require some deeper results related to the differentiability of such functions. In essence, the ultimate goal will be to derive the tangential bounds of Proposition IV.4.41 without assuming differentiability.

We begin by recalling Definition IV.4.39:

Definition 4.49 (Concave, convex functions) *A function $g(x)$ is **concave** on an interval $I = (a, b)$, which may be infinite, if for any $x, y \in I$:*

$$g(tx + (1-t)y) \geq tg(x) + (1-t)g(y) \quad \text{for } t \in [0,1], \tag{4.76}$$

*and is **convex** on I if for any $x, y \in I$:*

$$g(tx + (1-t)y) \leq tg(x) + (1-t)g(y) \quad \text{for } t \in [0,1]. \tag{4.77}$$

*When the inequalities are strict for $t \in (0,1)$, such functions are referred to as **strictly concave** and **strictly convex**, respectively.*

Since $g(x)$ is convex if and only if $-g(x)$ is concave, it is common to see many results stated only for convex functions, with the understanding that they are also true for concave functions by reversing the inequality. The same is true for strictly concave/convex, one changes the inequality to be strict.

Lemma 4.50 (Convex $\Rightarrow$ Slope function increases in each variable) *Let $g(x)$ be convex on $I = (a, b)$, and $\{z, y, z\} \subset I$ with $x < y < z$. Then:*

$$\frac{g(y) - g(x)}{y - x} \leq \frac{g(z) - g(x)}{z - x} \leq \frac{g(z) - g(y)}{z - y}. \tag{4.78}$$

In other words, the slope function $s(x, y) \equiv \frac{g(y)-g(x)}{y-x}$, is an increasing function in each variable.

Proof. *Expressing $y = (1-t)x + tz$ with $t = \frac{y-x}{z-x}$, and applying (4.77) yields:*

$$g(y) \leq g(x) + \frac{y - x}{z - x}\left[g(z) - g(x)\right],$$

which obtains the first inequality. Similarly expressing $y = tx + (1-t)z$ with $t = \frac{z-y}{z-x}$ obtains the second. ■

The significance of (4.78) is that it implies that convex functions are continuous, with left and right derivatives everywhere. Thus convex functions are differentiable at all but at most countably many points. This implies but is a little better than differentiable m-a.e., since recalling Example III.3.50 on the Cantor ternary set, a set can have Lebesgue measure 0 and be uncountable.

Proposition 4.51 (Convex $\Rightarrow$ Continuous; differentiable a.e.) *If $g(x)$ is convex on $I = (a, b)$, then $g(x)$ is:*

1. *Continuous on (a, b).*

2. *Left differentiable and right differentiable at every point, with both $D^-g(x)$ and $D^+g(x)$ increasing functions.*

3. *Differentiable at all but at most countably many points, meaning $g'(x) = D^-g(x) = D^+g(x)$, except for at most countably many points.*

Proof. *For $h, k > 0$, let $x = y - h$ and $z = y + k$ in (4.78):*

$$\frac{g(y) - g(y-h)}{h} \leq \frac{g(y+k) - g(y-h)}{h+k} \leq \frac{g(y+k) - g(y)}{k}. \tag{1}$$

As $k \to 0$, the right most term in (1) is decreasing by Lemma 4.50 and bounded below, so has a limit. Similarly, as $h \to 0$, the left most term is increasing, and bounded above, and again has a limit.

1. *Letting $h+k \to 0$, the middle term in (1) is bounded by these limits and thus $g(y+k) - g(y-h) \to 0$, and so for all y:*

   $$\lim_{x \to y-} g(x) = \lim_{x \to y+} g(x),$$

   meaning the left and right limits agree. This common limit equals $g(y)$, since otherwise, the outer terms in (1) could not have a limit as $h \to 0$ and $k \to 0$. Thus $g(x)$ is continuous on (a, b).

2. *The existence of $D^-g(y)$ follows from the existence of the limit of the left most term in (1), and similarly for the existence of $D^+g(y)$, so one-sided derivatives exist everywhere. That both derivatives are increasing functions again follows from (4.78). For example, if $y < w$:*

   $$\frac{g(y+k) - g(y)}{k} \leq \frac{g(w+k) - g(w)}{k},$$

   and so $D^+g(y) \leq D^+g(w)$.

3. *It follows by (1) that $D^-g(y) \leq D^+g(y)$. For every y with $D^-g(y) < D^+g(y)$, the open interval $(D^-g(y), D^+g(y))$ contains a rational number. And since each such interval contains a distinct rational, there can be at most countable many such intervals. Thus, $g'(x)$ exists except for at most countable many points.*

■

We are now ready for the main technical result needed for Jensen's inequality for conditional expectations. The following result generalizes Proposition IV.4.41, which assumed differentiability of $g(x)$. In particular, note that (1) in the proof below is identical to item 2 of that result when $g'(x) = c_x$ exists as is assumed there, and generalizes that result otherwise.

Proposition 4.52 ($g(x)$ convex $\Rightarrow g(x) = \sup_{cy+d \leq g(y)} \{cx + d\}$) *Let $g(x)$ be convex on $I = (a, b)$. Then:*

$$g(x) = \sup_{cy+d \leq g(y)} \{cx + d\}, \tag{4.79}$$

where the supremum is defined over all affine functions $A(y) = cy + d$ such that $A(y) \leq g(y)$ for all y.

Further, for each x this supremum can be defined over such $A(y)$ with countably many parameters $\{c, d\}$.

Proof. *We claim that for each x, there exists c_x so that for all y:*

$$g(y) \geq g(x) + c_x(y - x). \tag{1}$$

Letting $y = x + h$, then $y = x - k$, we require such c_x to satisfy for all $h, k > 0$:

$$\frac{g(x) - g(x-k)}{k} \leq c_x \leq \frac{g(x+h) - g(x)}{h}.$$

By Lemma 4.50, the upper bound increases with $h > 0$ and has a minimum $D^+g(x)$ in the limit as $h \to 0$, and the lower bound is also an increasing function that has a maximum $D^-g(x)$ in the limit as $k \to 0$. Letting:

$$c_x = \frac{1}{2}\left(D^-g(x) + D^+g(x)\right),$$

it follows that (1) is satisfied for all x, and by Proposition 4.51, $c_x = g'(x)$ for all but at most countably many x.

Define the affine function:

$$A_x(y) = g(x) + c_x(y - x), \tag{2}$$

so $A_x(y) \leq g(y)$ for all y, and $A_x(x) = g(x)$, by the above derivation.

Now define:

$$\mathcal{R} = \{A(y) = cy + d | A(y) \leq g(y) \text{ for all } y\}.$$

We have just proved that $\mathcal{R}$ is not empty. In addition, for all x:

$$g(x) = \sup_{A(y)\in\mathcal{R}} \{A(x)\}. \tag{3}$$

First, since $A(y) \leq g(y)$ for all y, it follows that $\sup_{A(y)\in\mathcal{R}} \{A(x)\} \leq g(x)$. On the other hand, choosing $A(y) = A_x(y)$ obtains $A_x(x) = g(x)$, so (3) is proved, which is (4.79).

Finally, given x, let $A_n(y) = d_n + c_x y$ with $d_n \equiv g(x) - c_x x - \frac{1}{n}$. Then $A_n(y) \in \mathcal{R}$ for all n, since $A_n(y) < A_x(y)$, and:

$$g(x) = \sup_n \{A_n(x)\}.$$

■

With the technical details on convex functions now settled, we are ready for Jensen's inequality for conditional expectations.

Proposition 4.53 (Conditional Jensen's inequality) *Let X be a μ-integrable random variable on a probability space $(\mathcal{S}, \mathcal{E}, \mu)$, and $\mathcal{F} \subset \mathcal{E}$ a sigma subalgebra.*

If φ is ***convex*** *and $\varphi(X)$ is μ-integrable, then μ-a.e.:*

$$\varphi\left(E\left[X|\mathcal{F}\right]\right) \leq E\left[\varphi(X)|\mathcal{F}\right]. \tag{4.80}$$

If φ is ***concave*** *and $\varphi(X)$ is μ-integrable, then μ-a.e.:*

$$\varphi\left(E\left[X|\mathcal{F}\right]\right) \geq E\left[\varphi(X)|\mathcal{F}\right]. \tag{4.81}$$

Proof. *Let φ be convex. If $ax + b \leq \varphi(x)$ for all x, then by items 2 and 3 of Proposition 4.46, we have μ-a.e.:*

$$aE\left[X|\mathcal{F}\right] + b = E\left[aX + b|\mathcal{F}\right] \leq E\left[\varphi(X)|\mathcal{F}\right].$$

Restricting to the countable collection $\{(a,b)\}$ *from Proposition 4.52, it follows that:*

$$\sup_{(a,b)} [aE[X|\mathcal{F}] + b] \leq E[\varphi(X)|\mathcal{F}], \tag{1}$$

and since this supremum is over countably many pairs (a,b), *this inequality is again true* μ*-a.e.*

By Proposition 4.52, for any convex function:

$$\varphi(x) = \sup_{(a,b)}\{ax + b\}, \tag{2}$$

where the supremum is over countably many (a,b) *with* $ay + b \leq \varphi(y)$ *for all* y. *Combining* (1) *and* (2), *the result follows* μ*-a.e.*

If φ *is concave then* $-\varphi$ *is convex, and thus* μ*-a.e.:*

$$-\varphi(E[X|\mathcal{F}]) \leq E[-\varphi(X)|\mathcal{F}],$$

and the result follows from linearity of item 2 of Proposition 4.46. ■

4.4.3 $L_p(\mathcal{S})$-Space Properties

For the results of this section, recall the $L_p(\mathcal{S})$-space of Definition V.9.16, which was earlier referenced in Remark 3.23. We restate this definition for a probability space given the dramatic change of notation.

Definition 4.54 ($L_p(\mathcal{S})$-space) *Given a probability space* $(\mathcal{S}, \mathcal{E}, \mu)$, *define* $L_p(\mathcal{S})$***-space,*** *and sometimes denoted* $L_p(\mathcal{S},\mu)$***-space,*** *as the space of all random variables* X *with finite* $L_p(\mathcal{S})$***-norm:***

$$\|X\|_p < \infty.$$

When $1 \leq p < \infty$, $\|X\|_p$ *is defined:*

$$\|X\|_p = \left(\int_X |X|^p \, d\mu\right)^{1/p}. \tag{4.82}$$

For $p = \infty$, $\|X\|_\infty$ *denotes the* ***essential supremum of*** X, *denoted* ess sup X, *and defined by:*

$$\|X\|_\infty = \text{ess sup}\, X \equiv \inf\{\alpha|\mu[\{|X| > \alpha\}] = 0\}. \tag{4.83}$$

Remark 4.55 *(*$X \in L_p(\mathcal{S})$ ***for*** $1 \leq p \leq \infty$ ***then*** $X \in L_1(\mathcal{S})$*) To make the statements of the following results more transparent, we state both that "*X *is a* μ*-integrable random variable," and "*$X \in L_p(\mathcal{S})$ *for* $1 \leq p \leq \infty$*."*

By Exercise V.9.26, the assumption that $X \in L_p(\mathcal{S})$ *is* μ*-integrable is redundant. Indeed for any finite measure space, if* $p \leq p'$ *then:*

$$L_{p'}(\mathcal{S}) \subset L_p(\mathcal{S}),$$

and thus all $L_p(\mathcal{S})$*-spaces are subsets of* $L_1(\mathcal{S})$, *and all* $L_p(\mathcal{S})$*-functions are* μ*-integrable.*

As noted in Exercise V.9.17, this is not true without finiteness of the measure space. In general, there are no predictable inclusions between $L_{p'}(\mathcal{S})$ *and* $L_p(\mathcal{S})$.

Proposition 4.56 (**$X \in L_p(\mathcal{S})$ then $E[X|\mathcal{F}] \in L_p(\mathcal{S})$**) *Let X be a μ-integrable random variable on a probability space $(\mathcal{S}, \mathcal{E}, \mu)$, and $\mathcal{F} \subset \mathcal{E}$ a sigma subalgebra.*

If $X \in L_p(\mathcal{S})$ for $1 \leq p \leq \infty$, then $E[X|\mathcal{F}] \in L_p(\mathcal{S})$ and:

$$\|E[X|\mathcal{F}]\|_p \leq \|X\|_p . \tag{4.84}$$

Proof. *For $1 \leq p < \infty$, let $\varphi(x) = |x|^p$ and note that $\varphi(x)$ is convex. Since $X \in L_p(\mathcal{S})$ assures integrability of $\varphi(X)$ by (4.82), Jensen's inequality obtains:*

$$|E[X|\mathcal{F}]|^p \leq E[|X|^p|\mathcal{F}].$$

Integrating this inequality by item 4 of Proposition V.3.42, then applying (4.56):

$$\int |E[X|\mathcal{F}]|^p d\mu \leq \int E[|X|^p|\mathcal{F}] d\mu \equiv \int |X|^p d\mu.$$

and the result is proved taking a pth root.

If $p = \infty$, let $\|X\|_\infty = \beta \geq 0$ and assume $\|E[X|\mathcal{F}]\|_\infty = \beta + \epsilon$ for $\epsilon > 0$. Since $E[X|\mathcal{F}]$ is $\mathcal{F}$-measurable by definition, define $B \in \mathcal{F}$ by:

$$B = \{\beta < E[X|\mathcal{F}] \leq \beta + \epsilon\}.$$

By item 5 of Proposition V.3.42:

$$\beta\mu(B) \leq \int_B E[X|\mathcal{F}] d\mu \leq (\beta + \epsilon)\mu(B). \tag{1}$$

Since $X = 0$ μ-a.e. on B by definition of $\|X\|_\infty$:

$$\int_B X d\mu = 0. \tag{2}$$

But by (1), this is a contraction to (4.56):

$$\int_B E[X|\mathcal{F}] d\mu = \int_B X d\mu,$$

unless $\mu(B) = 0$. Thus $\|E[X|\mathcal{F}]\|_\infty \leq \beta$ and the proof is complete. ■

Corollary 4.57 (**$\|E[X|\mathcal{F}_0]\|_p \leq \|E[X|\mathcal{F}]\|_p$ if $\mathcal{F}_0 \subset \mathcal{F}$**) *Let X be a μ-integrable random variable on a probability space $(\mathcal{S}, \mathcal{E}, \mu)$, and $\mathcal{F}_0 \subset \mathcal{F} \subset \mathcal{E}$ sigma subalgebras.*

If $X \in L_p(\mathcal{S})$ for $1 \leq p \leq \infty$, then $E[X|\mathcal{F}], E[X|\mathcal{F}_0] \in L_p(\mathcal{S})$ and:

$$|E[X]| \leq \|E[X|\mathcal{F}_0]\|_p \leq \|E[X|\mathcal{F}]\|_p \leq \|X\|_p . \tag{4.85}$$

Proof. *By Proposition 4.56, both $E[X|\mathcal{F}], E[X|\mathcal{F}_0] \in L_p(\mathcal{S})$. Applying this result again to $E[X|\mathcal{F}]$ obtains that $E[E[X|\mathcal{F}]|\mathcal{F}_0] \in L_p(\mathcal{S})$ and:*

$$\|E[E[X|\mathcal{F}]|\mathcal{F}_0]\|_p \leq \|E[X|\mathcal{F}]\|_p \leq \|X\|_p .$$

An application of the tower property obtains $E[E[X|\mathcal{F}]|\mathcal{F}_0] = E[X|\mathcal{F}_0]$, and the right two inequalities are proved.

Letting $\mathcal{F}_0 = \{\emptyset, \mathcal{S}\}$, $E[X|\mathcal{F}_0] = E[X]$ by item 1 of Example 4.41, and then $\|E[X|\mathcal{F}_0]\|_p = |E[X]|$ and the proof is complete. ■

The last result proves that conditional expectation preserves $L_p(\mathcal{S})$ convergence.

Proposition 4.58 (Conditional $L_p(\mathcal{S})$-convergence) *Let X, $\{X_n\}_{n=1}^{\infty}$ be μ-integrable random variables on a probability space $(\mathcal{S}, \mathcal{E}, \mu)$, and $\mathcal{F} \subset \mathcal{E}$ a sigma subalgebra.*

If $X_n \to X$ in $L_p(\mathcal{S})$, where $1 \leq p \leq \infty$:

$$\|X_n - X\|_p \to 0,$$

then $E[X_n|\mathcal{F}] \to E[X|\mathcal{F}]$ in $L_p(\mathcal{S})$:

$$\|E[X_n|\mathcal{F}] - E[X|\mathcal{F}]\|_p \to 0. \tag{4.86}$$

Proof. *By linearity of conditional expectations and (4.84):*

$$\begin{aligned} \|E[X_n|\mathcal{F}] - E[X|\mathcal{F}]\|_p &= \|E[X_n - X|\mathcal{F}]\|_p \\ &\leq \|X_n - X\|_p. \end{aligned}$$

■

4.5 Conditional Expectations in the Limit

In this section, we summarize several results on how conditional expectations work with various limiting operations. These results are closely related to the familiar "integration to the limit" results of Book V, and appear even more closely related than what in fact is true, due to the notational conventions of conditional expectations. See Remark 4.59 for example.

4.5.1 Conditional Monotone Convergence

We begin with **Lebesgue's monotone convergence theorem** of Proposition V.3.23, named for **Henri Léon Lebesgue** (1875–1941). We restate the somewhat generalized version of this result of Corollary V.3.25 in a probability space context to facilitate the discussion in Remark 4.59.

Recall that the statement of this result assumes only measurability, not integrability. As noted in Remark V.3.24, the conclusion in (4.87) cannot therefore assure that $\int_E X d\mu$ is finite, but only that whether finite or infinite, this integral equals the limit of the integral sequence.

Proposition V.3.23/Corollary V.3.25: *Let $\{X_n\}_{n=1}^{\infty}$ be an increasing sequence of nonnegative random variables defined on a probability space $(\mathcal{S}, \mathcal{E}, \mu)$,* and *$X$ a random variable so that $X = \lim_{n\to\infty} X_n$, μ-a.e. on $E \in \mathcal{E}$. Then:*

$$\int_E X d\mu = \lim_{n\to\infty} \int_E X_n d\mu. \tag{4.87}$$

Remark 4.59 (On conditional monotone convergence) *In the notation of Definition 4.1, and recalling Definition V.3.7 on the integral over a measurable set, (4.87) states that when integrals are finite:*

$$E[\chi_E X] = \lim_{n\to\infty} E[\chi_E X_n], \tag{1}$$

for all $E \in \mathcal{E}$. Here χ_E is the characteristic function of E, defined to equal 1 on E and 0 elsewhere. Thus when $E = \mathcal{S}$:

$$E[X] = \lim_{n\to\infty} E[X_n]. \tag{2}$$

Looking ahead, the result in (4.88) appears to generalize (2) *to conditional expectations.*

But this is an example of the potential for misunderstanding given the notational convention for conditional expectations. In reality, the next result asserts that under somewhat stronger conditions on the random variables $\{X_n\}_{n=1}^{\infty}$ *and* X, *that* $E[X_n|\mathcal{F}] \to E[X|\mathcal{F}]$, μ*-a.e. In other words, (4.88) asserts the* μ*-a.e. convergence of the* $\mathcal{F}$*-measurable random variables* $\{E[X_n|\mathcal{F}]\}_{n=1}^{\infty}$ *to the* $\mathcal{F}$*-measurable random variable* $E[X|\mathcal{F}]$, *and does not assert a result on limits of integrals as the notation can be misunderstood to imply.*

That said, (4.88) is an important result and not one, despite the notation, that is predictable from the construction of the conditional expectation. Indeed, recall that given X, *the* $\mathcal{F}$*-measurable random variable* $E[X|\mathcal{F}]$ *was derived as an application of the Radon-Nikodým theorem of Proposition 1.12. In the terminology of that result,* $E[X|\mathcal{F}] = \frac{dv}{d\mu}$, *the Radon-Nikodým derivative of* v, *a (signed) measure on* $\mathcal{F}$ *constructed with* X *as density function, and* μ. *As a quick review will confirm, the construction of* $\frac{dv}{d\mu}$ *in the Book V proof provides little insight on how properties of this function reflect properties of* X. *Indeed, the Book V proof does not even assume that* v *has a density.*

But note that there is a result on integration to the limit that is implicit in (4.88). Since $\{E[X_n|\mathcal{F}]\}_{n=1}^{\infty}$ *are random variables on* $(\mathcal{S}, \mathcal{F}, \mu)$ *by definition, and nonnegative and increasing by item 3 of Proposition 4.46, Lebesgue's monotone convergence theorem of Book V can be applied to assert that for all* $F \in \mathcal{F}$:

$$\int_F E[X|\mathcal{F}]\, d\mu = \lim_{n\to\infty} \int_F E[X_n|\mathcal{F}]\, d\mu. \tag{3}$$

This is indeed an integration result, but it is not a new one. By (4.56), (3) *simply restates* (1) *for all* $F \in \mathcal{F}$.

So the big idea of the next result is this:

If nonnegative, increasing, μ*-integrable random variables* $\{X_n\}_{n=1}^{\infty}$ *on* $(\mathcal{S}, \mathcal{E}, \mu)$ *converge* μ*-a.e. to a* μ*-integrable random variable* X, *then the* $\mathcal{F}$*-measurable version of these variates,* $\{E[X_n|\mathcal{F}]\}_{n=1}^{\infty}$, *also converge* μ*-a.e., and to* $E[X|\mathcal{F}]$, *the* $\mathcal{F}$*-measurable version of* X.

In contrast to the Book V statement noted above, in order for conditional expectations to be well-defined, we require μ-integrability of all random variables. Since $0 \leq X_n \leq X$ for all n, it is enough to assume only integrability of X.

Proposition 4.60 (Conditional monotone convergence theorem) *Let* $\{X_n\}_{n=1}^{\infty}$ *be an increasing sequence of nonnegative random variables on a probability space* $(\mathcal{S}, \mathcal{E}, \mu)$, *and* $\mathcal{F} \subset \mathcal{E}$ *a sigma subalgebra.*

If $X_n \to X$, μ*-a.e. for* μ*-integrable* X, *then:*

$$E[X_n|\mathcal{F}] \to E[X|\mathcal{F}], \ \mu\text{-a.e.} \tag{4.88}$$

Proof. *By assumption,* $X - X_n \geq 0$ *for all* n, $\{X - X_n\}_{n=1}^{\infty}$ *is decreasing, integrable as noted above, and* $X - X_n \to 0$ μ*-a.e. So for all* n *and* $A \in \mathcal{E}$:

$$\int_A (X - X_n)\, d\mu \leq \int_A (X - X_1)\, d\mu < \infty. \tag{1}$$

Lebesgue's dominated convergence theorem of Proposition V.3.45 applies due to (1), *and assures that as* $n \to \infty$,

$$\int_B (X - X_n)\, d\mu \to 0. \tag{2}$$

By the definition of conditional expectation, for all $B \in \mathcal{F}$:

$$\int_B E[X - X_n|\mathcal{F}]\, d\mu = \int_B (X - X_n)\, d\mu.$$

Thus by (2) *and linearity of conditional expectations from Proposition 4.46:*

$$\int_B (E[X_n|\mathcal{F}]\,d\mu - E[X|\mathcal{F}])\,d\mu \to 0, \text{ all } B \in \mathcal{F}. \tag{3}$$

Given $\epsilon > 0$, *define:*

$$B_n = \{E[X_n|\mathcal{F}]\,d\mu - E[X|\mathcal{F}] > \epsilon\}.$$

Then $B_n \in \mathcal{F}$ *by* $\mathcal{F}$*-measurability of conditional expectations, and thus so too is* $B^{(N)} \equiv \bigcap_{n=N}^{\infty} B_n$ *for any* N. *Using properties of integrals from Proposition V.3.42 and* (3)*:*

$$\epsilon\mu(B^{(N)}) \leq \int_{B^{(N)}} (E[X_n|\mathcal{F}]\,d\mu - E[X|\mathcal{F}])\,d\mu \to 0.$$

Thus $\mu(B^{(N)}) = 0$ *for all* N *and all* $\epsilon > 0$.

Letting A^c *denote the complement of the set* A*:*

$$\{E[X_n|\mathcal{F}]\,d\mu \to E[X|\mathcal{F}]\}^c = \bigcup_{\epsilon\in\mathbb{Q}} \bigcup_{N=1}^{\infty} B^{(N)},$$

a set of μ*-measure* 0, *so* $E[X_n|\mathcal{F}] \to E[X|\mathcal{F}]$ μ*-a.e.* ■

4.5.2 Conditional Fatou's Lemma

Fatou's lemma, named for **Pierre Fatou** (1878–1929), was seen in Proposition V.3.19. We restate the somewhat generalized version of this result of Corollary V.3.21 in a probability space context to facilitate the discussion in Remark 4.62.

As is the case for Lebesgue's monotone convergence theorem, this result assumes only measurability, and not integrability. Thus the conclusion in (4.91) cannot assure integrability of X, but implies that X is μ-integrable when the limit inferior of the integral sequence is finite.

Before beginning, we recall Definition V.2.16 in the context of random variables.

Definition 4.61 (Limits inferior/superior) *Given a sequence of random variables* $\{X_n\}_{n=1}^{\infty}$ *defined on a probability space* $(\mathcal{S}, \mathcal{E}, \mu)$, *the* ***limit inferior*** *and* ***limit superior*** *of the sequence are defined pointwise as follows:*

For each $s \in \mathcal{S}$*:*

$$\liminf_{n\to\infty} X_n(s) = \sup\nolimits_n \inf_{k\geq n} X_k(s), \tag{4.89}$$

$$\limsup_{n\to\infty} X_n(s) = \inf\nolimits_n \sup_{k\geq n} X_k(s). \tag{4.90}$$

When clear from the context, the subscript $n \to \infty$ *is often dropped from the* lim inf *and* lim sup *notation.*

That this definition actually reflects "limits" was verified in Exercise V.2.18, which proved in the above notation:

$$\liminf_{n\to\infty} X_n(x) = \lim_{n\to\infty} \inf_{k\geq n} X_k(x),$$
$$\limsup_{n\to\infty} X_n(x) = \lim_{n\to\infty} \sup_{k\geq n} X_k(x).$$

Proposition V.3.19/Corollary V.3.21: Let $\{X_n\}_{n=1}^{\infty}$ *be a sequence of nonnegative random variables on a probability space* $(\mathcal{S}, \mathcal{E}, \mu)$, *and* X is a *nonnegative random variable with* $X \equiv \liminf X_n$, μ*-a.e. on* $E \in \mathcal{E}$. *Then:*

$$\int_E X d\mu \leq \liminf \int_E X_n d\mu. \tag{4.91}$$

Remark 4.62 (On Fatou's lemma) *As in Remark 4.59, (4.91) states that when integrals are finite, that for all $E \in \mathcal{E}$:*

$$E[\chi_E X] \leq \liminf E[\chi_E X_n], \tag{1}$$

so when $E = \mathcal{S}$:

$$E[X] \leq \liminf E[X_n]. \tag{2}$$

Looking ahead, the result in (4.92) appears to generalize (2) *to conditional expectations.*

Repeating the key points made in Remark 4.59, (4.92) asserts that under somewhat stronger conditions on the random variables $\{X_n\}_{n=1}^{\infty}$ and $X = \liminf X_n$, μ-a.e., that $E[X|\mathcal{F}] \leq \liminf E[X_n|\mathcal{F}]$, μ-a.e. In other words, (4.92) bounds the $\mathcal{F}$-measurable random variable $E[X|\mathcal{F}]$ μ-a.e. by the limit inferior of the $\mathcal{F}$-measurable random variables $\{E[X_n|\mathcal{F}]\}_{n=1}^{\infty}$.

Again this result implies an integration to the limit result, but not a new one. Since $\{E[X_n|\mathcal{F}]\}_{n=1}^{\infty}$ are random variables on $(\mathcal{S}, \mathcal{F}, \mu)$ by definition, and nonnegative by item 3 of Proposition 4.46, Fatou's lemma of Book V can be applied to assert that for all $F \in \mathcal{F}$:

$$\begin{aligned} \int_F E[X|\mathcal{F}]\, d\mu &\leq \int_F \liminf E[X_n|\mathcal{F}]\, d\mu \\ &\leq \liminf \int_F E[X_n|\mathcal{F}]\, d\mu. \end{aligned} \tag{3}$$

By (4.56), (3) *restates* (1) *for all $F \in \mathcal{F}$.*

As noted for monotone convergence, in order for conditional expectations to be well-defined, we require μ-integrability of the random variable sequence and its limit inferior.

Proposition 4.63 (Conditional Fatou's lemma) *Let $\{X_n\}_{n=1}^{\infty}$ be a sequence of nonnegative μ-integrable random variables on a probability space $(\mathcal{S}, \mathcal{E}, \mu)$, and $\mathcal{F} \subset \mathcal{E}$ a sigma subalgebra.*

If $X = \liminf X_n$, μ-a.e. for μ-integrable X, then:

$$E[X|\mathcal{F}] \leq \liminf E[X_n|\mathcal{F}], \ \mu\text{-a.e.} \tag{4.92}$$

Proof. *Define:*

$$Y_n \equiv \inf_{k \geq n} X_k.$$

Then $\{Y_n\}_{n=1}^{\infty}$ are nonnegative and increasing, and as noted above:

$$\lim_{n\to\infty} Y_n \equiv \liminf X_n = X, \ \mu\text{-a.e.}$$

Since X is μ-integrable, conditional monotone convergence obtains:

$$E[Y_n|\mathcal{F}] \to E[X|\mathcal{F}], \ \mu\text{-a.e.} \tag{1}$$

Further, $Y_n \leq X_k$ for all $k \geq n$, and so by monotonicity of conditional expectations in item 3 of Proposition 4.46:

$$E[Y_n|\mathcal{F}] \leq \inf_{k \geq n} E[X_k|\mathcal{F}], \ \mu\text{-a.e.} \tag{2}$$

Combining (1) *and* (2)*:*

$$E[X|\mathcal{F}] = \lim_{n\to\infty} E[Y_n|\mathcal{F}] \leq \lim_{n\to\infty} \inf_{k\geq n} E[X_k|\mathcal{F}], \ \mu\text{-a.e.}$$

This upper bound equals $\liminf E[X_n|\mathcal{F}]$ *as noted above, and the proof is complete.* ■

4.5.3 Conditional Dominated Convergence

Lebesgue's dominated convergence theorem, named for **Henri Léon Lebesgue** (1875–1941), was seen in Proposition V.3.45. We restate a somewhat generalized version of this result of Corollary V.3.47 in a probability space context to facilitate the discussion in Remark 4.64. We assume only convergence μ-a.e., so to avoid requiring completeness of the probability space, must then also assume μ-measurability of X.

Proposition V.3.45: *Let $\{X_n\}_{n=1}^{\infty}$ be a sequence of random variables on a probability space $(\mathcal{S}, \mathcal{E}, \mu)$, and X a random variable with $X \equiv \lim_{n\to\infty} X_n$, μ-a.e. on $E \in \mathcal{E}$.*

If there exists a random variable Y, μ-integrable on E, so that for all n:

$$|X_n| \leq Y,$$

then X is μ-integrable on E, and:

$$\int_E X d\mu = \lim_{n\to\infty} \int_E X_n d\mu. \tag{4.93}$$

Further,

$$\lim_{n\to\infty} \int_E |X_n - X| \, d\mu = 0. \tag{4.94}$$

As is the case for Lebesgue's monotone convergence theorem and Fatou's lemma, this result assumes only μ-measurability of $\{X_n\}_{n=1}^{\infty}$ and X. However, the conclusion in (4.93) assures μ-integrability of X due to the assumption that $\{X_n\}_{n=1}^{\infty}$ is dominated by μ-integrable Y.

Remark 4.64 (On conditional dominated convergence) *Existence of the above integrals is part of the conclusion for this result, and thus (4.93) and (4.94) and be restated with $E = \mathcal{S}$:*

$$E[X] = \lim_{n\to\infty} E[X_n], \tag{1}$$

$$\lim_{n\to\infty} E[|X_n - X|] = 0. \tag{2}$$

Looking ahead, both (1) *and (4.95) assert the μ-integrability of $\{X_n\}_{n=1}^{\infty}$ and X so that expectations and conditional expectations are well-defined. However, as above, unlike the integration to the limit result of* (1), *(4.95) asserts μ-a.e. convergence of the $\mathcal{F}$-measurable random variables, $E[X_n|\mathcal{F}] \to E[X|\mathcal{F}]$.*

Similarly, while (2) *states an integration to the limit result, (4.96) asserts μ-a.e. convergence of the $\mathcal{F}$-measurable random variables, $E[|X_n - X|\,|\mathcal{F}] \to 0$.*

In contrast to the above results, we require only measurability of the random variable sequence and X, since μ-integrability follows from the assumption that $|X_n| \leq Y$ for μ-integrable Y.

Proposition 4.65 (Conditional dominated convergence theorem) *Let $\{X_n\}_{n=1}^{\infty}$ be a sequence of random variables on a probability space $(\mathcal{S}, \mathcal{E}, \mu)$, X a random variable with $X \equiv \lim_{n\to\infty} X_n$, μ-a.e., and $\mathcal{F} \subset \mathcal{E}$ a sigma subalgebra.*

If there exists a μ-integrable random variable Y so that for all n:

$$|X_n| \leq Y, \ \mu\text{-a.e.},$$

then:

$$E[X|\mathcal{F}] = \lim_{n\to\infty} E[X_n|\mathcal{F}], \ \mu\text{-a.e.}, \tag{4.95}$$

and:

$$\lim_{n\to\infty} E[|X_n - X|\,|\mathcal{F}] = 0, \ \mu\text{-a.e.} \tag{4.96}$$

Proof. *First,* $|X_n| \leq Y$ μ*-a.e. by assumption, so* $|X_n| \to |X| \leq Y$ *and* $|X_n - X| \leq 2Y$*, again* μ*-a.e. Thus all* X_n*,* $|X_n - X|$ *and* X *are* μ*-integrable, and the above conditional expectations are well-defined.*

For (4.95), consider the random variables sequences $\{Y + X_n\}_{n=1}^{\infty}$ *and* $\{Y - X_n\}_{n=1}^{\infty}$*, which by assumption converge* μ*-a.e. to* $Y \pm X$*. The assumption* $|X_n| \leq Y$ *assures that these sequences are nonnegative, and thus by conditional Fatou's lemma:*

$$\begin{aligned} E\left[Y + X|\mathcal{F}\right] &\leq \liminf E\left[Y + X_n|\mathcal{F}\right], \ \mu\text{-a.e.}, \\ E\left[Y - X|\mathcal{F}\right] &\leq \liminf E\left[Y - X_n|\mathcal{F}\right], \ \mu\text{-a.e.} \end{aligned}$$

By linearity of conditional expectations from Proposition 4.46, and then subtracting finite $E[Y]$ *from both inequalities obtains:*

$$\begin{aligned} E\left[X|\mathcal{F}\right] &\leq \liminf E\left[X_n|\mathcal{F}\right], \ \mu\text{-a.e.}, \\ E\left[-X|\mathcal{F}\right] &\leq \liminf E\left[-X_n|\mathcal{F}\right], \ \mu\text{-a.e.} \end{aligned} \tag{1}$$

By linearity and I.(3.13), this second inequality is equivalent to:

$$E\left[X|\mathcal{F}\right] \geq \limsup E\left[X_n|\mathcal{F}\right], \ \mu\text{-a.e.} \tag{2}$$

Combining (1) *and* (2)*:*

$$\limsup E\left[X_n|\mathcal{F}\right] \leq E\left[X|\mathcal{F}\right] \leq \liminf E\left[X_n|\mathcal{F}\right], \ \mu\text{-a.e.}$$

This statement can only be true if $\limsup E\left[X_n|\mathcal{F}\right] = \liminf E\left[X_n|\mathcal{F}\right]$*,* μ*-a.e. and by Corollary I.3.46, this common value equals* $\lim_{n\to\infty} E\left[X_n|\mathcal{F}\right]$*,* μ*-a.e., proving (4.95).*

For (4.96), $|X_n - X| \leq 2Y$ *obtains that* $2Y - |X_n - X|$ *is nonnegative for all n and has pointwise limit 2Y. Applying conditional Fatou's lemma of Proposition 4.63, and then linearity, obtains* μ*-a.e.:*

$$\begin{aligned} E\left[2Y|\mathcal{F}\right] &\leq \liminf E\left[2Y - |X_n - X|\,|\mathcal{F}\right] \\ &= E\left[2Y|\mathcal{F}\right] + \liminf E\left[-|X_n - X|\,|\mathcal{F}\right] \\ &= E\left[2Y|\mathcal{F}\right] - \limsup E\left[|X_n - X|\,|\mathcal{F}\right], \end{aligned}$$

where the last step is again I.(3.13). Subtracting $E\left[2Y|\mathcal{F}\right]$*, which is finite* μ*-a.e. by assumption, obtains:*

$$\limsup E\left[|X_n - X|\,|\mathcal{F}\right] \leq 0.$$

Since each conditional expectation is nonnegative μ*-a.e. by monotonicity, the limit superior of this sequence equals 0,* μ*-a.e. Thus the limit inferior of this sequence is also 0 since* $\liminf \leq \limsup$*, and thus so too is the limit by Corollary I.3.46, proving (4.96).* ■

5

The Characteristic Function

This chapter introduces and develops a number of properties of the characteristic function of a distribution or joint distribution function. As will be seen, characteristic functions provide a more general and thus more powerful tool for analysis than do moment generating functions of Chapter IV.4, since these exist for any distribution function. In contrast, it was seen in Proposition IV.4.25 that if a distribution function of a random variable has a moment generating function, then of necessity it must also have finite moments of all orders.

We begin with a section on the multivariate moment generating function to develop some of its properties and to complement the analogous one variable results of Chapter IV.4.

As preparation for the development of the characteristic function, which closely resembles the Fourier transform of Chapter V.7, we then add to the discussion of Section V.7.1 on integration of complex-valued functions. As seen there, the results already derived for real-valued functions will play a prominent and simplifying role in this discussion.

Characteristic functions are then formally defined in terms of a probability space integral. This integral is then transformed into various guises using the Book V results on change of variables, obtaining the more familiar definitions from discrete and continuous probability theory. Adding to the examples of Books II and IV, the next section illustrates the characteristic function for a variety of familiar distribution functions.

Properties of characteristic functions are then derived, first on $\mathbb{R}$ and requiring only a notational translation of the Book V results on the Fourier transform, and then generalized to characteristic functions on $\mathbb{R}^n$.

The next section derives Bochner's theorem, which identifies necessary and sufficient conditions for a function to be a characteristic function. The final section then proves a result on uniqueness of moments that was quoted and used in the Book IV investigation.

5.1 The Moment Generating Function

The multivariate moment generating function generalizes Definition IV.4.13 to joint distribution functions. In keeping with the restatement of expectations in terms of integrals defined on the probability space, we have the following.

Definition 5.1 (Moment generating function) *Let $X : \mathcal{S} \rightarrow \mathbb{R}^n$ be a random vector defined on a probability space $(\mathcal{S}, \mathcal{E}, \lambda)$ with joint distribution function $F(x)$.*

*The **moment generating function of** $X \equiv (X_1, X_2, ..., X_n)$, denoted $M_X(t)$, is defined on $t = (t_1, t_2, ..., t_n)$ by:*

$$M_X(t) \equiv E\left[e^{t \cdot X}\right] \equiv \int_{\mathcal{S}} e^{t \cdot X} d\lambda, \tag{5.1}$$

DOI: 10.1201/9781003275770-5

when this integral exists for $|t| < t_0$ *for some* $t_0 > 0$. *Here* $|t| = \left(\sum_{j=1}^n t_j^2\right)^{1/2}$ *is the standard norm in (2.14), and* $t \cdot X = \sum_{j=1}^n X_j t_j$ *is the* **dot product** *or* **inner product** *of the n-vectors t and X.*

The function $M_X(t)$ *is also called the* ***moment generating function of the joint distribution*** $F(x)$, *and denoted* $M_F(t)$.

Remark 5.2 (On Lebesgue-Stieltjes and other integrals for $M_X(t)$**)** *While defined as a* λ*-integral on* $\mathcal{S}$, $M_X(t)$ *is equivalently expressed as a* λ_F*-integral on* $\mathbb{R}^n$, *where* λ_F *is the probability measure on* $\mathbb{R}^n$ *induced by the distribution function* $F(x)$ *of* X.

To see this, consider X *as a measurable transformation of Definition 1.28:*

$$X : (\mathcal{S}, \mathcal{E}, \lambda) \to (\mathbb{R}^n, \mathcal{B}(\mathbb{R}^n), m),$$

where m *is Lebesgue measure. This transformation induces a measure* λ_X *of (1.21) on the range space, and* $\lambda_X = \lambda_F$ *as noted in Proposition 1.30. By the change of variables result of Proposition V.4.18, the integral on* $\mathcal{S}$ *can be expressed as a* ***Lebesgue-Stieltjes integral on*** $\mathbb{R}^n$:

$$M_X(t) \equiv \int_{\mathcal{S}} e^{t \cdot X} d\lambda = \int_{\mathbb{R}^n} e^{t \cdot x} d\lambda_F. \tag{5.2}$$

In the special case where $\lambda_F \ll m$, *meaning that* λ_F ***is absolutely continuous with respect to Lebesgue measure*** m *by Definition 1.11,* λ_F *has a density function* $f(x)$ *with respect to* m *in the sense of Definition 1.9. That is, by the Radon-Nikodým theorem of Proposition 1.12,* $f(x)$ *is Borel measurable and for all* $A \in \mathcal{B}(\mathbb{R}^n)$:

$$\lambda_F(A) = \int_A f(x) dm.$$

By Proposition 1.37, this representation of λ_F *can be extended to all Lebesgue measurable functions that equal* $f(x)$ *m-a.e., while by Proposition 1.10, any such density is also a density function associated with* $F(x)$.

It then follows from Proposition V.4.8, that the Lebesgue-Stieltjes integral of (5.2) can be expressed as a ***Lebesgue integral:***

$$M_X(t) = \int_{\mathbb{R}^n} e^{t \cdot x} d\lambda_F = \int_{\mathbb{R}^n} e^{t \cdot x} f(x) dm. \tag{1}$$

The integrals in (5.1) and (5.2) were developed in Chapter V.3. Thus the moment generating function of Book IV, as well as all moment-related notions, were there defined in terms of Riemann-Stieltjes integrals of Book III, and then transformed under various assumptions to other representations that were likely more familiar to the reader.

The Riemann-Stieltjes approach can be seen to be completely general as follows. Since $g(x) = e^{t \cdot x}$ *is continuous, Proposition V.3.66 obtains that given any bounded right semi-closed rectangle* $R \equiv \prod_{j=1}^n (a_j, b_j]$, *then with* $\bar{R} \equiv \prod_{j=1}^n [a_j, b_j]$:

$$\int_R e^{t \cdot x} d\lambda_F = \int_{\bar{R}} e^{t \cdot x} dF,$$

where the integral on the right in now the Riemann-Stieltjes integral of Section III.4.4 with respect to F, *the joint distribution function of* X. *Since the Lebesgue-Stieltjes integral on the left is well-defined over* $\mathbb{R}^n$, *letting all* $a_j \to -\infty$ *and* $b_j \to \infty$ *obtains the* ***Riemann-Stieltjes integral*** *representation as in Book IV:*

$$M_X(t) = \int_{\mathbb{R}^n} e^{t \cdot x} d\lambda_F \equiv \int_{\mathbb{R}^n} e^{t \cdot x} dF. \tag{2}$$

In the special case where $F(x)$ is continuously differentiable, then F has a continuous density function $f(x) = \frac{\partial^n F}{\partial x_1 \ldots \partial x_n}$ with:

$$F(x) = (\mathcal{R}) \int_{-\infty}^{x} f(y)dy,$$

defined as a Riemann integral in $\mathbb{R}^n$. This integral obtains F by the Fubini result of Corollary III.1.77 and the fundamental theorem of calculus of Proposition III.1.33, noting that the results there on bounded integration domains apply in the limit to these unbounded domains. Then by Proposition III.4.97, the Riemann-Stieltjes integral of (2) can be expressed as a ***Riemann integral:***

$$M_X(t) = \int_{\mathbb{R}^n} e^{t \cdot x} dF = (\mathcal{R}) \int_{\mathbb{R}^n} e^{t \cdot x} f(x)dx, \tag{3}$$

again extended from a bounded to unbounded domain.

If $F(x)$ is a discrete distribution, then (2) can be expressed as a summation as in item 2 of Proposition III.4.97.

Exercise 5.3 (MGF of $Y = AX + \mu$) *Let $X \equiv (X_1, X_2, ..., X_n)$ be a random vector defined on $(\mathcal{S}, \mathcal{E}, \lambda)$ with distribution function $F(x)$ and moment generating function $M_X(t)$ that exists for $|t| < t_0$ for some $t_0 > 0$.*

Given a matrix $A : \mathbb{R}^n \to \mathbb{R}^m$ and fixed $\mu \in \mathbb{R}^m$, let $Y : \mathcal{S} \to \mathbb{R}^m$ be defined by:

$$Y = AX + \mu.$$

Prove that $M_Y(s)$ exists for $s \in \mathbb{R}^m$ with $|s| < s_0$ for some $s_0 > 0$, and that:

$$M_Y(s) = e^{\mu \cdot s} M_X(A^T s), \tag{5.3}$$

where $A^T : \mathbb{R}^m \to \mathbb{R}^n$ is the transpose of A defined by $a_{ij}^T \equiv a_{ji}$. Hint: Justify that Y is a random vector on $(\mathcal{S}, \mathcal{E}, \lambda)$, meaning is appropriately measurable. Then using the standard norm of (2.14), since $|A^T s|$ is continuous $\mathbb{R}^m \to \mathbb{R}$, s_0 can be chosen so that $|s| < s_0$ implies $|A^T s| < t_0$. Finally, by keeping careful note of subscripts, prove that with the above notation:

$$s \cdot AX = A^T s \cdot X.$$

When $X \equiv (X_1, X_2, ..., X_n)$ and $\{X_i\}_{i=1}^n$ are independent variates, we have the following corollary to Proposition 4.7.

Corollary 5.4 ($M_X(t)$, independent $\{X_i\}_{i=1}^n$) *Let $X \equiv (X_1, X_2, ..., X_n)$ be a random vector defined on $(\mathcal{S}, \mathcal{E}, \lambda)$ with $\{X_i\}_{i=1}^n$ independent random variables. If $\{M_{X_i}(t_i)\}_{i=1}^n$ exist for $|t_i| < t_0$, and $M_X(t)$ exists for $|t| < t_0'$ where $t = (t_1, t_2, ..., t_n)$, then for $t \in C \equiv \{|t| < t_0'\} \bigcap \{|t_i| < t_0 \text{ all } i\}$:*

$$M_X(t) = \prod\nolimits_{i=1}^{n} M_{X_i}(t_i). \tag{5.4}$$

Proof. *Given $t = (t_1, t_2, ..., t_n) \in C$, let $Y_i = \exp(t_i X_i)$ and $Y = \prod_{i=1}^n Y_i$. Then $\{Y_i\}_{i=1}^n$ are independent random variables by Proposition II.3.56, and Y and $\{Y_i\}_{i=1}^n$ are λ-integrable by definition of C. Thus by Proposition 4.7:*

$$E[Y] = \prod\nolimits_{i=1}^{n} E[Y_i],$$

and the result follows by definition. ■

For the next result on the moment generating functions of marginal distributions, recall Definition 1.45.

Proposition 5.5 (MGF of marginal distributions) *Given a random vector* $X \equiv (X_1, X_2, ..., X_n)$ *defined on* $(\mathcal{S}, \mathcal{E}, \lambda)$, *assume that* $M_X(t)$ *exists for* $|t| < t_0$ *for some* $t_0 > 0$, *and let* $I \equiv \{i_1, ..., i_m\} \subset \{1, 2, ..., n\}$.

Then the moment generating function $M_{X_I}(t)$ *of the* ***marginal distribution function*** $F_I(x_{i_1}, x_{i_2}, ..., x_{i_m})$ *exists for* $s \in \mathbb{R}^m$ *with* $|s| < s_0$ *for some* $s_0 > 0$, *and:*

$$M_{X_I}(s) = M_X(s'), \tag{5.5}$$

where $s'_{i_j} \equiv s_j$ *for* $1 \leq j \leq m$ *and* $s'_k \equiv 0$ *otherwise.*

Proof. *Define an* $m \times n$ *matrix* $A : \mathbb{R}^n \to \mathbb{R}^m$ *as follows, noting that such* A *has* m *rows and* n *columns. For the* j*th row, let* $A_{j,i_j} = 1$ *and* $A_{j,k} = 0$ *for* $k \neq i_j$. *Then* $AX = X_I$, *and by Exercise 5.3:*

$$M_{X_I}(s) = M_X(A^T s).$$

To see that $A^T s = s'$ *defined above, recall* $A^T_{k,j} \equiv A_{j,k}$. *Thus for* $j \leq m$, $A^T_{i_j,j} \equiv 1$ *and* $A^T_{k,j} = 0$ *otherwise, while for* $m < j \leq n$, $A^T_{k,j} = 0$ *for all* k. *Hence* $\left(A^T s\right)_{i_j} = s_j$ *and* $\left(A^T s\right)_k = 0$ *otherwise, and the proof is complete.* ■

Given a random vector $X \equiv (X_1, X_2, ..., X_n)$ defined on $(\mathcal{S}, \mathcal{E}, \lambda)$ and Borel measurable $g : \mathbb{R}^n \to \mathbb{R}$, the general expectation $E[g(X)]$ is given in Definition 4.1. Next we introduce the notion of **multivariate moments** generalizing Section 4.1.2, with $g(X_1, X_2, ..., X_n) = X_1^{m_1} ... X_n^{m_n}$.

There is no single notational convention for such moments, and of necessity, any convention will be cumbersome to write, and virtually impossible to articulate.

Definition 5.6 (Multivariate moments $\mu'_{(m_1,...,m_n)}$) *Given the random vector* $X \equiv (X_1, X_2, ..., X_n)$ *defined on* $(\mathcal{S}, \mathcal{E}, \lambda)$, *and* n*-tuple of nonnegative integers* $m \equiv (m_1, ..., m_n)$, *define the associated* $(m_1, ..., m_n)$*th* ***multivariate moment of*** X, *denoted* $\mu'_{(m_1,...,m_n)}$, *by:*

$$\mu'_{(m_1,...,m_n)} \equiv E\left[X_1^{m_1} ... X_n^{m_n}\right] \equiv \int_{\mathcal{S}} X_1^{m_1} ... X_n^{m_n} d\lambda, \tag{5.6}$$

when this integral exists, meaning that:

$$E\left[|X_1|^{m_1} ... |X_n|^{m_n}\right] < \infty.$$

Multivariate central moments *are defined analogously when* $\mu_j \equiv E[X_j]$ *exists, by:*

$$\mu_{(m_1,...,m_n)} \equiv \int_{\mathcal{S}} (X_1 - \mu_1)^{m_1} ... (X_n - \mu_n)^{m_n} d\lambda, \tag{5.7}$$

again when $E\left[|X_1 - \mu_1|^{m_1} ... |X_n - \mu_n|^{m_n}\right] < \infty$.

Recalling Definition 3.36, the integrand in (5.6) is sometimes written as $X^m \equiv X_1^{m_1} ... X_n^{m_n}$, *with moment* $\mu'_{(m)} \equiv \mu'_{(m_1,...,m_n)}$, *and similarly for central moments.*

Remark 5.7 (On Lebesgue-Stieltjes and other integrals for $\mu'_{(m_1,...,m_n)}$) *By the same transformations as seen in Remark 5.2,* $\mu'_{(m_1,...,m_n)}$ *can be expressed as the Lebesgue-Stieltjes integral with probability measure* λ_F*:*

$$\mu'_{(m_1,...,m_n)} \equiv \int_{\mathbb{R}^n} x_1^{m_1} ... x_n^{m_n} d\lambda_F,$$

or as a Riemann-Stieltjes integral:

$$\mu'_{(m_1,...,m_n)} \equiv \int_{\mathbb{R}^n} x_1^{m_1}...x_n^{m_n} dF,$$

where $F(x)$ denotes the associated joint distribution function. In the special cases noted there, such moments can also be expressed as Lebesgue and Riemann integrals, or as sums.

We now derive two properties of multivariate moment generating functions that will be reminiscent of Book IV results where $n = 1$. The first result generalizes Proposition IV.4.25, that existence of $M_X(t)$ assures the existence of all moments. For this proof, recall Remark 3.23 which introduced the l_p-norms on $\mathbb{R}^n$. Needed here is the l_1-norm defined on $x = (x_1, ..., x_n)$ by:

$$\|x\|_1 \equiv \sum\nolimits_{j=1}^{n} |x_j|. \tag{5.8}$$

For this result and the next, note that if $m_j \neq 0$ and $m_k = 0$ for $k \neq j$ then:

$$\mu'_{(m_1,...,m_n)} = E\left[X_j^{m_j}\right] = \mu'_{m_j},$$

the m_j-moment of X_j. Similarly if $m_j \neq 0$ and $m_k \neq 0$, but $m_l = 0$ for $l \neq j, k$ then:

$$\mu'_{(m_1,...,m_n)} = E\left[X_j^{m_j} X_k^{m_k}\right],$$

and so forth.

Thus existence of $\mu'_{(m_1,...,m_n)}$ in the next result, and the formula for moments in the following result, also apply to the various moments of each X_j, etc.

Proposition 5.8 **($M_X(t) \Rightarrow \mu'_{(m_1,...,m_n)}$ all m)** *If $X \equiv (X_1, X_2, ..., X_n)$ is a random vector defined on $(\mathcal{S}, \mathcal{E}, \lambda)$ and $M_X(t)$ exists for $|t| < t_0$ for some $t_0 > 0$, then $\mu'_{(m)} \equiv \mu'_{(m_1,...,m_n)}$ exists for all n-tuples of nonnegative integers $m \equiv (m_1, ..., m_n)$.*

Proof. *We must prove that $E[|X_1|^{m_1} ... |X_n|^{m_n}] < \infty$ for all $m \equiv (m_1, ..., m_n)$. Since $|X_j| \leq \|X\|_1$ for all j by (5.8):*

$$|X_1|^{m_1} ... |X_n|^{m_n} \leq \|X\|_1^M,$$

with $M \equiv \sum_{j=1}^n m_j$. Hence by monotonicity of integrals (item 4 of Proposition V.3.42), the proof is complete by proving that for all nonnegative integers M:

$$\int_{\mathcal{S}} \|X\|_1^M \, d\lambda < \infty. \tag{1}$$

To this end, choose $t' \equiv (t, ..., t) \in \mathbb{R}^n$ with $|t'| < t_0$. Now $\|X\|_1^M \leq e^{t\|X\|_1}$ if $\|X\|_1 / \ln \|X\|_1 > M/t$, and since $x/\ln x$ is increasing and unbounded as $x \to \infty$, choose a so that $\|X\|_1 / \ln \|X\|_1 > M/t$ if $\|X\|_1 > a$. Then for $\|X\|_1 > a$:

$$\|X\|_1^M \leq e^{t\|X\|_1} = \exp\left[\sum\nolimits_{j=1}^{n} t\,|X_j|\right] \leq e^{t' \cdot X} + e^{-t' \cdot X}. \tag{2}$$

This obtains:

$$\begin{aligned}
\int_{\mathcal{S}} \|X\|_1^M \, d\lambda &= \int_{\|X\|_1 \leq a} \|X\|_1^M \, d\lambda + \int_{\|X\|_1} \|X\|_1^M \, d\lambda \\
&\leq ca^M + \int_{\|X\|_1 > a} (e^{t' \cdot X} + e^{-t' \cdot X}) d\lambda \\
&\leq ca^M + M_X(t') + M_X(-t').
\end{aligned}$$

Thus existence of $M_X(t)$ for $|t| < t_0$ for some $t_0 > 0$ proves (1) and the existence of $\mu'_{(m)}$ for all $m \equiv (m_1, ..., m_n)$. ■

The final result is a generalization of Proposition IV.4.27, which identifies a power series representation for $M_X(t)$ and derives the result that justifies the name "moment generating" function.

In preparation, it is necessary to develop some combinatorial results. Applying the Taylor series for e^x, and then the multinomial theorem of (4.32):

$$\begin{aligned} e^{t\cdot x} &= \sum\nolimits_{m=0}^{\infty}\frac{1}{m!}\left(\sum\nolimits_{j=1}^{n}t_jx_j\right)^m \\ &= \sum\nolimits_{m=0}^{\infty}\frac{1}{m!}\left[\sum\nolimits_{(m_1,m_2,..m_n)}\frac{m!}{m_1!m_2!...m_n!}(t_1x_1)^{m_1}...(t_nn_n)^{m_n}\right] \\ &= \sum\nolimits_{m=0}^{\infty}\sum\nolimits_{(m_1,m_2,..m_n)}\prod\nolimits_{j=1}^{n}\frac{t_j^{m_j}x_j^{m_j}}{m_j!}, \end{aligned}$$

where the inner summation is over all distinct n-tuples $(m_1, m_2, ..m_n)$ with $m_j \geq 0$ and $\sum_{j=1}^{n} m_j = m$.

Since this series is absolutely convergent, this summation can be arbitrarily rearranged without changing its value (Proposition 6.15, **Reitano** (2010)), and so equivalently expressed:

$$e^{t\cdot x} = \sum\nolimits_{(m_1,...,m_n)}\prod\nolimits_{j=1}^{n}\frac{t_j^{m_j}x_j^{m_j}}{m_j!}, \tag{5.9}$$

where this summation is over all n-tuples of nonnegative integers, meaning $m_j \geq 0$ for all j.

This latter expression then obtains that for all nonnegative integer n-tuples $\{m_j\}_{j=1}^{n}$ and $m \equiv \sum_{j=1}^{n} m_j$:

$$\left.\frac{\partial^m e^{t\cdot x}}{\partial t_1^{m_1}\cdots\partial t_n^{m_n}}\right|_{t=0} = x_1^{m_1}\cdots x_n^{m_n},$$

where $t = 0$ is shorthand for $t_j = 0$ for all j. As in the the one-variate Taylor series case, this follows as in Proposition 9.109 of the above text by justifying term-by-term differentiation, which in turn is justified by the uniform convergence of the associated series.

Then given $(m_1', ..., m_n')$ and $f(x) \equiv \prod_{j=1}^{n} t_j^{m_j'}x_j^{m_j'}/m_j'!$, consider $\frac{\partial^m f}{\partial t_1^{m_1}...\partial t_n^{m_n}}$. If at least one $m_j > m_j'$, then $\frac{\partial^m f}{\partial t_1^{m_1}...\partial t_n^{m_n}} \equiv 0$. If at least one $m_j' > m_j$, then this derivative will have at least one t_j-factor and thus equal 0 when evaluated at $t = 0$. The only remaining case is $(m_1', ..., m_n') = (m_1, ..., m_n)$, and then $\frac{\partial^m f}{\partial t_1^{m_1}...\partial t_n^{m_n}} = x_1^{m_1}\cdots x_n^{m_n}$.

Proposition 5.9 (Power series for $M_X(t)$) *If $X \equiv (X_1, X_2, ..., X_n)$ is a random vector defined on $(\mathcal{S}, \mathcal{E}, \lambda)$ and $M_X(t)$ exists for $|t| < t_0$ for some $t_0 > 0$, then:*

$$M_X(t) = \sum\nolimits_{(m_1,...,m_n)}\mu'_{(m_1,...,m_n)}\prod\nolimits_{j=1}^{n}\frac{t_j^{m_j}}{m_j!}, \tag{5.10}$$

where this summation is over all n-tuples of nonnegative integers $m \equiv (m_1, ..., m_n)$, and $\mu'_{(m_1,...,m_n)}$ is defined in (5.6).

Thus:

$$\mu'_{(m_1,...,m_n)} = \left.\frac{\partial^m M_X(t)}{\partial t_1^{m_1}\cdots\partial t_n^{m_n}}\right|_{t=0}. \tag{5.11}$$

Proof. *Let $t' \equiv (t, ..., t) \in \mathbb{R}^n$ with $0 < t < t_0/\sqrt{n}$ and thus $|t'| < t_0$. By (2) of the above proof:*

$$e^{t\|X\|_1} \leq e^{t' \cdot X} + e^{-t' \cdot X}, \tag{1}$$

and thus existence of $M_X(\pm t')$ assures that $e^{t\|X\|_1}$ is λ-integrable.

With $m \equiv \sum_{j=1}^{n} m_j$ and $|t_j| \leq t$ for all j, then for all N:

$$\begin{aligned} \left| \sum\nolimits_{m \leq N} \prod\nolimits_{j=1}^{n} \frac{t_j^{m_j} X_j^{m_j}}{m_j!} \right| &\leq \sum\nolimits_{m \leq N} \prod\nolimits_{j=1}^{n} \frac{|t_j|^{m_j} |X_j|^{m_j}}{m_j!} \\ &\leq \sum\nolimits_{m \leq N} \prod\nolimits_{j=1}^{n} \frac{t^{m_j} |X_j|^{m_j}}{m_j!}. \end{aligned}$$

On the other hand, using the calculations above:

$$\begin{aligned} e^{t\|X\|_1} &= \exp\left[\sum\nolimits_{j=1}^{n} t |X_j|\right] \\ &= \sum\nolimits_{(m_1,...,m_n)} \prod\nolimits_{j=1}^{n} \frac{t^{m_j} |X_j|^{m_j}}{m_j!}. \end{aligned}$$

Combining obtains that for all N:

$$\left| \sum\nolimits_{m \leq N} \prod\nolimits_{j=1}^{n} \frac{t_j^{m_j} X_j^{m_j}}{m_j!} \right| \leq e^{t\|X\|_1}. \tag{2}$$

The expression on the left is thus λ-integrable for all N by (1), and converges pointwise to $e^{t \cdot X}$.

By Lebesgue's dominated convergence theorem of Proposition V.3.45, and Proposition V.3.42 properties of the integral:

$$\begin{aligned} M_X(t) &= \lim_{N \to \infty} \sum\nolimits_{m \leq N} \int_{\mathcal{S}} \prod\nolimits_{j=1}^{n} \frac{t_j^{m_j} X_j^{m_j}}{m_j!} d\lambda \\ &= \lim_{N \to \infty} \sum\nolimits_{m \leq N} \prod\nolimits_{j=1}^{n} \frac{t_j^{m_j}}{m_j!} \int_{\mathcal{S}} \prod\nolimits_{j=1}^{n} X_j^{m_j} d\lambda \\ &= \lim_{N \to \infty} \sum\nolimits_{m \leq N} \prod\nolimits_{j=1}^{n} \frac{t_j^{m_j}}{m_j!} \mu'_{(m_1,...,m_n)} \\ &= \sum\nolimits_{(m_1,...,m_n)} \prod\nolimits_{j=1}^{n} \frac{t_j^{m_j}}{m_j!} \mu'_{(m_1,...,m_n)}. \end{aligned}$$

Since $M_X(t)$ is absolutely convergent for $|t| < t_0$, (5.11) follows by direct calculation as above. ■

Corollary 5.10 (Existence of $M_X(t)$) *If $X \equiv (X_1, X_2, ..., X_n)$ is a random vector defined on $(\mathcal{S}, \mathcal{E}, \lambda)$ and for some $s > 0$:*

$$E[e^{s\|X\|_1}] < \infty,$$

where $\|X\|_1 \equiv \sum_{j=1}^{n} |X_j|$, then $M_X(t)$ exists for $|t| < s$, and thus satisfies (5.10).

Proof. *As in (2) of the prior proof, if $|t_j| \leq s$ for all j, then for all N:*

$$\left| \sum\nolimits_{m \leq N} \prod\nolimits_{j=1}^{n} \frac{t_j^{m_j} X_j^{m_j}}{m_j!} \right| \leq e^{s\|X\|_1}, \tag{1}$$

and this upper bounding function is integrable by assumption.

Since from (5.9):

$$\begin{aligned} e^{t\cdot X} &= \sum\nolimits_{(m_1,\ldots,m_n)} \prod\nolimits_{j=1}^{n} \frac{t_j^{m_j} X_j^{m_j}}{m_j!} \\ &= \lim_{N\to\infty} \sum\nolimits_{m\leq N} \prod\nolimits_{j=1}^{n} \frac{t_j^{m_j} X_j^{m_j}}{m_j!}, \end{aligned}$$

Lebesgue's dominated convergence theorem of Proposition V.3.45 obtains by (1)*, that if* $|t_j| \leq s$ *for all* j*:*

$$M_X(t) \equiv \int_S e^{t\cdot X} d\lambda < \infty.$$

Hence $M_X(t)$ *exists if* $|t_j| \leq s$ *for all* j*, and this is assured if* $|t| < s$. ■

5.2 Integration of Complex-Valued Functions

Simply put, to transition from moment generating functions to characteristic functions is to transition from the mathematics of $M_X(t) \equiv E\left[e^{X\cdot t}\right]$ in the general case of a random vector X, to $C_X(t) \equiv E\left[e^{iX\cdot t}\right]$, where $i = \sqrt{-1}$ is the so-called imaginary unit. The significance of this transition is that $M_X : \mathbb{R}^n \to \mathbb{R}$ is a real-valued function, while $C_X : \mathbb{R}^n \to \mathbb{C}$ is complex-valued. Thus we must address the applicability of the needed Book V measure and integration theory of real-valued functions, to complex-valued functions.

In this section we continue the discussion of Section V.7.1, which introduced this investigation for the development of complex-valued Fourier transform $\hat{f} : \mathbb{R} \to \mathbb{C}$. There it was seen that results for integration of real-valued functions generalized naturally and usually easily to this complex-valued context. Here we recall a few details to set the stage, but largely focus on results needed for the current development of complex-valued functions defined on $\mathbb{R}^n$.

We start with a definition.

Definition 5.11 (z, $\bar{z}$, $\mathrm{Re}(z)$, $\mathrm{Im}(z)$, z) *The* ***complex conjugate*** $\bar{z}$ *of a complex number* $z \equiv a + bi$*, where* a*,* $b \in \mathbb{R}$ *and* $i = \sqrt{-1}$*, is defined by:*

$$\bar{z} = a - bi. \tag{5.12}$$

The number a *is called the* ***real part of*** z*, and* b *the* ***imaginary part of*** z*, denoted:*

$$a \equiv \mathrm{Re}(z) = (z + \bar{z})/2, \qquad b \equiv \mathrm{Im}(z) = (z - \bar{z})/2i. \tag{5.13}$$

The ***absolute value of*** z *is defined:*

$$|z| \equiv \sqrt{z\bar{z}} = \sqrt{a^2 + b^2}, \tag{5.14}$$

where the positive square root is taken by convention, and called the ***principal square root.*** *Thus for example,* $|i| = 1$.

Exercise 5.12 ($||x| - |y|| \leq |x - y|$ **for** $x, y \in \mathbb{C}$) *Given* $x, y \in \mathbb{C}$*, prove that:*

$$||x| - |y|| \leq |x - y|. \tag{5.15}$$

Hint: While apparent geometrically for $x, y \in \mathbb{R}$, this can be proved with (5.14) and some algebra, or by first proving the ***triangle inequality*** *for $u, v \in \mathbb{C}$:*

$$|u+v| \leq |u| + |v|,$$

and applying it creatively. The result in (5.15) is called the ***reverse triangle inequality.***

The Taylor series for the exponential function e^{iy} is well-defined for $y \in \mathbb{R}$:

$$e^{iy} \equiv \sum\nolimits_{j=0}^{\infty} \frac{(iy)^j}{j!}, \tag{5.16}$$

meaning this series is absolutely convergent for all $y \in \mathbb{R}$ since $|i| = 1$. Below, y will equal $X \cdot t$.

The summation in (5.16) can be split into even and odd powers:

$$\sum\nolimits_{j=0}^{\infty} \frac{(iy)^j}{j!} = \sum\nolimits_{j=0}^{\infty} \frac{(-1)^j y^{2j}}{(2j)!} + i \sum\nolimits_{j=0}^{\infty} \frac{(-1)^j y^{2j+1}}{(2j+1)j!}.$$

Recognizing the two Taylor series on the right obtains **Euler's formula:**

$$e^{iy} = \cos y + i \sin y, \tag{5.17}$$

named for **Leonhard Euler** (1707–1783).

A simple consequence of Euler's formula is that

$$\left|e^{iy}\right| = 1, \text{ all } y \in \mathbb{R}, \tag{5.18}$$

since $\cos^2 y + \sin^2 y = 1$ for all such y. Euler's formula also contains the beautiful result known as **Euler's identity:**

$$e^{i\pi} = -1, \tag{5.19}$$

while considering $e^{\pm iy}$ yields:

$$\cos y = \frac{e^{iy} + e^{-iy}}{2}, \qquad \sin y = \frac{e^{iy} - e^{-iy}}{2i}. \tag{5.20}$$

Remark 5.13 (Principal square root $\sqrt{z}$) *The notion of principal square root extends to complex numbers $z = a + bi$ by expressing z:*

$$z = re^{i\theta},$$

with $r = \sqrt{a^2 + b^2}$ and $-\pi < \theta \leq \pi$. Recall that θ is measured ***counterclockwise*** *from the x-axis which is $\theta = 0$, then the positive y-axis is $\theta = \pi/2$ and so forth. The* ***principal square root*** *is then defined:*

$$\sqrt{z} = \sqrt{r} e^{i\theta/2}.$$

Writing:

$$z = \sqrt{a^2 + b^2}\left(\frac{a}{\sqrt{a^2 + b^2}} + i\frac{b}{\sqrt{a^2 + b^2}}\right),$$

obtains by (5.17) that $\cos\theta = a/\sqrt{a^2 + b^2}$ and $\sin\theta = b/\sqrt{a^2 + b^2}$. Then by the half-angle formulas for $-\pi < \theta \leq \pi$:

$$\cos\frac{\theta}{2} = \sqrt{(1 + \cos\theta)/2}, \quad \sin\frac{\theta}{2} = sgn(\theta)\sqrt{(1 - \cos\theta)/2},$$

where $sgn(\theta) = 1$ if $\theta \geq 0$ and -1 otherwise.

Substituting, and noting that $sgn\theta = sgnb$, obtains the principal square root:

$$\sqrt{a+bi} = \sqrt{\left(\sqrt{a^2+b^2}+a\right)\Big/2} + isgn(b)\sqrt{\left(\sqrt{a^2+b^2}-a\right)\Big/2}, \tag{5.21}$$

where we take the positive square root of real numbers and define $sgn(b)$ as above.

Recalling Definition V.7.1, measurability and integrability of complex-valued functions $f : \mathbb{R} \to \mathbb{C}$ followed from the decomposition of $f(x) = u(x) + iv(x)$, where $u, v : \mathbb{R} \to \mathbb{R}$. We next generalize the Book V definition from the domain space $(\mathbb{R}, \sigma(\mathbb{R}), \mu)$ to a general measure space $(\mathcal{S}, \mathcal{E}, \lambda)$. This includes general $(\mathbb{R}^n, \sigma(\mathbb{R}^n), \mu)$ as well as all probability spaces, the context of the current investigation.

Definition 5.14 (Measurability, integrability of $f : \mathcal{S} \to \mathbb{C}$) *Given a measure space $(\mathcal{S}, \mathcal{E}, \lambda)$, a function $f : \mathcal{S} \to \mathbb{C}$ is defined to be* ***measurable,*** *and sometimes* ***$\mathcal{E}$-measurable,*** *if $f(x) = u(x)+iv(x)$ and both $u(x)$ and $v(x)$ are $\mathcal{E}$-measurable. That is, $f(x)$ is measurable if $u^{-1}(A) \in \mathcal{E}$ and $v^{-1}(A) \in \mathcal{E}$ for all Borel sets $A \in \mathcal{B}(\mathbb{R})$.*

If $u(x)$ and $v(x)$ are λ-integrable, we define the ***λ-integral of*** *$f(x)$ by:*

$$\int f(x)d\lambda \equiv \int u(x)d\lambda + i\int v(x)d\lambda. \tag{5.22}$$

Recalling Definition V.3.7, if $A \in \mathcal{E}$:

$$\int_A f(x)d\lambda \equiv \int \chi_A(x)u(x)d\lambda + i\int \chi_A(x)v(x)d\lambda.$$

As was seen in Remark V.7.2, defining integrability of $f(x)$ in terms of the integrability of $u(x)$ and $v(x)$ is consistent with Definition V.3.38, which requires integrability of $|f(x)|$. In addition, properties of integrals of real-valued functions summarized in Proposition V.3.42 that are well-defined for integrals of complex-valued functions are readily seen to apply by the above definition. Usually the proof involves no more than splitting the integral by (5.22), applying the Proposition V.3.42 result to the component integrals, and reassembling. While the triangle inequality also applied, the proof was more subtle and seen in Proposition V.7.3.

An example of a property that does not extend is monotonicity, since $f(x) \leq g(x)$ is not well-defined for complex valued functions.

Turning to some of the powerful integration tools of Book V, **Lebesgue's dominated convergence theorem** was seen to extend to complex-valued functions in Proposition V.7.4. The proof was largely to decompose, apply the real-valued result, and then reassemble. For the Fourier analysis of Book V, this was the primary tool needed.

For the current chapter, we will also need the change of variables result of Proposition V.4.18, and Fubini's theorem of Proposition V.5.20.

Proposition V.4.18 addressed change of variables under measurable transformations, reflecting the associated induced measures of Definition 1.28 as stated in (1.21).

Proposition 5.15 ($\int_{A'} g(x')d\lambda_T$ as a λ-integral; $g : X \to \mathbb{C}$) *Let $T : (X, \sigma(X), \lambda) \to (X', \sigma(X'), \lambda')$ be a measurable transformation and λ_T defined on $\sigma(X')$ by (1.20).*

Then for any complex-valued, measurable function $g : X' \to \mathbb{C}$:

$$\int_{X'} g(x')d\lambda_T = \int_X g(Tx)d\lambda, \tag{5.23}$$

though both integrals may be infinite.

More generally, a complex-valued, measurable function $g(x')$ defined on X' is λ_T-integrable if and only if $g(Tx)$ is λ-integrable, and when integrable:

$$\int_{A'} g(x')d\lambda_T = \int_{T^{-1}A'} g(Tx)d\lambda, \tag{5.24}$$

for all $A \in \sigma(X')$.

Proof. *We leave the proof as an exercise in verifying that the above strategy to decompose, apply the earlier result, then reassemble, provides the needed result.* ■

For Fubini's theorem of Proposition V.5.20, recall that $\sigma'(X \times Y)$ is the smallest sigma algebra that contains the measurable rectangles $A \times B$, where $A \in \sigma(X)$ and $B \in \sigma(Y)$.

Proposition 5.16 (Fubini's theorem on finite $(X \times Y, \sigma'(X \times Y), \mu \times v); f : X \times Y \to \mathbb{C}$) *Let $(X, \sigma(X), \mu)$ and $(Y, \sigma(Y), v)$ be finite measure spaces, and $f(x, y)$ a complex-valued integrable function on the finite measure space $(X \times Y, \sigma'(X \times Y), \mu \times v)$. Then:*

1. *For all x, $f_x(y)$ is $\sigma(Y)$-measurable, and,*

1'. *For all y, $f_y(x)$ is $\sigma(X)$-measurable.*

2. *$\int_Y f_x(y)dv \equiv \int_Y f(x, y)dv$ is $\sigma(X)$-measurable, and,*

2'. *$\int_X f_y(x)d\mu \equiv \int_X f(x, y)d\mu$ is $\sigma(Y)$-measurable.*

3. *The $\mu \times v$-integral of $f(x, y)$ can be evaluated as an iterated integral:*

$$\int_X \left[\int_Y f(x, y)dv\right] d\mu = \int_{X \times Y} f(x, y)d(\mu \times v) = \int_Y \left[\int_X f(x, y)d\mu\right] dv. \tag{5.25}$$

Proof. *Left as an exercise in deploying the above strategy.* ■

5.3 The Characteristic Function

The characteristic function of a random variable or random vector defined on a probability space $(\mathcal{S}, \mathcal{E}, \lambda)$ is formulaically related to the respective moment generating function definitions as noted above. But the mathematics of characteristic functions is most closely related to the Chapter V.7 development of Fourier transforms, applied in the context of probability theory.

We begin with a definition, and then return to this connection. As above, by complex valued is meant that $C_X : \mathbb{R} \to \mathbb{C}$, respectively, $C_X : \mathbb{R}^n \to \mathbb{C}$.

Definition 5.17 (Characteristic function of X) *If X is a random variable defined on a probability space $(\mathcal{S}, \mathcal{E}, \lambda)$ with distribution function $F(x)$, the* ***characteristic function of*** X *is defined as the complex valued function of $t \in \mathbb{R}$:*

$$C_X(t) \equiv E\left[e^{iXt}\right] = \int_{\mathcal{S}} e^{iXt} d\lambda, \tag{5.26}$$

where $i \equiv \sqrt{-1}$.

If $X \equiv (X_1, X_2, ..., X_n)$ is a random vector defined on $(\mathcal{S}, \mathcal{E}, \lambda)$, the ***characteristic function of*** X *is defined as the complex valued function of $t = (t_1, t_2, ..., t_n) \in \mathbb{R}^n$:*

$$C_X(t) \equiv E\left[e^{iX\cdot t}\right] = \int_{\mathcal{S}} e^{iX\cdot t} d\lambda, \tag{5.27}$$

where $x \cdot t = \sum_{j=1}^{n} x_j t_j$ is the ***dot product*** *or* ***inner product*** *of t and X.*

The function $C_X(t)$ is also called the ***characteristic function of the distribution*** *$F(x)$, and denoted $C_F(t)$.*

Remark 5.18 (On Lebesgue-Stieltjes and other integrals for $C_X(t)$) *While defined as a λ-integral on $\mathcal{S}$ as in (5.22), $C_X(t)$ is equivalently expressed as a λ_F-integral on $\mathbb{R}$ or $\mathbb{R}^n$, where λ_F is the probability measure on $\mathbb{R}$ or $\mathbb{R}^n$ induced by the distribution function $F(x)$ of X of Proposition 1.27.*

For example, we can interpret the random vector X as a measurable transformation of Definition 1.28:

$$X : (\mathcal{S}, \mathcal{E}, \lambda) \to (\mathbb{R}^n, \mathcal{B}(\mathbb{R}^n), m),$$

where m is Lebesgue measure. This transformation induces a measure λ_X on the range space by (1.21), and as noted in Proposition 1.30, $\lambda_X = \lambda_F$, the Borel measure on $\mathbb{R}^n$ induced by the distribution function F.

By the change of variables result of Proposition 5.15, the integral on $\mathcal{S}$ in (5.27) can be expressed as a ***Lebesgue-Stieltjes integral on*** $\mathbb{R}^n$*:*

$$C_X(t) = \int_{\mathcal{S}} e^{it\cdot X} d\lambda = \int_{\mathbb{R}^n} e^{it\cdot x} d\lambda_X. \tag{5.28}$$

The integral in (5.26) can be equivalently expressed as an integral on $\mathbb{R}$.

Thus, the function $C_X(t)$ is also called the ***characteristic function of the probability measure*** λ_X.

As noted in Remark 5.7, these integrals can then also be represented under various assumptions as Riemann-Stieltjes, Lebesgue and Riemann integrals, or sums. For example by Proposition III.4.28, if F is a discrete distribution function, the Riemann-Stieltjes integral representation reduces to:

$$C_X(t) = \sum\nolimits_{i=1}^{\infty} e^{ix_i t} f(x_i). \tag{5.29}$$

Throughout this chapter, we will alternate between the above integral representations of $C_X(t)$, as λ-integrals on $\mathcal{S}$ or λ_X-integrals on $\mathbb{R}/\mathbb{R}^n$, depending on which better suites the given analysis.

Perhaps the most remarkable thing about the above definition is that this is the first definition in Books IV and VI involving an expectation that omits the mandatory qualification, "*when this integral exists.*" This omission is deliberate, because $C_X(t)$ always exists.

Proposition 5.19 (On existence of $C_X(t)$) *If X is a random variable or random vector defined on a probability space $(\mathcal{S}, \mathcal{E}, \lambda)$, the characteristic function $C_X(t)$ of Definition 5.17 always exists, and:*

$$|C_X(t)| \leq 1. \tag{5.30}$$

Proof. *By (5.18):*

$$\left|e^{iXt}\right| = \left|e^{iX\cdot t}\right| = 1,$$

and since $\left|e^{iX\cdot t}\right|$ is bounded and thus λ-integrable, $e^{iX\cdot t}$ is λ-integrable by Remark V.7.2 and $C_X(t)$ exists.

The triangle inequality of Proposition V.7.3 for integrals of complex-valued, λ-integrable functions then obtains:

$$|C_X(t)| = \left|\int_S e^{iX\cdot t} d\lambda\right| \leq \int_S d\lambda = 1,$$

and the proof is complete. ■

While appearing similar formulaically to the analogously defined moment generating functions of Definition 5.1, the seemingly simple placement of i in the exponential has made a stark difference for existence.

By Proposition IV.4.25, the existence of $M_X(t)$ on an open interval $(-t_0, t_0)$ for a random variable required all moments of the random variable X to be finite, and this was generalized in Proposition 5.8 to random vectors. In the case of random variables, while existence of all moments is necessary, it is not sufficient for the existence of $M_X(t)$ as seen in Example IV.4.32 on the lognormal distribution. Defining the random vector X with independent lognormal variates, it is an exercise to check that again all moments exist, but the moment generating function does not.

Reinforcing this formulaic connection, we will see below that when $M_X(t)$ exists, then $C_X(t) = M_X(it)$, for random variables or random vectors. This result makes the calculation of $C_F(t)$ easy when $M_F(t)$ exists, while otherwise, $C_F(t)$ must be calculated directly and requires complex variable integrations or summations.

Remark 5.20 ($C_X(t)$ and the Fourier transform) *By Definition V.7.6 for $n = 1$, $C_X(t)$ as expressed in (5.28) is the* ***Fourier-Stieltjes transform of the finite measure*** *λ_F on $\mathbb{R}$. For general n, the same terminology is used for the finite measure λ_F on $\mathbb{R}^n$. Thus as defined above, the characteristic function of X is the Fourier transform of the measure λ_F on $\mathbb{R}$ or $\mathbb{R}^n$ induced by F, the distribution function of X.*

In the special case where λ_F ***is absolutely continuous with respect to Lebesgue measure*** *m by Definition 1.11, denoted $\lambda_F \ll m$, it follows that λ_F has a density function $f(x)$ with respect to m in the sense of Definition 1.9. That is, by the Radon-Nikodým theorem of Proposition 1.12, there exists Borel measurable $f(x)$ so that for all $A \in \mathcal{B}(\mathbb{R}^n)$:*

$$\lambda_F(A) = \int_A f(x) dm.$$

By Proposition 1.37, this representation of λ_F can be extended to all Lebesgue measurable functions that equal $f(x)$ m-a.e., while by Proposition 1.10, any such density is also a density function associated with $F(x)$.

In this special case, the Lebesgue-Stieltjes integral of (5.28) can be expressed as a ***Lebesgue integral*** *by Proposition V.4.8:*

$$C_X(t) = \int_{\mathbb{R}^n} e^{it\cdot x} d\lambda_F = \int_{\mathbb{R}^n} e^{it\cdot x} g_\lambda(x) dm. \tag{5.31}$$

Again by Definition V.7.6 for $n = 1$, $C_X(t)$ is the ***Fourier transform of*** *$g_\lambda(x)$, denoted $\widehat{g}_\lambda(t)$. For general n, the same terminology is used for integrable functions $g_\lambda(x)$ on $\mathbb{R}^n$.*

5.4 Examples of Characteristic Functions

In this section we identify the characteristic functions of many of the probability densities of random variables introduced in previous books. Some derivations are left to the reader as exercises. As noted above and proved below in (5.52), $C_X(t) = M_X(it)$ for $t \in (-t_0, t_0)$ if $M_X(t)$ exists on this interval. When this interval is $\mathbb{R}$, then this result obtains the functional form of $C_X(t)$ for all t. When this interval is bounded, the functional form of $C_X(t)$ must be independently verified outside of this interval.

In cases where $M_F(t)$ does not exist, one obtains $C_F(t)$ by complex variable summations or integration. In such cases, perhaps unsurprisingly, a closed formula for $C_X(t)$ is sometimes unknown or very complicated, requiring the use of special functions which again admit no closed form representation.

Moment generating functions and other results on many of these distribution functions can be found in Sections IV.4.2.6 and IV.4.2.7.

5.4.1 Discrete Distributions

1. **Discrete Rectangular Distribution on** $[0, 1]$

   The discrete rectangular distribution is defined by the density function:

$$f_R\left(\frac{j}{n}\right) = \frac{1}{n}, \qquad j = 1, 2, .., n.$$

   Applying (5.29) or (5.52) to $M_R(t)$ in IV.(4.58):

$$C_R(t) = \frac{\exp[i(1 + 1/n)t] - \exp[it/n]}{n(\exp[it/n] - 1)}. \tag{5.32}$$

   This result generalizes to an analogously defined discrete rectangular distribution on $[a, b]$, and then:

$$C_{R_{a,b}}(t) = e^{iat} C_R([b - a]t). \tag{5.33}$$

   A special case of this distribution with $n = 1$ and all probability mass at x_0 is sometimes called a **delta function** and denoted δ_{x_0}. By definition: $\delta_{x_0}(x_0) = 1$, and $\delta_{x_0}(x) = 0$ for $x \neq x_0$. Thus:

$$C_\delta(t) = e^{itx_0}. \tag{5.34}$$

2. **Binomial Distribution**

   The binomial distribution has parameters n and $0 < p < 1$ and is defined by the density:

$$f_B(j) = \binom{n}{j} p^j (1 - p)^{n-j}, \qquad j = 0, 1, .., n.$$

   Applying (5.29) or (5.52) to $M_B(t)$ in IV.(4.63):

$$C_B(t) = \left(1 + p(e^{it} - 1)\right)^n. \tag{5.35}$$

3. **Geometric Distribution**

   The geometric distribution has parameter $0 < p < 1$ and is defined by the density:

$$f_G(j) = p(1 - p)^j, \qquad j = 0, 1, 2, ..$$

   The geometric summation in (5.29) is absolutely convergent for all p and thus:

$$C_G(t) = \frac{p}{1 - (1 - p)e^{it}}. \tag{5.36}$$

   In contrast, $M_G(t)$ of IV.(4.64) exists only for t with $(1 - p)e^t < 1$, or $t < -\ln(1 - p)$.

4. **Negative Binomial Distribution**

The negative binomial distribution has parameters $k \in \mathbb{N}$ and $0 < p < 1$ and is defined by the density:

$$f_{NB}(j) = \binom{j+k-1}{k-1} p^k (1-p)^j, \qquad j = 0, 1, 2, ..$$

The summation in (5.29) is again absolutely convergent for all p and thus:

$$C_{NB}(t) = \left(\frac{p}{1-(1-p)e^{it}} \right)^k, \tag{5.37}$$

while in contrast, $M_{NB}(t)$ of (4.66) exists only for $t < -\ln(1-p)$.

5. **Poisson Distribution**

The Poisson distribution has parameter $\lambda > 0$ and is defined by the density:

$$f_P(j) = e^{-\lambda} \frac{\lambda^j}{j!}, \qquad j = 0, 1, 2, ...$$

Applying (5.29) or (5.52) to $M_P(t)$ in IV.(4.68):

$$C_P(t) = \exp[\lambda(e^{it} - 1)]. \tag{5.38}$$

5.4.2 Continuous Distributions

1. **Continuous Uniform Distribution**

The continuous uniform distribution is defined on $x \in [a, b]$ by density:

$$f_U(x) = \begin{cases} 1/(b-a), & x \in [a, b], \\ 0, & x \notin [a, b]. \end{cases}$$

Applying (5.31) or (5.52) to $M_U(t)$ in IV.(4.71):

$$C_U(t) = \frac{e^{ibt} - e^{iat}}{t(b-a)}. \tag{5.39}$$

2. **Exponential Distribution** and **Gamma Distribution**

The exponential distribution has parameter $\lambda > 0$ and is defined by the density:

$$f_E(x) = \begin{cases} 0, & x < 0, \\ \lambda e^{-\lambda x}, & x \geq 0. \end{cases}$$

The exponential is a special case of the more general gamma distribution defined with parameters $\lambda > 0$ and $\alpha > 0$ by:

$$f_\Gamma(x) = \begin{cases} 0, & x < 0, \\ \lambda^\alpha x^{\alpha-1} e^{-\lambda x}/\Gamma(\alpha), & x \geq 0, \end{cases}$$

where the gamma function $\Gamma(\alpha)$ is defined in (2.30).

Applying (5.31) or (5.52) to $M_\Gamma(t)$ in IV.(4.73), the characteristic functions are then defined for all t by:

$$C_\Gamma(t) = \left(1 - \frac{it}{\lambda}\right)^{-\alpha}, \qquad C_E(t) = \left(1 - \frac{it}{\lambda}\right)^{-1}. \tag{5.40}$$

In contrast, $M_\Gamma(t)$ and $M_E(t)$ exist only for $t < \lambda$.

3. **Beta Distribution**

The beta distribution has parameters $v > 0$, $w > 0$, and is defined on the interval $[0, 1]$ by the density function:

$$f_\beta(x) = \frac{x^{v-1}(1-x)^{w-1}}{B(v,w)},$$

where the beta function $B(u, v)$ is defined:

$$B(v,w) = \int_0^1 y^{v-1}(1-y)^{w-1}dy. \tag{5.41}$$

Applying (5.52) to $M_\beta(t)$ in IV.(4.76), or (5.31) using the Book IV derivation for $M_\beta(t)$, the characteristic function is given by:

$$C_\beta(t) = 1 + \sum_{n=1}^{\infty}\prod_{j=0}^{n-1}\left(\frac{v+j}{v+w+j}\right)\frac{(it)^n}{n!}. \tag{5.42}$$

4. **Cauchy Distribution**

The Cauchy density has parameters $\gamma > 0$ and x_0, and is defined by:

$$f_C(x) = \frac{1}{\pi\gamma}\frac{1}{1+\left(\frac{x-x_0}{\gamma}\right)^2},$$

while the **standard Cauchy distribution** is parameterized with $x_0 = 0$ and $\gamma = 1$. The Cauchy distribution has no finite moments and hence no moment generating function.

However, the characteristic function can be calculated based on contour integration and an important result called **Cauchy's integral theorem,** named for **Augustin-Louis Cauchy** (1789–1857), which is outside the scope of our analysis. Alternatively, we can use Fourier inversion as seen in the next section to produce:

$$C_C(t) = \exp\left(ix_0t - \gamma\,|t|\right). \tag{5.43}$$

The Fourier inversion approach using (5.56) is left as an exercise.

Note that $C_C(t)$ is not differentiable at $t = 0$, consistent with Proposition 5.25 for a distribution with no finite moments.

5. **Normal Distribution**

The normal distribution has parameters $\sigma > 0$ and μ and is defined by the density:

$$f_N(x) = \frac{1}{\sigma\sqrt{2\pi}}\exp\left(-\frac{(x-\mu)^2}{2\sigma^2}\right),$$

while the **standard normal distribution** is parameterized with $\mu = 0$ and $\sigma = 1$.

Applying (5.52) to $M_N(t)$ in IV.(4.78), or (5.31) using the Book IV derivation for $M_N(t)$, the characteristic function is given by:

$$C_N(t) = \exp\left(i\mu t - \frac{1}{2}\sigma^2t^2\right). \tag{5.44}$$

6. **Multivariate Normal Distribution**

   The density function of the multivariate normal distribution, when it exists, is defined in (6.6) with symmetric, positive definite $n \times n$ matrix C, and vector $\mu = (\mu_1, \mu_2, ..., \mu_n)$. By positive definite is meant that $x^T Cx \geq 0$ and $x^T Cx = 0$ if and only if $x = 0$. As a function of $x = (x_1, x_2, ..., x_n)$, this density is given by:

$$f_X(x) = (2\pi)^{-n/2} [\det C]^{-1/2} \exp\left[-\frac{1}{2}(x-\mu)^T C^{-1}(x-\mu)\right].$$

   By (6.9), C is the covariance matrix of the random vector $(X_1, X_2, ..., X_n)$ and defined by $C_{ij} = E[(X_i - \mu_i)(X_j - \mu_j)]$, and μ is the vector of first moments, $\mu_j = E[X_j]$.

   Generalizing (5.44), it will be proved in Proposition 6.14 by a change of variables that:

$$C_X(t) = \exp\left[i\mu \cdot t - \frac{1}{2}t^T Ct\right]. \tag{5.45}$$

   Here $t^T Ct$ as above denotes the matrix product with t^T the row vector transpose of the column vector t.

   As in the one-dimensional case, $C_X(t) = M_X(it)$ where $M_X(t)$ denotes the moment generating function of X in (6.7). We prove this general result in (5.67) of Proposition 5.35.

7. **Lognormal Distribution** The lognormal distribution has parameters $\sigma > 0$ and μ and is defined on $[0, \infty)$ by:

$$f_L(x) = \frac{1}{\sigma x\sqrt{2\pi}} \exp\left(-(\ln x - \mu)^2 / (2\sigma^2)\right). \tag{5.46}$$

   Although $C_L(t)$ exists in theory, there is no known closed form version of this function.

Example 5.21 (Multivariate normal distribution) *Since we will not study the multivariate normal distribution until Chapter 6, we consider a simpler example here. This result will be needed for the proof of Bochner's theorem, and will hopefully make the general result more accessible. The reader is encouraged to review the summary of matrix results in Section 6.1.1 if needed.*

Let $\mu = 0$ and $C = \lambda^{-1}I$, with I the $n \times n$ identity matrix. Then $C^{-1} = \lambda I$, $\det C = \lambda^{-n}$, and $x^T C^{-1}x = \sum_{i=1}^n \lambda x_i^2$, so the above multivariate normal density function is:

$$\begin{aligned} f_N(x) &= \frac{\lambda^{n/2}}{(2\pi)^{n/2}} \exp\left[-\sum\nolimits_{i=1}^n \lambda x_i^2 \Big/ 2\right] \\ &= \prod\nolimits_{i=1}^n \frac{\lambda^{1/2}}{(2\pi)^{1/2}} \exp\left[-\lambda x_i^2/2\right] \\ &= \prod\nolimits_{i=1}^n f_N(x_i), \end{aligned}$$

where $f_N(x_i)$ is the density of the 1-dimensional normal variate in item 5 above, with $\mu_i = 0$ and $\sigma_i = \lambda^{-1/2}$ for all i.

By Corollary 1.57, $f_N(x)$ is the density function of a random vector X with independent normal variates, $X = (X_1, X_2, ..., X_n)$, each with $\mu_i = 0$ and $\sigma_i = \lambda^{-1/2}$. Thus by

Proposition 5.33 and (5.44):

$$\begin{aligned} C_X(t) &= \prod\nolimits_{i=1}^{n} \exp\left(-\frac{1}{2}\lambda^{-1}t_i^2\right) \\ &= \exp\left[-\frac{1}{2}t^T Ct\right], \end{aligned}$$

as in (5.45).

5.5 Properties of Characteristic Functions on $\mathbb{R}$

As noted in Remark 5.20 for $n = 1$, the characteristic function of X is identical to the Fourier transform of the measure λ_F on $\mathbb{R}$ induced by F, the distribution function of X. In cases where λ_F has a density function $f(x)$ in the sense of Definition 1.9, or equivalently by Proposition 1.34, when $F(x)$ has a density function $f(x)$, the characteristic function of X is identical to $\widehat{f}(t)$, the Fourier transform of $f(x)$. Thus all the hard work on Fourier transforms done in Book V makes it possible to simply quote the most important properties of this transform, adapting the language to the context of characteristic functions.

One result on characteristic functions on $\mathbb{R}$ will be deferred to the end of this chapter. There we return to the as yet unproved Proposition IV.4.61. That result stated that if a distribution function $F(x)$ on $\mathbb{R}$ has moments of all orders, and the associated power series converges absolutely on an open interval about 0, then $F(x)$ is uniquely determined by these moments.

We begin with an exercise:

Exercise 5.22 (Affine transformations of X and $C_X(t)$) *Given a random variable X defined on $(\mathcal{S}, \mathcal{E}, \lambda)$ with distribution function F, let $Y = aX + b$ for $a, b \in \mathbb{R}$. Prove that:*

$$C_Y(t) = e^{ibt}C_X(at), \tag{5.47}$$

using two approaches:

1. *Use properties from Proposition V.3.42 of the defining integral on $\mathcal{S}$ in (5.26).*
2. *Letting G denote the distribution function of Y, define $C_Y(t)$ and $C_X(t)$ in terms of the induced measures λ_F and λ_G as in (5.28), and then use change of variable results from Section V.4.2. Hint: Consider the transformation $T : (\mathbb{R}, \mathcal{B}(\mathbb{R}), \lambda_F) \to (\mathbb{R}, \mathcal{B}(\mathbb{R}), \lambda_G)$ defined by $T : x \to ax + b$. Check that $\lambda_G = (\lambda_F)_T$ in the notation of Definition 1.28, then apply Proposition 5.15.*

The next result has no counterpart in Book V on Fourier transforms, since independence is strictly a probability space notion. See Exercise 5.24 for an alternative proof of this result that utilizes Proposition 4.7, and see the proof of Proposition 5.33 that generalizes this alternative proof.

Proposition 5.23 (Characteristic functions and independent sums) *Let $\{X_i\}_{i=1}^n$ be independent random variables on a probability space $(\mathcal{S}, \mathcal{E}, \lambda)$. Then with $X = \sum_{i=1}^n X_i$:*

$$C_X(t) = \prod\nolimits_{i=1}^{n} C_{X_i}(t). \tag{5.48}$$

Proof. *By induction it is only necessary to prove this result for* $n = 2$, *and simplifying notation let* $Z = X + Y$. *With* λ_Z *the Borel measure on* $\mathbb{R}$ *of (1.21) induced by* Z, *then by (5.28):*

$$C_Z(t) \equiv \int_{\mathbb{R}} e^{izt} d\lambda_Z. \tag{1}$$

Define the random vector $W = (X, Y)$ *on* $(\mathcal{S}, \mathcal{E}, \lambda)$*:*

$$W : (\mathcal{S}, \mathcal{E}, \lambda) \to \left(\mathbb{R}^2, \mathcal{B}\left(\mathbb{R}^2\right), \lambda_X \times \lambda_Y\right),$$

so $W(s) = (X(s), Y(s))$, *where the Borel space on the right is the product space of Chapter I.7 of* $(\mathbb{R}, \mathcal{B}(\mathbb{R}), \lambda_X)$ *and* $(\mathbb{R}, \mathcal{B}(\mathbb{R}), \lambda_Y)$. *Now define the continuous and hence by Proposition V.2.12, Borel measurable transformation* $T(x, y) = x + y$*:*

$$T : \left(\mathbb{R}^2, \mathcal{B}\left(\mathbb{R}^2\right), \lambda_X \times \lambda_Y\right) \to (\mathbb{R}, \mathcal{B}(\mathbb{R}), \lambda_Z).$$

By Definition 1.28, the transformation T *induces a measure* $(\lambda_X \times \lambda_Y)_T$ *on the range space. For* $A \in \mathcal{B}(\mathbb{R})$, *since* $(TW)^{-1} = W^{-1}T^{-1}$ *and* $TW = Z$*:*

$$(\lambda_X \times \lambda_Y)_T (A) \equiv \lambda_X \times \lambda_Y \left[T^{-1}(A)\right] \equiv \lambda \left[W^{-1}T^{-1}(A)\right] \equiv \lambda_Z(A),$$

and so:

$$(\lambda_X \times \lambda_Y)_T = \lambda_Z. \tag{2}$$

Hence by Proposition 5.15, the integral in (1) *can be expressed using* (2)*:*

$$C_Z(t) \equiv \int_{\mathbb{R}} e^{izt} d\lambda_Z = \int_{\mathbb{R}^2} e^{i(x+y)t} d\left(\lambda_X \times \lambda_Y\right).$$

As $\left|e^{i(x+y)t}\right| = 1$ *by (5.18),* $\left|e^{i(x+y)t}\right|$ *is integrable over* $\mathbb{R}^2$ *since* $\lambda_X \times \lambda_Y$ *is a probability measure, and thus* $e^{i(x+y)t}$ *is integrable over* $\mathbb{R}^2$ *by Remark V.7.2. Fubini's theorem of Proposition 5.16 then applies to obtain:*

$$C_Z(t) = \int_{\mathbb{R}} e^{ixt} d\lambda_X \int_{\mathbb{R}} e^{iyt} d\lambda_Y = C_X(t) C_Y(t).$$

■

Exercise 5.24 (Characteristic functions and independent sums) *Prove Proposition 5.23 using Proposition 4.7, by proving that:*

$$C_Z(t) \equiv E\left[e^{iXt}e^{iYt}\right] = E\left[e^{iXt}\right] E\left[e^{iYt}\right] \equiv C_X(t)C_Y(t).$$

Hint: Since complex-valued, we cannot by Proposition II.3.56 assert independence of e^{iXt} *and* e^{iYt} *from independence of* X *and* Y, *nor can we assert that such independence would assure the middle equality above by Proposition 4.7. Instead use Euler's formula in (5.17), so:*

$$e^{iXt}e^{iYt} = \cos X \cos Y - \sin X \sin Y + i \left[\cos X \sin Y + \sin X \cos Y\right].$$

Take expectations, apply Proposition 4.7 after justifying its applicability, and reassemble. See also the proof of Proposition 5.33.

We now summarize properties of $C_X(t)$ in a series of propositions that are directly obtainable from Book V. To make the connections with the Book V results more transparent, we express $C_X(t)$ as a Lebesgue-Stieltjes integral on $\mathbb{R}$ as in (5.28), rather than as an integral on $\mathcal{S}$ as in (5.26). Then by (1.22), this result can equivalently be stated in terms of λ_F.

The first result addresses smoothness of the characteristic function $C_X(t)$, which is related to the existence of moments.

Proposition 5.25 (Smoothness of $C_X(t)$) *Given a random variable X defined on a probability space $(\mathcal{S}, \mathcal{E}, \lambda)$ with induced Borel measure λ_X by (1.21).*

1. *$C_X(t)$ is uniformly continuous on $\mathbb{R}$, and $C_X(0) = 1$.*
2. *If $\mu'_n \equiv E[X^n]$ exists for positive integer n:*

$$\int_{-\infty}^{\infty} |x^n| \, d\lambda_X < \infty,$$

then $C_X(t)$ is differentiable up to order n.

Further for $1 \leq k \leq n$:

$$C_X^{(k)}(t) = i^k \int_{-\infty}^{\infty} x^k e^{itx} d\lambda_X, \tag{5.49}$$

$C_X^{(k)}(t)$ is uniformly continuous on $\mathbb{R}$, and:

$$C_X^{(k)}(0) = i^k \mu'_k. \tag{5.50}$$

3. *If $\int_{-\infty}^{\infty} e^{|sx|} d\mu_X < \infty$ for any $s \neq 0$, then $C_X(t)$ is infinitely differentiable, and for $|t| \leq |s|$:*

$$C_X(t) = \sum\nolimits_{j=0}^{\infty} \frac{(it)^j}{j!} \mu'_k. \tag{5.51}$$

Thus for $|t| \leq |s|$:

$$C_X(t) = M_X(it). \tag{5.52}$$

Proof. *This is a restatement of Proposition V.7.10, with two additions. First, $C_X(0) = 1$ is true by definition.*

For (5.52), note that (5.51) asserts that this series is absolutely convergent for $|t| \leq |s|$, and thus since $|i| = 1$, so too is the series:

$$\sum\nolimits_{j=0}^{\infty} \frac{t^j}{j!} \mu'_k.$$

Proposition IV.4.59 then assures that $M_X(t)$ exists on this interval, and is given by this series. The identity in (5.52) now follows by substitution. ■

Turning to the behavior of $C_X(t)$ at infinity, the associated rate of decay is related to the existence of a density function for $\lambda_F = \lambda_X$, and its smoothness. The first result is the **Riemann-Lebesgue lemma,** named for **Bernhard Riemann** (1826–1866) and **Henri Lebesgue** (1875–1941), and states that if $F(x)$ has a density function, equivalently by Proposition 1.34, if λ_F has a density function, then $|C_X(t)| \to 0$ as $t \to \pm\infty$.

For the second result, the expression:

$$f(x) = o(g(x)) \text{ as } x \to \infty,$$

for $g(x) > 0$ means by Definition 7.2 that:

$$\frac{|f(x)|}{g(x)} \to 0 \text{ as } x \to \infty.$$

In words, this is read "$f(x)$ **is little-o of** $g(x)$ **as** $x \to \infty$."

Proposition 5.26 (Behavior of $C_X(t)$ at $\pm\infty$) *Given a random variable X defined on a probability space $(\mathcal{S}, \mathcal{E}, \lambda)$ with distribution function $F(x)$ and induced Borel measure λ_X:*

1. ***(Riemann-Lebesgue Lemma)*** *If λ_X has a density function $f(x)$ in the sense of Definition 1.9, or equivalently by Proposition 1.34, if $F(x)$ has a density function $f(x)$, then:*
$$|C_X(t)| \to 0 \text{ as } t \to \pm\infty\ . \tag{5.53}$$

2. *If the density function $f(x)$ is differentiable up to order n and $f^{(k)}(x)$ is integrable for $k \leq n$, then:*
$$|C_X(t)| = o(|t|^{-n}) \text{ as } t \to \pm\infty\ . \tag{5.54}$$

Proof. *This is a restatement of Proposition V.7.15.* ■

The next two results address invertibility of $C_X(t)$, by which is meant to recover λ_X from $C_X(t)$, or equivalently, to recover $F(x)$ from $C_X(t)$ by (1.18).

The first result is completely general, and of theoretical interest. The second result constructs a continuous density function $f(x)$ for λ_X under the assumption that $C_X(t)$ is Lebesgue integrable. By Proposition 1.34, this is also a density function for $F(x)$.

Proposition 5.27 (Inversion of $C_X(t)$) *Let λ_X be a probability measure on $\mathbb{R}$ and $C_X(t)$ the associated characteristic function.*

For $b > a$:

$$\lim_{T\to\infty} (2\pi)^{-1} \int_{-T}^{T} \frac{e^{-iat} - e^{-ibt}}{it} C_X(t)dm = \lambda_X[(a,b)] + \frac{1}{2}\lambda_X[\{a,b\}], \tag{5.55}$$

where $\lambda_X[\{a,b\}]$ denotes the measure of these points.
Proof. *This is a restatement of Proposition V.7.18.* ■

Proposition 5.28 (Inversion of integrable $C_X(t)$) *If $C_X(t)$ is Lebesgue integrable, then λ_X has a continuous density function $f(x)$:*

$$f(x) = (2\pi)^{-1} \int_{-\infty}^{\infty} C_X(t)e^{-ixt}dt. \tag{5.56}$$

Hence for all $A \in \mathcal{B}(\mathbb{R})$:

$$\lambda_X[A] = \int_A f(x)dm. \tag{5.57}$$

Proof. *This is a restatement of Proposition V.7.31.* ■

Turning to uniqueness, the next result states that the distribution function of a random variable is uniquely identified by its characteristic function. For this statement we use the alternate notation of $C_F(t)$.

Proposition 5.29 (Uniqueness of $C_X(t)$: If $C_F(t) = C_G(t)$ then $F = G$) *Let $F(x)$ and $G(x)$ be distribution functions on $\mathbb{R}$ with induced Borel measures λ_F and λ_G.*

If $C_F(t) = C_G(t)$ for all t, then for all Borel sets $A \in \mathcal{B}(\mathbb{R})$:

$$\lambda_F(A) = \lambda_G(A). \tag{5.58}$$

Thus for all x:

$$F(x) = G(x). \tag{5.59}$$

Proof. *This is a restatement of Corollary V.7.23, except for (5.59), which follows from (5.58) with $A = (-\infty, x]$.* ■

The final result addresses the connection between weak convergence of finite measures and pointwise convergence of the associated characteristic functions, and states that these are equivalent. This result is known as **Lévy's continuity theorem,** named for **Paul Lévy** (1886–1971).

We state the result in terms of weak convergence of distribution functions, but recall that $F_n \Rightarrow F$ and $\lambda_{F_n} \Rightarrow \lambda_F$ and $\lambda_{X_n} \Rightarrow \lambda_X$ are equivalent statements by Definition 3.22 and (1.22), and are all equivalent to convergence in distribution, $X_n \rightarrow_d X$.

Proposition 5.30 (Lévy's continuity theorem for $C_F(t)$) *Let* $\{F_n\}_{n=1}^{\infty}$, F *be distribution functions on* $\mathbb{R}$, *and* $\{C_{F_n}\}_{n=1}^{\infty}$, C_F *the associated characteristic functions.*

Then $F_n \Rightarrow F$ *if and only if* $C_{F_n}(t) \rightarrow C_F(t)$ *for all* t.

Further, given $\{F_n\}_{n=1}^{\infty}$ *and associated characteristic functions* $\{C_{F_n}\}_{n=1}^{\infty}$, *if* $C_{F_n}(t) \rightarrow \phi(t)$ *for all* t, *where* $\phi(t)$ *is continuous at* $t = 0$, *then there exists a distribution function* F *so that* $F_n \Rightarrow F$ *and* $\phi(t) = C_F(t)$.

Proof. *This is a restatement of Proposition V.7.35 and Corollary V.7.36, which should have been attributed to Paul Lévy in that book.* ■

5.6 Properties of Characteristic Functions on $\mathbb{R}^n$

In this section, we largely follow the outline of the previous section on $\mathbb{R}$, but of necessity will need to do more work since we have no Book V Fourier transform theory on $\mathbb{R}^n$ to restate. Fortunately, with the aid of the one-dimensional theory, a number of essential results of the n-dimensional theory can be derived with the aid of change of variables of Proposition 5.15, and/or Fubini's theorem of Proposition 5.16.

We begin with a result on the characteristic functions of the marginal distributions of a given distribution function.

Proposition 5.31 ($C_{X_I}(s)$ from $C_X(t)$) *Let* $X = (X_1, X_2, ..., X_n)$ *be a random vector defined on a probability space* $(\mathcal{S}, \mathcal{E}, \lambda)$ *with joint distribution function* $F(x_1, x_2, ..., x_n)$, *and for* $I = \{i_1, ..., i_m\} \subset \{1, 2, ..., n\}$, *let* $F_I(x_I) \equiv F_I(x_{i_1}, x_{i_2}, ..., x_{i_m})$ *denote the associated marginal distribution function defined on* $\mathbb{R}^m$ *by (1.34).*

Then the characteristic function $C_{F_I}(s)$ *of* F_I *is defined on* $s \in \mathbb{R}^m$ *in terms of the characteristic function* $C_F(t)$ *of* X *defined on* $t \in \mathbb{R}^n$ *by:*

$$C_{F_I}(s) = C_F(\hat{s}), \tag{5.60}$$

where $\hat{s} \in \mathbb{R}^n$ *is given by:*

$$\hat{s}_{i_j} = s_j, \quad \hat{s}_{i_k} = 0, \text{ otherwise.}$$

Proof. *Define the random vector* $X_I \equiv (X_{i_1}, X_{i_2}, ..., X_{i_m})$, *and recall that by Proposition II.3.36,* $F_I(x_I)$ *is the distribution function of* X_I.

Then by (5.27):

$$C_F(\hat{s}) \equiv E\left[e^{i\hat{s}\cdot X}\right] = E\left[e^{is\cdot X_I}\right] \equiv C_{F_I}(s),$$

and the proof is complete. ■

Turning to affine transformations, the next result is analogous to Exercise 5.3 for the moment generating function. For variety, we provide an alternative proof in terms of Lebesgue-Stieltjes integrals.

Proposition 5.32 (Affine Transforms of X and $C_X(t)$) *Given a random vector $X \equiv (X_1, X_2, ..., X_n)$ defined on a probability space $(\mathcal{S}, \mathcal{E}, \lambda)$, a matrix $A : \mathbb{R}^n \to \mathbb{R}^m$, and fixed $b \in \mathbb{R}^m$, define the random vector $Y : \mathcal{S} \to \mathbb{R}^m$ by $Y = AX + b$. Then for $t \in \mathbb{R}^m$:*

$$C_Y(t) = e^{ib\cdot t} C_X(A^T t), \tag{5.61}$$

where $A^T : \mathbb{R}^m \to \mathbb{R}^n$ is the transpose of A defined by $a_{ij}^T \equiv a_{ji}$.

Proof. *Since $T : (\mathbb{R}^n, \mathcal{B}(\mathbb{R}^n)) \to (\mathbb{R}^m, \mathcal{B}(\mathbb{R}^m))$ defined by $Tx = Ax + b$ is continuous, it is Borel measurable by Proposition V.2.12, and thus $Y = TX$ is λ-measurable and a random vector. In detail:*

$$Y^{-1}\left[\mathcal{B}(\mathbb{R}^m)\right] = X^{-1}\left[T^{-1}\mathcal{B}(\mathbb{R}^m)\right] \subset X^{-1}\left[\mathcal{B}(\mathbb{R}^n)\right] \subset \mathcal{E}.$$

As in (1.21), let λ_X and λ_Y denote the probability measures on $\mathcal{B}(\mathbb{R}^n)$ and $\mathcal{B}(\mathbb{R}^m)$ induced by X and Y respectively, and consider the transformation T on probability spaces:

$$T : (\mathbb{R}^n, \mathcal{B}(\mathbb{R}^n), \lambda_X) \to (\mathbb{R}^m, \mathcal{B}(\mathbb{R}^m), \lambda_Y).$$

Then T induces a measure $(\lambda_X)_T$ on the range space by Definition 1.28, and we claim that:

$$(\lambda_X)_T = \lambda_Y.$$

If $A \in \mathcal{B}(\mathbb{R}^m)$, then by definition:

$$(\lambda_X)_T(A) \equiv \lambda_X\left[T^{-1}(A)\right] \equiv \lambda\left[X^{-1}T^{-1}(A)\right] \equiv \lambda_Y(A),$$

since $X^{-1}T^{-1} = (TX)^{-1}$ and $TX = Y$.

Thus by Proposition 5.15:

$$C_Y(t) \equiv \int_{\mathbb{R}^n} e^{iy\cdot t} d\lambda_Y(y) = \int_{\mathbb{R}^n} e^{iTx\cdot t} d\lambda_X(x).$$

Since $Tx = Ax + b$:

$$C_Y(t) = e^{ib\cdot t} \int_{\mathbb{R}^n} e^{iAx\cdot t} d\lambda_X(x).$$

The final step is to recall that $Ax \cdot t = x \cdot A^T t$ from Exercise 5.3, and this obtains (5.61). ■

The next result on the characteristic function of a sum of independent random vectors generalizes Proposition 5.23, but the reader can confirm that the proof there works here with only a change of dimension in the notation. Instead we prove this result by generalizing the proof in Exercise 5.24.

Proposition 5.33 (Characteristic functions and independent sums) *Let $\{X_i\}_{i=1}^m$ be independent random n-vectors defined on a probability space $(\mathcal{S}, \mathcal{E}, \lambda)$. Then with $X \equiv \sum_{i=1}^m X_i$ and $t = (t_1, .., t_n)$:*

$$C_X(t) = \prod\nolimits_{i=1}^m C_{X_i}(t). \tag{5.62}$$

Proof. *By induction, it is only necessary to prove this result for $n = 2$, and simplifying notation let $Z = X + Y$ where $X \equiv (X_1, X_2, ..., X_n)$ and $Y \equiv (Y_1, Y_2, ..., Y_n)$.*

Using (5.17):

$$\begin{aligned} C_Z(t) &= E\left[e^{iX\cdot t} e^{iY\cdot t}\right] \\ &= E\left[\prod\nolimits_{j=1}^n (\cos X_j t_j + i \sin X_j t_j) \prod\nolimits_{j=1}^n (\cos Y_j t_j + i \sin Y_j t_j)\right]. \end{aligned}$$

Every term of this product of $2n$ *factors reflects two partitions of* $(1, ..., n)$ *into* $A \bigcup \tilde{A}$ *and* $B \bigcup \tilde{B}$. *Letting* $|A|$ *denote the number of elements of* A, *etc., the general term in this summation is:*

$$W \equiv i^{|\tilde{A}|+|\tilde{B}|} \prod\nolimits_{j\in A} \cos X_j t_j \prod\nolimits_{j\in \tilde{A}} \sin X_j t_j \prod\nolimits_{j\in B} \cos Y_j t_j \prod\nolimits_{j\in \tilde{B}} \sin Y_j t_j.$$

Independence of X *and* Y *assures by Proposition II.3.56 the independence of the* X_j*-product terms and the* Y_j*-product terms, and thus by Proposition 4.7:*

$$E[W] = i^{|\tilde{A}|+|\tilde{B}|} E\left[\prod\nolimits_{j\in A} \cos X_j t_j \prod\nolimits_{j\in \tilde{A}} \sin X_j t_j\right] E\left[\prod\nolimits_{j\in B} \cos Y_j t_j \prod\nolimits_{j\in \tilde{B}} \sin Y_j t_j\right].$$

By linearity of expectations, which follows from linearity of the integral by Proposition V.3.42:

$$\begin{aligned} C_Z(t) &= \sum\nolimits_{A,B} i^{|\tilde{A}|+|\tilde{B}|} E\left[\prod\nolimits_{j\in A} \cos X_j t_j \prod\nolimits_{j\in \tilde{A}} \sin X_j t_j\right] \\ &\quad \times E\left[\prod\nolimits_{j\in B} \cos Y_j t_j \prod\nolimits_{j\in \tilde{B}} \sin Y_j t_j\right] \\ &= E\left[\prod\nolimits_{j=1}^{n} (\cos X_j t_j + i \sin X_j t_j)\right] E\left[\prod\nolimits_{j=1}^{n} (\cos Y_j t_j + i \sin Y_j t_j)\right] \\ &= C_X(t) C_Y(t). \end{aligned}$$

■

When $X \equiv (X_1, X_2, ..., X_n)$ and $\{X_i\}_{i=1}^n$ are independent random variates, we have the following corollary.

Corollary 5.34 ($C_X(t)$**, independent** $\{X_i\}_{i=1}^n$**)** *Let* $X \equiv (X_1, X_2, ..., X_n)$ *be a random vector defined on a probability space* $(\mathcal{S}, \mathcal{E}, \lambda)$ *with* $\{X_i\}_{i=1}^n$ *independent random variables. Then with* $t = (t_1, t_2, ..., t_n)$*:*

$$C_X(t) = \prod\nolimits_{i=1}^{n} C_{X_i}(t_i). \tag{5.63}$$

Proof. *Define the random vector* $Y_i \equiv (Y_{i1}, Y_{i2}, ..., Y_{in})$ *by* $Y_{ii} = X_i$, *and* $Y_{ik} = 0$ *for* $k \neq i$. *It is an exercise to check that* $\{Y_i\}_{i=1}^n$ *are independent random vectors. Hint: Recall Proposition II.3.51, that one only needs to check Definition 1.53 for relatively simple sets in* $\mathcal{B}(\mathbb{R}^n)$.

Thus by Proposition 5.33, with $t = (t_1, .., t_n)$*:*

$$C_X(t) = \prod\nolimits_{i=1}^{n} C_{Y_i}(t),$$

and the proof is complete since $C_{Y_i}(t) = C_{X_i}(t_i)$ *by definition of* Y_i. ■

The next result addresses smoothness properties of $C_X(t)$, generalizing Proposition 5.25. Again, the degree of smoothness is determined by the existence of moments.

Proposition 5.35 (Smoothness of $C_X(t)$**)** *Given a random vector* $X \equiv (X_1, X_2, ..., X_n)$ *defined on a probability space* $(\mathcal{S}, \mathcal{E}, \lambda)$ *with induced Borel measure* λ_X *by (1.21).*

1. $C_X(t)$ *is uniformly continuous on* $\mathbb{R}^n$ *and* $C_X(0) = 1$.

2. *If* $E[\|X\|_1^k] < \infty$ *for* $\|X\|_1 \equiv \sum_{j=1}^n |X_j|$ *and positive integer* k*:*

$$\int_{\mathbb{R}^n} \|x\|_1^k \, d\lambda_X < \infty,$$

   then $C_X(t)$ *is differentiable up to order* k.

Further, given $(m_1, ..., m_n)$ *with* $\sum_{j=1}^{n} m_j = m \leq k$*:*

$$\frac{\partial^m C_X(t)}{\partial t_1^{m_1} \cdots \partial t_n^{m_n}} = i^m \int_{\mathbb{R}^n} \prod\nolimits_{j=1}^{n} x_j^{m_j} e^{it \cdot x} d\lambda_X, \tag{5.64}$$

is uniformly continuous, and:

$$\left. \frac{\partial^m C_X(t)}{\partial t_1^{m_1} \cdots \partial t_n^{m_n}} \right|_{t=0} = i^m \mu'_{(m_1,...,m_n)}, \tag{5.65}$$

where $t = 0$ *means* $t_j = 0$ *for all* j*, and* $\mu'_{(m_1,...,m_n)}$ *is defined in (5.6).*

3. *If* $E[e^{s\|X\|_1}] < \infty$ *for any* $s > 0$*, then* $C_X(t)$ *is infinitely differentiable, and for* $|t| \leq s$*:*

$$C_X(t) = \sum\nolimits_{(m_1,...,m_n)} \mu'_{(m_1,...,m_n)} \prod\nolimits_{j=1}^{n} \frac{(it_j)^{m_j}}{m_j!}, \tag{5.66}$$

where the summation is over all nonnegative integer n-tuples $(m_1, ..., m_n)$*.*

Thus for $|t| < s$*:*

$$C_X(t) = M_X(it). \tag{5.67}$$

Proof. *We address these statements in turn.*

1. *First,* $C_X(0) = 1$ *by definition. Then by the triangle inequality of Proposition V.7.3:*

$$|C_X(t+h) - C_X(t)| \leq \int_{\mathbb{R}^n} \left| e^{ih \cdot x} - 1 \right| d\lambda_X, \tag{1}$$

noting that $\left| e^{ix} \right| = 1$ *for all* x *by (5.18) obtains* $\left| e^{i(t+h) \cdot x} - e^{it \cdot x} \right| = \left| e^{ih \cdot x} - 1 \right|$ *independent of* t*. Now by continuity,* $e^{ih \cdot x} - 1 \to 0$ *as* $h = (h_1, ..., h_n) \to 0$ *for all* $x \in \mathbb{R}^n$*, and* $\left| e^{ih \cdot x} - 1 \right| \leq 2$ *which is integrable, so Lebesgue's dominated convergence theorem of Proposition V.7.4 obtains as* $h \to 0$*:*

$$|C_X(t+h) - C_X(t)| \to 0.$$

This convergence is independent of t *by (1), and this proves uniform continuity.*

2. *First, by (4.32):*

$$\|x\|_1^k \equiv \sum\nolimits_{m=k} c_{\bar{m}} \prod\nolimits_{j=1}^{n} |x_j|^{m_j},$$

where the summation is over all n-tuples $\bar{m} = (m_1, ..., m_n)$ *with* $m \equiv \sum_{j=1}^{n} m_j$*, and* $\{c_{\bar{m}}\}$ *are positive constants. Now given* $(m_1, ..., m_n)$ *with* $m < k$*:*

$$\prod\nolimits_{j=1}^{n} |x_j|^{m_j} \leq \max\left(1, \prod\nolimits_{j=1}^{n} |x_j|^{m'_j}\right),$$

where $(m'_1, ..., m'_n) \geq (m_1, ..., m_n)$ *componentwise and* $m' \equiv \sum_{j=1}^{n} m'_j = k$*. Thus* $\int_{\mathbb{R}^n} \|x\|_1^k d\lambda_X < \infty$ *assures that for all* $m \leq k$*:*

$$\int_{\mathbb{R}^n} \prod\nolimits_{j=1}^{n} |x_j|^{m_j} d\lambda_X < \infty. \tag{2}$$

In other words, $\mu'_{(m_1,...,m_n)}$ *exists for all* $(m_1, ..., m_n)$ *with* $\sum_{j=1}^{n} m_j \leq k$*.*

We prove (5.64) and uniform continuity by induction, noting that this result is true by item 1 when $m = 0$. *So assume that (5.64) is satisfied for given* $(m_1, ..., m_n)$ *with associated* $m < k$, *and that* $\frac{\partial^m C_F(t)}{\partial t_1^{m_1} ... \partial t_n^{m_n}}$ *is uniformly continuous. Defining* $G(t) \equiv \frac{\partial^m C_F(t)}{\partial t_1^{m_1} ... \partial t_n^{m_n}}$ *for notational simplicity, we prove that* $\frac{\partial G}{\partial t_l}$ *satisfies (5.64) and is uniformly continuous. Let* t' *be defined with* $t'_l = t_l + h$, *and* $t'_j = t_j$ *otherwise. Then by the induction assumption:*

$$\frac{G(t') - G(t)}{h} = i^m \int_{\mathbb{R}^n} \prod_{j=1}^{n} x_j^{m_j} e^{it \cdot x} \frac{e^{ihx_l} - 1}{h} d\lambda_X. \tag{3}$$

By the bound proved in Proposition V.7.5:

$$\left| \frac{e^{ihx_l} - 1}{h} \right| \leq |x_l|,$$

and since $m + 1 \leq k$, *the triangle inequality of Proposition V.7.3 obtains by* (2)*:*

$$\left| \frac{G(t') - G(t)}{h} \right| \leq \int_{\mathbb{R}^n} |x_l| \prod_{j=1}^{n} |x_j|^{m_j} d\lambda_X < \infty.$$

Lebesgue's dominated convergence theorem now applies to (3)*, and since* $\frac{e^{ihx_l} - 1}{h} \to ix_l$ *pointwise, (5.64) is true for* $(m_1, ..., m_l + 1, ..., m_n)$.

We leave as an exercise that this partial derivative is uniformly continuous, using the approach of item 1, while (5.65) follows from (5.64).

3. *Since* $|i| = 1$, (2) *of the proof of Proposition 5.9 obtains that if* $|t_j| \leq s$ *for all* j, *then for all* N:

$$\left| \sum_{m \leq N} \prod_{j=1}^{n} \frac{(it_j)^{m_j} X_j^{m_j}}{m_j!} \right| \leq e^{s\|X\|_1}. \tag{4}$$

Similarly, the derivation of (5.9) obtains:

$$\begin{aligned} e^{it \cdot X} &= \sum_{(m_1, ..., m_n)} \prod_{j=1}^{n} \frac{(it_j)^{m_j} X_j^{m_j}}{m_j!} \\ &= \lim_{N \to \infty} \sum_{m \leq N} \prod_{j=1}^{n} \frac{(it_j)^{m_j} X_j^{m_j}}{m_j!}. \end{aligned}$$

Since the upper bounding function in (4) *is integrable by assumption, Lebesgue's dominated convergence theorem of Proposition V.7.4 obtains that if* $|t_j| \leq s$ *for all* j:

$$\begin{aligned} C_X(t) &\equiv \lim_{N \to \infty} \sum_{m \leq N} \int_{\mathcal{S}} \prod_{j=1}^{n} \frac{(it_j)^{m_j} X_j^{m_j}}{m_j!} d\lambda \\ &= \lim_{N \to \infty} \sum_{m \leq N} \mu'_{(m_1, ..., m_n)} \prod_{j=1}^{n} \frac{(it_j)^{m_j}}{m_j!} \\ &= \sum_{(m_1, ..., m_n)} \mu'_{(m_1, ..., m_n)} \prod_{j=1}^{n} \frac{(it_j)^{m_j}}{m_j!}. \end{aligned}$$

Finally, $M_X(t)$ *exists for* $|t| < s$ *by Corollary 5.10, so (5.67) then follows by (5.10).*

■

The next result is the **Riemann-Lebesgue lemma,** generalizing Proposition 5.26. Named for **Bernhard Riemann** (1826–1866) and **Henri Lebesgue** (1875–1941), it states that if $F(x)$ has a density function, equivalently by Proposition 1.34, if $\lambda_X = \lambda_F$ has a density function, then $|C_X(t)| \to 0$ as $t = (t_1, t_2, ..., t_n) \to \pm\infty$. As always, such limits are shorthand for $t_j \to \pm\infty$ for all j.

Proposition 5.36 (Riemann-Lebesgue Lemma: Behavior of $C_X(t)$ at $\pm\infty$) *Let $X \equiv (X_1, X_2, ..., X_n)$ be a random vector defined on $(\mathcal{S}, \mathcal{E}, \lambda)$ with joint distribution function $F(x)$ and induced Borel measure λ_X.*

If λ_X has a density function $f(x)$ in the sense of Definition 1.9, or equivalently by Proposition 1.34, if $F(x)$ has a density function $f(x)$, then:

$$|C_X(t)| \to 0 \text{ as } t \to \pm\infty . \tag{5.68}$$

Proof. *By Definition 1.9 and Proposition 1.37, such $f(x)$ is nonnegative m-a.e. and Lebesgue integrable on $\mathbb{R}^n$. Since $(\mathbb{R}^n, \mathcal{M}_L(\mathbb{R}^n), m)$ is a complete measure space, we can assume that $f(x)$ is nonnegative, and then Proposition V.2.33 obtains an increasing sequence of simple functions $\{\varphi_n(x)\}_{j=1}^{\infty}$, each defined on a finite collection of disjoint right semi-closed rectangles $\prod_{i=1}^n (a_i, b_i]$, each equal to 0 outside a set of finite Lebesgue measure, and $\varphi_n(x) \to f(x)$ m-a.e.*

By Lebesgue's dominated convergence theorem of Corollary V.3.47:

$$\int_{\mathbb{R}^n} \varphi_n(x)dm \to \int_{\mathbb{R}^n} f(x)dm,$$

and for any $\epsilon > 0$ there is an M so that for $n \geq M$:

$$\int_{\mathbb{R}^n} |f(x) - \varphi_n(x)|\, dm < \epsilon/2. \tag{1}$$

Using the reverse triangle inequality in (5.15), then the triangle inequality of Proposition V.7.3, (5.18), and (1), obtains that for any $n \geq M$ and all t:

$$\begin{aligned} &\left| \left| \int_{\mathbb{R}^n} f(x)e^{it\cdot x}dm \right| - \left| \int_{\mathbb{R}^n} \varphi_n(x)e^{it\cdot x}dm \right| \right| \\ &\leq \left| \int_{\mathbb{R}^n} f(x)e^{it\cdot x}dm - \int_{\mathbb{R}^n} \varphi_n(x)e^{it\cdot x}dm \right| \\ &< \epsilon/2. \end{aligned} \tag{2}$$

For fixed $n \geq M$, let $\varphi_n(x) = \sum_{j=1}^{N_n} c_j \chi_{R_j}(x)$ where $R_j = \prod_{i=1}^n (a_i^{(j)}, b_i^{(j)}]$. We apply Fubini's theorem of Proposition 5.16, with the iterated integrals evaluated as Riemann by splitting $e^{it_j x_j}$ by (5.17) and applying Proposition III.2.18:

$$\begin{aligned} \int_{\mathbb{R}^n} \varphi_n(x)e^{it\cdot x}dm &= \sum\nolimits_{j=1}^{N_n} c_j \prod\nolimits_{i=1}^{n} \int_{a_i^{(j)}}^{b_i^{(j)}} e^{it_j x_j}dx_j \\ &= \sum\nolimits_{j=1}^{N_n} c_j \prod\nolimits_{i=1}^{n} \frac{e^{it_j a_j} - e^{it_j b_j}}{it_j}. \end{aligned}$$

Then since $\left|e^{it_j a_j} - e^{it_j b_j}\right| \leq 2$ by (5.18):

$$\left| \int_{\mathbb{R}^n} \varphi_n(x)e^{it\cdot x}dm \right| \leq 2^n \sum\nolimits_{j=1}^{N_n} \frac{|c_j|}{\prod_{i=1}^n |t_j|}. \tag{3}$$

Thus $\left|\int_{\mathbb{R}^n} \varphi_n(x)e^{it\cdot x}dm\right|$ converges to 0 as $t = (t_1, t_2, ..., t_n) \to \pm\infty$, while by (2), $\left|\int_{\mathbb{R}^n} f(x)e^{it\cdot x}dm\right|$ can be made arbitrarily close uniformly in t, and this proves (5.68). ■

The next two results address invertibility of $C_X(t)$, by which is meant to recover λ_X from $C_X(t)$, or equivalently, to recover $F(x)$ from $C_X(t)$ by (1.18).

The first result is completely general, and of theoretical interest. We will use this result in the proof of uniqueness in Proposition 5.39. The second result constructs a continuous density function $f(x)$ for λ_X under the assumption that $C_X(t)$ is Lebesgue integrable. By Proposition 1.34, this is also a density function for $F(x)$.

Recall Definition 3.4, that for a given Borel set $A \subset \mathbb{R}^n$, the **boundary of** A, denoted $\partial(A)$, is defined as:

$$\partial(A) \equiv \{x | x \text{ is a limit point of } A \text{ and } \widetilde{A}\},$$

where $\widetilde{A}$ denotes the complement of A. When $A = \prod_{j=1}^n (a_j, b_j]$ it is an exercise to check that:

$$\partial(A) = \prod\nolimits_{j=1}^n [a_j, b_j] - \prod\nolimits_{j=1}^n (a_j, b_j).$$

Proposition 5.37 (Inversion of $C_X(t)$) *Let F be a distribution function defined on $\mathbb{R}^n$, $\lambda_F = \lambda_X$ the induced probability measure, and $C_X(t)$ the associated characteristic function.*

Given bounded $A = \prod_{j=1}^n (a_j, b_j]$ with $\lambda_X[\partial A] = 0$, and $R_T \equiv \{t | |t_j| \leq T \text{ all } j\}$, then defined as Lebesgue integral:

$$\lambda_X[A] = \lim\nolimits_{T\to\infty} \frac{1}{(2\pi)^n} \int_{R_T} \prod\nolimits_{j=1}^n \frac{e^{-ia_jt_j} - e^{-ib_jt_j}}{it_j} C_X(t) dm. \tag{5.69}$$

Proof. *By (5.28),*

$$\begin{aligned} &\frac{1}{(2\pi)^n} \int_{R_T} \prod\nolimits_{j=1}^n \frac{e^{-ia_jt_j} - e^{-ib_jt_j}}{it_j} C_F(t) dm \qquad (1) \\ &= \frac{1}{(2\pi)^n} \int_{R_T} \int_{\mathbb{R}^n} \prod\nolimits_{j=1}^n \frac{\exp\left[i\left(x_j - a_j\right)t_j\right] - \exp\left[i\left(x_j - b_j\right)t_j\right]}{it_j} d\lambda_X(x) dm(t). \end{aligned}$$

Now:

$$\frac{\exp\left[i\left(x_j - a_j\right)t_j\right] - \exp\left[i\left(x_j - b_j\right)t_j\right]}{it_j} = \int_{(x_j-b_j)}^{(x_j-a_j)} e^{iyt_j} dy,$$

and since $\left|e^{ix}\right| = 1$ for all $x \in \mathbb{R}$ by (5.18), the absolute value of each term of the integrand in (1) is bounded by $b_j - a_j$ by the triangle inequality of Proposition V.7.3.

This integrand is continuous in x_i and t_i, bounded in absolute value by $\prod_{j=1}^n (b_j - a_j)$, and thus integrable. This follows because the Lebesgue integral is over bounded R_T, while the Lebesgue-Stieltjes integral is with respect to a probability measure. So by Fubini's theorem of Proposition 5.16, this iterated integral can be reversed, and the Lebesgue integral reduced to iterated integrals. Denoting this integral by I_T:

$$\begin{aligned} I_T &= \frac{1}{(2\pi)^n} \int_{\mathbb{R}^n} \left[\int_{R_T} \prod\nolimits_{j=1}^n \frac{\exp\left[i\left(x_j - a_j\right)t_j\right] - \exp\left[i\left(x_j - b_j\right)t_j\right]}{it_j} dm(t) \right] d\lambda_X(x) \\ &= \int_{\mathbb{R}^n} \prod\nolimits_{j=1}^n \left[\frac{1}{2\pi} \int_{-T}^{T} \frac{\exp\left[i\left(x_j - a_j\right)t_j\right] - \exp\left[i\left(x_j - b_j\right)t_j\right]}{it_j} dt_j \right] d\lambda_x(x). \end{aligned}$$

Each inner integral is now identical to the one-dimensional expression in the proof of Proposition V.7.18.

Using the same steps and notation there:

$$I_T = \int_{\mathbb{R}^n} \prod\nolimits_{j=1}^n \left[\frac{1}{\pi} \int_{-\infty}^{\infty} \left[S_{(x_j-a_j)}(T) - S_{(x_j-b_j)}(T) \right] \right] d\lambda_X(x).$$

Recall:

$$S_\theta(t) \equiv sgn(\theta)S(t\,|\theta|),$$

*the **sign function** or **signum function** $sgn(\theta)$ is defined:*

$$sgn(\theta) = \begin{cases} 1, & \theta > 0, \\ 0, & \theta = 0, \\ -1, & \theta < 0, \end{cases}$$

and:

$$S(t) \equiv \int_0^t \frac{\sin x}{x}dx.$$

As in the Book V proof, it now follows that:

$$\lim\nolimits_{T\to\infty} I_T = \int_{\mathbb{R}^n} \prod\nolimits_{j=1}^n \lambda_{a_j,b_j}(x_j)d\lambda_X(x), \tag{2}$$

where:

$$\lambda_{a_j,b_j}(x) = \begin{cases} 0, & x < a_j, \\ \frac{1}{2}, & x = a_j, \\ 1, & a_j < x < b_j, \\ \frac{1}{2}, & x = b_j, \\ 0, & x > b_j. \end{cases}$$

Letting $\mathring{A} = \prod_{j=1}^n (a_j, b_j)$ and $\overline{A} = \prod_{j=1}^n [a_j, b_j]$, it follows that with $\chi_B(x)$ the characteristic function of the set B, that for all x:

$$\chi_{\mathring{A}}(x) \leq \prod\nolimits_{j=1}^n \lambda_{a_j,b_j}(x_j) \leq \chi_{\overline{A}}(x),$$

and thus by monotonicity of the integral (item 4, Proposition V.3.42):

$$\int_{\mathbb{R}^n} \chi_{\mathring{A}}(x)d\lambda_X(x) \leq \int_{\mathbb{R}^n} \prod\nolimits_{j=1}^n \lambda_{a_j,b_j}(x_j)d\lambda_X(x) \leq \int_{\mathbb{R}^n} \chi_{\overline{A}}(x)d\lambda_X(x).$$

Since $\lambda_X[\partial A] = \lambda_X\left[\overline{A} - \mathring{A}\right] = 0$ by assumption, these bounding integrals agree by item 6 of Proposition V.3.42. But then since $\chi_{\mathring{A}}(x) \leq \chi_A(x) \leq \chi_{\overline{A}}(x)$, this and (2) prove that:

$$\lim\nolimits_{T\to\infty} I_T = \int_{\mathbb{R}^n} \chi_A(x)d\lambda_X(x) \equiv \lambda_X[A].$$

■

In the special case of integrable $C_X(t)$, we can say more about the inverse transform as was the case for $n = 1$ in Proposition 5.28.

Proposition 5.38 (Inversion of integrable $C_X(t)$) *If $C_X(t)$ is Lebesgue integrable on $\mathbb{R}^n$, then λ_X has a continuous density function $f(x)$:*

$$f(x) = (2\pi)^{-n} \int_{\mathbb{R}^n} C_X(t)e^{-ix\cdot t}dt. \tag{5.70}$$

Hence for all $A \in \mathcal{B}(\mathbb{R}^n)$:

$$\lambda_X[A] = \int_A f(x)dm. \tag{5.71}$$

Proof. *Rather than here prove the existence of a density function* $f(x)$ *that satisfies (5.70), we defer the proof to Remark 5.52 following Bochner's theorem in Proposition 5.51. The first part of that theorem's somewhat long proof will accomplish the needed result.*

Given the existence of the density function $f(x)$ *and associated distribution function* $F(x)$*, Proposition 1.34 obtains that* $f(x)$ *is also the density for* λ_X *as in Definition 1.9, and (5.71) follows.* ■

As in the one-dimensional case of Proposition 5.29, we now address the question of the **uniqueness of the n-dimensional Fourier-Stieltjes transform,** or in the current context, the **uniqueness of the characteristic function.**

Specifically, given distribution functions F and G, the next result proves that if $C_F(t) = C_G(t)$ for all $t \in \mathbb{R}^n$, then $\lambda_F(A) = \lambda_G(A)$ for all Borel sets $A \in \mathcal{B}(\mathbb{R}^n)$, and in particular, $F(x) = G(x)$ for all $x \in \mathbb{R}^n$.

Proposition 5.39 (Uniqueness of $C_X(t)$: If $C_F(t) = C_G(t)$ then $\lambda_F = \lambda_G$) *Let* $F(x)$ *and* $G(x)$ *be distribution functions on* $\mathbb{R}^n$ *with induced Borel measures* λ_F *and* λ_G*.*

If $C_F(t) = C_G(t)$ *for all* t*, then for all Borel sets* $A \in \mathcal{B}(\mathbb{R}^n)$*:*

$$\lambda_F(A) = \lambda_G(A). \tag{5.72}$$

Thus for all x*:*

$$F(x) = G(x). \tag{5.73}$$

Proof. *By Proposition 5.37:*

$$\lambda_F(A) = \lambda_G(A), \tag{1}$$

for all bounded right semi-closed rectangles $A = \prod_{j=1}^{n}(a_j, b_j]$ *with* $\lambda_F[\partial A] = \lambda_G[\partial A] = 0$.

Assume that we can extend the identity in (1) to the semi-algebra $\mathcal{A}'$ *of all right semi-closed rectangles* $\{\prod_{j=1}^{n}(a_j, b_j]\}$*, without any restriction on the measures of the boundaries, recalling that this collection is indeed a semi-algebra by Corollary I.7.3. Then this result would extend from* $\mathcal{A}'$ *to the associated algebra* $\mathcal{A}$ *of all finite disjoint unions of* $\mathcal{A}'$*-sets by the Carathéodory extension theorem 2 of Proposition I.6.13, and then to the smallest sigma algebra* $\sigma(\mathcal{A})$ *that contains* $\mathcal{A}$ *by the uniqueness of extensions result of Proposition I.6.14. But then* $\sigma(\mathcal{A}) = \mathcal{B}(\mathbb{R}^n)$ *by Proposition I.8.1, completing the proof that (1) is true for all* $A \in \mathcal{B}(\mathbb{R}^n)$*.*

To prove that (1) applies in $\mathcal{A}'$*, we recall a technical detail from the proof part a of the portmanteau theorem of Proposition 3.17. There it was demonstrated that there is a dense and uncountable set of reals,* $D_F \subset \mathbb{R}$*, with countable complement, so that for any* k*, and any* $r \in D_F$*:*

$$\lambda_F[\{x \in \mathbb{R}^n | x_k = r\}] = 0. \tag{2}$$

With D_G *analogously defined, let* $D = D_F \bigcap D_G$*. Then* D *is a set of reals that satisfies property (2) for both measures.*

To see that D *is dense, by construction* $D_F = \mathbb{R} - E_F$ *with* E_F *countable, and similarly* $D_G = \mathbb{R} - E_G$*. Using De Morgan's laws of Exercise I.2.2, and recalling that* $A - B \equiv A \bigcap \widetilde{B}$*:*

$$D_F \bigcap D_G = \mathbb{R} - \left(E_F \bigcup E_G\right).$$

Hence D *has a countable complement, and it is an exercise to confirm that* D *is therefore dense and uncountable. Hint: Proof by contradiction.*

Define the class of rectangles $\mathcal{R}$ *by* $\mathcal{R} = \left\{\prod_{j=1}^{n}(a_j, b_j] | \, a_j, b_j \in D\right\}$*. By the proof of part a of Proposition 3.17, every vertex of the* $\mathcal{R}$*-rectangles is a continuity point of both distribution functions* F *and* G *by construction. Since the bounding hyperplanes of such* *rectangles have both λ_F-measure and λ_G-measure 0, so too do the boundaries of all rectangles in $\mathcal{R}$. Thus by the Proposition 5.37,* (1) *is satisfied for all $A \in \mathcal{R}$.*

To extend this identity from $\mathcal{R}$ to $\mathcal{A}'$, let $A = \prod_{j=1}^{n}(\widetilde{a}_j, b_j]$, with all $b_j \in D$ and $\widetilde{a}_j$ finite but arbitrary. For each j, let $\{a_j^{(m)}\}_{m=1}^{\infty} \subset D$ be a decreasing sequence with $a_j^{(m)} \to \widetilde{a}_j$. Such sequences exist by density of D. Then $A^{(m)} \equiv \prod_{j=1}^{n}(a_j^{(m)}, b_j] \in \mathcal{R}$ for all m, so $\lambda_F(A^{(m)}) = \lambda_G(A^{(m)})$ by the previous paragraph. Also, $\{A^{(m)}\}_{m=1}^{\infty}$ is nested collection of sets, $A^{(m)} \subset A^{(m+1)}$, and since $A = \bigcup_{m=1}^{\infty} A^{(m)}$, it follows by continuity from below of Proposition I.2.45, that $\lambda_F(A) = \lambda_G(A)$.

Now define $A = \prod_{j=1}^{n}(\widetilde{a}_j, \widetilde{b}_j]$ with $\widetilde{b}_j$ and $\widetilde{a}_j$ arbitrary, and for each j let $\{b_j^{(m)}\}_{m=1}^{\infty} \subset D$ be a decreasing sequence with $b_j^{(m)} \to \widetilde{b}_j$. Then by the previous paragraph, if $A^{(m)} \equiv \prod_{j=1}^{n}(\widetilde{a}_j, b_j^{(m)}]$, then $\lambda_F(A^{(m)}) = \lambda_G(A^{(m)})$. Also, $\{A^{(m)}\}_{m=1}^{\infty}$ is nested, $A^{(m+1)} \subset A^{(m)}$, and since $A = \bigcap_{m=1}^{\infty} A^{(m)}$, it follows by continuity from above that $\lambda_F(A) = \lambda_G(A)$.

Hence (1) *is satisfied for all bounded $A \in \mathcal{A}'$. Since a rectangle $A = \prod_{j=1}^{n}(-\infty, b_j]$ can be decomposed into countably many disjoint bounded rectangles, the result in* (1) *follows by countable additivity of measures. The same conclusion applies to $A = \prod_{j=1}^{n}(a_j, \infty)$. Thus $\lambda_F(A) = \lambda_G(A)$ for all $A \in \mathcal{A}'$, and the proof is complete.* ■

The final investigation is into the generalization of **Lévy's continuity theorem** of Proposition 5.30, but here split into two parts. As noted there, these results are named for **Paul Lévy** (1886–1971). For this proof, recall that $F_m \Rightarrow F$ and $\lambda_{F_m} \Rightarrow \lambda_F$ and $\lambda_{X_m} \Rightarrow \lambda_X$ are equivalent statements by Definition 3.8 and (1.22), and all are equivalent to convergence in distribution, $X_m \to_d X$.

Proposition 5.40 (Lévy's continuity theorem) *Let $\{F_m\}_{m=1}^{\infty}$, F be a collection of distribution functions on $\mathbb{R}^n$ with associated probability measures $\{\lambda_m\}_{m=1}^{\infty}$, λ and characteristic functions $\{C_{F_m}\}_{m=1}^{\infty}$, C_F.*

Then $F_m \Rightarrow F$ if and only if $C_{F_m}(t) \to C_F(t)$ for all t.

Proof. *One direction of this result is a consequence of the portmanteau theorem of Proposition 3.17. By item 5 of that result, $\lambda_m \Rightarrow \lambda$ if and only if $\int g(x)d\lambda_m \to \int g(x)d\lambda$ for every bounded, continuous real-valued function g defined on $\mathbb{R}^n$. By (5.17), for any $t \in \mathbb{R}^n$:*

$$e^{ix\cdot t} = (\cos(x \cdot t) + i\sin(x \cdot t)) = g_1(x) + ig_2(x),$$

with $g_1(x)$ and $g_2(x)$ continuous and bounded real valued functions. Thus by linearity of the integral, $F_m \Rightarrow F$ obtains $C_{F_m}(t) \to C_F(t)$ for all t.

If $C_{F_m}(t) \to C_F(t)$ for all t, define $t_k \in \mathbb{R}^n$ to have 1 in the kth component and 0s elsewhere, and let $s \in \mathbb{R}$. Then by Proposition 5.31, $C_{F_m}(st_k) = C_{F_{m,k}}(s)$ with $F_{m,k}$ the kth marginal distribution function of F_m. Similarly, $C_F(st_k) = C_{F_k}(s)$ where F_k is the kth marginal of F. It then follows by hypothesis that for each k, $C_{F_{m,k}}(s) \to C_{F_k}(s)$ as $m \to \infty$ for all $s \in \mathbb{R}$.

By Lévy's continuity theorem on $\mathbb{R}$ of Proposition 5.30, $F_{m,k} \Rightarrow F_k$ for all k. In other words, each of the n one-variate marginal distribution functions of the distribution function sequence $\{F_m\}_{m=1}^{\infty}$ converges weakly to the respective marginal distribution of F.

Thus for each k, the sequence of marginal distributions $\{F_{m,k}\}_{m=1}^{\infty}$ is tight by Proposition II.8.18. So by Definition 3.37 applied to the associated probability measures $\{\lambda_{m,k}\}_{m=1}^{\infty}$, given $\epsilon > 0$, there is a T_k so that for all m:

$$\lambda_{m,k}\left[(-T_k, T_k]\right] > 1 - \epsilon/n.$$

Since:

$$\lambda_{m,k}\left[(-T_k, T_k]\right] = \lambda_m\left[\mathbb{R}^{k-1} \times (-T_k, T_k] \times \mathbb{R}^{n-k}\right],$$

it follows that for each k*, there is a* T_k *so that for all* m*:*

$$\lambda_m[\{x \in \mathbb{R}^n | -T_k < x_k \leq T_k\}] > 1 - \epsilon/n. \tag{1}$$

Let $T = \max\{T_k\}_{k=1}^n$ *and note that:*

$$\mathbb{R}^n - (-T, T]^n \subset \bigcup\nolimits_{k=1}^n \{x | x_k \leq -T_k \text{ or } x_k > T_k\}.$$

This follows since if $x \in \mathbb{R}^n - (-T, T]^n$*, then for each* k*,* $x_k \leq -T \leq -T_k$ *or* $x_k \geq T \geq T_k$*.*

Each set in this union has λ_m*-measure less than* ϵ/n *by (1), and so* $\lambda_m[\mathbb{R}^n - (-T, T]^n] < \epsilon$ *by subadditivity of measures, and then finite additivity obtains for all* m*:*

$$\mu_m[(-T, T]^n] > 1 - \epsilon.$$

Thus $\{\mu_m\}_{m=1}^\infty$ *is tight, and we can apply Prokhorov's theorem of Proposition 3.49 and Corollary 3.51.*

Proposition 3.49 asserts existence of a weakly convergent subsequence $\{\lambda_{m_j}\}_{j=1}^\infty$ *and a limit probability measure* υ *such that* $\lambda_{m_j} \Rightarrow \upsilon$*. By the first part of this proposition it follows that* $C_{\lambda_{m_j}}(t) \to C_\upsilon(t)$ *for all* t*. But* $C_{\lambda_m}(t) \to C_\lambda(t)$ *for all* t *by hypothesis and thus also* $C_{\lambda_{m_j}}(t) \to C_\lambda(t)$*. Hence* $C_\upsilon(t) = C_\lambda(t)$ *for all* t*, and by the above uniqueness theorem of Proposition 5.39,* $\upsilon = \lambda$*.*

By the same argument, every weakly convergent subsequence of $\{\lambda_m\}_{m=1}^\infty$ *converges to* λ*, and Corollary 3.51 obtains that* $\lambda_m \Rightarrow \lambda$*.* ■

Corollary 5.41 (Lévy's continuity theorem) *Let* $\{F_m\}_{m=1}^\infty$ *be a collection of distribution functions on* $\mathbb{R}^n$ *with associated probability measures* $\{\lambda_m\}_{m=1}^\infty$ *and characteristic functions* $\{C_{F_m}\}_{m=1}^\infty$*.*

If $C_{F_m}(t) \to \varphi(t)$ *for all* t *with* φ *continuous at* $t = 0$*, then there exists a probability measure* λ *and associated distribution function* F *so that* $C_F(t) = \varphi(t)$ *and* $\lambda_m \Rightarrow \lambda$*.*

Proof. *Since* $C_{F_m}(t) \to \varphi(t)$ *for all* t*, let* $t = (0, ..., 0, s, 0, ..., 0)$ *with* $s \in \mathbb{R}$ *in the* k*th component. Then by Proposition 5.31,* $C_{F_{m,k}}(s) \to \varphi_k(s)$ *for all* s*, with* $F_{m,k}$ *the* k*th marginal distribution of* F_m *and* $\varphi_k(s) \equiv \varphi(0, .., s, 0, .., 0)$*. Since each* $\varphi_k(s)$ *is continuous at* $s = 0$ *by assumption, Proposition 5.30 obtains the existence of probability measures* $\{\lambda_k\}_{k=1}^n$ *on* $\mathbb{R}$ *so that* $\lambda_{m,k} \Rightarrow \lambda_k$*, where* $\lambda_{m,k}$ *is the probability measure associated with* $F_{m,k}$*, and* $\varphi_k(s)$ *is the characteristic function of* λ_k*.*

By Proposition II.8.18, each such sequence of marginal distributions $\{F_{m,k}\}_{m=1}^\infty$ *is tight, and using the same steps as in the previous proof can conclude that* $\{\lambda_m\}_{m=1}^\infty$ *is tight.*

Given any weakly convergent subsequence $\{\lambda_{m_j}\}_{j=1}^\infty$ *and limit probability measure* λ*, noting that at least one such subsequence and limit exist by Prokhorov's theorem, the conclusion that* $\lambda_{m_j} \Rightarrow \lambda$ *assures by Proposition 5.40 that* $C_{\lambda_{m_j}}(t) \to C_\lambda(t)$ *for all* t*. But by hypothesis,* $C_{\lambda_m}(t) \to \varphi(t)$ *and thus also* $C_{\lambda_{m_k}}(t) \to \varphi(t)$*, and so* $C_\lambda(t) = \varphi(t)$ *and* $\varphi(t)$ *is the characteristic function of* λ*.*

An application of Corollary 3.51 then yields that $\lambda_m \Rightarrow \lambda$*.* ■

5.6.1 The Cramér-Wold Theorem

The **Cramér-Wold theorem**, also called the **Cramér-Wold device,** was introduced in Proposition 3.28. It is named for a 1936 result of **Harald Cramér** (1893–1985) and **Herman Wold** (1908–1992), and addresses the following question.

If $\{X_m\}_{m=1}^\infty$*,* X*, are random vectors defined on* $(\mathcal{S}, \mathcal{E}, \lambda)$ *with range in* $\mathbb{R}^n$*, what is the relationship between* $X_m \Rightarrow X$ *and* $t \cdot X_m \Rightarrow t \cdot X$ *for* $t \in \mathbb{R}^n$*?*

The first convergence, $X_m \Rightarrow X$, is a statement about weak convergence of the associated joint distributions, while the latter convergence, $t \cdot X_m \Rightarrow t \cdot X$, is a statement about the weak convergence of one-dimensional distributions defined relative to half-spaces in $\mathbb{R}^n$.

To see this, for given t let $X'_m \equiv t \cdot X_m$ and $X' \equiv t \cdot X$. Then since $\{X'_m \leq \alpha\} = \{t \cdot X_m \leq \alpha\}$, the distribution function of X'_m is defined by:

$$F_{X'_m}(\alpha) = \lambda\left[(X'_m)^{-1}(-\infty, \alpha]\right] = \lambda\left[(X_m)^{-1}(H_t(\alpha))\right].$$

Here $H_t(\alpha) \subset \mathbb{R}^n$ is the half space defined:

$$H_t(\alpha) \equiv \{t \cdot x \leq \alpha\}.$$

Hence given $t \in \mathbb{R}^n$, the weak convergence $X'_m \Rightarrow X'$ provides a highly summarized statement about the random vectors X_m and X. Specifically, if α is a continuity point of $F_{X'}$, then $F_{X'_m}(\alpha) \to F_{X'}(\alpha)$, which to say only that

$$\lambda\left[(X_m)^{-1}(H_t(\alpha))\right] \to \lambda\left[(X)^{-1}(H_t(\alpha))\right].$$

Remarkably, the Cramér-Wold theorem states that $X_m \Rightarrow X$ if and only if $t \cdot X_m \Rightarrow t \cdot X$ **for all** $t \in \mathbb{R}$. Proposition 3.28 provided the simpler half of the proof, the "only if" result.

The deeper "if" half of the proof requires the powerful tools of Fourier analysis, here interpreted in the context of characteristic functions. That results on one-dimensional "marginal" distributions imply results in the multivariate context has already been seen in the various proofs of the prior section.

For an application of this result, see the proof of the n-dimensional **central limit theorem** of Proposition 7.23.

Proposition 5.42 (Cramér-Wold theorem) *Let* $\{X_m\}_{m=1}^{\infty}$, X *be random vectors on* $(\mathcal{S}, \mathcal{E}, \lambda)$ *with range in* $\mathbb{R}^n$. *Then* $X_m \Rightarrow X$ *if and only if* $t \cdot X_m \Rightarrow t \cdot X$ *for all* $t \in \mathbb{R}^n$.

Proof. *That* $X_m \Rightarrow X$ *assures* $t \cdot X_m \Rightarrow t \cdot X$ *for all* $t \in \mathbb{R}^n$ *was proved in Proposition 3.28 as an application of the mapping theorem on* $\mathbb{R}^j/\mathbb{R}^k$. *To prove the converse, we use Lévy's continuity theorem in one dimension.*

For given $t \in \mathbb{R}^n$, *define random variables* $X'_m \equiv t \cdot X_m$ *and* $X' \equiv t \cdot X$. *Then* $X'_m \Rightarrow X'$ *by assumption, and Lévy's continuity theorem obtains that* $C_{X'_m}(s) \to C_{X'}(s)$ *for all* $s \in \mathbb{R}$. *By definition:*

$$C_{X'_m}(s) \equiv \int_{-\infty}^{\infty} e^{iys} d\lambda_{X'_m}(y),$$

with $C_{X'}(s)$ *similarly defined. Thus for all* s:

$$\int_{-\infty}^{\infty} e^{iys} d\lambda_{X'_m}(y) \to \int_{-\infty}^{\infty} e^{iys} d\lambda_{X'}(y), \tag{1}$$

and this is true for any $t \in \mathbb{R}^n$.

For $t \in \mathbb{R}^n$, *define* $T_t : \mathbb{R}^n \to \mathbb{R}$ *by* $T_t : x \to t \cdot x$, *noting that* T_t *is continuous and thus Borel measurable by Proposition V.2.12. Considered as a measurable transformation of Definition 1.28:*

$$T_t : (\mathbb{R}^n, \mathcal{B}(\mathbb{R}^n), \lambda_{X_m}) \to (\mathbb{R}, \mathcal{B}(\mathbb{R}), m),$$

the measure λ_{X_m} *and transformation* T_t *induce a measure* $(\lambda_{X_m})_{T_t}$ *on the range space defined on* $A \in \mathcal{B}(\mathbb{R})$ *by:*

$$(\lambda_{X_m})_{T_t}(A) \equiv \lambda_{X_m}\left[T_t^{-1}(A)\right].$$

We claim that:

$$(\lambda_{X_m})_{T_t} = \lambda_{X'_m}. \tag{2}$$

To see this, let $A \in \mathcal{B}(\mathbb{R})$. By definition of $(\lambda_{X_m})_{T_t}$ above, then λ_{X_m} in (1.21), and recalling that $f^{-1}g^{-1} = (gf)^{-1}$ for transformations:

$$\begin{aligned}
(\lambda_{X_m})_{T_t}(A) &\equiv \lambda_{X_m}\left[T_t^{-1}(A)\right] \\
&= \lambda\left[X_m^{-1}T_t^{-1}(A)\right] \\
&= \lambda[\{X_m \cdot t \in A\}] \\
&= \lambda\left[(X'_m)^{-1}(A)\right] \\
&\equiv \lambda_{X'_m}(A).
\end{aligned}$$

Similarly defining T_t on $(\mathbb{R}^n, \mathcal{B}(\mathbb{R}^n), \lambda_X)$ obtains $(\lambda_X)_{T_t} = \lambda_{X'}$.

Using the change of variables of Proposition 5.15 with $g(z) = e^{izs}$:

$$\int_{\mathbb{R}^n} e^{i(t\cdot x)s} d\lambda_{X_m}(x) = \int_{-\infty}^{\infty} e^{iys} d\lambda_{X'_m}(y) \equiv C_{X'_m}(s), \tag{3}$$

and analogously:

$$\int_{\mathbb{R}^n} e^{i(t\cdot x)s} d\lambda_X(x) = \int_{-\infty}^{\infty} e^{iys} d\lambda_{X'}(y) \equiv C_{X'}(s). \tag{4}$$

Now $C_{X'_m}(s) \to C_{X'}(s)$ for all $s \in \mathbb{R}$, and all $t \in \mathbb{R}^n$ by (1). Letting $s = 1$, this can be restated by (3) and (4), that for all $t \in \mathbb{R}^n$:

$$\int_{\mathbb{R}^n} e^{i(t\cdot x)} d\lambda_{X_m} \to \int_{\mathbb{R}^n} e^{i(t\cdot x)} d\lambda_X.$$

Thus $C_{X_m}(t) \to C_X(t)$ for all $t \in \mathbb{R}^n$, and by Lévy's continuity theorem this is equivalent to $X_m \Rightarrow X$. ■

5.7 Bochner's Theorem

Bochner's theorem, named for **Salomon Bochner** (1899–1982), provides a necessary and sufficient condition on a complex valued function $\varphi : \mathbb{R}^n \to \mathbb{C}$ to ensure that there exists a probability measure λ, or equivalently, a distribution function F, so that $\varphi(t) = C_F(t)$. Published in two papers in 1932 and 1933, the first developed the results on $\mathbb{R}$, while the second generalized these results to $\mathbb{R}^n$.

To start, any such φ must possess the properties common to all characteristic functions. Proposition 5.45 summarizes several properties, three of which will look quite natural. The fourth property is a surprising one, and not because it will be difficult to derive. The surprise is that it is not a property that has been observed or indeed utilized in any of the above results.

First a definition:

Definition 5.43 (Positive semidefinite $\varphi : \mathbb{R}^n \to \mathbb{C}$) *A complex valued function $\varphi : \mathbb{R}^n \to \mathbb{C}$ is said to* ***positive semidefinite****, and sometimes* ***nonnegative definite****, if for any m, any $\{t_j\}_{j=1}^m \subset \mathbb{R}^n$, and any $\{z_j\}_{j=1}^m \subset \mathbb{C}$:*

$$\sum\nolimits_{j=1}^{m} \sum\nolimits_{k=1}^{m} \varphi(t_j - t_k) z_j \bar{z}_k \geq 0, \tag{5.74}$$

where $\bar{z}_k$ is the complex conjugate of z_k of Definition 5.11.

Remark 5.44 (On positive semidefinite) *The double sum in (5.74) can also be expressed in matrix notation:*

$$z^T\Phi\bar{z} \geq 0,$$

or:

$$\bar{z}^T\Phi z \geq 0,$$

where Φ is the $m \times m$ matrix:

$$\Phi_{jk} = \varphi(t_j - t_k), \tag{5.75}$$

$z \equiv (z_1, ..., z_m)$ and $\bar{z} \equiv (\bar{z}_1, ..., \bar{z}_m)$. By linear algebra convention, vectors are taken as column matrices, and thus z^T, the ***transpose of*** *z, is the associated row matrix.*

Thus the terminology "positive semidefinite" is reminiscent of that used for a matrix A in linear algebra, as will be seen in the discussion on the multivariate normal distribution. But there is an important difference between these two notions.

The matrix A in this later context is a real matrix, and $x^TAx \geq 0$ is a statement about $x \in \mathbb{R}^m$. For such A and x, x^TAx is a real number by definition, and the positive semidefinite requirement is that this real number must be nonnegative.

In contrast, here Φ is a complex matrix, and $z^T\Phi\bar{z} \geq 0$ is a statement about $z \in \mathbb{C}^m$. So first and foremost, the above definition states that $z^T\Phi\bar{z}$ is always a real number. In addition, as in the real matrix case, this real number must be nonnegative.

Proposition 5.45 (Properties of a characteristic function φ) *If $\varphi : \mathbb{R}^n \to \mathbb{C}$ is the characteristic function of a distribution function F defined on $\mathbb{R}^n$, then φ has the following properties:*

1. *φ is uniformly continuous on $\mathbb{R}^n$ and $\varphi(0) = 1$, where $0 \equiv (0, ..., 0)$.*
2. *$|\varphi(t)| \leq 1$ for all t.*
3. *$\varphi(-t) = \overline{\varphi(t)}$ for all t.*
4. *φ is positive semidefinite.*

Proof. *Item 1 is Proposition 5.35, while item 2 is (5.18) and the triangle inequality of Proposition V.7.3. In detail, if $\varphi(t) \equiv C_F(t)$:*

$$|C_F(t)| \leq \int_{\mathbb{R}^n} \left|e^{ix\cdot t}\right| d\lambda_F(x) = 1.$$

For item 3, (5.17) obtains that:

$$\overline{e^{iy}} = \cos y - i \sin y = e^{-iy}.$$

Hence:

$$C_F(-t) = \int_{\mathbb{R}^n} e^{-ix\cdot t} d\lambda_F(x) = \int_{\mathbb{R}^n} \overline{e^{ix\cdot t}} d\lambda_F(x) = \overline{\int_{\mathbb{R}^n} e^{ix\cdot t} d\lambda_F(x)} = \overline{C_F(t)},$$

noting that the third equality follows from (5.17) and linearity of the integral from Proposition V.3.42.

For item 4, again by linearity, and (5.14) in the last step:

$$\begin{aligned}\sum_{j=1}^{m}\sum_{k=1}^{m} C_F(t_j-t_k)z_j\bar{z}_k &= \int_{\mathbb{R}^n}\sum_{j=1}^{m}\sum_{k=1}^{m} e^{ix\cdot(t_j-t_k)}z_j\bar{z}_k d\lambda_F(x)\\ &= \int_{\mathbb{R}^n}\sum_{j=1}^{m}\sum_{k=1}^{m} e^{ix\cdot t_j}z_j e^{-ix\cdot t_k}\bar{z}_k d\lambda_F(x)\\ &= \int_{\mathbb{R}^n}\sum_{j=1}^{m} e^{ix\cdot t_j}z_j \sum_{k=1}^{m}\overline{e^{ix\cdot t_k}z_k} d\lambda_F(x)\\ &= \int_{\mathbb{R}^n}\left|\sum_{j=1}^{m} e^{ix\cdot t_j}z_j\right|^2 d\lambda_F(x) \geq 0.\end{aligned}$$

■

Exercise 5.46 (**$C_F(t)$** ***is positive semidefinite.***) *Using $C_F(0)=1$ of item 1 and item 3, prove that the double sum in item 4 is always a real number by showing that:*

$$\sum_{j=1}^{m}\sum_{k=1}^{m} C_F(t_j-t_k)z_j\bar{z}_k = \sum_{j=1}^{m}|z_j|^2 + 2\sum_{k<j}\mathrm{Re}\left[C_F(t_j-t_k)z_j\bar{z}_k\right],$$

where $\mathrm{Re}[z]$ *is the real part of z as in Definition 5.11. Hint: Split the sum into $j=k$, $j<k$ and $k<j$.*

5.7.1 Positive Semidefinite Functions

In this section we develop additional properties of positive semidefinite functions that will be important in the proof of Bochner's theorem. There positive semidefiniteness will be an important assumption on $\varphi(t)$, the function we hope to prove is a characteristic function, so it will be helpful to know what other properties of $\varphi(t)$ this implies.

Remark 5.47 (Hermitian matrix) *A complex matrix Φ that satisfies (5.76) below is said to be a* ***Hermitian matrix,*** *and this identity is sometimes expressed:*

$$\Phi = \bar{\Phi}^T.$$

In other words, a Hermitian matrix Φ equals its ***conjugate transpose****. A Hermitian matrix is also called* ***self-adjoint.***

Named for ***Charles Hermite*** *(1822–1901), who proved that such matrices have only real eigenvalues, and, generalizing the result for real symmetric matrices for which $A=A^T$, he proved that such matrices have n linearly independent eigenvectors.*

Recall that given a real matrix $A:\mathbb{R}^n\to\mathbb{R}^n$, that $\lambda\in\mathbb{R}$ is a (real) ***eigenvalue*** *of A if there exists an* ***eigenvector*** *$x\in\mathbb{R}^n$ so that $Ax=\lambda x$. In general, a real matrix A need not have n linearly independent eigenvectors, but this is the case if A is symmetric.*

Similarly, given a complex matrix $B:\mathbb{C}^n\to\mathbb{C}^n$, then $\lambda\in\mathbb{C}$ is a (complex) eigenvalue of B if there exists an eigenvector $z\in\mathbb{C}^n$ so that $Bz=\lambda z$. In general, a complex matrix can have real and/or complex eigenvalues, and need not have n linearly independent eigenvectors. Hermite proved that if $B=\bar{B}^T$, then B has only real eigenvalues, and n linearly independent eigenvectors.

Another important property of Hermitian matrices, and one used in the proof of item 4 below, is that the determinant of a Hermitian matrix is real and nonnegative:

$$\det\Phi \geq 0.$$

For more on Hermitian matrices, see ***Hoffman and Kunze*** *(1971), or,* ***Strang*** *(2009).*

Exercise 5.48 (**$(AB)^T = B^T A^T$ and $\overline{AB} = \bar{A}\bar{B}$**) *Given complex $n \times n$ matrices A and B, prove that:*

$$(AB)^T = B^T A^T, \text{ and, } \overline{AB} = \bar{A}\bar{B}.$$

Proposition 5.49 (Properties of positive semidefinite $\varphi(t)$) *If $\varphi : \mathbb{R}^n \to \mathbb{C}$ is positive semidefinite then:*

1. *$\varphi(0) \in \mathbb{R}$ and $\varphi(0) \geq 0$, where $0 \equiv (0, ..., 0)$.*

2. *Given $\{t_j\}_{j=1}^m \subset \mathbb{R}^n$, the matrix Φ defined in (5.75) is Hermitian:*

$$\bar{\Phi} = \Phi^T, \tag{5.76}$$

where Φ^T denotes the transpose of the matrix Φ. In other words, $\bar{\Phi}_{jk} = \Phi_{kj}$. Equivalently as a statement on φ, for all $t \in \mathbb{R}^n$:

$$\overline{\varphi(t)} = \varphi(-t). \tag{5.77}$$

3. *$|\varphi(t)| \leq \varphi(0)$ for all $t \in \mathbb{R}^n$.*

4. *If continuous at $t = 0$, then $\varphi(t)$ is uniformly continuous on $\mathbb{R}^n$.*

Proof. *By Exercise 5.48 it follows that:*

$$\left(z^T \Phi \bar{z}\right)^T = \bar{z}^T \Phi^T z, \text{ and, } \overline{\left(z^T \Phi \bar{z}\right)} = \bar{z}^T \bar{\Phi} z.$$

However, since $z^T \Phi \bar{z} \in \mathbb{R}$ for all $z = (z_1, ..., z_m)$ by (5.74):

$$\left(z^T \Phi \bar{z}\right)^T = z^T \Phi \bar{z} = \overline{\left(z^T \Phi \bar{z}\right)}.$$

Combining obtains that for all $z = (z_1, ..., z_m)$:

$$\bar{z}^T \Phi^T z = \bar{z}^T \bar{\Phi} z. \tag{1}$$

Let $e_j \in \mathbb{R}^n$ have 1 in the jth component and 0s elsewhere.

a. *If $z \equiv e_j$ in (1), this obtains by (5.74) and (5.75) that:*

$$0 \leq \varphi(0) = \overline{\varphi(0)}, \tag{2a}$$

which proves item 1.

b. *Letting $z \equiv e_j + e_k$ in (1) and using (2a) yields:*

$$\varphi(t_j - t_k) + \varphi(t_k - t_l) = \overline{\varphi(t_j - t_k)} + \overline{\varphi(t_k - t_j)},$$

which by (5.13) obtains:

$$\operatorname{Im}\left[\varphi(t_j - t_k)\right] = -\operatorname{Im}\left[\varphi(t_k - t_j)\right]. \tag{2b}$$

c. *Letting $z \equiv e_j + ie_k$ in (1) and (2a) yields again by (5.13):*

$$\operatorname{Re}\left[\varphi(t_j - t_k)\right] = \operatorname{Re}\left[\varphi(t_k - t_l)\right]. \tag{2c}$$

Combining items (2a)-(2c) obtains (5.77), which proves item 2.

Next recall (5.14), that for any complex number $w = a + bi$ that $w\bar{w} = a^2 + b^2 \equiv |w|^2$. Letting $m = 2$ in (5.74) obtains that for all $\{z_j\}_{j=1}^2 \subset \mathbb{C}$ and all $t \equiv t_2 - t_1 \in \mathbb{R}^n$ that:

$$\left(|z_1|^2 + |z_2|^2\right)\varphi(0) + z_1\bar{z}_2\varphi(-t) + \bar{z}_1 z_2\varphi(t) \geq 0. \tag{3}$$

If $\varphi(t) = 0$ for all $t \in \mathbb{R}^n$ then item 3 is satisfied, so assume there exists $t \in \mathbb{R}^n$ with $\varphi(t) \neq 0$, and then let $z_1 = \bar{z}_2 = \sqrt{-\varphi(t)}$. If $-\varphi(t) \equiv a + bi$, $\sqrt{-\varphi(t)}$ is defined as the principal square root of (5.21):

$$\sqrt{-\varphi(t)} = \sqrt{\left(\sqrt{a^2+b^2}+a\right)\Big/\,2} + isgn(b)\sqrt{\left(\sqrt{a^2+b^2}-a\right)\Big/\,2},$$

where we take the positive square root of real numbers and define $sgn(b) = 1$ if $b \geq 0$ and -1 otherwise.

A little algebra obtains:

$$|z_1|^2 = |z_2|^2 = |\varphi(t)|, \qquad z_1\bar{z}_2 = -\varphi(t), \qquad \bar{z}_1 z_2 = -\overline{\varphi(t)},$$

and then from (3) and (5.77):

$$2|\varphi(t)|\varphi(0) - 2|\varphi(t)|^2 \geq 0.$$

Since $|\varphi(t)| > 0$ by assumption, item 3 follows.

For item 4 let $m = 3$, and $t = (t_1, t_2, t_3) \equiv (t, s, 0)$, and so by (5.77):

$$\Phi = \begin{pmatrix} \varphi(0) & \varphi(t-s) & \varphi(t) \\ \overline{\varphi(t-s)} & \varphi(0) & \varphi(s) \\ \overline{\varphi(t)} & \overline{\varphi(s)} & \varphi(0) \end{pmatrix}.$$

Since Φ is Hermitian by item 2, $\det\Phi \geq 0$ as noted in Remark 5.47, and we evaluate this determinant using the so-called ***diagonal method.***

For this method, the matrix is first augmented with the first two columns:

$$\Phi^{Aug} = \begin{pmatrix} \varphi(0) & \varphi(t-s) & \varphi(t) & \varphi(0) & \varphi(t-s) \\ \overline{\varphi(t-s)} & \varphi(0) & \varphi(s) & \overline{\varphi(t-s)} & \varphi(0) \\ \overline{\varphi(t)} & \overline{\varphi(s)} & \varphi(0) & \overline{\varphi(t)} & \overline{\varphi(s)} \end{pmatrix}.$$

The determinant then equals the sum of the three down-diagonal products, less the sum of the three up-diagonal products:

$$\begin{aligned} \det\Phi &= \varphi(0)^3 + \varphi(t-s)\varphi(s)\overline{\varphi(t)} + \overline{\varphi(t-s)\varphi(s)}\varphi(t) \\ &\quad - \varphi(0)\left[|\varphi(t)|^2 + |\varphi(s)|^2 + |\varphi(t-s)|^2\right]. \end{aligned}$$

Using (5.13) and $|z|^2 + |w|^2 = |z - w|^2 + 2\operatorname{Re}[\bar{z}w]$ obtains:

$$\begin{aligned} \det\Phi &= \varphi(0)^3 - \varphi(0)\left[|\varphi(t) - \varphi(s)|^2 + |\varphi(t-s)|^2\right] - 2\operatorname{Re}\left[\varphi(s)\overline{\varphi(t)}\left(\varphi(0) - \varphi(t-s)\right)\right] \\ &\leq \varphi(0)^3 - \varphi(0)\left[|\varphi(t) - \varphi(s)|^2 + |\varphi(t-s)|^2\right] + 2\varphi(0)^2|\varphi(0) - \varphi(t-s)|, \end{aligned} \tag{4}$$

where in the last step we use $-\operatorname{Re}(zw) \leq |z||w|$, then item 3.

If $\varphi(0) = 0$ then $\varphi(t) = 0$ for all t by item 3 and thus item 4 is assured, so assume that $\varphi(0) > 0$. Since $\det\Phi \geq 0$, it follows from (4):

$$0 \leq \varphi(0)^2 - \left[|\varphi(t) - \varphi(s)|^2 + |\varphi(t-s)|^2\right] + 2\varphi(0)|\varphi(0) - \varphi(t-s)|.$$

Thus:

$$\begin{aligned}|\varphi(t)-\varphi(s)|^2 &\leq \varphi(0)^2-|\varphi(t-s)|^2+2\varphi(0)\,|\varphi(0)-\varphi(t-s)| \\ &\leq 2\varphi(0)\,|\varphi(0)-\varphi(t-s)|,\end{aligned}$$

since $|\varphi(t-s)|^2 \leq \varphi(0)^2$ *by item 3. Hence continuity at* 0 *yields uniform continuity on* $\mathbb{R}^n$, *and the proof is complete.* ■

In summary, the above proposition assures that if $\varphi:\mathbb{R}^n \to \mathbb{C}$ is positive semidefinite, $\varphi(0)=1$, and φ is continuous at $t=0$, then φ has all of the properties of a characteristic function of a distribution function noted in Proposition 5.49. Bochner's theorem states that in fact this is enough to ensure that such φ is indeed a characteristic function on $\mathbb{R}^n$.

Before stating and proving this result, we require an additional result on positive semidefinite functions.

Proposition 5.50 (More properties of positive semidefinite $\varphi(t)$) *If* $\varphi:\mathbb{R}^n \to \mathbb{C}$ *is positive semidefinite and continuous at* $0 \equiv (0,...,0)$, *then:*

1. *The function* $\varphi_y(t) \equiv \varphi(t)e^{it\cdot y}$ *is positive semidefinite for any* $y \in \mathbb{R}^n$.
2. *If* $h:\mathbb{R}^n \to \mathbb{R}$ *is nonnegative, continuous and Riemann integrable, then defined as a Riemann integral:*

$$\psi(t) \equiv \int_{\mathbb{R}^n} h(y)\varphi_y(t)dy,$$

is positive semidefinite.

Proof. *Let* m, $\{t_j\}_{j=1}^m \subset \mathbb{R}^n$, *and* $\{z_j\}_{j=1}^m \subset \mathbb{C}$ *be given. Then:*

$$\begin{aligned}\sum_{j=1}^m\sum_{k=1}^m \varphi_y(t_j-t_k)z_j\bar{z}_k &= \sum_{j=1}^m\sum_{k=1}^m \varphi(t_j-t_k)e^{i(t_j-t_k)\cdot y}z_j\bar{z}_k \\ &= \sum_{j=1}^m\sum_{k=1}^m \varphi(t_j-t_k)e^{it_j\cdot y}z_je^{-it_k\cdot y}\bar{z}_k \\ &= \sum_{j=1}^m\sum_{k=1}^m \varphi(t_j-t_k)w_j\bar{w}_k\end{aligned}$$

with $w_j \equiv e^{it_j\cdot y}z_j$. *This expression is nonnegative by assumption on* φ, *proving item 1.*

For item 2, $h(y)\varphi_y(t)$ *is continuous by item 4 of Proposition 5.49. Further,* $\left|e^{it\cdot y}\right|=1$ *by (5.18) and* $\left|\varphi_y(t)\right| \leq \varphi(0)$ *by item 3 above, so* $\left|h(y)\varphi_y(t)\right| \leq \varphi(0)h(y)$ *and* $h(y)\varphi_y(t)$ *is Riemann integrable. As in Definition 5.14, we define such integrability in terms of the associated real and imaginary parts of* $h(y)\varphi_y(t)$, *which we notationally suppress below.*

Thus by Proposition III.1.62, $\int_{\mathbb{R}^n} h(y)\varphi_y(t)dy$ *is the limit of Riemann sums of Definition III.1.52. For a given Riemann sum of mesh size* δ, *with* $v_i > 0$ *denoting the volume of the ith cell* C_i *and* $y_i \in C_i$, *define:*

$$\psi_\delta(t) \equiv \sum\nolimits_i h(y_i)\varphi_{y_i}(t)v_i.$$

Since this summation is absolutely convergent, we can rearrange sums in the following:

$$\begin{aligned}\sum_{j=1}^m\sum_{k=1}^m \psi_\delta(t_j-t_k)z_j\bar{z}_k &= \sum_{j=1}^m\sum_{k=1}^m\sum\nolimits_i h(y_i)\varphi_{y_i}(t_j-t_k)z_j\bar{z}_k \\ &= \sum\nolimits_i h(y_i)v_i\sum_{j=1}^m\sum_{k=1}^m \varphi_{y_i}(t_j-t_k)z_j\bar{z}_k.\end{aligned}$$

Because $h(y_i)v_i \geq 0$, *it follows from item 1 that for any mesh-size* δ:

$$\sum_{j=1}^m\sum_{k=1}^m \psi_\delta(t_j-t_k)z_j\bar{z}_k \geq 0, \tag{1}$$

and each $\psi_\delta(t)$ *is positive semidefinite.*

By definition, $\psi_\delta(t) \to \psi(t)$ pointwise as $\delta \to 0$, and thus by continuity this convergence is uniform on compact sets by Exercise III.1.14. As any finite collection $\{t_j - t_k\}_{j,k=1}^m$ is contained in a compact set, it follows by uniform convergence that:

$$\sum\nolimits_{j=1}^m \sum\nolimits_{k=1}^m \psi_\delta(t_j - t_k) z_j \bar{z}_k \to \sum\nolimits_{j=1}^m \sum\nolimits_{k=1}^m \psi(t_j - t_k) z_j \bar{z}_k,$$

and this last sum is nonnegative by (1). ■

5.7.2 Bochner's Theorem

With the prepatory work complete, we are finally ready to prove Bochner's theorem, which characterizes all characteristic functions. The proof below was adapted from online lecture notes of **Rongfeng Sun** of the National University of Singapore.

It will be noted that the proof of this result is still quite long, even with the above preliminary development. The majority of the proof will address the special case where φ is absolutely Lebesgue integrable on $\mathbb{R}^n$. In this case, we will be able to identify a density function $f(x)$ for the distribution function $F(x)$ we seek, and then to prove that $\varphi(t) = C_F(t)$. Once accomplished, the general case will be surprisingly easy to establish.

Proposition 5.51 (Bochner's theorem) *A function $\varphi : \mathbb{R}^n \to \mathbb{C}$ is the characteristic function of a distribution function defined on $\mathbb{R}^n$ if and only if φ is positive semidefinite, continuous at $t = 0$, and $\varphi(0) = 1$.*

Proof. *Since characteristic functions have the stated properties by Proposition 5.45, the essence of this result is the sufficiency of these conditions.*

As noted above, the proof of sufficiency will first address the case where φ is absolutely integrable on $\mathbb{R}^n$. Recalling Proposition 5.38, which proved that an integrable characteristic function assures that the underlying distribution function has an associated density function, this portion of the proof seeks to create such a density from φ.

***1. Integrable** φ : Assume that φ is absolutely Lebesgue integrable on $\mathbb{R}^n$, meaning that $|\varphi(t)|$ is Lebesgue integrable. Since continuous by Proposition 5.49, φ is Riemann integrable over bounded rectangles by Proposition III.1.63. Further, the Lebesgue and Riemann integrals agree by Proposition III.2.18 over bounded rectangles, and then in the limit over $\mathbb{R}^n$. Thus we alternate below between these integration frameworks.*

***a. Identifying the candidate density** $f(x)$:*

Define $f : \mathbb{R}^n \to \mathbb{C}$ by:

$$f(x) = \frac{1}{(2\pi)^n} \int_{\mathbb{R}^n} \varphi(t) e^{-ix \cdot t} dt. \tag{1}$$

Since $\left|e^{-ix\cdot t}\right| = 1$, it follows by the triangle inequality of Proposition V.7.3 and integrability of $|\varphi(t)|$ that $f(x)$ is well-defined and absolutely bounded. In addition, $f(x)$ is continuous:

$$|f(x) - f(x')| \le \frac{1}{(2\pi)^n} \int_{\mathbb{R}^n} |\varphi(t)| \left|e^{-ix\cdot t} - e^{-ix'\cdot t}\right| dt.$$

Since $|\varphi(t)| \left|e^{-ix\cdot t} - e^{-ix'\cdot t}\right| \le 2|\varphi(t)|$ and $\left|e^{-ix\cdot t} - e^{-ix'\cdot t}\right| \to 0$ pointwise as $x' \to x$, Lebesgue's dominated convergence theorem of Proposition V.7.4 obtains that $|f(x) - f(x')| \to 0$.

We claim that $f(x)$ is real-valued and nonnegative. To see this, fix x, define $R_T^+ = [0,T]^n$, and let:

$$f_T(x) \equiv \frac{1}{(2\pi)^n} \frac{1}{T^n} \int_{R_T^+ \times R_T^+} \varphi(t - s) e^{-ix\cdot(t-s)} ds dt, \tag{2}$$

which as above, has the same value as a Riemann or Lebesgue integral. The Riemann integral is then the limit of Riemann sums by Proposition III.1.62, and indeed Riemann sums with finitely many terms. Any such Riemann sum is real and nonnegative by the positive semidefinite property of φ with $m = 1$, and thus in the limit, this integral is real and nonnegative:

$$f_T(x) \geq 0. \tag{3}$$

We next perform a change of variables on the integral in (2), using the invertible linear transformation on $\mathbb{R}^{2n}$ defined by $T : (s,t) \to (u,v)$ where $u \equiv t - s$, $v \equiv s$. This linear transformation has $2n \times 2n$ diagonal matrix:

$$A \equiv \begin{pmatrix} -I & I \\ Z & I \end{pmatrix},$$

there I denotes the $n \times n$ identity matrix, and Z the $n \times n$ zero matrix. The determinant of A equal to $(-1)^n$, so $|\det A| = 1$.

As a Lebesgue integral, the integral in (2) can be defined over $\mathbb{R}^n \times \mathbb{R}^n$ by multiplying the integrand by the characteristic function $\chi_{R_T^+ \times R_T^+}(s,t)$ recalling Definition III.2.9. Applying the change of variables result of Proposition V.4.32:

$$\int_{\mathbb{R}^n \times \mathbb{R}^n} \chi_{R_T^+ \times R_T^+}(s,t)\varphi(t-s)e^{-ix\cdot(t-s)}dsdt = \int_{\mathbb{R}^n \times \mathbb{R}^n} \chi_{R_T^+ \times R_T^+}(v, u+v)\varphi(u)e^{-ix\cdot u}dudv.$$

Reversing the order of integration using Fubini's theorem of Proposition 5.16, the dv-integral can be explicitly evaluated. By definition $\chi_{R_T^+ \times R_T^+}(v, u+v) = 1$ if for all i both $0 \leq v_i \leq T$ and $-u_i \leq v_i \leq T - u_i$. These v_i-sets intersect only when $-T \leq u_i \leq T$ for all i, and then this intersection set is either $0 \leq v_i \leq T - u_i$ for $u_i \geq 0$ or $-u_i \leq v_i \leq T$ for $u_i < 0$, and in both cases this set has measure $T - |u_i|$. Thus with $R_T = [-T,T]^n$:

$$\begin{aligned} f_T(x) &\equiv \frac{1}{(2\pi)^n}\frac{1}{T^n}\int_{\mathbb{R}^n} \chi_{R_T}(u)\prod\nolimits_{i=1}^n (T - |u_i|)\varphi(u)e^{-ix\cdot u}du \\ &= \frac{1}{(2\pi)^n}\int_{\mathbb{R}^n}\prod\nolimits_{i=1}^n\left(1 - \frac{|u_i|}{T}\right)\chi_{R_T}(u)\varphi(u)e^{-ix\cdot u}du. \end{aligned}$$

Denoting this integrand by $\psi_T(u)$, note that $|\psi_T(u)| \leq |\varphi(u)|$, which is integrable, and that $\psi_T(u) \to \varphi(u)e^{-ix\cdot u}$ pointwise as $T \to \infty$. Thus by Lebesgue's dominated convergence theorem of Proposition V.7.4, it follows that $f_T(x) \to f(x)$ in (1), and hence by (3), $f(x) \geq 0$ as claimed.

While $f(x)$ is real, continuous and nonnegative, to be a density function we must prove integrability and that this function integrates to 1.

b. $\varphi(t) = C_F(t)$, ***the characteristic function of*** $F(x)$:

We now show that φ is the characteristic function of $f(x)$ in the sense of (5.31), with the function $g_\lambda(x)$ there replaced by $f(x)$. Then once proved integrable, this will prove that $f(x)$ is a density function since $1 = \varphi(0) = \int f(x)dx$, where this integral can be defined as a Riemann integral since f is continuous.

To prove that f is integrable, we first approximate f with integrable functions. For $\lambda > 0$, and $|x|^2 \equiv \sum_{i=1}^n x_i^2$ by (2.14), define:

$$h_\lambda(x) \equiv f(x)\exp\left[-\frac{\lambda|x|^2}{2}\right].$$

Since $f(x)$ is bounded, $h_\lambda(x)$ is integrable for all $\lambda > 0$, and also $h_\lambda(x) \to f(x)$ pointwise as $\lambda \to 0$. Since $h_\lambda(x)$ need not be a density, we formally take the Fourier transform $\widehat{h}_\lambda(y)$ of $h_\lambda(x)$, recalling that this is identical to the characteristic function definition by Remark 5.20.

To this end:

$$\widehat{h}_\lambda(y) \equiv \int_{\mathbb{R}^n} h_\lambda(s)e^{iy\cdot s}ds = \frac{1}{(2\pi)^n}\int_{\mathbb{R}^n}\int_{\mathbb{R}^n}\varphi(t)\exp\left[-\lambda\left|s\right|^2/2\right]e^{-is\cdot t}e^{iy\cdot s}dtds. \tag{4}$$

By the integrability assumption on φ, this integrand is integrable over $\mathbb{R}^n \times \mathbb{R}^n$, and thus Fubini's theorem of Proposition 5.16 obtains:

$$\widehat{h}_\lambda(y) = \frac{\lambda^{-n/2}}{(2\pi)^{n/2}}\int_{\mathbb{R}^n}\varphi(t)\left[\frac{\lambda^{n/2}}{(2\pi)^{n/2}}\int_{\mathbb{R}^n}\exp\left[-\lambda\left|s\right|^2/2\right]e^{is\cdot(y-t)}ds\right]dt.$$

Recalling Example 5.21, the inner integral is seen to be the characteristic function of a multivariate normal density evaluated at $y-t$. By that example and a change of variables $y-t \to s$:

$$\begin{aligned}\widehat{h}_\lambda(y) &= \frac{\lambda^{-n/2}}{(2\pi)^{n/2}}\int_{\mathbb{R}^n}\varphi(t)\exp\left[-\left|y-t\right|^2/2\lambda\right]dt \\ &= \frac{\lambda^{-n/2}}{(2\pi)^{n/2}}\int_{\mathbb{R}^n}\varphi(y+s)\exp\left[-\left|s\right|^2/2\lambda\right]ds.\end{aligned}$$

Letting $y = 0$ in (4) and above, and noting the multivariate normal density of Example 5.21 in the above integrand, but this time with $C = \lambda I$, this obtains with item 3 of Proposition 5.49:

$$\int_{\mathbb{R}^n} h_\lambda(s)ds = \widehat{h}_\lambda(0) \leq \sup\left|\varphi(s)\right| = \varphi(0) = 1. \tag{5}$$

Since $h_\lambda(x) \to f(x)$ monotonically for each x, and all functions are nonnegative, Lebesgue's monotone convergence theorem of Proposition III.2.37 assures that f is integrable and that $\int_{\mathbb{R}^n} f(s)ds \leq 1$.

Combining results above:

$$\int_{\mathbb{R}^n} h_\lambda(s)e^{iy\cdot s}ds = \frac{\lambda^{-n/2}}{(2\pi)^{n/2}}\int_{\mathbb{R}^n}\varphi(y+s)\exp\left[-\left|s\right|^2/2\lambda\right]ds. \tag{6}$$

Since $h_\lambda(s)e^{iy\cdot s} \to f(s)e^{iy\cdot s}$ pointwise as $\lambda \to 0$, and the integrals of $h_\lambda(s)e^{iy\cdot s}$ are bounded by the integral of $f(s)$, Lebesgue's dominated convergence theorem of Proposition V.7.4 assures that:

$$\int_{\mathbb{R}^n} h_\lambda(s)e^{iy\cdot s}ds \to \int_{\mathbb{R}^n} f(s)e^{iy\cdot s}ds. \tag{7}$$

By a change of variables in the integral on the right in (6):

$$\frac{\lambda^{-n/2}}{(2\pi)^{n/2}}\int_{\mathbb{R}^n}\varphi(y+s)\exp\left[-\left|s\right|^2/2\lambda\right]ds = \frac{1}{(2\pi)^{n/2}}\int_{\mathbb{R}^n}\varphi(y+t\sqrt{\lambda})\exp\left[-\left|t\right|^2/2\right]dt.$$

The integrand converges pointwise as $\lambda \to 0$ to $\varphi(y)\exp\left[-\left|t\right|^2/2\right]$, and since φ is bounded, Lebesgue's dominated convergence theorem applies to conclude that this integral converges to $\varphi(x)$. Thus:

$$\frac{\lambda^{-n/2}}{(2\pi)^{n/2}}\int_{\mathbb{R}^n}\varphi(y+s)\exp\left[-\left|s\right|^2/2\lambda\right]ds \to \frac{\varphi(y)}{(2\pi)^{n/2}}\int_{\mathbb{R}^n}\exp\left[-\left|t\right|^2/2\right]dt = \varphi(y), \tag{8}$$

noting another multivariate normal from Example 5.21, this time with $C = I$.

Combining (6)–(8)*:*

$$\varphi(x) = \int_{\mathbb{R}^n} f(s)e^{ix\cdot s}ds.$$

Letting $x = 0$ *confirms that* $\int_{\mathbb{R}^n} f(s)ds = \varphi(0) = 1$*, and thus* f *is a density function, and* φ *is its characteristic function.*

2. General φ

In the general case, we approximate φ *with integrable functions and apply part 1. Let:*

$$\begin{aligned}\varphi_\lambda(x) &\equiv \varphi(x)\exp\left[-\lambda |x|^2/2\right] \\ &= \frac{1}{\lambda^{n/2}(2\pi)^{n/2}}\int_{\mathbb{R}^n}\varphi(x)\exp\left[-|s|^2/2\lambda\right]e^{is\cdot x}ds,\end{aligned}$$

where the last step follows as in Example 5.21, that $\exp\left[-\lambda |x|^2/2\right]$ *is the characteristic function of the multivariate normal with* $C = \lambda I$*.*

Now $\varphi_\lambda(x)$ *is continuous at* $x = 0$ *with* $\varphi(0) = 1$*, and is positive semidefinite by Proposition 5.50. Thus there exists a density function* $f_\lambda(x)$ *by part 1 such that* $\varphi_\lambda(x)$ *is the characteristic function of* $f_\lambda(x)$*. But* $\varphi_\lambda(x) \to \varphi(x)$ *pointwise as* $\lambda \to 0$*, and so by Corollary 5.41 to Lévy's continuity theorem, there exists a distribution function* F *such that* $\varphi(x) = C_F(x)$*.* ∎

Remark 5.52 (Inversion of integrable $C_X(t)$) *The proof of (5.70) of Proposition 5.38 was deferred to this point.*

To this end, let $C_X(t)$ *be the characteristic function of a probability measure* λ_X*. By definition,* $C_X(t)$ *is also the characteristic function of the associated distribution function* $F(x)$ *of (1.18), and* $C_X(t)$ *is assumed integrable by the hypothesis of Proposition 5.38. Since* $\varphi(t) \equiv C_X(t)$ *satisfies the assumptions of Bochner's theorem, part 1 of the above proof assures the existence of a continuous density function* $f(x)$ *so that* $C_X(t)$ *is the characteristic function of* $f(x)$*.*

By the uniqueness theorem of Proposition 5.39, it follows that λ_X *is the probability measure associated with* $f(x)$*, and equivalently,* $F(x)$ *is the distribution function associated with* $f(x)$*.*

5.8 A Uniqueness of Moments Result

Proposition IV.4.61 stated, without proof, the key result needed for that section's investigation into the uniqueness of moments and the moment generating function. Perhaps ironically, the proof of this result below takes place in the more general environment of characteristic functions, where we prove that moments uniquely determine the characteristic function. The last step is then to apply the uniqueness theorem of characteristic functions, which converts this uniquely defined characteristic function into a unique probability measure and distribution function.

Proposition IV.4.61 was stated in terms of the distribution function F, and made no mention of the associated probability measure λ_F. While studied in Book II, and then acknowledged in the context of moments in Section IV.4.1.2, probability measures could play no formal role in the development of moments and moment generating functions until the integration theory of Book V was developed.

Below we restate this result from the current perspective. But recalling Proposition IV.4.59, the assumptions below assure the existence of a moment generation function for F, and that this function is defined by the given series.

Proposition 5.53 (Uniqueness of moments) *Let λ_F be a Borel measure on $\mathbb{R}$ induced by a probability distribution function F and assume that $\mu'_n \equiv \int_{-\infty}^{\infty} x^n d\lambda_F$ exists for all n.*

If the power series $\sum_{n=0}^{\infty} \mu'_n t^n/n!$ converges absolutely on $(-t_0, t_0)$ for some $t_0 > 0$, then F is the only distribution function with these moments.

Proof. *Denoting by $\mu'_{|n|}$ the absolute moments:*

$$\mu'_{|n|} = \int_{-\infty}^{\infty} |x|^n \, d\lambda_F,$$

we first show that $\mu'_{|n|} s^n/n! \to 0$ as $n \to \infty$ for some s with $0 < s < t_0$.

Given any t with $0 < t < t_0$, $\sum_{n=0}^{\infty} \mu'_n t^n/n!$ converges by assumption. Hence for such t, $\mu'_n t^n/n! \to 0$ as $n \to \infty$, so $\mu'_{|n|} t^n/n! \to 0$ for even n.

For odd indexes, the inequality $\mu'_{|2n+1|} \leq \left(\mu'_{2n}\mu'_{2n+2}\right)^{1/2}$ of Example IV.4.49 obtains:

$$\mu'_{|2n+1|} \leq \mu'_{(2n^+)} \equiv \max[\mu'_{2n}, \mu'_{2n+2}].$$

With $s \equiv \lambda t$ for $0 < t < t_0$ and arbitrary $0 < \lambda < 1$:

$$\frac{\mu'_{|2n+1|} |s|^{2n+1}}{(2n+1)!} \leq c_{(2n^+)} \frac{\mu'_{(2n^+)} |t|^{2n^+}}{(2n^+)!}, \tag{1}$$

where:

$$c_{(2n^+)} = \begin{cases} (2n+2)\,\lambda^{2n+1}/|t|, & \text{if } 2n^+ = 2n+2, \\ |t|\,\lambda^{2n+1}/(2n+1), & \text{if } 2n^+ = 2n. \end{cases}$$

Hence since both $c_{(2n^+)} \to 0$ and $\mu'_{(2n^+)} |t|^{2n^+} / (2n^+)! \to 0$ as noted above, it follows from (1) that $\mu'_{|2n+1|} |s|^{2n+1} / (2n+1)! \to 0$ if $s = \lambda t < t$.

Next, since $\left|e^{itx}\right| = 1$ by (5.18), Proposition V.7.5 obtains:

$$\left| e^{itx} \left(e^{ihx} - \sum\nolimits_{j=0}^{n} \frac{(ihx)^j}{j!} \right) \right| \leq \frac{|hx|^{n+1}}{(n+1)!}.$$

Applying the triangle inequality of Proposition V.7.3:

$$\begin{aligned} &\left| C_F(t+h) - \sum\nolimits_{j=0}^{n} \frac{h^j}{j!} \int_{-\infty}^{\infty} (ix)^j \, e^{itx} d\lambda_F \right| \\ &= \left| \int_{-\infty}^{\infty} e^{i(h+t)x} d\lambda_F - \int_{-\infty}^{\infty} \sum\nolimits_{j=0}^{n} \frac{(ihx)^j}{j!} e^{itx} d\lambda_F \right| \\ &\leq \int_{-\infty}^{\infty} \left| e^{itx} \left(e^{ihx} - \sum\nolimits_{j=0}^{n} \frac{(ihx)^j}{j!} \right) \right| d\lambda_F \\ &\leq \frac{|h|^{n+1} \mu'_{|n|}}{(n+1)!}. \end{aligned}$$

Denoting by $C_F^{(j)}(t)$ the jth derivative of $C_F(t)$ and applying (5.49) obtains:

$$\left| C_F(t+h) - \sum\nolimits_{j=0}^{n} \frac{h^j}{j!} C_F^{(j)}(t) \right| \leq \frac{|h|^{n+1} \mu'_{|n|}}{(n+1)!}.$$

Letting $n \to \infty$, it follows that for $|h| \leq s$ defined above, and all t, that:

$$C_F(t+h) = \sum\nolimits_{j=0}^{\infty} \frac{h^j}{j!} C_F^{(j)}(t). \tag{2}$$

Assume that G is another distribution function with the same moments as F. Then as above it will follow that for all t:

$$C_G(t+h) = \sum\nolimits_{j=0}^{\infty} \frac{h^j}{j!} C_G^{(j)}(t), \tag{3}$$

for $|h| \leq s$, using the same definition of s as above, since it depended only on these moments.

Letting $t = 0$ and applying (5.49) proves that for all j:

$$C_F^{(j)}(0) = i^j \mu_j' = C_G^{(j)}(0),$$

and thus by (2) and (3):

$$C_F(h) = C_G(h), \ \ |h| \leq s.$$

Hence $C_F^{(j)}(h) = C_G^{(j)}(h)$ for $|h| \leq s$ and all j.

Now letting $t = \pm(s - \epsilon)$ with $\epsilon = .1s$ for example, then since $C_F^{(j)}(t) = C_G^{(j)}(t)$ as just noted, it follows that:

$$C_F(h) = C_G(h), \ \ |h| < 2s - \epsilon,$$

and so too, $C_F^{(j)}(h) = C_G^{(j)}(h)$ for $|h| \leq 2s - \epsilon$ and all j. Repeating with $t = \pm 2(s - \epsilon)$, and so forth, we then conclude that for every m, that:

$$C_F(h) = C_G(h), \ \ |h| < (m+1)s - m\epsilon.$$

Hence $C_F(h) = C_G(h)$ for all h.

By the uniqueness property of Proposition 5.39, this implies that $\mu_F(A) = \mu_G(A)$ for all Borel sets $A \in \mathcal{B}(\mathbb{R})$. Taking $A = (-\infty, x]$ produces $F(x) = G(x)$ for all x. ■

6

Multivariate Normal Distribution

This chapter investigates the **multivariate normal distribution**, which as in the one-dimensional case discussed in earlier books, is also called the **multivariate Gaussian distribution,** named for **Carl Friedrich Gauss** (1777–1855).

In the first section we motivate the definition of this distribution function by first investigating affine transformations of random vectors of independent normal variates, and then generalizing. While the initial construction obtains a density function for this distribution, the general definition, framed in terms of characteristic or moment generating functions, encompasses distributions without densities. The second section investigates when a density function exists.

Continuing the discussion of Chapter IV.5, simulation of multivariate normal vectors is discussed next, for which the Cholesky decomposition of a positive definite matrix plays a central role.

The last section investigates properties of the multivariate normal distribution function, starting with moments, and then develops the interesting and oft-misquoted property that the component variates of a multivariate normal are independent if and only if they are uncorrelated. Samples of normal variates are then investigated, proving that the sample mean and sample variance are independent random variables, with distribution functions that are then derived.

6.1 Derivation and Definition

In this section we derive various results relating to affine transformations of a random vector of independent normal variates, which will motivate the final definition of what it means for a random vector to have a multivariate normal distribution.

Studied in Books II and IV, the **normal density function** $f_N(x)$ depends on a location parameter $\mu \in \mathbb{R}$ and a scale parameter $\sigma > 0$, and is defined by:

$$f_N(x) = \frac{1}{\sigma\sqrt{2\pi}} \exp\left(-\frac{(x-\mu)^2}{2\sigma^2}\right), \tag{6.1}$$

where $\exp y \equiv e^y$ to simplify notation. This is often call the **Gaussian probability density** after **Carl Friedrich Gauss** (1777–1855), one of the codiscoverers of this formula.

When $\mu = 0$ and $\sigma = 1$, this density is known as the **unit normal** or **standard normal density,** and often denoted $\phi(x)$:

$$\phi(x) = \frac{1}{\sqrt{2\pi}} \exp\left(-\frac{x^2}{2}\right). \tag{6.2}$$

Turning to a random vector $X \equiv (X_1, X_2, ..., X_n)$ on a probability space $(\mathcal{S}, \mathcal{E}, \lambda)$, under what criteria should we declare that X has a multivariate normal distribution?

DOI: 10.1201/9781003275770-6

- If $\{X_i\}_{i=1}^n$ are independent normal variates with parameters $\{(\mu_i, \sigma_i)\}_{i=1}^n$, one would certainly expect X to be an example of a multivariate normal. By (6.1), independence and Corollary 1.57, it would follow that X had a density function defined on $x \equiv (x_1, ..., x_n)$:

$$f(x) = \frac{1}{(2\pi)^{n/2}\prod_{i=1}^n \sigma_i} \exp\left(-\frac{1}{2}\sum\nolimits_{i=1}^n \frac{(x_i - \mu_i)^2}{\sigma_i^2}\right).$$

Framed in terms of the moment generating function of X, Corollary 5.4 and $M_N(t)$ in IV.(4.78) obtain for $t = (t_1, t_2, ..., t_n)$:

$$\begin{aligned} M_X(t) &= \prod\nolimits_{i=1}^n \exp\left(\mu_i t_i + \frac{1}{2}\sigma_i^2 t_i^2\right) \\ &= \exp\left(\mu \cdot t + \frac{1}{2}\sum\nolimits_{i=1}^n \sigma_i^2 t_i^2\right), \end{aligned}$$

recalling that $\mu \cdot t = \sum_{i=1}^n \mu_i t_i$ denotes the dot product or inner product of $\mu \equiv (\mu_1, ..., \mu_n)$ and $t \equiv (t_1, ..., t_n)$.

A similar result is obtained for $C_X(t)$ using Corollary 5.34 and (5.44):

$$C_X(t) = \exp\left(i\mu \cdot t - \frac{1}{2}\sum\nolimits_{i=1}^n \sigma_i^2 t_i^2\right)$$

- If $\{X_i\}_{i=1}^n$ are independent normal variates with parameters $\{(\mu_i, \sigma_i)\}_{i=1}^n$, define $Y \equiv (Y_1, Y_2, ..., Y_n)$ by $Y = AX + \alpha$, where A is a linear transformation on $\mathbb{R}^n$ and $\alpha = (\alpha_1, \alpha_2, ..., \alpha_n)$ a constant vector. Again it seems compelling to assert that Y be declared multivariate normal.

Indeed, X in the first bullet is created just this way from $Z \equiv (Z_1, Z_2, ..., Z_n)$:

$$X = A^{(\sigma)}Z + \mu, \tag{1}$$

with independent **standard** normal variates $\{Z_i\}_{i=1}^n$, $A^{(\sigma)}$ a diagonal matrix with $A_{ii}^{(\sigma)} = \sigma_i$, and $\alpha = \mu \equiv (\mu_1, \mu_2, ..., \mu_n)$.

It is an exercise to check that if $Y = AX + \alpha$ with $X = A^{(\sigma)}Z + \mu$, then $Y = A'Z + \alpha'$ with $A' = AA^{(\sigma)}$ and α' constant. Thus there is no advantage to transforming anything by standard normal variates.

We will see that such Y need not have a density function as in the first bullet, but we can identify $M_Y(t)$ and $C_Y(t)$ using the transformation results of Exercise 5.3 and Proposition 5.32, and $M_X(t)$ and $C_X(t)$ derived above.

With this introduction, we now investigate a more detailed development of these observations.

6.1.1 Density Function Approach

The first result investigates the existence and form of the density function for the random vector $Y = AZ + \mu$, with $Z \equiv (Z_1, Z_2, ..., Z_n)$ a vector of independent standard normals, and A an $n \times n$ **nonsingular**, or **invertible** matrix. See Section 6.2 for a discussion of these results when A is $n \times n$ but not invertible, or is an $m \times n$ matrix with $m \neq n$.

We begin with some definitions. The reader is referred to Section V.4.3.2 for a short review of linear transformations $T : \mathbb{R}^n \to \mathbb{R}^m$, and Sections III.4.3.1 and V.9.1 for a more general summary related to vector spaces. For more details on a variety of topics in linear algebra, see **Strang** (2009) or **Hoffman and Kunze** (1971).

Definition 6.1 (Transpose; Inverse of matrix A) *Given a transformation $T : \mathbb{R}^n \to \mathbb{R}^m$:*

- *T is called a **linear transformation** if for all $x, y \in \mathbb{R}^n$ and $a, b \in \mathbb{R}$:*

$$T(ax + by) = aTx + bTy.$$

- *An $m \times n$ matrix A induces a **linear transformation** $T : \mathbb{R}^n \to \mathbb{R}^m$ by matrix multiplication, meaning that if $x \in \mathbb{R}^n$ then $Tx \equiv Ax \in \mathbb{R}^m$.*

- *The **null space** $\mathcal{N}(A) \subset \mathbb{R}^n$ and **range space** $\mathcal{R}(A) \subset \mathbb{R}^m$ are defined:*

$$\mathcal{N}(A) = \{x \in \mathbb{R}^n | Ax = 0 \in \mathbb{R}^m\},$$
$$\mathcal{R}(A) = \{Ax | x \in \mathbb{R}^n\}.$$

- *A is **nondegenerate** if there exists $x \in \mathbb{R}^n$ with $Ax \neq 0$. Otherwise A is **degenerate**, $\mathcal{N}(A) = \mathbb{R}^n$ and $A = Z$, the zero matrix.*

- *The **transpose** of A, denoted A^T, is an $m \times n$ matrix, so $A^T : \mathbb{R}^m \to \mathbb{R}^n$, defined componentwise by:*

$$A^T_{ij} = A_{ji}.$$

  Equivalently, the rows of A are the columns of A^T, in order.

- *If $n = m$, the **inverse** of A, denoted A^{-1} when it exists, is an $n \times n$ matrix that satisfies:*

$$AA^{-1} = A^{-1}A = I, \tag{6.3}$$

  *where I is the **identity matrix**, defined as $I_{jj} = 1$ and $I_{jk} = 0$ for $j \neq k$.*

  *When A^{-1} exists, we say that A is **nonsingular** or **invertible**, and otherwise, A is **singular** or **not invertible**.*

Remark 6.2 (On $\mathcal{N}(A)$ and $\mathcal{R}(A)$ as spaces) *Note that $\mathcal{N}(A)$ and $\mathcal{R}(A)$ are called "spaces" because they are vector subspaces of $\mathbb{R}^n$, respectively, $\mathbb{R}^m$. Recalling Definition III.4.33, this means that if $x, y \in \mathcal{N}(A)$ and $a, b \in \mathbb{R}$, then $ax + by \in \mathcal{N}(A)$, and similarly for $\mathcal{R}(A)$.*

An important theorem relates the dimensions of these spaces for $A : \mathbb{R}^n \to \mathbb{R}^m$. Even though $\mathcal{N}(A) \subset \mathbb{R}^n = Dom(A)$, the domain space of A, and $\mathcal{R}(A) \subset \mathbb{R}^m$, this result states:

$$\dim[\mathcal{N}(A)] + \dim[\mathcal{R}(A)] = \dim[Dom(A)]. \tag{6.4}$$

Example 6.3 ($T \Leftrightarrow A$) *Linear transformations T and matrices A are intimately related. As noted above, given an $m \times n$ matrix A, an the associated $T : \mathbb{R}^n \to \mathbb{R}^m$ is defined by matrix multiplication.*

*Conversely, any linear transformation $T : \mathbb{R}^n \to \mathbb{R}^m$ induces a unique $m \times n$ matrix A relative to the standard bases of $\mathbb{R}^n$ and $\mathbb{R}^m$. Specifically, if $x = (x_1, ..., x_n)$, then $x = \sum_{j=1}^n x_j e_j$ where $\{e_j\}_{j=1}^n \subset \mathbb{R}^n$ is the **standard basis** for $\mathbb{R}^n$ and defined by $e_j = (e_{j_1}, ..., e_{j_n})$ with $e_{j_j} = 1$ and $e_{j_k} = 0$ for $k \neq j$.*

Given T, define the columns $\{A^{(j)}\}_{j=1}^n$ of A by $A^{(j)} = Te_j$. Then by linearity:

$$Ax = \sum\nolimits_{j=1}^n x_j A^{(j)} = \sum\nolimits_{j=1}^n x_j Te_j = Tx.$$

In general, the matrix representation A for T depends on the bases used in $\mathbb{R}^n$ and $\mathbb{R}^m$, but we always assume the standard basis unless otherwise specified.

Exercise 6.4 **($(AB)^T = B^T A^T$; $(AB)^{-1} = B^{-1}A^{-1}$; $\left(A^T\right)^{-1} = \left(A^{-1}\right)^T$)** *Derive several properties of matrix manipulations.*

1. *Given matrices A, B so that AB is well-defined, show that:*

$$(AB)^T = B^T A^T.$$

Hint: To be well-defined, A is $n \times p$ and B is $p \times m$, so AB is $n \times m$. Verify that $B^T A^T$ is well-defined and an $m \times n$ matrix, the dimension of $(AB)^T$. Pay close attention to subscripts in the multiplications.

2. *Given $n \times n$ invertible matrices A, B, show that AB is invertible and:*

$$(AB)^{-1} = B^{-1}A^{-1}.$$

Hint: For X^{-1} to be the inverse of X requires (6.3) to be satisfied.

3. *Given an $n \times n$ invertible matrix A, show that:*

$$\left(A^T\right)^{-1} = \left(A^{-1}\right)^T \tag{6.5}$$

and thus A^T is also invertible.

4. *By induction, show that these results generalize to products of any finite number of matrices, where for the transposition result of item 1, we only require that the product is well-defined.*

Recall that by convention we interpret vectors as column matrices (or column vectors), so if A is an $n \times n$ matrix and X and μ are n-vectors, then AX is well-defined and $AX + \mu$ is a (column) n-vector.

Proposition 6.5 (Density of $Y = AX + \mu$) *Let $X \equiv (X_1, X_2, ..., X_n)$ be a random vector of independent standard normal variables with density functions as in (6.2). Let $Y \equiv (Y_1, Y_2, ..., Y_n)$ be defined by:*

$$Y \equiv AX + \mu,$$

where $A : \mathbb{R}^n \to \mathbb{R}^n$ is an invertible matrix and $\mu = (\mu_1, \mu_2, ..., \mu_n)$ a constant vector.

Then the probability density function $f_Y(y)$ of Y is defined on $y = (y_1, y_2, ..., y_n)$ by:

$$f_Y(y) = (2\pi)^{-n/2} \left[\det C\right]^{-1/2} \exp\left[-\frac{1}{2}(y-\mu)^T C^{-1}(y-\mu)\right], \tag{6.6}$$

where $C = AA^T$, and $\det C$ denotes the determinant of C.

Proof. *We first express the density function of X in a format suitable to apply a transformation. Since $f(x_1, x_2, ..., x_n) = \prod_{i=1}^n \phi(x_i)$ by Corollary 1.57, with $\phi(x)$ given in (6.2):*

$$f(x) = (2\pi)^{-n/2} \exp\left[-\frac{1}{2}x^T I x\right],$$

where I denotes the identity matrix, x^T the row matrix transpose of the column matrix x, and thus:

$$x^T I x = \sum\nolimits_{i=1}^n x_i^2.$$

The inverse of the transformation $g(X) \equiv AX + \mu$ is the linear transformation $g^{-1}(Y) = A^{-1}(Y - \mu)$, and so the Jacobian determinant of g^{-1} as in (2.28) is the determinant of the *matrix* A^{-1}. *Recall that* $\det A^{-1} = 1/\det A$, *and this is well-defined since* A *is invertible and hence* $\det A \neq 0$.

Applying (2.29) and Exercise 6.4:

$$\begin{aligned} f_Y(y) &= (2\pi)^{-n/2}\left|\det A^{-1}\right| \exp\left[-\frac{1}{2}\left(A^{-1}(y-\mu)\right)^T I \left(A^{-1}(y-\mu)\right)\right] \\ &= \frac{1}{(2\pi)^{n/2}\left|\det A\right|}\exp\left[-\frac{1}{2}(y-\mu)^T C^{-1}(y-\mu)\right], \end{aligned} \tag{1}$$

where $C^{-1} = \left(A^{-1}\right)^T I A^{-1}$.

Using various matrix manipulations from Exercise 6.4:

$$C^{-1} = \left(A^{-1}\right)^T A^{-1} = \left(A^T\right)^{-1} A^{-1} = \left(AA^T\right)^{-1},$$

and so $C \equiv AA^T$. *Then by Remark V.4.29,* $\det C = \det AA^T = (\det A)^2 > 0$, *so* $\left|\det A\right|^{-1} = [\det C]^{-1/2}$ *is well-defined, and the result follows from* (1). ■

Remark 6.6 (Generalizing the density function $f_Y(y)$) *The above result derives the density function* $f_Y(y)$ *of* $Y \equiv AX + \mu$, *which assures that:*

$$\int f_Y(y)dy = 1. \tag{1}$$

Since continuous, this integral exists as a Riemann or Lebesgue integral, but it is of interest to determine how one would verify this integral value directly.

In addition, while $C \equiv AA^T$ *above for an invertible matrix* A, *it is also of interest to know if* (1) *is true for other matrices* C *that do not arise in this way. To be sure,* C *in (6.6) cannot be completely general:*

1. *Any matrix* C *of interest would have to be nonsingular, so* $\det C \neq 0$, *and indeed given the* $[\det C]^{-1/2}$*-term in (6.6), we would of necessity require* $\det C > 0$.

2. *While* $C \equiv AA^T$ *is symmetric by Exercise 6.4, we can always assume that any given* $n \times n$ *matrix* C *is symmetric, meaning* $C = C^T$, *which assures the same for* C^{-1} *by (6.5). This follows because for any square matrix* B:

$$x^T Bx = x^T\left[\frac{1}{2}\left(B + B^T\right)\right]x,$$

and $\frac{1}{2}\left(B + B^T\right)$ *is symmetric. Thus there is no loss of generality in requiring symmetry of* C *in the more general formula for* $f_Y(y)$. *Indeed, there is a major benefit as will be seen with the spectral theorem below.*

Before continuing, it is timely to formalize additional matrix notions that are important for the emerging discussion. Note that in contrast to Remark 5.47, here A is a real matrix defined on the real space $\mathbb{R}^m$.

Definition 6.7 (Symmetric; positive semidefinite; positive definite) *An* $m \times m$ *real matrix* B *is:*

- ***Symmetric*** *if* $B^T = B$.
- ***Positive semidefinite*** *if* B *is symmetric and* $x^T Bx \geq 0$ *for all* $x \in \mathbb{R}^m$.

- ***Positive definite*** *if B is positive semidefinite and $x^T Bx = 0$ if and only if $x = 0$.*
- ***Orthogonal*** or ***orthonormal*** *if $B^{-1} = B^T$.*

Remark 6.8 (On determinants and eigenvalues) *As noted in Book V, while we need various results from linear algebra in these books, we cannot derive them in any detail and instead reference **Strang** (2009) or **Hoffman and Kunze** (1971) for details. Adding to the discussion in Section V.4.3.2, we make the following observations:*

1. *If B is **positive definite,** then of necessity B is nonsingular so $\det B \neq 0$. This follows because if B were singular, there would exist $x \neq 0$ with $Bx = 0$ and thus $x^T Bx = 0$.*

   Moreover, B is positive definite if and only if $\det B > 0$. This follows from the important result that such a matrix has n (counting multiplicities) positive eigenvalues $\{\lambda_j\}_{j=1}^n$, and then $\det A = \prod_{j=1}^n \lambda_j$.

   To see that eigenvalues are positive, if λ is an eigenvalue there exists $x \neq 0$ by nonsingularity, so that $Bx = \lambda x$. But then by positive definiteness, $0 < x^T Bx = \lambda |x|^2$.

2. *If B is **positive semidefinite** and not positive definite, then there exists $x \neq 0$ with $Bx = 0$. Hence $\det B = 0$ and B is not invertible. Then B has both positive eigenvalues like positive definite matrices, but also the eigenvalue $\lambda_0 = 0$.*

3. *If B is **orthogonal,** then $\det B = \pm 1$. This follows from $B^T B = I$, and thus $\det B^T \det B = 1$. Since $\det B = \det B^T$, the result follows.*

   The term orthogonal (or orthonormal) means that the columns $\{B^{(j)}\}_{j=1}^n$ of B are orthogonal (or orthonormal):

   $$B^{(j)} \cdot B^{(k)} = \begin{cases} 0, & j \neq k, \\ 1, & j = k, \end{cases}$$

   and this follows from $B^T B = I$.

Example 6.9 **($C \equiv AA^T$, $A : \mathbb{R}^n \to \mathbb{R}^n$)** *When $Y = AX + \mu$ as above, with $A : \mathbb{R}^n \to \mathbb{R}^n$ an $n \times n$ invertible matrix, then $C \equiv AA^T$ is always **symmetric** since by Exercise 6.4:*

$$C^T = \left(AA^T\right)^T = \left(A^T\right)^T A^T = C.$$

*Further, such C is **positive definite,** since if $x \in \mathbb{R}^n$:*

$$x^T Cx = x^T AA^T x = \left(A^T x\right)^T A^T x = \left|A^T x\right|^2,$$

where $\left|A^T x\right|$ denotes the standard norm of the n-vector $A^T x$ as defined in (2.14).

Thus $x^T Cx \geq 0$ for all x and $x^T Cx = 0$ if and only if $A^T x = 0$. Since A^T is nonsingular by Exercise 6.4, it follows that $x^T Cx = 0$ if and only if $x = 0$.

Remark 6.10 **($C \equiv AA^T$, $A : \mathbb{R}^n \to \mathbb{R}^m$)** *In the next section we will be considering $Y = AX + \mu$ with $A : \mathbb{R}^n \to \mathbb{R}^m$ an $m \times n$ matrix, so the matrix $C \equiv AA^T$ of interest is an $m \times m$ matrix. Then C is again **symmetric** exactly as above, and if $x \in \mathbb{R}^m$:*

$$x^T Cx = \left|A^T x\right|^2.$$

*It then follows that any such C is at least **positive semidefinite.** Further, C is **positive definite** if and only if $x = 0$ is the only solution of $A^T x = 0$. Thus C is positive definite if and only if A^T has null space $\mathcal{N}(A) = \{0\}$.*

Since $A^T : \mathbb{R}^m \to \mathbb{R}^n$ is an $n \times m$ matrix, if $m > n$ then by (6.4), there always exist nonzero solutions of $A^T x = 0$. If $n > m$, then there may exist such solutions or not.

For the next proof on the integral of $f(x)$, we need the **spectral theorem**, and for this, symmetry of B is required.

Proposition 6.11 (Spectral theorem) *If $B : \mathbb{R}^n \to \mathbb{R}^n$ is a symmetric $n \times n$ matrix, then B has n (counting multiplicities) real eigenvalues $\{\lambda_j\}_{j=1}^n$, orthogonal eigenvectors $\{Q^{(j)}\}_{j=1}^n$, and B can be factored:*

$$B = Q\Lambda Q^T,$$

where Λ is a diagonal matrix of eigenvalues, and Q has eigenvectors as columns. Hence, Q is orthogonal and $Q^T = Q^{-1}$.

Proof. *See* **Strang** *(2009) or* **Hoffman and Kunze** *(1971).* ■

We are now ready for our general result on when $f_Y(y)$ in (6.6) is a density function. Here we require C to be positive definite, but by the above discussion one can equivalently require C to be a symmetric matrix with $\det C > 0$.

Proposition 6.12 ($\int f_Y(y)dy = 1$) *Let $f_Y(y)$ be defined as in (6.6) with C an $n \times n$ positive definite matrix and given $\mu \in \mathbb{R}^n$. Then:*

$$\int f_Y(y)dy = 1,$$

and $f_Y(y)$ is a density function.

Proof. *By definition $f_Y(y) > 0$ for all y, so to be a density function we only need to verify integrability and the value of the integral. By a change of variables $y-\mu \to x$ and Proposition V.4.44:*

$$\int f_Y(y)dy = (2\pi)^{-n/2} [\det C]^{-1/2} \int_{\mathbb{R}^n} \exp\left[-\frac{1}{2}x^T C^{-1} x\right] dx. \tag{1}$$

The spectral theorem states that:

$$C^{-1} = Q\Lambda Q^T, \tag{2}$$

with Q orthogonal and thus $Q^T = Q^{-1}$. Hence $Q^T C^{-1} Q = \Lambda$, so making the change of variables $x = Qz$ in (1), and applying Proposition V.4.32 obtains:

$$\begin{aligned}\int f_Y(y)dy &= (2\pi)^{-n/2} [\det C]^{-1/2} |\det Q| \int_{\mathbb{R}^n} \exp\left[-\frac{1}{2}(Qz)^T C^{-1}(Qz)\right] dz \\ &= (2\pi)^{-n/2} [\det C]^{-1/2} \int_{\mathbb{R}^n} \exp\left[-\frac{1}{2}z^T \Lambda z\right] dz \\ &= (2\pi)^{-n/2} [\det C]^{-1/2} \int_{\mathbb{R}^n} \exp\left[-\frac{1}{2}\sum\nolimits_{j=1}^n \lambda_j z_j^2\right] dz,\end{aligned}$$

recalling that $|\det Q| = 1$ and Λ is diagonal.

Finally, (2) obtains that:

$$\det C^{-1} = \det \Lambda = \prod\nolimits_{j=1}^n \lambda_j,$$

and since $\det C^{-1} = 1/\det C$, this yields $[\det C]^{-1/2} = \prod_{j=1}^n \sqrt{\lambda_j}$.

Thus by Fubini's theorem of Proposition V.5.16, recognizing the density functions of the normal variate in (6.1):

$$\int f_Y(y)dy = \prod\nolimits_{j=1}^n \left[\sqrt{\frac{\lambda_j}{2\pi}} \int_{\mathbb{R}} \exp\left(-\frac{1}{2}\lambda_j z_j^2\right) dz\right] = 1.$$

■

Remark 6.13 (On $\int f_Y(y)dy = 1$) *If $f_Y(y)$ is defined as in (6.6) with C an $n \times n$ positive definite matrix and given $\mu \in \mathbb{R}^n$, there is an alternative way to see that such $f_Y(y)$ is a density function besides the direct integration of Proposition 6.12.*

By the spectral theorem, $C = Q\Lambda Q^T$, where the eigenvalues $\{\lambda_j\}_{j=1}^n$ of C in the diagonal matrix Λ satisfy $\lambda_j > 0$ for all j by Remark 6.8. Letting $\sqrt{\Lambda}$ denote the diagonal matrix with elements $\{\sqrt{\lambda_j}\}_{j=1}^n$:

$$\begin{aligned} C &= Q\sqrt{\Lambda}\sqrt{\Lambda}Q^T \\ &\equiv AA^T, \end{aligned}$$

where $A = Q\sqrt{\Lambda}$ is an $n \times n$ matrix. Further, A is nonsingular since Q is orthogonal and thus nonsingular, and all $\lambda_j > 0$.

If $X \equiv (X_1, X_2, ..., X_n)$ is a vector of independent standard normal variables, define $Y = AX + \mu$. Then by Proposition 6.5, $f_Y(t)$ in (6.6) is the density function for Y, and the derivation is complete.

In summary, all functions $f_Y(y)$ as defined as in (6.6), with C an $n \times n$ positive definite matrix and given μ, are density functions because they are the density functions of $Y = AX + \mu$ with A obtainable as above.

See also Proposition 6.34.

In the next section we take a characteristic function approach to identifying properties of the random vector $Y = AX + \mu$, where X is again a random vector of independent standard normals, but where we no longer assume that A is invertible, or even square. Thus we end this section identifying the moment generating and characteristic functions of Y from the density function in Proposition 6.5.

By Remark 6.13, this is equivalent to the assumption that Y is defined by the transformation $Y = AX + \mu$, with A invertible and $C \equiv AA^T$.

Proposition 6.14 ($M_Y(t)$, $C_Y(t)$) *Let $f_Y(y)$ be defined as in (6.6) with C an $n \times n$ positive definite matrix and given μ. Then:*

$$\begin{aligned} M_Y(t) &= \exp\left(\mu \cdot t + \tfrac{1}{2}t^T Ct\right), \\ C_Y(t) &= \exp\left(i\mu \cdot t - \tfrac{1}{2}t^T Ct\right). \end{aligned} \tag{6.7}$$

Proof. *We derive $M_Y(t)$, and then the result for $C_Y(t)$ follows from (5.67).*

Using Definition 5.17 as transformed to the Lebesgue integral in (5.31) of Remark 5.20, obtains with $c \equiv (2\pi)^{-n/2}[\det C]^{-1/2}$ and a change of variables $y - \mu \to x$ by Proposition V.4.44:

$$\begin{aligned} M_Y(t) &= c\int_{\mathbb{R}^n} \exp[t \cdot y] \exp\left[-\frac{1}{2}(y-\mu)^T C^{-1}(y-\mu)\right] dy \\ &= c\exp[t \cdot \mu]\int_{\mathbb{R}^n} \exp[t \cdot x] \exp\left[-\frac{1}{2}x^T C^{-1}x\right] dx. \end{aligned}$$

Rewriting:

$$M_Y(t) = c\exp\left(\mu \cdot t + \frac{1}{2}t^T Ct\right)\int_{\mathbb{R}^n} \exp[g(x)]\,dx,$$

where:

$$g(x) = t \cdot x - \frac{1}{2}t^T Ct - \frac{1}{2}x^T C^{-1}x. \tag{1}$$

We claim that for a vector a to be identified that:

$$g(x) = -\frac{1}{2}(x-a)^T C^{-1}(x-a). \tag{2}$$

Expanding (2), *this requires:*

$$g(x) = -\frac{1}{2}x^T C^{-1}x + a^T C^{-1}x - \frac{1}{2}a^T C^{-1}a.$$

Comparing with (1), *this obtains:*

$$a^T C^{-1} = t^T,$$
$$a^T C^{-1}a = t^T Ct.$$

Recalling that C is symmetric obtains $C^{-1}a = t$ from the first identity, and thus $a = Ct$, and this satisfies the second identity.

Thus:

$$M_Y(t) = c\exp\left(\mu \cdot t + \frac{1}{2}t^T Ct\right)\int_{\mathbb{R}^n} \exp\left[-\frac{1}{2}(x - Ct)^T C^{-1}(x - Ct)\right] dx,$$

and a change of variables $x - Ct \to z$ obtains the density in (6.6), which integrates to 1 by Proposition 6.12. ■

6.1.2 Characteristic Function Approach

In this section we consider more general transformations: $Y \equiv AX + \mu$, than those possible with the density function approach. Specifically, we no longer assume that A is invertible, nor even that it is $n \times n$. In this general setting, we use moment generating and characteristic functions, noting that moment generating functions will be seen to always exist within this model.

For general $m \times n$ matrix $A : \mathbb{R}^n \to \mathbb{R}^m$ below, the matrix $C \equiv AA^T$ is well-defined and symmetric, but is now an $m \times m$ matrix. As noted in Remark 6.10, C will generally not be positive definite, but will always be positive semidefinite. That said, the formulas for $M_Y(t)$ and $C_Y(t)$ below are identical to the respective formulas of Proposition 6.14, which applied when A was invertible.

Proposition 6.15 **($M_Y(t)$, $C_Y(t)$ for $Y = AX + \mu$)** *Let $X \equiv (X_1, X_2, ..., X_n)$ be a vector of independent standard normal variables, $A : \mathbb{R}^n \to \mathbb{R}^m$ an $m \times n$ matrix, and $\mu = (\mu_1, \mu_2, ..., \mu_m) \in \mathbb{R}^m$. If $Y \equiv (Y_1, Y_2, ..., Y_m)$ is defined by $Y \equiv AX + \mu$, then for all $t \equiv (t_1, t_2, ..., t_m)$:*

$$\begin{aligned} M_Y(t) &= \exp\left(\mu \cdot t + \tfrac{1}{2}t^T Ct\right), \\ C_Y(t) &= \exp\left(i\mu \cdot t - \tfrac{1}{2}t^T Ct\right), \end{aligned} \tag{6.8}$$

where $C \equiv AA^T$.

Proof. *Recalling that $M_{X_i}(s_i) = \exp\left(\frac{1}{2}s_i^2\right)$ from IV.(4.79), it follows by independence and Corollary 5.4 that with $s \equiv (s_1, s_2, ..., s_n)$:*

$$M_X(s) = \prod\nolimits_{i=1}^{n} M_{X_i}(s_i) = \exp\left(\frac{1}{2}s^T Is\right),$$

where I is the identity matrix and thus $s^T Is = \sum_{i=1}^n s_i^2$.

Then from (5.3) of Exercise 6.4:

$$\begin{aligned} M_Y(t) &= e^{\mu \cdot t} M_X(A^T t) \\ &= \exp\left(\mu \cdot t + \frac{1}{2}\left(A^T t\right)^T I \left(A^T t\right)\right) \\ &= \exp\left(\mu \cdot t + \frac{1}{2}t^T Ct\right), \end{aligned}$$

with $C \equiv AA^T$ as above.

The same approach works for $C_Y(t)$, now using (5.44) and Corollary 5.34. More simply, we note that $C_Y(t) = M_Y(it)$ by (5.67). ■

The following exercise identifies the significance of μ and C in terms of the moments of Y. These moments can also be derived in the more limited context of Proposition 6.5 using the density function $f_Y(y)$, where $C = AA^T$ with A an $n \times n$ invertible matrix.

Besides being more general, the reader will undoubtedly see that the following approach is far simpler.

Exercise 6.16 (μ, C, as moments of Y) *Let Y be defined by $Y \equiv AX + \mu$, with $A : \mathbb{R}^n \to \mathbb{R}^m$ an $m \times n$ matrix and $\mu = (\mu_1, \mu_2, ..., \mu_m) \in \mathbb{R}^m$. Apply (5.65) of Proposition 5.35 to $C_Y(t)$ to prove:*

$$\mu_i = E[Y_i], \ C_{ii} = Var[Y_i], \ C_{ij} = Cov[Y_i, Y_j]. \tag{6.9}$$

Thus μ_i is the mean of Y_i, C_{ii} the variance of Y_i, and for $i \neq j$, C_{ij} is the covariance of Y_i and Y_j:

$$Var[Y_i] \equiv E\left[(Y_i - \mu_i)^2\right], \qquad Cov[Y_i, Y_j] \equiv E[(Y_i - \mu_i)(Y_j - \mu_j)]. \tag{6.10}$$

6.1.3 Multivariate Normal Definition

Given the affine transformation $Y = AX + \mu$, where $X \equiv (X_1, X_2, ..., X_n)$ are independent standard normal variables and μ is a constant vector, the above sections highlight the significance of the matrix $C \equiv AA^T$. When A is invertible, Y has a density function by Proposition 6.5 in which C plays a prominent role, while in that case and the more general case of Proposition 6.15, the matrix C plays a prominent role in the characteristic and moment generating functions of Y.

With the above results, we are ready to define when a random vector $Y \equiv (Y_1, ..., Y_n)$ will be said to have a **multivariate normal distribution.**

We state that C is nondegenerate to avoid the case where $C = Z$, the zero matrix, and then the characteristic function $C_Y(t)$ in (6.11) is that of a constant random vector, $Y \equiv \mu$. Such random variables or random vectors are also called **degenerate.**

Definition 6.17 (Multivariate normal random vector) *A random vector $Y \equiv (Y_1, ..., Y_n)$ is said to have a **multivariate normal distribution,** or a **multivariate Gaussian distribution,** if there exists a nondegenerate $n \times n$ positive semidefinite matrix C and an n-vector μ so that for $t = (t_1, ..., t_n)$:*

$$C_Y(t) = \exp\left[i\mu \cdot t - \frac{1}{2}t^T Ct\right]. \tag{6.11}$$

*We will express this a $Y \sim MN_n(\mu, C)$, and say that Y is **multivariate normally distributed with μ and C.***

The next exercise proves that when C is positive definite, there is a close link between multivariate normal random vectors $Y \equiv (Y_1, ..., Y_n) \sim MN_n(\mu, C)$, and the normality of any linear combination of the component variates. See also Proposition 6.23 and Example 6.24.

Exercise 6.18 ($Y \sim MN_n(\mu, C)$ versus $\sum_{i=1}^n \lambda_i Y_i$ is normal) *Let C be positive definite. Prove that $Y \equiv (Y_1, ..., Y_n)$ is multivariate normal with μ and C if and only if $Z \equiv \sum_{i=1}^n \lambda_i Y_i$ is a normal random variable for all $\{\lambda_i\}_{i=1}^n \subset \mathbb{R}$ with at least one λ_i*

nonzero. Hint: Use characteristic functions in (6.11) and (5.44), and the uniqueness results for characteristic functions. Recall the general formulas for the mean and variance of Z from IV.(4.39) and IV.(4.46).

The mathematical form of the characteristic function in (6.11) was derived in (6.7) for random vectors Y with a density function, where C was positive definite. It was also derived for random vectors $Y = AX + \mu$ with associated positive semidefinite C. However, it is by no means obvious that so defined, that the functional form of $C_Y(t)$ above always represents a characteristic function. Stated as a question:

Given any nondegenerate positive semidefinite matrix C and n-vector μ, is $C_Y(t)$ as defined in (6.11) the characteristic function of a random vector?

Bochner's theorem of Proposition 5.51 addresses this question, and states that $C_Y(t) : \mathbb{R}^n \to \mathbb{C}$ is the characteristic function of a distribution function defined on $\mathbb{R}^n$, if and only if $C_Y(t)$ is a positive semidefinite function, continuous at $t = 0$, and $C_Y(0) = 1$. This result is applicable by Proposition 1.24, which states that given any distribution function defined on $\mathbb{R}^n$, there exists a random vector with this distribution function. Thus such $C_Y(t)$ is the characteristic function of a distribution function defined on $\mathbb{R}^n$ if and only if it is the characteristic function of a random vector.

Proposition 6.19 (exp $\left[i\mu \cdot t - \frac{1}{2}t^TCt\right]$ **is a characteristic function)** *For any nondegenerate $n \times n$ positive semidefinite matrix C and n-vector μ, the function $C_Y(t) = \exp\left[i\mu \cdot t - \frac{1}{2}t^TCt\right]$ of (6.11) is the characteristic function of a distribution function on $\mathbb{R}^n$.*

Proof. *We prove this by verifying the requirements of Bochner's theorem.*

Continuity at $t = 0$ is apparently satisfied, as is $C_Y(0) = 1$. For positive semidefiniteness, given m, $\{t_j\}_{j=1}^m \subset \mathbb{R}^n$, and $\{z_j\}_{j=1}^m \subset \mathbb{C}$, we prove that:

$$Q \equiv \sum\nolimits_{j=1}^{m} \sum\nolimits_{k=1}^{m} C_Y(t_j - t_k) z_j \bar{z}_k \geq 0, \tag{1}$$

where $\bar{z}_k$ is the complex conjugate of z_k.

First note that:

$$-\frac{1}{2}(t_j - t_k)^T C(t_j - t_k) = -\frac{1}{2}{t_j}^T C t_j - \frac{1}{2}{t_k}^T C t_k + {t_j}^T C t_k, \tag{2}$$

since ${t_j}^T C t_k = {t_k}^T C t_j$. This follows from Exercise 6.4 and the observation that since ${t_j}^T C t_k$ is a number, $\left({t_j}^T C t_k\right)^T = {t_j}^T C t_k$.

Using (2)*, the definition of $C_Y(t)$, and some algebra:*

$$Q = \bar{w}^T A w, \tag{3}$$

where A is an $m \times m$ matrix with $A_{jk} = {t_j}^T C t_k$, and $w = (w_1, ..., w_m)$ with:

$$w_j = z_j \exp\left[i\mu \cdot t_j - \frac{1}{2}{t_j}^T C t_j\right].$$

To complete the proof, let $w = a + bi$ where a, $b \in \mathbb{R}^m$. Then with $\bar{w} = a - bi$:

$$\begin{aligned}(a - bi)^T A\,(a + bi) &= a^T A a - i^2 b^T A b + i\left[a^T A b - b^T A a\right] \\ &= a^T A a + b^T A b,\end{aligned} \tag{4}$$

since $a^T A b = b^T A a$ as noted above.

Since C is a positive semidefinite matrix and $A_{jk} = t_j{}^T C t_k$, given $a \in \mathbb{R}^m$:

$$a^T A a = \sum\nolimits_{j=1}^{m} \sum\nolimits_{k=1}^{m} (a_j t_j)^T C (a_k t_k) \geq 0,$$

where $a_j t_j \equiv (a_j t_{j,1}, ..., a_j t_{j,n})$. Similarly, $b^T A b \geq 0$ for all $b \in \mathbb{R}^m$. Thus by (3) and (4), Q is real and nonnegative, and the proof is complete. ■

The next result sharpens the conclusion of Bochner's theorem, and identifies the distribution function with the characteristic function in (6.11).

Proposition 6.20 **(exp $\left[i\mu \cdot t - \frac{1}{2}t^T Ct\right]$ is the CF of $Y = AX + \mu$)** *Given any nondegenerate positive semidefinite matrix C and n-vector μ, $C_Y(t)$ as defined in (6.11) is the characteristic function of the random vector $Y = AX + \mu$, with $C = AA^T$ for nondegenerate $n \times n$ matrix A, and $X \equiv (X_1, X_2, ..., X_n)$ a vector of independent standard normal variables.*

Proof. *Since C is symmetric by definition of positive semidefinite, it follows from the spectral theorem of Proposition 6.11 that:*

$$C = Q\Lambda Q^T.$$

Recall Λ is a diagonal matrix of eigenvalues $\{\lambda_j\}_{j=1}^n$, Q has orthogonal eigenvectors $\{Q^{(j)}\}_{j=1}^n$ as columns, and $Q^T = Q^{-1}$. By Remark 6.8, positive semidefiniteness of C assures that all $\lambda_j \geq 0$.

Let $\sqrt{\Lambda}$ denote the diagonal matrix with elements $\{\sqrt{\lambda_j}\}_{j=1}^n$. Recalling Exercise 6.4:

$$\begin{aligned} C &= Q\sqrt{\Lambda}\sqrt{\Lambda}Q^T \\ &= AA^T, \end{aligned}$$

with $A \equiv Q\sqrt{\Lambda}$ an $n \times n$ matrix. Thus A will be nonsingular if C is positive definite and then all $\lambda_j > 0$, and will be singular if any $\lambda_j = 0$. Since C is nondegenerate, at least one $\lambda_j \neq 0$, and thus A is a nondegenerate matrix.

Let $X \equiv (X_1, X_2, ..., X_n)$ be a vector of independent standard normal variables, and define $Y = AX + \mu$. Then by Proposition 6.15, $C_Y(t)$ is given as in (6.11), and the proof is complete. ■

The next result consolidates conclusions now derivable from the above investigations. The key point is that despite the abstraction of Definition 6.17, a random vector is multivariate normal if and only if it equals the affine transformation of a vector of independent standard normal variates.

Proposition 6.21 **($Y \sim MN_n(\mu, C)$ if and only if $Y = AX + \mu$)** *A random vector $Y \equiv (Y_1, ..., Y_n)$ has a multivariate normal distribution by Definition 6.17, $Y \sim MN_n(\mu, C)$, if and only if there exists a nondegenerate $n \times n$ matrix A so that $C = AA^T$ and:*

$$Y = AX + \mu,$$

where $X \equiv (X_1, X_2, ..., X_n)$ is a vector of independent standard normal variables.

In addition, any such random vector has a moment generating function given by:

$$M_Y(t) = \exp\left[\mu \cdot t + \frac{1}{2}t^T Ct\right], \tag{6.12}$$

and when A is invertible, Y has a density function given in (6.6).

Proof. *The first statement combines the results of Propositions 6.15 and 6.20.*

The moment generating function in (6.12) follows from this characterization and Proposition 6.15, while the existence of a density is Proposition 6.5. ■

Remark 6.22 (On $Y = AX + \mu$) *It is worth a few moments to contemplate the relationship between the earlier result in Proposition 6.15, and that in Proposition 6.21, and for this we change notation a bit.*

In Proposition 6.15, if $X \equiv (X_1, X_2, ..., X_n)$ is a vector of independent standard normal variables, $B : \mathbb{R}^n \to \mathbb{R}^m$ an $m \times n$ matrix, and $\mu = (\mu_1, \mu_2, ..., \mu_m)$ a constant vector, then $Y \equiv (Y_1, Y_2, ..., Y_m)$ defined by $Y \equiv BX + \mu$ has characteristic function $C_Y(t)$ as in Definition 6.17, with $C = BB^T$ an $m \times m$ matrix. In other words, $Y \sim MN_m(\mu, C)$.

Proposition 6.21 then states that for any such Y there exists an $m \times m$ matrix $A : \mathbb{R}^m \to \mathbb{R}^m$, so that $C = AA^T$ and $Y = AX' + \mu$, where $X' \equiv (X_1', X_2', ..., X_m')$ is a vector of independent standard normal variables.

It thus appears that $Y \sim MN_m(\mu, C)$ can be generated by either an n-vector or an m-vector of independent standard normal variables, which perhaps seems odd when $n \neq m$.

We will see in Corollaries 6.31 and 6.32, that:

$$\mathit{rank}(C) = \mathit{rank}(A) = \mathit{rank}(B) \leq \min(n, m),$$

*where **rank**$(\cdot)$ is the dimension of the range space $\mathcal{R}(\cdot)$ of these matrices. While $A, C : \mathbb{R}^m \to \mathbb{R}^m$ and $B : \mathbb{R}^n \to \mathbb{R}^m$, the range in $\mathbb{R}^m$ of these matrices is always the same, and never exceeds* $\min(n, m)$.

Thus AX' and BX are contained in spaces of the same dimension, and of dimension m or less, whether we start with an m-vector X' or an n-vector X.

The final result proves that affine transformations of multivariate normal vectors obtain multivariate normal vectors. This then provides a simple corollary on the marginal distributions of multivariate normal vectors.

Proposition 6.23 (If $Y \sim MN_n(\mu, C)$ then $BY + \nu \sim MN_m(B\mu + \nu, BCB^T)$) *Let $Y \equiv (Y_1, Y_2, ..., Y_n)$ be multivariate normally distributed by Definition 6.17 with μ and C, and let $B : \mathbb{R}^n \to \mathbb{R}^m$ be a nondegenerate $m \times n$ matrix and ν and m-vector.*

- *If C is positive definite, then $Z \equiv BY + \nu$ is multivariate normally distributed on $\mathbb{R}^m$ with $\mu_Z = B\mu + \nu$ and $C_Z = BCB^T$.*
- *If C is positive semidefinite, the same result is true unless C_Z is degenerate.*

Proof. *By (5.61) and (6.11), and recalling Exercise 6.4:*

$$\begin{aligned} C_Z(s) &= \exp[i\nu \cdot s]\, C_Y(B^T s) \\ &= \exp[i\nu \cdot s] \exp\left[i\mu \cdot B^T s - \frac{1}{2}\left(B^T s\right)^T C B^T s\right] \\ &= \exp\left[i\left(B\mu + \nu\right) \cdot s - \frac{1}{2} s^T B C B^T s\right]. \end{aligned}$$

Now BCB^T is symmetric by Exercise 6.4 and the symmetry of C:

$$\left(BCB^T\right)^T = \left(B^T\right)^T (C)^T (B)^T = BCB^T.$$

Further, BCB^T is positive semidefinite because C has this property. If $x \in \mathbb{R}^m$:

$$x^T BCB^T x = \left(B^T x\right)^T C \left(B^T x\right) \geq 0.$$

In summary, Z has the characteristic function as in (6.11), with symmetric, positive semidefinite $C_Z = BCB^T$, and vector μ_Z, so Z is multivariate normal unless C_Z is degenerate.

- *If C is positive definite, then nondegeneracy of B implies nondegeneracy of B^T by (6.14), so there exists x with $y = B^T x \neq 0$ and thus $x^T BCB^T x = y^T Cy > 0$, so BCB^T is nondegenerate.*
- *If C is positive semidefinite, then BCB^T can be degenerate if $y^T Cy = 0$ for all $y = B^T x$. This happens for example if, recalling Definition 6.1, $\mathcal{R}(B^T) \subset \mathcal{N}(C)$.*

■

Example 6.24 ($Y \sim MN_n(\mu, C)$ and $B \cdot Y$) *Recalling Exercise 6.18, if $Y \sim MN_n(\mu, C)$ and $B = (b_1, ..., b_n)$ is nondegenerate, $B : \mathbb{R}^n \to \mathbb{R}$, then considered as a vector or a $1 \times n$ row matrix:*

$$B \cdot Y = BY = \sum\nolimits_{j=1}^{n} b_j Y_j.$$

When C is positive definite, then $Z = BY$ has a normal distribution with $\mu_Z = B\mu = \sum_{j=1}^{n} b_j \mu_j$, and $C_Z = BCB^T > 0$ by positive definiteness. In this case $C_Z = \sigma_Z^2$.

If C is positive semidefinite, then $x^T Cx = 0$ for some $x \neq 0$. Thus if $B = x^T$ then $\sigma_Z^2 = 0$ and Z is a degenerate random variable.

Proposition 6.23 provides a simple approach to proving that if $Y \sim MN_n(\mu, C)$, then each marginal distribution of Y is multivariate normal by this definition, or degenerate.

For this, recall Definition 1.45, that given a distribution function $F(x_1, x_2, ..., x_n)$ and $I = \{i_1, ..., i_m\} \subset \{1, 2, ..., n\}$, the **marginal distribution function** $F_I(x_I) \equiv F_I(x_{i_1}, x_{i_2}, ..., x_{i_m})$ is defined on $\mathbb{R}^m$ by:

$$F_I(x_I) \equiv \lim_{x_J \to \infty} F(x_1, x_2, ..., x_n),$$

where $x_J \equiv (x_{j_1}, x_{j_2}, ..., x_{j_{n-m}})$ for $j_k \in J \equiv \widetilde{I}$.

It was seen in Proposition II.3.36 that $F_I(x_I)$ is the joint distribution function of $X_I \equiv (X_{i_1}, X_{i_2}, ..., X_{i_m})$.

Corollary 6.25 (On marginals of $Y \sim MN_n(\mu, C)$) *Let $Y \equiv (Y_1, Y_2, ..., Y_n)$ be multivariate normally distributed with μ and C, and $I = \{i_1, ..., i_m\} \subset \{1, 2, ..., n\}$. Then:*

- *If C is positive definite, then $Y_I \equiv (Y_{i_1}, Y_{i_2}, ..., Y_{i_m})$ is multivariate normally distributed with $\mu_I = (\mu_{i_1}, \mu_{i_2}, ..., \mu_{i_m})$ and C_I the submatrix of covariances between all Y_{i_j} and Y_{i_k}.*
- *If C is positive semidefinite, then the same result is true unless C_I is degenerate.*

Proof. *Define the matrix $B_I : \mathbb{R}^n \to \mathbb{R}^m$ by $B_I : (x_1, x_2, ..., x_n) \to (x_{i_1}, x_{i_2}, ..., x_{i_m})$. Thus if $B_I = \{b_{ji}\}$ is $m \times n$, then $b_{ji_j} = 1$ and $b_{ji} = 0$ for $i \neq i_j$.*

Then by the above result, Y_I is multivariate normal with:

$$\mu_I = B_I \mu = (\mu_{i_1}, \mu_{i_2}, ..., \mu_{i_m}),$$

while:

$$\begin{aligned} C_I &= B_I C B_I^T \\ &= \{C_{i_j, i_k}\}_{j,k=1}^{m}. \end{aligned}$$

In other words, C_I is the submatrix of elements of C with indexes i_j and i_k, and by Exercise 6.16, C_{i_j, i_k} equals the covariance of Y_{i_j} and Y_{i_k}.

The conclusions for positive definite C, respectively, positive semidefinite C, now follow Proposition 6.23. ■

Example 6.26 ($m = 1, 2$) *For $m = 1$ and simplifying notation, let $B_j : \mathbb{R}^n \to \mathbb{R}$ be defined by $B_j : (Y_1, Y_2, ..., Y_n) \to Y_j$. Formally, B_j is given by a $1 \times n$ matrix: $B = (0, .., 0, 1, 0, ..., 0)$ with 1 in the jth component. Then Y_j has a normal distribution with parameters μ_j and σ_{jj}, which are the mean and variance of Y_j. As in Example 6.24, the variance of Y_j is positive if C is positive definite, and could be 0 in the general case.*

For $m = 2$, let $B_{ij} : \mathbb{R}^n \to \mathbb{R}^2$ with $B_{ij} : (Y_1, Y_2, ..., Y_n) \to (Y_i, Y_j)$, so here, B_{ij} is given by a $2 \times n$ matrix with a 1 in the ith column of row 1, and the jth column of row 2. Then (Y_i, Y_j) is multivariate normal with vector $\mu_{ij} \equiv (\mu_i, \mu_j)$ and covariance matrix $C_{ij} \equiv \begin{pmatrix} c_{ii} & c_{ij} \\ c_{ij} & c_{jj} \end{pmatrix}$ when C is positive definite, but can be degenerate in the general case.

6.2 Existence of Densities

In this section, we investigate when a multivariate normal random vector by Definition 6.17, $Y \sim MN_n(\mu, C)$, has an associated density function.

We begin by summarizing results in the case where $Y = AX + \mu$ with associated $C = AA^T$. While this seemed to be a special case for multivariate normals in Section 6.1.1, it follows from Proposition 6.21 that this is the general case.

Example 6.27 ($Y = AX + \mu$; $C = AA^T$) *Let $X \equiv (X_1, X_2, ..., X_n)$ be a vector of n independent unit normals, and $Y \equiv (Y_1, Y_2, ..., Y_m)$ defined by $Y = AX + \mu$, where $A : \mathbb{R}^n \to \mathbb{R}^m$ is a nondegenerate $m \times n$ matrix and $\mu \equiv (\mu_1, \mu_2, ..., \mu_m)$ is a constant vector. By Proposition 6.15, Y has characteristic function $C_Y(t)$ that satisfies Definition 6.17 with $C = AA^T$, an $m \times m$ matrix.*

When does Y have a density function?

1. *$m = 1$:*

   Now $\mu \in \mathbb{R}$, $A = (a_1, ..., a_n)$ is a $1 \times n$ matrix, and $C = \sum_{j=1}^{n} a_j^2 > 0$ since A is assumed nondegenerate. Then $C_Y(t)$ of Proposition 6.15 equals the characteristic function in (5.44) of a normal variate Y' with given μ and $\sigma^2 \equiv C$, and density function given in (6.1) with these parameters.

   By the uniqueness theorem of Proposition 5.29, the variates Y and Y' have the same distribution function, and thus Y will always have a density given by the density of Y'.

2. *$m = n > 1$* ***and A invertible:***

   In this case, Y has density function $f_Y(y)$ defined in (6.6) by Proposition 6.5, and associated characteristic function in (6.7) of Proposition 6.14 with $C = AA^T$. This characteristic function satisfies Definition 6.17, since C is positive definite by Remark 6.10.

   If Y' is any multivariate normal vector by Definition 6.17 with this characteristic function, then by the uniqueness theorem of Proposition 5.39, Y' and Y have the same distribution function. In particular, any such Y' has a density function.

3. *$m = n > 1$* ***and A is not invertible, or,*** *$m > 1$* ***and*** *$m \neq n$:*

   Then Y has the characteristic function in (6.8) of Proposition 6.15 with $C = AA^T$, and thus satisfies Definition 6.17 since AA^T is positive semidefinite by Remark 6.10.

   On the question of density functions:

(a) *If* $m = n > 1$ *and* A *is not invertible, it then follows that* C *is not invertible. We will prove in Corollary 6.32 that:*

$$rank(C) = rank(A),$$

where rank is the dimension of the range space. Thus C *is never positive definite.*

(b) *If* $m > 1$ ***and*** $m \neq n$*, then it is not immediately apparent if* $C = AA^T$ *can be positive definite.*

We next exemplify the possibilities for the matrix $C = AA^T$.

Example 6.28 (On $C = AA^T$) *Here we consider examples with* $A : \mathbb{R}^n \to \mathbb{R}^m$*:*

1. $n = m$, A ***not invertible:***

   Let $A : \mathbb{R}^3 \to \mathbb{R}^3$ *be defined by* $A : (X_1, X_2, X_3) \to (X_1, X_2, X_1 + X_2)$*. Then:*

$$A = \begin{pmatrix} 1 & 0 & 0 \\ 0 & 1 & 0 \\ 1 & 1 & 0 \end{pmatrix}, \qquad AA^T = \begin{pmatrix} 1 & 0 & 1 \\ 0 & 1 & 1 \\ 1 & 1 & 2 \end{pmatrix},$$

   so $\det(AA^T) = 0$ *and the matrix* AA^T *is singular.*

   This is the general case for a singular $n \times n$ *matrix* A*, since* $\det(A^T) = \det(A)$*, and so:*

$$\det(AA^T) = [\det(A)]^2 = 0.$$

2. $n < m$*:*

   Let $A : \mathbb{R}^2 \to \mathbb{R}^3$ *be defined by* $A : (X_1, X_2) \to (X_1, X_2, X_1 + X_2)$*. Then:*

$$A = \begin{pmatrix} 1 & 0 \\ 0 & 1 \\ 1 & 1 \end{pmatrix}, \qquad AA^T = \begin{pmatrix} 1 & 0 & 1 \\ 0 & 1 & 1 \\ 1 & 1 & 2 \end{pmatrix},$$

   and $\det(AA^T) = 0$.

   This is the general case for an $m \times n$ *matrix* A*, with* $n < m$*. Since* $A^T : \mathbb{R}^m \to \mathbb{R}^n$ *and* $m > n$*, this assures by (6.4) that the dimension of the null space of* A^T *is at least* $m - n$*. Thus there exists* $x \neq 0$ *with* $A^T x = 0$*, and then also* $AA^T x = 0$*. Hence* AA^T *is singular and* $\det(AA^T) = 0$.

3. $n > m$*:*

   Let $A : \mathbb{R}^3 \to \mathbb{R}^2$ *be defined by* $A : (X_1, X_2, X_3) \to (X_1 + X_2, X_2 + X_3)$*. Then:*

$$A = \begin{pmatrix} 1 & 1 & 0 \\ 0 & 1 & 1 \end{pmatrix}, \qquad AA^T = \begin{pmatrix} 2 & 1 \\ 1 & 2 \end{pmatrix},$$

   and AA^T *is invertible with* $\det A > 0$*, so* AA^T *is positive definite.*

   On the other hand, defining $A : (X_1, X_2, X_3) \to (X_1 + X_2 + X_3, X_1 + X_2 + X_3)$ *leads to the conclusion that* $\det(AA^T) = 0$.

   Thus when $n > m$*,* AA^T *can be positive definite, but can also be singular and thus positive semidefinite.*

These examples illustrate some general results from linear algebra for a matrix $A : \mathbb{R}^n \to \mathbb{R}^m$ where the elements of A are real. We summarize the needed results below. For more detail on these and related results the reader is referred to **Hoffman and Kunze** (1971) or **Strang** (2009).

Definition 6.29 (Column/row rank; Rank) *Given a real $m \times n$ matrix $A : \mathbb{R}^n \to \mathbb{R}^m$:*

- ***Column rank***(A)*: The dimension of the subspace of $\mathbb{R}^m$ spanned by the n column vectors of A;*
- ***Row rank***(A)*: The dimension of the subspace of $\mathbb{R}^n$ spanned by the m row vectors of A;*
- ***Rank***(A)*: The dimension of $\mathcal{R}(A)$, the range space of A.*

The following result is known as the **rank theorem** which we state without proof. Quite remarkably, it states that the column rank and row rank agree.

We then investigate corollaries to be applied below.

Proposition 6.30 (Rank theorem) *Given a real matrix $A : \mathbb{R}^n \to \mathbb{R}^m$:*

$$\text{column rank}(A) = \text{row rank}(A). \tag{6.13}$$

Proof. *See for example, Theorem 3 of Chapter 3 in Hoffman and Kunze.* ■

Corollary 6.31 (Rank theorem) *Given a real matrix $A : \mathbb{R}^n \to \mathbb{R}^m$:*

$$\text{rank}(A) = \text{rank}\left(A^T\right), \tag{6.14}$$

and thus:

$$\text{rank}(A) \leq \min(n, m). \tag{6.15}$$

Proof. *Given $x = (x_1, ..., x_n)$:*

$$Ax = \sum\nolimits_{j=1}^{n} A^{(j)} x_j, \tag{1}$$

where $\{A^{(j)}\}_{j=1}^n$ are the columns of A. Hence the range of A is spanned by the columns of A and so:

$$\text{rank}(A) = \text{column rank}(A). \tag{2}$$

Similarly,

$$\text{rank}\left(A^T\right) = \text{column rank}(A^T) = \text{row rank}(A). \tag{3}$$

Combining (2) and (3) obtains (6.14) by the rank theorem.

By definition, $\text{rank}(A) \leq m$, while $\text{rank}(A^T) \leq n$, so (6.15) follows from (6.14). ■

Corollary 6.32 (Rank theorem) *Given a real matrix $A : \mathbb{R}^n \to \mathbb{R}^m$:*

$$\text{rank}(A) = \text{rank}\left(AA^T\right) = \text{rank}\left(A^T A\right). \tag{6.16}$$

Proof. *Since $\left(AA^T\right)^T = A^T A$ by Exercise 6.4, the second equality in (6.16) is (6.14). For the first equality, $\text{rank}(AA^T) = \text{column rank}(AA^T)$ by (2) of the prior proof, and we show that:*

$$\text{column rank}(AA^T) = \text{column rank}(A). \tag{1}$$

Let:

$$A^T = \left(A^T_{(1)}, ...A^T_{(m)}\right),$$

where $\{A^T_{(i)}\}_{i=1}^m \subset \mathbb{R}^n$ *are the m-columns of* A^T*. The m-columns of* AA^T *are then:*

$$AA^T = \left(AA^T_{(1)}, ...AA^T_{(m)}\right).$$

By (1) *of the prior proof, for any i and* $A^T_{(i)} = (x_{1i}, ..., x_{ni})$*:*

$$AA^T_{(i)} = \sum\nolimits_{j=1}^n A^{(j)} x_{ji}.$$

In other words, every column of the matrix AA^T *is a linear combination of the columns of* A*, and* (1) *is proved.* ■

Example 6.33 (On 2, 3 of Example 6.28) *Corollary 6.32 provides another perspective on items 2 and 3 of Example 6.28, where* $A : \mathbb{R}^n \to \mathbb{R}^m$*.*

2. AA^T ***is always singular when*** $n < m$*:*

   Combining (6.16) with (6.15) obtains:

   $$\text{rank}\left(AA^T\right) \leq \min(n, m) = n.$$

   Since $AA^T : \mathbb{R}^m \to \mathbb{R}^m$*, this states that the range space* $\mathcal{R}(AA^T)$ *has dimension at most* n*, and thus by (6.4), the null space* $\mathcal{N}(AA^T)$ *has dimension at least* $m - n$*.*

3. AA^T ***can be singular or nonsingular when*** $n > m$*:*

   By the same application of the above results:

   $$\text{rank}\left(AA^T\right) \leq \min(n, m) = m.$$

   Since $AA^T : \mathbb{R}^m \to \mathbb{R}^m$*,* AA^T *will be nonsingular when rank*$(AA^T) = m$*, and singular otherwise.*

We are finally ready for the main result on existence of density functions.

Proposition 6.34 (Existence of a density for $Y \sim MN_n(\mu, C)$**)** *Given a multivariate normally distributed random vector* $Y \sim MN_n(\mu, C)$*, then* Y *has a density function* $f_Y(y)$ *if and only if* C *is positive definite, and in this case,* $f_Y(y)$ *is given in (6.6).*

Proof. *Given* $Y \sim MN_n(\mu, C)$*, then since* C *is symmetric by definition of positive semidefinite, the spectral theorem of Proposition 6.11 obtains:*

$$C = Q\Lambda Q^T,$$

where Λ *is a diagonal matrix of eigenvalues* $\{\lambda_j\}_{j=1}^n$*, and* Q *has orthogonal eigenvectors* $\{Q^{(j)}\}_{j=1}^n$ *as columns. Hence,* Q *is orthogonal and* $Q^T = Q^{-1}$*. Since* C *is nondegenerate, at least one* $\lambda_j \neq 0$*.*

By Remark 6.8, positive semidefiniteness of C *assures that all* $\lambda_j \geq 0$*, so let* $\sqrt{\Lambda}$ *denote the diagonal matrix with elements* $\{\sqrt{\lambda_j}\}_{j=1}^n$*. Recalling Exercise 6.4:*

$$\begin{aligned} C &= Q\sqrt{\Lambda}\sqrt{\Lambda}Q^T \\ &= AA^T, \end{aligned}$$

where $A = Q\sqrt{\Lambda}$ *is an* $n \times n$ *matrix.*

If C *is positive definite, then for all* $x \neq 0$*:*

$$0 < x^T C x = x^T AA^T x = \left(A^T x\right)^T A^T x = \left|A^T x\right|^2. \tag{1}$$

Thus A^T is invertible, as is A by Exercise 6.4, and so Y has the density function $f_Y(y)$ in (6.6) by Proposition 6.5.

Conversely, assume that C is positive semidefinite and not positive definite. Let ν_Y denote the probability measure on $\mathbb{R}^n$ of Proposition 1.27 induced by the distribution function $F_Y(y)$ of Y. If $F_Y(y)$ had a density function $f_Y(y)$, or equivalently if ν_Y had a density function $f_Y(y)$ by Proposition 1.34, then for all Borel sets $H \in \mathcal{B}(\mathbb{R}^n)$:

$$\nu_Y(H) = \int_H f_Y(y)dm, \tag{2}$$

defined as a Lebesgue integral. From (2) and item 9 of Proposition V.3.42, if $m(H) = 0$, then $\nu_Y(H) = 0$. In other words, ν_Y is absolutely continuous with respect to m, denoted $\nu_Y \ll m$.

We now prove that there exists a Borel set $H \subset \mathbb{R}^n$ with $m(H) = 0$ but $\nu_Y(H) = 1$, and thus by contradiction, this proves that such $f_Y(y)$ does not exist.

To this end, if $X \equiv (X_1, X_2, ..., X_n)$ is a vector of independent unit normals, the random vector $Y' \equiv AX + \mu$ has the same characteristic function as Y by Proposition 6.15, and then by uniqueness of Proposition 5.39, $Y = Y'$ in distribution. Thus we can assume with the above notation that:

$$Y = Q\sqrt{\Lambda}X + \mu.$$

Since C is not positive definite, there exists at least one x_0 so that $x_0^T C x_0 = 0$, and thus by (1), $A^T x = 0$ and A^T is singular. By Exercise 6.4, A is singular, and since $A = Q\sqrt{\Lambda}$, it follows that at least one $\lambda_j = 0$. Assume that $\lambda_n = 0$ for notational simplicity. Then by matrix multiplication, $A = Q\sqrt{\Lambda}$ is an $n \times n$ matrix with columns $\{\sqrt{\lambda_j}Q^{(j)}\}_{j=1}^n$, and hence the nth column of A is a column of 0s. This implies that for any vector x, Ax has a zero for the nth component.

If the random vector X is defined on the probability space $(\mathcal{S}, \mathcal{E}, \lambda)$, then since $Y = AX + \mu$:

$$Y(\mathcal{S}) \subset H \equiv \{x \in \mathbb{R}^n | x_n = \mu_n\}. \tag{3}$$

Then:

$$\nu_Y(H) \equiv \lambda\left[Y^{-1}(H)\right] = 1,$$

yet $m(H) = 0$, since H has at most $n - 1$ dimensions. ■

Exercise 6.35 ($m(H) = 0$) *With H defined in (3) of the above proof, show that $m(H) = 0$. Specifically, prove that for any $\epsilon > 0$, there exists a countable disjoint collection of rectangles $\left\{\prod_{j=1}^n (a_j^{(k)}, b_j^{(k)}]\right\}_{k=1}^\infty$ such that $H \subset \bigcup_{k=1}^\infty \prod_{j=1}^n (a_j^{(k)}, b_j^{(k)}]$, and $\sum_{k=1}^\infty m\left(\prod_{j=1}^n (a_j^{(k)}, b_j^{(k)}]\right) < \epsilon$. Hint: For $j < n$ choose all $b_j^{(k)} = a_j^{(k)} + 1$, then choose $a_n^{(k)}$ and $b_n^{(k)}$ to contain μ_n and produce the measures needed. Note: For more details on how this characterization actually proves $m(H) = 0$, see Definition III.1.64 and the paragraph preceding that definition.*

Given $Y \sim MN_n(\mu, C)$, the final result of this section identifies a necessary and sufficient condition on an $m \times n$ matrix $B : \mathbb{R}^n \to \mathbb{R}^m$ so that $Z \equiv BY + \nu$ has a density function. For this result, recall that by Proposition 6.21, $C = AA^T$, where by Example 6.27, A is singular for general positive semidefinite C, and nonsingular for positive definite C.

Proposition 6.36 (Existence of a density for $Z \equiv BY + \nu$; $Y \sim MN_n(\mu, C)$) *Let $Y \sim MN_n(\mu, C)$ be a multivariate normally distributed random vector where $C = AA^T$ with*

$A : \mathbb{R}^n \to \mathbb{R}^n$, *and let* $B : \mathbb{R}^n \to \mathbb{R}^m$ *be an* $m \times n$ *matrix and* ν *an* m*-vector. Then* $Z \equiv BY + \nu$ *has a density function if and only if:*

$$\begin{aligned} \mathcal{N}(B^T) &= \{0\} \in \mathbb{R}^m, \text{ and,} \\ \mathcal{R}(B^T) \bigcap \mathcal{N}(A^T) &= \{0\} \in \mathbb{R}^n. \end{aligned} \tag{6.17}$$

Proof. *By Proposition 6.23,* $C_Z = BCB^T$. *Given* $x \in \mathbb{R}^m$ *with* $x \neq 0$, *then by Exercise 6.4:*

$$\begin{aligned} x^T C_Z x &= x^T B A A^T B^T x \\ &= \left(A^T B^T x\right)^T \left(A^T B^T x\right) \\ &= \left|A^T B^T x\right|^2. \end{aligned} \tag{1}$$

If (6.17) is satisfied, then $B^T x \neq 0$ *since* $\mathcal{N}(B^T) = \{0\}$. *Further,* $B^T x \in \mathcal{R}(B^T)$ *assures that* $B^T x \notin \mathcal{N}(A^T)$ *and hence* $A^T B^T x \neq 0$. *Thus* C_Z *is positive definite and* Z *has a density function by Proposition 6.34.*

If (6.17) is not satisfied, then either there exists $x \neq 0$ *with* $x \in \mathcal{N}(B^T)$, *or there exists* $x \neq 0$ *with* $x \in \mathcal{R}(B^T) \bigcap \mathcal{N}(A^T)$, *or both. In the first case,* $x^T C_Z x = 0$ *by (1). In the second case, there exists* y *with* $B^T y = x$, *and thus* $y^T C_Z y = 0$ *by (1). In either case,* C_Z *is not positive definite and thus has no density function by Proposition 6.34.* ■

Corollary 6.37 (Existence of a density for $Z \equiv BY + \nu$; $Y \sim MN_n(\mu, C)$**)** *Let* $Y \sim MN_n(\mu, C)$ *be a multivariate normally distributed random vector with positive definite* C, *and let* $B : \mathbb{R}^n \to \mathbb{R}^m$ *be an* $m \times n$ *matrix and* ν *an* m*-vector. Then* $Z \equiv BY + \nu$ *has a density function if and only if:*

$$\mathcal{N}(B^T) = \{0\} \in \mathbb{R}^m.$$

Proof. *Since* $C = AA^T$, *positive definiteness assures that* A^T *is nonsingular and* $\mathcal{N}(A^T) = \{0\}$. *Hence the second condition in (6.17) is always satisfied.* ■

We end this section with an example that investigates the implications of Proposition 6.36 and Corollary 6.37 given the relationship between n and m. The reader is encouraged to substitute actual numbers for the various parameters to make this more concrete.

Example 6.38 (Existence of a density for $Z \equiv BY + \nu$, **with** $Y \sim MN_n(\mu, C)$**)** *There are six cases to consider for* $Y \sim MN_n(\mu, C)$ *and* $B : \mathbb{R}^n \to \mathbb{R}^m$.

1. *C* ***Positive Definite:*** *This is the setting of Corollary 6.37, where* $C = AA^T$ *with* A^T *nonsingular and* $\mathcal{N}(A^T) = \{0\}$. *In this case* Y *has a density function by Proposition 6.5.*

   To have $\mathcal{N}(B^T) = \{0\}$ *for* $B^T : \mathbb{R}^m \to \mathbb{R}^n$, *or equivalently* $\dim \mathcal{N}(B^T) = 0$:

   (a) $m < n$:

   Here $B : \mathbb{R}^n \to \mathbb{R}^m$ *reduces dimension, and* B^T *creates an embedding of* $\mathbb{R}^m$ *into* $\mathbb{R}^n$.

   If B^T *has* ***full dimension,*** *meaning:*

   $$\dim \mathcal{R}(B^T) = m,$$

   then $\dim \mathcal{N}(B^T) = 0$ *by (6.4) and* $BY + \nu$ *has a density function.*

   If $\dim \mathcal{R}(B^T) < m$, *then* $\dim \mathcal{N}(B^T) > 0$ *and then* $BY + \nu$ *does not have a density function.*

(b) $m = n$:

Now by (6.4), $\mathcal{N}(B^T) = \{0\}$ if and only if $\mathcal{R}(B^T) = \mathbb{R}^n$. This happens if and only if B^T, and thus B, is invertible. Hence $BY + \nu$ has a density function if and only if B is invertible.

(c) $m > n$:

Here B increases dimension, and $\dim \mathcal{N}(B^T) > 0$ by (6.4). In this case, $BY + \nu$ does not have a density function.

2. *C* ***Positive Semidefinite:*** *This is the general setting of Proposition 6.36, where $C = AA^T$ with A^T singular and thus $\dim \mathcal{N}(A^T) = r > 0$. In this case, Y does not have a density function by Proposition 6.34.*

To have $\mathcal{R}(B^T) \bigcap \mathcal{N}(A^T) = \{0\}$ and $\mathcal{N}(B^T) = \{0\}$ for $B^T : \mathbb{R}^m \to \mathbb{R}^n$:

(a) $m < n$:

Now $\mathcal{R}(B^T) \subset \mathbb{R}^n$, so if $\mathcal{N}(B^T) = \{0\}$ then $\dim \mathcal{R}(B^T) = m$ by (6.4). Further, $\mathcal{N}(A^T) \subset \mathbb{R}^n$ with $\dim \mathcal{N}(A^T) = r > 0$. Thus to have $\mathcal{R}(B^T) \bigcap \mathcal{N}(A^T) = \{0\}$ it is necessary that $m + r \leq n$, so that the sum of the dimensions of these spaces does not exceed n. In other words, from a dimensional argument we must have:

$$r \leq n - m. \tag{1}$$

While a necessary condition, (1) is not sufficient to ensure that $BY + \nu$ has a density. This can be exemplified in $\mathbb{R}^n = \mathbb{R}^3$ with $\mathcal{R}(B^T)$ a plane through the origin, and $\mathcal{N}(A^T)$ a line on that plane, again through the origin. Then the dimensions work but the intersection restriction $\mathcal{R}(B^T) \bigcap \mathcal{N}(A^T) = \{0\}$ is violated.

In summary, $BY + \nu$ does not have a density function if (1) is violated, and may or may not have a density function otherwise.

(b) $m = n$:

Now $\dim \mathcal{R}(B^T) = n$ by (6.4) if $\mathcal{N}(B^T) = \{0\}$, and so the dimensionality argument in (1) is never satisfied since $r > 0$. Thus $BY + \nu$ does not have a density function.

(c) $m > n$:

As in item 1.c, $\dim \mathcal{N}(B^T) > 0$, so $BY + \nu$ does not have a density function.

6.3 The Cholesky Decomposition

In this section we address the following question, which continues the discussion of Chapter IV.5.

How can one simulate N-samples of normal vectors:

$$\{Y^{(j)}\}_{j=1}^N \equiv \{(Y_1^{(j)}, Y_2^{(j)}, ..., Y_n^{(j)})\}_{j=1}^N,$$

with prescribed mean vector μ and covariance matrix C?

One answer from Proposition 6.21 is that if we can simulate vectors of independent unit normal variables:

$$\{X^{(j)}\}_{j=1}^N \equiv \{(X_1^{(j)}, X_2^{(j)}, ..., X_n^{(j)})\}_{j=1}^N,$$

and factor the positive semidefinite (or definite) matrix $C = AA^T$ with A nondegenerate, then $Y^{(j)} \equiv AX^{(j)} + \mu$ has the appropriate distribution.

As it turns out, it is relatively easy to generate independent unit normal variates using a variety of mathematical programming languages as well as Microsoft Excel. Alternatively, using the Box-Muller transform of Proposition IV.5.4, pairs of such variates can be generated with pairs of independent continuous uniform variates, the latter again easily obtained with computer software. But the factorization of C involves an obvious challenge.

One approach to factoring a positive semidefinite (or definite) matrix, which is a symmetric matrix by definition, is provided by the spectral theorem of Proposition 6.11. For this we need to identify the n (counting multiplicities) real eigenvalues $\{\lambda_j\}_{j=1}^n$ and n orthogonal eigenvectors $\{Q^{(j)}\}_{j=1}^n$ of C, and this obtains:

$$B = Q\Lambda Q^T.$$

Here Λ is a diagonal matrix with $\{\lambda_j\}_{j=1}^n$ on the diagonal, and Q is an orthogonal matrix with columns $\{Q^{(j)}\}_{j=1}^n$.

Since C is positive semidefinite, $\lambda_j \geq 0$ for all j, and thus $\sqrt{\Lambda}$ is well-defined. Defining $A = Q\sqrt{\Lambda}$ obtains the needed factorization by Exercise 6.4 if at least one $\lambda_j > 0$.

While theoretically possible, identifying the eigenvalues and eigenvectors of a symmetric matrix can be computationally intensive. The eigenvalues of C are the n solutions of an nth degree polynomial, called the characteristic polynomial of C, which is given by $\det(C - \lambda I) = 0$. Eigenvectors are then found by determining a basis for the null space of $C - \lambda_j I$ for each such eigenvalue.

However, when C is **positive definite**, the following proposition provides an especially convenient unique factorization with $A = L$, a lower triangular matrix with positive elements on the diagonal. By **lower triangular** is meant that:

$$l_{ij} = 0, \text{ for } j > i,$$

while positive diagonal elements means that $l_{ii} > 0$ for all i. Thus L has $n(n+1)/2$ nonzero components.

This representation of C is called the **Cholesky decomposition of** C, named for **André-Louis Cholesky** (1875–1918), who developed this approach for real matrices, the current application. This decomposition is in fact valid more generally for positive semidefinite C and for certain complex matrices. For this generalization, the real matrix requirement of symmetric, where $C = C^T$ or $c_{ij} = c_{ji}$, is replaced by the complex matrix requirement of **self-adjoint,** where $C = \bar{C}^T$ or $c_{ij} = \bar{c}_{ji}$, with $\bar{c}_{ji}$ denoting the complex conjugate of c_{ji}. The matrix C is then also called **Hermitian** as noted in Remark 5.47.

Proposition 6.39 (Cholesky decomposition) *Given a positive definite matrix C, there exists a unique, invertible,* ***lower triangular*** *matrix L with positive diagonal so that $C = LL^T$.*

Proof. *The existence proof is by mathematical induction. Example 6.40 then derives an algorithm for constructing such decompositions iteratively, and it will be seen by that construction that the solution is unique.*

Existence is true for $n = 1$ where $C \equiv (c)$, since positive definite implies $c > 0$ and thus $L \equiv (\sqrt{c})$. So assume the existence result is true for symmetric matrices that are $(n-1) \times (n-1)$, and let C be a given $n \times n$ positive definite matrix. Then:

$$C = \begin{pmatrix} C' & y \\ y^T & c_{nn} \end{pmatrix},$$

where C' is $(n-1) \times (n-1)$ and symmetric, y is an $(n-1)$-vector, and c_{nn} a constant.

By considering x^TCx with $x = (x_1, x_2, ..., x_{n-1}, 0)$ and $x = (0, 0, ..., 0, x_n)$, respectively, positive definiteness of C assures that C' is positive definite and $c_{nn} > 0$. Also by assumption, $C' = LL^T$ where L is an $(n-1) \times (n-1)$ invertible, lower triangular matrix.

To factor C, we require:

$$\begin{pmatrix} C' & y \\ y^T & c_{nn} \end{pmatrix} = \begin{pmatrix} L & 0 \\ z^T & b \end{pmatrix} \begin{pmatrix} L^T & z \\ 0 & b \end{pmatrix},$$

where z and 0 denote $(n-1)$-vectors as columns, and $b > 0$ is a constant. A block matrix multiplication on the right obtains:

$$\begin{pmatrix} C' & y \\ y^T & c_{nn} \end{pmatrix} = \begin{pmatrix} LL^T & Lz \\ (Lz)^T & |z|^2 + b^2 \end{pmatrix},$$

where $|z|$ denotes the standard norm in $\mathbb{R}^{n-1}$ in (2.14).

Now $C' = LL^T$ by construction, so the proof is complete if we can solve for z and b:

$$Lz = y, \qquad |z|^2 + b^2 = c_{nn}.$$

Since L is invertible, $z \equiv L^{-1}y$, and the second equation for b requires:

$$c_{nn} > |z|^2. \tag{1}$$

We claim that this inequality is true by positive definiteness of C.

To this end, let C be positive definite with:

$$C = \begin{pmatrix} LL^T & Lz \\ (Lz)^T & c_{nn} \end{pmatrix},$$

and consider the vector $x = \begin{pmatrix} (L^T)^{-1} z \\ -1 \end{pmatrix} \neq 0$. Since $(L^T)^{-1} = (L^{-1})^T$ by Exercise 6.4, it follows that $Cx = \begin{pmatrix} 0 \\ |z|^2 - c_{nn} \end{pmatrix}$, and so by positive definiteness:

$$0 < x^TCx = c_{nn} - |z|^2,$$

and (1) is proved.

Combining:

$$\begin{pmatrix} C' & y \\ y^T & c_{nn} \end{pmatrix} = \begin{pmatrix} L & 0 \\ (L^{-1}y)^T & \sqrt{c_{nn} - |L^{-1}y|^2} \end{pmatrix} \begin{pmatrix} L^T & L^{-1}y \\ 0 & \sqrt{c_{nn} - |L^{-1}y|^2} \end{pmatrix}, \tag{2}$$

and the induction step is complete for proving the existence of lower triangular L.

To prove that the derived lower triangular matrix in (2) is invertible, recall that by an expansion into cofactors, that the determinant of a triangular matrix is the product of the diagonal elements. The determinant in (2) therefore equals the product of the diagonal elements of L, which is nonzero since L is invertible, and $\sqrt{c_{nn} - |L^{-1}y|^2} > 0$. Thus this derived matrix has nonzero determinant and is invertible.

The proof of existence is now complete, and uniqueness is addressed in Example 6.40.

■

Example 6.40 (Generating L) *Proposition 6.39 provides an existence result on an invertible, lower diagonal matrix L such that $C = LL^T$, so we now investigate the derivation of L given a positive definite matrix C. Because this derivation will be shown to be iterative, this assures the uniqueness of L. In short, L contains $n(n+1)/2$ nonzero components as noted above, while solving $C = LL^T$ obtains exactly $n(n+1)/2$ equations in these components.*

To simplify notation, let L_i denote the ith row of L so $L_i \equiv (L_{i1}, L_{i2}, ..., L_{in})$, noting that since lower triangular, $L_{ij} = 0$ for $j > i$. Then $C = LL^T$ reduces to a system of $n(n+1)/2$ equations, which using dot product notation becomes:

$$L_i \cdot L_j = c_{ij}, \qquad 1 \leq j \leq i \leq n. \tag{1}$$

The first equation is $L_1 \cdot L_1 = c_{11}$, or $L_{11}^2 = c_{11}$. Since $c_{11} > 0$ by positive definiteness, and $L_{1j} = 0$ for $j > 1$, define $L_1 = (\sqrt{c_{11}}, 0, ..., 0)$, taking the positive square root by definition of L.

For L_2 there are two equations:

$$L_2 \cdot L_1 = c_{21}, \qquad L_2 \cdot L_2 = c_{22}. \tag{2}$$

Since L_1 has only one nonzero component, the first equation in (2) has one unknown L_{21}, and then the second equation has only L_{22} unknown. This second calculation involves a square root, but see below to justify that this obtains a real number. This determines L_2 since $L_{2j} = 0$ for $j > 2$.

For general L_i there will be i equations:

$$L_i \cdot L_1 = c_{i1}, \quad L_i \cdot L_2 = c_{i2}, \quad ... \quad L_i \cdot L_i = c_{ii}.$$

The first $i-1$ equations involve L_j for $j < i$, all of which are known from previous steps. Since each L_j has only j nonzero components, each successive equation contains only one unknown and thus each is readily solvable.

In detail, $L_i \cdot L_1 = c_{i1}$ has one unknown, L_{i1}, since L_1 contains only one nonzero component L_{11}. Next, $L_i \cdot L_2 = c_{i2}$ contains L_{i1} and L_{i2}, but only the latter is unknown by the previous step. And this proceeds to $L_i \cdot L_i = c_{ii}$, which has only L_{ii} unknown, since $L_{i1}, ..., L_{i,i-1}$ were derived in the previous steps.

But there is one potential question for the existence of a real solution to this last equation, $L_i \cdot L_i = c_{ii}$, since this reduces to:

$$L_{ii}^2 = c_{ii} - \sum\nolimits_{k=1}^{i-1} L_{ik}^2,$$

and involves a square root. That the right hand expression is positive follows from the existence proof, since this equals $c_{nn} - \left|L^{-1}y\right|^2 = c_{nn} - |z|^2$, where in the notation of that proof, $z = (L_{i1}, ..., L_{i,i-1}, 0..., 0)$.

Exercise 6.41 *Recalling (6.9), if $Y \sim MN_n(\mu, C)$, then:*

$$\mu_i = E[Y_i], \; C_{ii} = Var[Y_i], \; C_{ij} = Cov[Y_i, Y_j].$$

For C positive definite and 3×3, implement the Cholesky decomposition to find L.

6.4 Properties of Multivariate Normal

In this section we develop three interesting results on the multivariate normal distribution. The first is the general formula for the higher central moments of the random vector

$Y \sim MN_n(\mu, C)$, which generalizes (6.9), and which provides an important special case known as **Isserlis' theorem.**

The second result states that the component normal variates of multivariate normal $Y \equiv (Y_1, Y_2, ..., Y_n)$ are independent if and only if they are uncorrelated. Independence of random variables always implies that these variates are uncorrelated, but the reverse implication is in general not true. We provide an example where this reverse implication fails, even for normal random variables. An apparent paradox? See below.

The third result has applications in sampling theory. Specifically, given a sample of normal random variables, the sample mean and sample variance are proved to be independent random variables, and with distribution functions identified below. While this is a result about normal random variables, we require multivariate normal results for the proof.

6.4.1 Higher Moments

If the random variable Y has a normal distribution with parameters μ and σ^2, it was seen in IV.(4.77) that for all integers $m \geq 0$:

$$E\left[(Y-\mu)^{2m+1}\right] = 0,$$

$$E\left[(Y-\mu)^{2m}\right] = \frac{\sigma^{2m}(2m)!}{2^m m!},$$

and thus in particular:

$$E[Y] = \mu, \quad E\left[(Y-\mu)^2\right] = \sigma^2.$$

In this section we generalize these results to a multivariate normal context.

Let $Y \equiv (Y_1, Y_2, ..., Y_n)$ be multivariate normally distributed by Definition 6.17, and thus by Proposition 6.21:

$$E\left[e^{t\cdot(Y-\mu)}\right] = \exp\left[\frac{1}{2}t^T Ct\right].$$

Expanding $e^{t\cdot(Y-\mu)}$ in a Taylor series and applying Lebesgue's dominated convergence theorem to $E\left[e^{t\cdot(Y-\mu)}\right]$, then equating to the Taylor series for $\exp\left[\frac{1}{2}t^T Ct\right]$, obtains:

$$\sum\nolimits_{l=0}^{\infty} \frac{1}{l!} E\left[(t\cdot(Y-\mu))^l\right] = \sum\nolimits_{m=0}^{\infty} \frac{\left(t^T Ct\right)^m}{2^m m!}, \tag{6.18}$$

recalling that $0! \equiv 1$.

The next proposition formalizes this identity, and then uses it to generalize the above results to multivariate moments. Admittedly, the results are less transparent.

Proposition 6.42 (Moments of $Y \sim MN_n(\mu, C)$) *Let $Y \equiv (Y_1, Y_2, ..., Y_n)$ be multivariate normally distributed by Definition 6.17, $Y \sim MN_n(\mu, C)$, with positive semidefinite matrix $C \equiv (c_{ij})_{1\leq i,j\leq n}$ and n-vector $\mu \equiv (\mu_1, \mu_2, ..., \mu_n)$. Given nonnegative integers $(M_1, M_2, ..., M_n)$, define $M \equiv \sum_{j=1}^n M_j$.*

If $M = 2m+1$, an odd integer:

$$E\left[\prod\nolimits_{j=1}^{n} (Y_j - \mu_j)^{M_j}\right] = 0. \tag{6.19}$$

If $M = 2m$, and even integer:

$$E\left[\prod\nolimits_{j=1}^{n} (Y_j - \mu_j)^{M_j}\right] = \frac{1}{2^m} \prod\nolimits_{j=1}^{n} M_j! \sum\nolimits_{(m_{ij})} \frac{\prod_{i,j=1}^n c_{i,j}^{m_{i,j}}}{\prod_{i,j=1}^n m_{i,j}!}, \tag{6.20}$$

where the summation is over all nonnegative integer n^2-tuples $(m_{ij})_{1\leq i,j\leq n}$ such that:

$$\sum\nolimits_{i,j=1}^{n} m_{i,j} = m,$$

and for each j :

$$\sum\nolimits_{i=1}^{n} m_{i,j} + \sum\nolimits_{i=1}^{n} m_{j,i} = M_j.$$

Proof. *To begin, we justify (6.18). Assume that the random vector $Y \sim MN_n(\mu, C)$ is defined on the probability space $(\mathcal{S}, \mathcal{E}, \lambda)$. Since the distribution function of Y is continuous from above and n-increasing by Proposition 1.23, such a space and random vector Y' always exists by Proposition 1.24, where $Y' = Y$ in distribution. As this proof only uses distributional properties of Y, we can denote Y' by Y for simplicity and assume Y is defined on this space.*

Now $e^{|t\cdot(Y-\mu)|} \leq e^{t\cdot(Y-\mu)} + e^{-t\cdot(Y-\mu)}$, so with expectations defined on $\mathcal{S}$, integral properties of Proposition V.3.29 obtain:

$$E\left[e^{|t\cdot(Y-\mu)|}\right] \leq e^{-t\cdot\mu} M_Y(t) + e^{t\cdot\mu} M_Y(-t),$$

and thus $e^{|t\cdot(Y-\mu)|}$ is integrable.

Defining $g_L(Y)$ by:

$$g_L(Y) = \sum\nolimits_{l=0}^{L} \frac{1}{l!} \left(t \cdot (Y - \mu)\right)^l,$$

it follows from Proposition V.3.42 that:

$$E\left[g_L(Y)\right] = \sum\nolimits_{l=0}^{L} \frac{1}{l!} E\left[\left(t \cdot (Y - \mu)\right)^l\right].$$

Further $g_L(Y) \to e^{t\cdot(Y-\mu)}$ pointwise as $L \to \infty$, and for all L:

$$E\left[|g_L(Y)|\right] \leq E\left[\sum\nolimits_{l=0}^{L} \frac{1}{l!} \left|\left(t \cdot (Y - \mu)\right)^l\right|\right] \leq E\left[e^{|t\cdot(Y-\mu)|}\right].$$

Thus by Lebesgue's dominated convergence theorem of Proposition V.3.45:

$$E\left[e^{t\cdot(Y-\mu)}\right] = \lim_{L\to\infty} \sum\nolimits_{l=0}^{L} \frac{1}{l!} E\left[\left(t \cdot (Y - \mu)\right)^l\right].$$

This plus the Taylor series for $\exp\left[\frac{1}{2} t^T C t\right]$ obtain (6.18).

Applying the multinomial theorem of (4.32) to $t \cdot (Y - \mu) \equiv \sum_{j=1}^{n} t_j (Y_j - \mu_j)$:

$$\frac{E\left[\left(t \cdot (Y - \mu)\right)^l\right]}{l!} = \sum\nolimits_{(l_1,\dots,l_n)} \frac{1}{\prod_{j=1}^{n} l_j!} E\left[\prod\nolimits_{j=1}^{n} (Y_j - \mu_j)^{l_j}\right] \prod\nolimits_{j=1}^{n} t_j^{l_j}, \tag{1}$$

where the summation is over all nonnegative integer n-tuples $(l_1, ..., l_n)$ with $\sum_{j=1}^{n} l_j = l$. Similarly applying this result to $t^T C t \equiv \sum_{i=1}^{n} \sum_{j=1}^{n} c_{i,j} t_i t_j$:

$$\frac{\left(t^T C t\right)^m}{2^m m!} = \sum\nolimits_{(m_{ij})} \frac{1}{2^m \prod_{i,j=1}^{n} m_{i,j}!} \prod\nolimits_{i,j=1}^{n} c_{i,j}^{m_{i,j}} \prod\nolimits_{i,j=1}^{n} (t_i t_j)^{m_{i,j}}, \tag{2}$$

where the summation is over all nonnegative integer n^2-tuples $(m_{ij})_{1\leq i,j\leq n}$ with $\sum_{i,j=1}^{n} m_{i,j} = m$.

Taylor series expansions define analytic functions which are then unique on their domain of convergence, say $|t| < t_0$. *Thus if we substitute the expressions in* (1) *and* (2) *into (6.18), we can equate the coefficients of the various t-products from these expansions. From* (2) *it follows that every t-product of the power series on the right in (6.18) has even degree, and hence the coefficients of all odd degree t-products for the power series on the left must be* 0. *From the expression for* $E\left[(t\cdot(Y-\mu))^l\right]$ *in* (1), *it follows that if* $\sum_{j=1}^n l_j = l$ *is odd, the coefficient of* $\prod_{j=1}^n t_j^{l_j}$ *must be* 0, *and this proves (6.19).*

By the same argument, the coefficients of the t-products in (2) *for given* m *must equal the coefficients of the t-products in* (1) *for* $l = M = 2m$. *This obtains that for each* m, *switching notation in* (1) *to* $l = M$:

$$\sum\nolimits_{(m_{ij})} \frac{1}{2^m \prod_{i,j=1}^n m_{i,j}!} \prod\nolimits_{i,j=1}^n c_{i,j}^{m_{i,j}} \prod\nolimits_{i,j=1}^n (t_i t_j)^{m_{i,j}} \tag{3}$$
$$= \sum\nolimits_{(M_1,\dots,M_n)} \frac{1}{\prod_{j=1}^n M_j!} E\left[\prod\nolimits_{j=1}^n (Y_j-\mu_j)^{M_j}\right] \prod\nolimits_{j=1}^n t_j^{M_j},$$

where $\sum_{i,j=1}^n m_{i,j} = m$ *and* $\sum_{j=1}^n M_j = M = 2m$.

To compare coefficients of t-products in (3), *note that for each* j:

$$\prod\nolimits_{i=1}^n (t_i t_j)^{m_{i,j}} = \prod\nolimits_{i=1}^n t_j^{m_{i,j}} \prod\nolimits_{i=1}^n t_i^{m_{i,j}} = t_j^{\Sigma_i m_{i,j}} \prod\nolimits_{i=1}^n t_i^{m_{i,j}},$$

and thus, changing notation in the last step:

$$\begin{aligned}\prod\nolimits_{j=1}^n \prod\nolimits_{i=1}^n (t_i t_j)^{m_{i,j}} &= \prod\nolimits_{j=1}^n t_j^{\Sigma_i m_{i,j}} \cdot \prod\nolimits_{i=1}^n \prod\nolimits_{j=1}^n t_i^{m_{i,j}} \\ &= \prod\nolimits_{j=1}^n t_j^{\Sigma_i m_{i,j}} \prod\nolimits_{i=1}^n t_i^{\Sigma_j m_{i,j}} \\ &= \prod\nolimits_{j=1}^n t_j^{\Sigma_i m_{i,j} + \Sigma_i m_{j,i}}.\end{aligned}$$

To finish the proof, fix the $(M_1, \dots, M_n)$*-term in the sum on the right in* (3). *By the above calculation, the coefficient of* $\prod_{j=1}^n t_j^{M_j}$ *must equal the sum of the coefficients on the left for all* (m_{ij}) *so that* $\Sigma_i m_{i,j} + \Sigma_i m_{j,i} = M_j$. *This obtains (6.20) and the proof is complete.* ■

Remark 6.43 (On $\sum_{(m_{ij})}$**)** *Note that for given* $(M_1, \dots, M_n)$, *the* (m_{ij})*-summation in (6.20) has potentially many repetitions of the same value.*

For example, given n^2*-tuple* (m_{ij}), *if there exists* m_{kl} *so that* $m_{kl} \neq m_{lk}$, *then simply switching these values obtains another term in the summation with the same value as the given* (m_{ij}) *because* C *is symmetric and thus* $c_{kl} = c_{lk}$. *Indeed if there are* L *such pairs in the given* n^2*-tuple* (m_{ij}), *there will be* 2^L *terms in the summation with this same value.*

This observation is at the heart of the proof of the following result, known as ***Isserlis' theorem,*** *and named for* ***Leon Isserlis*** *(1881–1966). It is often stated under the assumption that* $\mu = 0$.

Given $(Y_1, Y_2, \dots, Y_n)$, Isserlis' theorem focuses on expectations of the form:

$$E\left[\prod\nolimits_{k=1}^K (Y_{j_k} - \mu_{j_k})\right].$$

In the notation of Definition 5.6, this is the multivariate central moment $\mu_{(m_1,\dots,m_n)}$, where $m_{j_k} = 1$ for $k = 1, \dots, K$ and $m_j = 0$ otherwise. By Proposition 6.42, this expectation is 0 if K is odd, and nonzero otherwise.

To simplify notation and avoid double subscripts, the next result is stated with the full vector $Y = (Y_1, Y_2, ..., Y_n)$ and $n = 2m$. For subvectors $(Y_{j_1}, ..., Y_{j_{2m}})$, this result applies by only using the elements of C related to these subscripts. Alternatively, this result can be applied to the marginal vector Y_I of (1.35), using the associated matrix C_I of Corollary 6.25 and $I = (j_1, ..., j_{2m})$.

In the following result we do not repeat (6.19), but instead focus on the restatement of (6.20).

Proposition 6.44 (Isserlis' theorem) *Let $Y \equiv (Y_1, Y_2, ..., Y_n)$ be multivariate normally distributed by Definition 6.17, $Y \sim MN_n(\mu, C)$, with positive semidefinite matrix $C \equiv (c_{ij})_{1 \leq i,j \leq n}$ and vector $\mu \equiv (\mu_1, \mu_2, ..., \mu_n)$.*

If $n = 2m$:

$$E\left[\prod_{j=1}^{2m} (Y_j - \mu_j)\right] = \sum\nolimits'_{\{(i_k,j_k)\}_{k=1}^m} \prod_{k=1}^m c_{i_k,j_k}, \tag{6.21}$$

*where the $\sum'$-summation is over all **distinct** m-pairs $\{(i_k, j_k)\}_{k=1}^m$ so that $\bigcup_{k=1}^m (i_k, j_k) = (1, 2, ..., 2m)$.*

The number of terms in this summation is:

$$N_m = \frac{(2m)!}{2^m m!}. \tag{6.22}$$

Proof. *If $n = 2m = M$ in Proposition 6.42, then $M_j = 1$ for all j and thus for all j:*

$$\begin{array}{c} \sum_{i=1}^n m_{i,j} + \sum_{i=1}^n m_{j,i} = 1, \\ \sum_{i,j=1}^n m_{i,j} = m. \end{array} \tag{1}$$

The first constraint in (1) implies that all $m_{j,j} = 0$, and that for each j, there is exactly one k_j so that either $m_{k_j,j} = 1$ or $m_{j,k_j} = 1$. Identifying $(m_{ij})_{1 \leq i,j \leq n}$ as a matrix, this implies that for every j, there is exactly one nonzero element in either the jth row or jth column of this matrix, and this element cannot be on the diagonal. The second constraint in (1) assures that in total, there are m such elements with $m_{i,j} = 1$.

We claim that if $\{(i_k, j_k)\}_{k=1}^m$ are m arbitrarily chosen (i, j)-index pairs, and we define $m_{i_k,j_k} = 1$ and $m_{i,j} = 0$, otherwise, then $\{m_{i_k,j_k}\}_{k=1}^m$ satisfy the constraints in (1) if and only if:

$$\bigcup_{k=1}^m (i_k, j_k) = (1, 2, ..., 2m). \tag{2}$$

First, if $\{(i_k, j_k)\}_{k=1}^m$ do not satisfy (2), then there exists $l \in (1, 2, ..., 2m)$ with $l \notin \bigcup_{k=1}^m (i_k, j_k)$. Thus there exists no $m_{i_k,j_k} = 1$ with either $i_k = l$ or $j_k = l$, so:

$$\sum_{i=1}^n m_{i,l} + \sum_{i=1}^n m_{l,i} = 0,$$

and $\{(i_k, j_k)\}_{k=1}^m$ do not satisfy (1).

Conversely, if $\{(i_k, j_k)\}_{k=1}^m$ satisfy (2), we claim that $\{m_{i_k,j_k}\}_{k=1}^m$ satisfy (1). The second constraint in (1) is true by definition, so consider for given l:

$$\sum_{i=1}^n m_{i,l} + \sum_{i=1}^n m_{l,i}. \tag{3}$$

Since (2) is satisfied, there exists either (i_k, l) or (l, j_k) in $\{(i_k, j_k)\}_{k=1}^m$, but this collection cannot contain both. By a counting argument, $\bigcup_{k=1}^m (i_k, j_k)$ cannot contain $2m$ different numbers to satisfy (2) if the digit l is repeated. Thus the sum in (3) is 1, and the claim is proved.

It now follows from (6.20) with all $m_{i_k,j_k} = 1$ *and* $m_{i,j} = 0$*, otherwise, that:*

$$\prod_{i,j=1}^{2m} c_{i,j}^{m_{i,j}} = \prod_{k=1}^{m} c_{i_k,j_k},$$

and so with all $M_j = 1$*:*

$$E\left[\prod_{j=1}^{2m} (Y_j - \mu_j)\right] = \frac{1}{2^m} \sum_{\{(i_k,j_k)\}_{k=1}^m} \prod_{k=1}^{m} c_{i_k,j_k}. \tag{4}$$

Here the summation is over all index collections $\{(i_k, j_k)\}_{k=1}^m$ *which satisfy* (2).

We next show how this summation can be simplified. Given any index collection $\{(i_k, j_k)\}_{k=1}^m$ *which satisfies* (2) *and* $A \subset \{1, ..., m\}$*, define collections* $\{(i'_k, j'_k)\}_{k=1}^m$ *by:*

$$(i'_k, j'_k) = \begin{cases} (i_k, j_k), & k \in A, \\ (j_k, i_k), & k \in \tilde{A}, \end{cases} \tag{5}$$

where $\tilde{A}$ *is the complement of* A*. Considering all subsets* $A \subset \{1, ..., m\}$*, for any collection* $\{(i_k, j_k)\}_{k=1}^m$*, there are* 2^m *collections* $\{(i'_k, j'_k)\}_{k=1}^m$ *defined by* (5)*. Any such collection* $\{(i'_k, j'_k)\}_{k=1}^m$ *also satisfies* (2)*, and by symmetry of* C*:*

$$\prod_{k=1}^{m} c_{i_k,j_k} = \prod_{k=1}^{m} c_{i'_k,j'_k}. \tag{6}$$

Thus all index collections $\{(i_k, j_k)\}_{k=1}^m$ *which satisfy* (2) *can be partitioned into groups of* 2^m *index collections, for which all the members of each group are related by* (5)*, and all members of each group satisfy* (6)*. Thus* (4) *can be rewritten as in (6.21):*

$$E\left[\prod_{j=1}^{n} (Y_j - \mu_j)\right] = \sum'_{\{(i_k,j_k)\}_{k=1}^m} \prod_{k=1}^{m} c_{i_k,j_k}, \tag{7}$$

where the summation is now over one index collection from each group.

The proof is complete by noting that two collections $\{(i_k, j_k)\}_{k=1}^m$ *and* $\{(i'_k, j'_k)\}_{k=1}^m$ *are related as in* (5) *if and only if they contain the same* m*-pairs of indexes, where we identify* (i, j) *and* (j, i) *as pairs. Thus the summation in* (7) *is over all distinct collections of* m *pairs that satisfy* (2)*, where each pair is defined independent of order.*

For (6.22), note that there are $(2m)!$ *orderings of the* $(1, 2, ..., 2m)$*, which by sequentially grouping into pairs can be initially identified with* $(2m)!$ *collections of indexes* $\{(i_k, j_k)\}_{k=1}^m$*. By construction,* $\bigcup_{k=1}^m (i_k, j_k) = (1, 2, ..., 2m)$ *for any such ordering. These collections are not all distinct, since the ordering of the* m *pairs, and the index order within the* m*-pairs, does not change the given collection by the above identification.*

Hence for any given collection, there will be an additional $2^m - 1$ *collections with the* m *pairs* $\{(i_k, j_k)\}_{k=1}^m$ *in the same order, but with from* 1 *to* m *index-pairs in reverse order. Then given any of these in total* 2^m *collections, there will be* $m!$ *collections with the same pairs in some order. Thus the number of distinct collections of* m*-pairs, where we define distinct to omit reordering of the collections, or interchanges of indexes within pairs, is* $N_m = (2m)!/(2^m m!)$*, and the proof is complete.* ■

Example 6.45 *With* $m = 2$*,* $N_2 = 3$*, and:*

$$E\left[\prod_{j=1}^{4} (Y_j - \mu_j)\right] = c_{12}c_{34} + c_{13}c_{24} + c_{14}c_{23}.$$

With $m = 3$*,* $N_3 = 15$*, and we leave* $E\left[\prod_{j=1}^{6} (Y_j - \mu_j)\right]$ *as an exercise. Hint: As for* $m = 2$*, the first factors of the sum are* $\{c_{1j}\}_{j=1}^6$*, but then for each there is more than one second factor.*

Remark 6.46 (On N_m) *As suggested by Example 6.45, the number of terms in the summation in (6.21) grows very fast with* m :

$$N_1 = 1,\ N_2 = 3,\ N_3 = 15,\ N_4 = 105, ...,$$

and in general:

$$N_{m+1} = (2m+1)N_m.$$

Thus:

$$\begin{aligned} N_{m+1} &= (2m+1)(2m-1)(2m-3)...3 \\ &\equiv (2m+1)!!. \end{aligned}$$

The ***double factorial*** $n!!$ *is defined as the product of* n *and all smaller integers with the same parity as* n*, where by parity is meant in terms of even/odd.*

Stirling's formula*, also known as* ***Stirling's approximation,*** *is named for* ***James Stirling*** *(1692–1770) and was seen in (3.29) in the context of the gamma function. In terms of the factorial function, it states that as* $n \to \infty$*:*

$$\frac{n!}{\sqrt{2\pi}n^{n+1/2}e^{-n}} \approx e^{1/12n} \to 1. \tag{6.23}$$

Using this result in (6.22), it follows that as $m \to \infty$*:*

$$N_m \approx \sqrt{2}\left(\frac{2m}{e}\right)^m.$$

6.4.2 Independent vs. Uncorrelated Normals

In this section, we prove as a corollary of Proposition 6.34 that for a multivariate normal random vector $Y = (Y_1, ..., Y_n)$ with C positive definite, that the component normal variates are independent if and only if they are uncorrelated.

Given random variables Y_i and Y_j with means μ_i and μ_j and variances σ_i^2 and σ_j^2, recall from (4.39) that the **correlation** between Y_i and Y_j is defined:

$$corr[Y_i, Y_j] \equiv \frac{cov[Y_i, Y_j]}{\sigma_i \sigma_j},$$

with the **covariance** $cov[Y_i, Y_j]$ defined in (4.38):

$$cov(Y_i, Y_j) \equiv E[(Y_i - \mu_i)(Y_j - \mu_j)].$$

Combining the above formulas obtains an intuitive formula for correlation:

$$corr[Y_i, Y_j] \equiv E\left[\left(\frac{Y_i - \mu_i}{\sigma_i}\right)\left(\frac{Y_j - \mu_j}{\sigma_j}\right)\right].$$

The transformation $Y \to \frac{Y-\mu}{\sigma}$ is sometimes called "normalizing" the random variable Y. Though often applied to normal variates, here the term **normalize** reflects that we have centered and scaled Y to have a mean of 0 and a variance of 1.

The correlation is often denoted ρ_{Y_i,Y_j}, or sometimes ρ_{ij} when the context is clear. It follows from Corollary IV.4.50 that:

$$-1 \leq \rho_{ij} \leq 1.$$

Exercise 6.47 *The correlation between variates is not affected by affine transformations. Check that:*

$$corr[Y_i, Y_j] = corr[a_i Y_i + b_i, a_j Y_j + b_j],$$

as long as $a_i, a_j \neq 0$.

We recall Definition 1.53 to contrast independent with uncorrelated.

Definition 6.48 (Independent; uncorrelated) *Let* X, Y *be random variables on a probability space* $(\mathcal{S}, \mathcal{E}, \lambda)$. *We say that:*

- X *and* Y *are* ***independent*** *if for all* A, $B \in \mathcal{B}(\mathbb{R})$*:*

$$\lambda\left[X^{-1}(A)\bigcap Y^{-1}(B)\right] = \lambda\left[X^{-1}(A)\right]\lambda\left[Y^{-1}(B)\right].$$

- X *and* Y *are* ***uncorrelated*** *if:*

$$\rho_{X,Y} = 0,$$

or equivalently:

$$Cov[X, Y] = 0.$$

To compare these notions, first note that for random variables to be uncorrelated requires that these random variables have two moments. Hence the qualification in the following exercise.

Exercise 6.49 (Independent with moments ⇒ uncorrelated) *Prove that if* X *and* Y *are independent random variables on a probability space* $(\mathcal{S}, \mathcal{E}, \lambda)$, *and each with two finite moments, then* X *and* Y *are uncorrelated. Hint: Proposition 4.7 and Proposition V.3.56.*

In general, the opposite implication fails, even for general normal variates, as is seen below. However, when these normal variates are the components of a multivariate normal vector, remarkably, independent and uncorrelated are equivalent.

Proposition 6.50 (Independent ⇔ Uncorrelated for $Y \sim MN_n(\mu, C)$**)** *Let* $Y \equiv (Y_1, Y_2, ..., Y_n)$ *be multivariate normally distributed by Definition 6.17,* $Y \sim MN_n(\mu, C)$, *with* C *positive definite.*

Then $\{Y_i\}_{i=1}^n$ *are independent if and only if they are uncorrelated.*

Proof. *We prove that* $\{Y_i\}_{i=1}^n$ *are independent if and only if* $c_{ij} = 0$ *for* $i \neq j$, *where* $\{c_{ij}\}$ *are the components of* C.

By Proposition 6.34, Y *has a density function* $f_Y(y)$ *given in (6.6). If* $\{Y_i\}_{i=1}^n$ *are uncorrelated and thus* $c_{ij} \equiv Cov[Y_i, Y_j] = 0$ *for* $i \neq j$ *by Exercise 6.16, then* C *is diagonal and hence* $f_Y(y) = \prod_{i=1}^n f_{Y_i}(y_i)$, *where* $f_{Y_i}(y_i)$ *is a normal density function with mean* μ_i *and variance* c_{ii}. *By Corollary 1.57, this implies that* $\{Y_i\}_{i=1}^n$ *are independent.*

Conversely, if $\{Y_i\}_{i=1}^n$ *are independent then* $f_Y(y) = \prod_{i=1}^n f_{Y_i}(y_i)$. *Rewriting,* Y *is multivariate normal with a diagonal covariance matrix* C, *so* $c_{ij} = 0$ *for* $i \neq j$, *and* $\{Y_i\}_{i=1}^n$ *are uncorrelated.* ■

Example 6.51 (On the importance of $Y \sim MN_n(\mu, C)$**)** *It is important to emphasize that in general,* ***uncorrelated normals need not be independent.***

The above proposition specifies that this equivalence applies to normal variates $\{Y_i\}_{i=1}^n$ *that are the components of a multivariate normally distributed random vector* $Y \sim MN_n(\mu, C)$. *However, given a collection of uncorrelated normal variates, they need not be components of* $Y \sim MN_n(\mu, C)$, *and thus need not be independent.*

As a general example, let $\varphi(x) = \frac{1}{\sqrt{2\pi}} e^{-x^2/2}$ *and define:*

$$g(c) = \int_{-c}^{c} x^2 \varphi(x) dx.$$

Then $g(0) = 0$, g *is continuous and strictly increasing, and* $g(c) \to 1$ *as* $c \to \infty$ *since* $g(\infty)$ *is the variance of the standard normal. Thus by the intermediate value theorem of Exercise III.1.13, there exists* c' *with* $g(c') = 1/2$. *Numerically one obtains* $c' \approx 0.67449$.

Let X *have a standard normal distribution, and define:*

$$Y = \begin{cases} X, & |X| \leq c', \\ -X, & |X| > c'. \end{cases}$$

By symmetry, Y *has a standard normal distribution, meaning that* $F_Y(x) = F_X(x)$ *for all* x. *See Exercise 6.52.*

Further, X *and* Y *are uncorrelated normals:*

$$E[XY] = \int_{|x| \leq c'} x^2 \varphi(x) dx - \int_{|x| > c'} x^2 \varphi(x) dx = 0,$$

but X *and* Y *are not independent:*

$$\Pr[|X| \leq c', |Y| \leq c'] = \Pr[|X| \leq c'] \neq \Pr[|X| \leq c'] \Pr[|Y| \leq c'].$$

Exercise 6.52 *Prove that* Y *defined in Example 6.51 has a standard normal distribution function, that* $F_Y(y) = F_X(y)$ *for all* y.

6.4.3 Sample Mean and Variance

The notion of an N**-sample** was introduced in Definition II.4.1. Recall that $\mathcal{B}(\mathbb{R})$ is the Borel sigma algebra of Definition 1.2.

Definition 6.53 (N**-Sample)** *Let a probability space* $(\mathcal{S}, \mathcal{E}, \lambda)$ *and random variable* $X : \mathcal{S} \longrightarrow \mathbb{R}$ *be given. With* N *finite or infinite, a collection of random variables* $\{X_j\}_{j=1}^N$ *defined on a probability space* $(\mathcal{S}', \mathcal{E}', \lambda')$ *is said to be an* N***-sample of*** X, *or a* ***sample of*** X *when* N *is implied, if this collection is* ***independent, and identically distributed with*** X ***(i.i.d.-***X***):***

1. $\{X_j\}_{j=1}^N$ *are* ***independent*** *if given a finite collection* $(i_1, ..., i_m) \subset (1, 2, ..., N)$ *and* $\{A_j\}_{j=1}^m \subset \mathcal{B}(\mathbb{R})$*:*
$$\lambda' \left[\bigcap\nolimits_{j=1}^{m} X_{i_j}^{-1}(A_j) \right] = \prod\nolimits_{j=1}^{m} \lambda'[X_{i_j}^{-1}(A_j)]. \tag{6.24}$$

2. $\{X_j\}_{j=1}^N$ *are* ***identically distributed with*** X *if for all* j, *and all* $A \in \mathcal{B}(\mathbb{R})$*:*
$$\lambda'[X_j^{-1}(A)] = \lambda[X^{-1}(A)]. \tag{6.25}$$

As noted in Definition II.4.3, items 1 and 2 can be restated. If F_j denotes the distribution function of X_j and F the distribution function of X:

1. $\{X_j\}_{j=1}^N$ are **independent** if given a finite collection $(i_1, ..., i_m) \subset (1, 2, ..., N)$ and $F(x_{i_1}, ..., x_{i_m})$ the joint distribution function of $\{X_{i_j}\}_{j=1}^m$:
$$F(x_{i_1}, ..., x_{i_m}) = \prod\nolimits_{j=1}^{m} F_{i_j}(x_{i_j}). \tag{6.26}$$

2. $\{X_j\}_{j=1}^N$ are **identically distributed with** X if for all j, and all x:

$$F_j(x) = F(x). \tag{6.27}$$

Given a probability space $(\mathcal{S}, \mathcal{E}, \lambda)$ and a random variable $X : \mathcal{S} \longrightarrow \mathbb{R}$, Proposition II.4.4 proves the existence of $\{X_j\}_{j=1}^N$ and $(\mathcal{S}', \mathcal{E}', \lambda')$ for any N, finite or infinite, while Propositions II.4.13 and II.4.17 derive alternative constructions.

Returning to the application at hand, let X be a normal variate with mean μ and variance σ^2, and $\{X_j\}_{j=1}^N$ an N-sample of X.

Definition 6.54 (Statistic) *Given an N-sample $\{X_j\}_{j=1}^N$ defined on a probability space $(\mathcal{S}', \mathcal{E}', \lambda')$ and $n \leq N$, a* ***statistic*** *is a random variable S defined on $(\mathcal{S}', \mathcal{E}', \mu')$ by:*

$$\mathsf{S} = g(X_1, ..., X_n),$$

where $g : \mathbb{R}^n \to \mathbb{R}$ is Borel measurable.

Example 6.55 *Examples of statistics of interest for this section, and undoubtedly familiar to the reader, are:*

1. The ***sample mean,*** *denoted m (for μ) or $\bar{X}$:*

$$m \equiv \frac{1}{n}\sum\nolimits_{j=1}^n X_j. \tag{6.28}$$

2. The ***sample variance,*** *denoted s^2 (for σ^2):*

$$s^2 \equiv \frac{1}{n}\sum\nolimits_{j=1}^n (X_j - m)^2. \tag{6.29}$$

3. The ***unbiased sample variance,*** *denoted (here) $\hat{s}^2$:*

$$\hat{s}^2 \equiv \frac{1}{n-1}\sum\nolimits_{j=1}^n (X_j - m)^2. \tag{6.30}$$

Expectations of these statistics are summarized next, recalling Definition 4.1.

Item 2 below implies that s^2 is a **biased estimator for** σ^2, since $E\left[s^2\right] \neq \sigma^2$, while $\hat{s}^2$ in item 3, as the name implies, is an **unbiased estimator for** σ^2.

Proposition 6.56 (Expectation of m, s^2, and $\hat{s}^2$) *Let $\{X_j\}_{j=1}^n$ be an n-sample of a random variable X defined on a probability space $(\mathcal{S}', \mathcal{E}', \lambda')$.*

If $\mu = E[X]$ exists for item 1, and $\sigma^2 = E\left[(X-\mu)^2\right]$ exists for items 2 and 3, then:

1. m is an ***unbiased estimator for*** *μ:*

$$E[m] = \mu.$$

2. s^2 is a ***biased estimator for*** *σ^2:*

$$E\left[s^2\right] = \left(\frac{n-1}{n}\right)\sigma^2.$$

3. $\hat{s}^2$ is an ***unbiased estimator for*** *σ^2:*

$$E\left[\hat{s}^2\right] = \sigma^2.$$

Proof. *Item 1 follows from linearity of the λ'-integral from Proposition V.3.42:*

$$E\left[\frac{1}{n}\sum\nolimits_{j=1}^{n} X_j\right] = \frac{1}{n}\sum\nolimits_{j=1}^{n} E\left[X_j\right] = \mu,$$

since $E\left[X_j\right] = \mu$ for all j by definition of n-sample.

For items 2 and 3:

$$\begin{aligned}(X_j - m)^2 &= \left(X_j - \frac{1}{n}\sum\nolimits_{k=1}^{n} X_k\right)^2 \\ &= \left(\frac{n-1}{n}X_j - \frac{1}{n}\sum\nolimits_{k\neq j} X_k\right)^2 \\ &= \left(\frac{n-1}{n}\right)^2 X_j^2 - \frac{2(n-1)}{n^2}\sum\nolimits_{k\neq j} X_j X_k \\ &+ \frac{1}{n^2}\sum\nolimits_{k\neq j} X_k^2 + \frac{1}{n^2}\sum\nolimits_{k\neq j, l\neq j; k\neq l} X_l X_k.\end{aligned} \tag{1}$$

Taking expectations:

$$E\left[X_j^2\right] = \sigma^2 + \mu^2,$$

and by independence and Proposition 4.7:

$$E\left[X_l X_k\right] = \mu^2, \ k \neq l.$$

Noting that the three summations in (1) *have $n-1$, $n-1$, and $(n-1)(n-2)$ terms, respectively, obtains:*

$$\begin{aligned}E\left[(X_j - m)^2\right] &= \left[\left(\frac{n-1}{n}\right)^2 + \frac{n-1}{n^2}\right]\left(\sigma^2 + \mu^2\right) \\ &\quad + \left[\frac{(n-1)(n-2)}{n^2} - \frac{2(n-1)(n-1)}{n^2}\right]\mu^2 \\ &= \left(\frac{n-1}{n}\right)\sigma^2.\end{aligned}$$

Items 2 and 3 now follow by linearity of the integral. ■

The next result provides an interesting and somewhat surprisingly conclusion that for a normal n-sample, m and s^2 are independent random variables. We then determine the distribution functions of these variates.

Proposition 6.57 (Normal samples: m and s^2 are independent) *If $\{X_j\}_{j=1}^{n}$ is an n-sample from a normal distribution with mean μ and variance σ^2, then m and s^2 are independent random variables.*

Proof. *Since:*

$$\sum\nolimits_{k=1}^{n} (X_k - m) = 0, \tag{1}$$

it follows that:

$$s^2 = \frac{1}{n}\left[\sum\nolimits_{j=2}^{n} (X_j - m)^2 + \left(\sum\nolimits_{j=2}^{n} (X_j - m)\right)^2\right]. \tag{2}$$

Assume that we can prove that the random variable m is independent of the random vector $(X_2 - m, ..., X_n - m)$. Then by Proposition II.3.56, m will be independent from s^2 since $s^2 = g(X_2 - m, ..., X - m)$ with Borel measurable $g(x)$ defined in (2).

To this end, define a random vector Y on the probability space $(\mathcal{S}', \mathcal{E}', \mu')$ by:

$$Y : \mathcal{S}' \to (m, X_2 - m, ..., X_n - m).$$

Note that $Y = BX$ with $X \equiv (X_1, X_2, ..., X_n)$ and B an $n \times n$ matrix given by:

$$B = \begin{pmatrix} 1/n & 1/n & 1/n & 1/n & \cdots & 1/n \\ -1/n & 1-1/n & -1/n & -1/n & \cdots & -1/n \\ -1/n & -1/n & 1-1/n & -1/n & \cdots & -1/n \\ \vdots & \vdots & \vdots & \vdots & \vdots & \vdots \\ -1/n & -1/n & -1/n & -1/n & \cdots & 1-1/n \end{pmatrix}.$$

By definition of n-sample, X is multivariate normal with $\mu = (\mu, ..., \mu)$ and $C = \sigma^2 I$, where I is the identity matrix. Then by Proposition 6.23, Y is multivariate normally distributed with $\mu_B = B\mu$ and $C_B = BCB^T$. Hence $\mu_B = (\mu, 0, ..., 0)$ and $C_B = \sigma^2 BB^T$. Adding the first row of B to all the other rows does not change $\det B$, *and one obtains from this that* $\det(B) = 1/n$ *and B is invertible. Thus C_B is positive definite by Example 6.9, and Proposition 6.34 obtains the density function for Y as in (6.6):*

$$f_Y(y) = (2\pi)^{-n/2} [\det C_B]^{-1/2} \exp\left[-\frac{1}{2}(y - \mu_B)^T C_B^{-1}(y - \mu_B)\right]. \tag{3}$$

The final steps are to prove that $(y - \mu_B)^T C_B^{-1}(y - \mu_B)$ in (3) is a sum of y_1^2, and an expression involving $(y_2, ..., y_n)$. This then assures that $f_Y(y) = f(y_1)f(y_2, ..., y_n)$, and the independence proof will be complete by Corollary 1.57.

To this end, recalling Exercise 6.4:

$$\begin{aligned} (y - \mu_B)^T C_B^{-1}(y - \mu_B) &= \sigma^2 (y - \mu_B)^T \left(B^{-1}\right)^T B^{-1}(y - \mu_B) \\ &= \sigma^2 \left[B^{-1}(y - \mu_B)\right]^T B^{-1}(y - \mu_B) \\ &= \sigma^2 \left|B^{-1}(y - \mu_B)\right|^2. \end{aligned}$$

For the last step, recall that for a vector x, that $x^T x = |x|^2 \equiv \sum_{j=1}^n x_j^2$, the squared length of x in (2.14).

The final step is to evaluate $\left|B^{-1}(y - \mu_B)\right|^2$, for which B^{-1} is needed. Using row reduction one obtains the following, or at least confirms this as an exercise:

$$B^{-1} = \begin{pmatrix} 1 & -1 & -1 & -1 & \cdots & -1 \\ 1 & 1 & 0 & 0 & \cdots & 0 \\ 1 & 0 & 1 & 0 & \cdots & 0 \\ \vdots & \vdots & \vdots & \vdots & \vdots & \vdots \\ 1 & 0 & 0 & 0 & \cdots & 1 \end{pmatrix}.$$

In other words, B^{-1} is the identity matrix, modified with a first column of 1s, and a first row of -1s starting in the second column.

Letting $z \equiv y - \mu_B$ for notational convenience:

$$B^{-1}z = (z_1 - \sum\nolimits_{j=2}^n z_j, z_1 + z_2, ..., z_1 + z_n),$$

so

$$\begin{aligned} \left|B^{-1}z\right|^2 &= \left(z_1 - \sum\nolimits_{j=2}^n z_j\right)^2 + \sum\nolimits_{j=2}^n (z_1 + z_j)^2 \\ &= nz_1^2 + \left(\sum\nolimits_{j=2}^n z_j\right)^2 + \sum\nolimits_{j=2}^n z_j^2 \\ &= n(y_1 - \mu)^2 + \left(\sum\nolimits_{j=2}^n y_j\right)^2 + \sum\nolimits_{j=2}^n y_j^2. \end{aligned}$$

Thus by (3), $f_Y(y) = f(y_1)f(y_2, ..., y_n)$, and the proof is complete. ■

The final detail is to determine the distribution functions of m and s^2.

Proposition 6.58 (Distribution functions of m and s^2) *If $\{X_j\}_{j=1}^n$ is an n-sample from a normal distribution with mean μ and variance σ^2, then:*

1. *m has a* ***normal distribution*** *with mean μ and variance σ^2/n.*

2. *$ns^2/\sigma^2 = (n-1)\hat{s}^2/\sigma^2$ has a* ***chi-squared distribution*** *with $n-1$ degrees of freedom.*

Proof. *The distributional result in item 1 is assigned as Exercise 6.59.*

For item 2, since $(X_j - \mu)/\sigma$ is standard normal, Example IV.2.4 obtains that $(X_j - \mu)^2/\sigma^2$ has gamma distribution with parameters $\lambda = \alpha = 1/2$, and equivalently, a chi-squared distribution with 1 degree of freedom. A summation of n independent chi-squared variates is by item 5 of Section IV.4.4.1, gamma with $\lambda = 1/2$ and $\alpha = n/2$, and equivalently, a chi-squared distribution with n degrees of freedom, denoted $\chi^2_{n\ d.f.}$. The apparent discrepancy of 1 degree of freedom will be fixed by switching from $\sum_{j=1}^n (X_j - \mu)^2/\sigma^2$ to $\sum_{j=1}^n (X_j - m)^2/\sigma^2$ as follows.

First:

$$\begin{aligned}\sum_{j=1}^n \left(\frac{X_j - \mu}{\sigma}\right)^2 &= \sum_{j=1}^n \left(\frac{X_j - m}{\sigma} + \frac{m-\mu}{\sigma}\right)^2 \\ &= \sum_{j=1}^n \left(\frac{X_j - m}{\sigma}\right)^2 + n\left(\frac{m-\mu}{\sigma}\right)^2,\end{aligned}$$

by (1) of the preceding proof. Thus:

$$\sum_{j=1}^n \left(\frac{X_j - \mu}{\sigma}\right)^2 = \frac{ns^2}{\sigma^2} + \left(\frac{m-\mu}{\sigma/\sqrt{n}}\right)^2.$$

From the above discussion, the term on the left is $\chi^2_{n\ d.f.}$. Since $(m-\mu)/[\sigma/\sqrt{n}]$ is standard normal by Exercise 6.59, the second term on the right is $\chi^2_{1\ d.f.}$ by the above discussion. Then by independence of m and s^2, this obtains independence of the terms on the right by Proposition II.3.56.

Evaluating the moment generating function of both sides obtains by independence and (4.40), as well the gamma moment generating function formula in IV.(4.73), that for $t < \lambda = 1/2$:

$$(1-2t)^{-n/2} = M_{ns^2/\sigma^2}(t)(1-2t)^{-1/2}.$$

Thus:

$$M_{ns^2/\sigma^2}(t) = (1-2t)^{-(n-1)/2},$$

so ns^2/σ^2 is gamma with $\lambda = 1/2$ and $\alpha = (n-1)/2$ by the uniqueness of moments result of Proposition IV.4.61. Then again by definition, ns^2/σ^2 is $\chi^2_{n-1\ d.f.}$. ■

Exercise 6.59 *If $\{X_j\}_{j=1}^n$ is an n-sample from a normal distribution with mean μ and variance σ^2, use Proposition 4.20 to show that m has a normal distribution with mean μ and variance σ^2/n. Hint: Proposition II.3.56.*

7

Applications of Characteristic Functions

In this chapter, we explore a variety of topics of interest which provide applications of characteristic functions.

First, the central limit theorem of Proposition IV.6.13 can now be significantly generalized in various ways from the earlier context of distribution functions with moment generating functions. Because every distribution function has a characteristic function, and moreover is uniquely "characterized" by this characteristic function, it is the perfect tool to use for such generalizations. For example the first generalization applies to distribution functions with only two moments, and then we consider results on independent but not identically distributed variates, and finally derive a central limit theorem in $\mathbb{R}^n$.

The second investigation focuses on distribution functions of sums of random variables, and in particular, distribution families related under addition. The approach used is similar to that in Section IV.4.4.1, but now replacing moment generating functions with characteristic functions. Consequently, the results will extend far beyond the earlier special situations. We then apply these methods to investigate so-called "infinitely divisible" distributions and some of their properties.

The final section considers the distribution function of products of random variables within the limited framework needed for the next chapter.

7.1 Central Limit Theorems

The central limit theorem of Proposition IV.6.13 required the very strong assumption that the distribution function $F(x)$ of the independent and identically distributed random variables had an associated moment generating function $M_X(t)$ for $t \in (-t_0, t_0)$ with $t_0 > 0$. With characteristic functions, we can effectively reproduce the earlier proof, but now require only the assumption that this distribution function has two moments. This is the "classical" statement of the central limit theorem, from which many generalizations exist and some are investigated in this chapter.

The idea of using characteristic functions to prove central limit theorems and other weak convergence results was introduced by **Paul Lévy** (1886–1971). Consequently, one often finds Lévy's name attached to statements of various versions of this theorem when the proof is based on an analysis of characteristic functions.

We present four generalizations of the earlier Book IV result. For background on probability spaces for independent, identically distributed random variables, the reader is referred to Chapter II.4, while various results on weak convergence of distribution functions can be found in Chapters II.8 and IV.6, as well as in this book's Chapter 3.

Before beginning, we prove a couple of needed technical estimates related to complex numbers.

DOI: 10.1201/9781003275770-7

Lemma 7.1 *1. If $z \in \mathbb{C}$, then:*

$$\left|e^z - (1+z)\right| \leq |z|^2 e^{|z|}. \tag{7.1}$$

2. Given $\{a_j\}_{j=1}^n, \{b_j\}_{j=1}^n \subset \mathbb{C}$ with $|a_j| \leq \lambda$ and $|b_j| \leq \lambda$:

$$\left|\prod_{j=1}^n a_j - \prod_{j=1}^n b_j\right| \leq \lambda^{n-1} \sum_{j=1}^n |a_j - b_j|. \tag{7.2}$$

Proof.

1. *As noted in Section 5.2, e^z is well-defined for $z \in \mathbb{C}$ by the absolutely convergent series*

$$e^z = \sum_{j=0}^\infty \frac{z^j}{j!}. \tag{7.3}$$

Thus:

$$\left|e^z - (1+z)\right| = \left|\sum_{j=2}^\infty \frac{z^j}{j!}\right| \leq |z|^2 \sum_{j=0}^\infty \frac{|z|^j}{(j+2)!} \leq |z|^2 e^{|z|}.$$

2. *We prove this result by induction, noting that this inequality is apparently true for $n = 1$. Assuming that this result is true for $n - 1$:*

$$\begin{aligned}\left|\prod_{j=1}^n a_j - \prod_{j=1}^n b_j\right| &\leq \left|a_1\left(\prod_{j=2}^n a_j - \prod_{j=2}^n b_j\right)\right| + \left|a_1 \prod_{j=2}^n b_j - b_1 \prod_{j=2}^n b_j\right| \\ &\leq \lambda\left[\lambda^{n-2} \sum_{j=2}^n |a_j - b_j|\right] + |a_1 - b_1|\,\lambda^{n-1} \\ &= \lambda^{n-1} \sum_{j=1}^n |a_j - b_j|.\end{aligned}$$

■

7.1.1 The Classical Central Limit Theorem

The term "classical" is typically applied to the next statement of the central limit theorem. It generalizes Proposition IV.6.13, now requiring only the existence of two moments. In contrast, the earlier result assumed the existence of a moment generating function, and thus by Proposition IV.4.25 required the existence of all moments, and by Proposition IV.4.27, a convergent power series in these moments.

For this convergence result, we again **normalize** the sum or average of independent, identically distributed random variables to produce a random variable Y_n with:

$$E[Y_n] = 0, \qquad Var[Y_n] = 1.$$

A calculation obtains that this normalization of a random variable W is achieved by the transformation:

$$W \to \frac{W - E[W]}{\sqrt{Var[W]}}. \tag{7.4}$$

Recall the "big-O" and "little-o" terminology from Definition V.7.13:

Definition 7.2 (big O, little o) *Let $f(x)$ and $g(x) > 0$ be two functions defined on $\mathbb{R}$. The expression:*

$$f(x) = O(g(x)) \text{ as } x \to \infty,$$

or in words, "$f(x)$ ***is big-O of*** *$g(x)$* ***as*** *$x \to \infty$," means that there exists x_0 and a positive constant K so that:*

$$|f(x)| \leq Kg(x) \text{ for all } x \geq x_0.$$

The expression:

$$f(x) = o(g(x)) \text{ as } x \to \infty,$$

or in words, "$f(x)$ ***is little-o of*** *$g(x)$* ***as*** *$x \to \infty$," means that:*

$$\frac{|f(x)|}{g(x)} \to 0 \text{ as } x \to \infty.$$

When $|f(x)| \to 0$ as $x \to \infty$, it is common to express this as $f(x) = o(1)$ as $x \to \infty$.

Exercise 7.3 (On Y_n) *Confirm that Y_n below is of the form in (7.4) with $W = \sum_{j=1}^{n} X_j$ or $W = \frac{1}{n}\sum_{j=1}^{n} X_j$.*

Proposition 7.4 (Central limit theorem 2) *Let F_X denote the distribution function of a random variable X with mean μ and variance σ^2, and let Y_n denote the normalized sum or average of an n-sample $\{X_j\}_{j=1}^{n}$:*

$$Y_n \equiv \frac{\sum_{j=1}^{n} X_j - n\mu}{\sqrt{n}\sigma} = \frac{\frac{1}{n}\sum_{j=1}^{n} X_j - \mu}{\sigma/\sqrt{n}}.$$

By n-sample is meant that these X_j-variates are independent and identically distributed with X.

Then as $n \to \infty$, the distribution function F_{Y_n} of Y_n converges in distribution to Φ, the distribution function of the standard normal:

$$F_{Y_n} \Rightarrow \Phi. \tag{7.5}$$

Proof. *Letting $Y = X - \mu$ and $Y_n = \sum_{j=1}^{n}\left(\frac{X_j-\mu}{\sqrt{n}\sigma}\right)$, it follows from (5.47) and (5.48) that:*

$$C_{Y_n}(t) = \left[C_Y\left(\frac{t}{\sqrt{n}\sigma}\right)\right]^n. \tag{1}$$

By Proposition V.7.5:

$$\left|e^{iy} - \sum_{j=0}^{2} (iy)^j / j!\right| \leq \min\left[\frac{|y|^3}{6}, |y|^2\right].$$

and so with the triangle inequality of Proposition V.7.3 and $y = Yt/\sqrt{n}\sigma$:

$$\begin{aligned}
&\left|E\left[e^{iYt/\sqrt{n}\sigma}\right] - E\left[\sum_{j=0}^{2} (it)^j \left(Y/\sqrt{n}\sigma\right)^j / j!\right]\right| \\
&\leq E\left|e^{iYt/\sqrt{n}\sigma} - \sum_{j=0}^{2} (it)^j \left(Y/\sqrt{n}\sigma\right)^j / j!\right| \\
&\leq E\left[\min\left(\left(|Yt|/\sqrt{n}\sigma\right)^3/6, \left(|Yt|/\sqrt{n}\sigma\right)^2\right)\right] \\
&= \frac{t^2}{n} E\left[\min\left(\frac{|t|}{6\sqrt{n}}\left(\frac{|Y|}{\sigma}\right)^3, \left(\frac{Y}{\sigma}\right)^2\right)\right].
\end{aligned} \tag{2}$$

Fix t and let:

$$f_n(Y) = \min\left(\frac{|t|}{6\sqrt{n}}\left(\frac{|Y|}{\sigma}\right)^3, \left(\frac{Y}{\sigma}\right)^2\right).$$

Then $f_n(Y) \leq f(Y) \equiv (Y/\sigma)^2$ *for all* n, *and* f *is integrable by the existence of* σ^2. *Since* $f_n(Y) \to 0$ *pointwise as* $n \to \infty$, *Lebesgue's dominated convergence theorem of Proposition V.3.45 obtains that* $E[f_n(Y)] \to 0$ *as* $n \to \infty$.

Thus, the upper bound in (2) *is* $o(n^{-1})$. *Since* $E\left[e^{iyt/\sqrt{n}\sigma}\right] \equiv C_Y(t/[\sqrt{n}\sigma])$, $E[Y] = 0$ *and* $E[Y^2] = \sigma^2$, *this yields:*

$$C_Y(t/[\sqrt{n}\sigma]) = 1 - t^2/2n + o(n^{-1}),$$

and so by (1):

$$C_{Y_n}(t) = \left(1 - t^2/2n + o(n^{-1})\right)^n.$$

If it can be proved that for all t:

$$\left(1 - t^2/2n + o(n^{-1})\right)^n \to \exp\left(-\frac{1}{2}t^2\right), \tag{3}$$

then (5.44) and Lévy's continuity theorem of Proposition 5.30 will complete the proof.

For the limit in (3), *the error term of* $o(n^{-1})$ *is complex, and hence we must prove this as a result in* $\mathbb{C}$. *To this end:*

$$\begin{aligned} &\left|\exp(-t^2/2) - \left(1 - t^2/2n + o(n^{-1})\right)^n\right| \\ &\leq \left|\exp(-t^2/2) - \exp\left[-t^2/2 + n\left[o(n^{-1})\right]\right]\right| \\ &\quad + \left|\exp\left[-t^2/2 + n\left[o(n^{-1})\right]\right] - \left[1 - t^2/2n + o(n^{-1})\right]^n\right|. \end{aligned} \tag{4}$$

Since $n\left[o(n^{-1})\right] \to 0$ *by definition, the first expression on the right in* (4) *converges to zero as* $n \to \infty$ *by the continuity of the exponential function.*

For the second expression, let $w = \exp(z)$ *where* $z = -t^2/2n + o(n^{-1})$. *Applying (7.2):*

$$|w^n - (1+z)^n| \leq n\lambda^{n-1}|w - (1+z)|,$$

where:

$$\lambda = \max\left[|1+z|, |\exp z|\right] \leq \max\left[1 + |z|, \exp|z|\right].$$

Then with (7.1):

$$|w^n - (1+z)^n| \leq n\lambda^{n-1}|z|^2 e^{|z|}.$$

As $n \to \infty$, $|z| = O(n^{-1})$, $\lambda^{n-1} = O(\exp(-t^2/2))$, *and so:*

$$|w^n - (1+z)^n| = O(n^{-1})\, O(\exp(-t^2/2)).$$

Putting the pieces together obtains:

$$\left|\exp\left(-\frac{1}{2}t^2 + o(n^{-1})\right) - \left(1 - \frac{1}{2}t^2/n + o(n^{-1})\right)^n\right| \to 0,$$

completing the proof. ■

7.1.2 Lindeberg's Central Limit Theorem

The classical central limit theorem can be generalized from sums of independent, identically distributed random variables:

$$S_n \equiv \sum\nolimits_{j=1}^{n} X_j,$$

to sums of independent, but not necessarily identically distributed random variables. Specifically, we now define:

$$S_n \equiv \sum\nolimits_{j=1}^{r_n} X_{n,j},$$

where $r_n \to \infty$ as $n \to \infty$, and for each n, $\{X_{n,j}\}_{j=1}^{r_n}$ are only assumed to be independent.

This general setup applies to two common situations:

1. For all n, $X_{n,j} = X_j$ and $r_n = n$. In this case, $S_n \equiv \sum_{j=1}^{n} X_j$ and the generalization relative to Proposition 7.4 is that while independent, the collection $\{X_j\}_{j=1}^{\infty}$ is not assumed to be identically distributed.

2. More generally, $\{X_{n,j}\}_{j=1}^{r_n}$ are assumed independent for each n, but not necessarily identically distributed.

In the general setting of item 2, the collection $\left\{\{X_{n,j}\}_{j=1}^{r_n}\right\}_{n=1}^{\infty}$ is called a **triangular array of random variables** because of the shape of the array that is created by placing the set of variates $\{X_{n,j}\}_{j=1}^{r_n}$ into the nth row of an array. Thus each row of the array contains independent random variables by assumption, while each column contains independent or dependent random variables. That columns may contain dependent random variables should not surprise. In item 1, not only are the random variables $\{X_{n,j}\}_{n=f(j)}^{\infty}$ in the jth column dependent, where $f(j) = \min\{n : r_n \geq j\}$, but they are identical.

The first generalization of this type was formulated in 1922 by **J. W. (Jarl Waldemar) Lindeberg** (1876–1932), and applied in the context in item 1 above. That is, $S_n \equiv \sum_{j=1}^{n} X_j$ and $\{X_j\}_{j=1}^{\infty}$ are independent but not assumed to be identically distributed. To prove this result, it is necessary to impose some condition that ensures that S_n is not dominated by one or a few random variables.

Example 7.5 (CLT Failure: General independent random variables) *Let $\{X_j\}_{j=1}^{\infty}$ be independent random variables where X_j has mean 0 and variance $j\,(j!)$. Then $S_n \equiv \sum_{j=1}^{n} X_j$ has mean 0, and mathematical induction obtains that S_n has a variance of $(n+1)! - 1$.*

Normalizing, it follows that:

$$\frac{S_n}{\sqrt{(n+1)!-1}} = \frac{X_n}{\sqrt{(n+1)!-1}} + \frac{S_{n-1}}{\sqrt{(n+1)!-1}}, \tag{1}$$

where:

$$Var\left[\frac{X_n}{\sqrt{(n+1)!-1}}\right] = \frac{n(n!)}{(n+1)!-1},$$

$$Var\left[\frac{S_{n-1}}{\sqrt{(n+1)!-1}}\right] = \frac{n!-1}{(n+1)!-1}.$$

The first variance approaches 1 as $n \to \infty$, while the second approaches 0. An application of Chebyshev's inequality of Proposition IV.4.34 obtains that for every $s > 0$:

$$\Pr\left[\left|\frac{S_{n-1}}{\sqrt{(n+1)!-1}}\right| \geq s\right] \leq \frac{1}{s^2}\frac{n!-1}{(n+1)!-1} \to 0.$$

Then by (1):

$$\Pr\left[\left|\frac{S_n}{\sqrt{(n+1)!-1}}\right| \geq s\right] \geq \Pr\left[\left|\frac{S_{n-1}}{\sqrt{(n+1)!-1}}\right| \geq \frac{s}{2}\right] + \Pr\left[\left|\frac{X_n}{\sqrt{(n+1)!-1}}\right| \geq \frac{s}{2}\right]$$
$$\approx \Pr\left[\left|\frac{X_n}{\sqrt{(n+1)!-1}}\right| \geq \frac{s}{2}\right], \text{ as } n \to \infty.$$

Hence, while the normalized summation $S_n/\sqrt{(n+1)!-1}$ has mean 0 and variance 1 by construction, it has little hope of achieving a normal distribution as $n \to \infty$. This normalized variate approaches the distribution of X_n, normalized to have a variance that approaches 1 as $n \to \infty$.

The restriction imposed by Lindeberg to avoid this anomalous behavior has come to be known as **Lindeberg's condition**, or **the Lindeberg condition.** We note that:

- The domain of integration of $|X_j - \mu_j| \geq ts_n$ in (7.6) is sometimes expressed as $|X_j - \mu_j| > ts_n$, and similarly for (7.7). It will be seen in the proof that either formulation works equally well.
- The integrals in this definition are sometimes expressed as Lebesgue-Stieltjes integrals on $\mathbb{R}$, reflecting the Borel measures induced by the respective random variables X_j or $X_{n,j}$. These measures are defined in (1.21), and the transformation of these integrals is an application of Proposition V.4.18, as seen in Section 4.1.

Definition 7.6 (Lindeberg's Condition) *If $\{X_j\}_{j=1}^{\infty}$ are independent random variables defined on the probability space $(\mathcal{S}, \mathcal{E}, \lambda)$ with $\mu_j = E[X_j]$, $\sigma_j^2 = Var[X_j]$, and $s_n^2 \equiv \sum_{j=1}^{n} \sigma_j^2$,* ***Lindeberg's condition*** *is satisfied if for every $t > 0$:*

$$\lim_{n\to\infty} \frac{1}{s_n^2} \sum\nolimits_{j=1}^{n} \int_{|X_j-\mu_j|\geq ts_n} (X_j - \mu_j)^2 \, d\lambda = 0. \tag{7.6}$$

More generally, let $\left\{\{X_{n,j}\}_{j=1}^{r_n}\right\}_{n=1}^{\infty}$ be a triangular array of random variables defined on $(\mathcal{S}, \mathcal{E}, \lambda)$, and thus $\{X_{n,j}\}_{j=1}^{r_n}$ are independent random variables for each n. If $\mu_{n,j} = E[X_{n,j}]$, $\sigma_{n,j}^2 = Var[X_{n,j}]$ and $s_n^2 \equiv \sum_{j=1}^{r_n} \sigma_{n,j}^2$, ***Lindeberg's condition*** *is satisfied if for every $t > 0$:*

$$\lim_{n\to\infty} \frac{1}{s_n^2} \sum\nolimits_{j=1}^{r_n} \int_{|X_{n,j}-\mu_{n,j}|\geq ts_n} (X_{n,j} - \mu_{n,j})^2 \, d\lambda = 0. \tag{7.7}$$

Remark 7.7 (On Lindeberg's condition) *Note that in the absence of the restriction on the domains of integral, meaning $|X_j - \mu_j| \geq ts_n$ or $|X_{n,j} - \mu_{n,j}| \geq ts_n$, that the expressions above would be identically 1 for all n. For example, by definition of s_n^2:*

$$\frac{1}{s_n^2} \sum\nolimits_{j=1}^{n} \int (X_j - \mu_j)^2 \, d\lambda = 1,$$

and similarly for (7.7). Thus the intuition of the Lindeberg condition is that as $n \to \infty$, all of the variability of these random variables comes from regions near the means, where "near" is defined relative to s_n, and not from regions far from the means.

To formalize this in the notation of (7.6):

$$\sigma_j^2 \equiv \int_{|X_j-\mu_j|<ts_n} (X_j - \mu_j)^2 \, d\lambda + \int_{|X_j-\mu_j|\geq ts_n} (X_j - \mu_j)^2 \, d\lambda. \tag{1}$$

Thus:

$$s_n^2 = \sum\nolimits_{j=1}^{n} \int_{|X_j-\mu_j|<ts_n} (X_j-\mu_j)^2 \, d\lambda + \sum\nolimits_{j=1}^{n} \int_{|X_j-\mu_j|\geq ts_n} (X_j-\mu_j)^2 \, d\lambda,$$

and the Lindeberg condition can be equivalently stated that for every $t>0$:

$$\lim_{n\to\infty} \frac{1}{s_n^2} \sum\nolimits_{j=1}^{n} \int_{|X_j-\mu_j|<ts_n} (X_j-\mu_j)^2 \, d\lambda = 1. \tag{7.8}$$

In other words, for any $t>0$, the regions $\{|X_j-\mu_j|<ts_n\}_{j=1}^{n}$ contain asymptotically all of the variance of $\{X_j\}_{j=1}^{n}$ as $n\to\infty$.

Similar restatements apply to (7.7).

The next result provides another interesting implication of Lindeberg's condition which precludes the result of Example 7.5. It states that the maximum contribution of σ_j^2 to s_n^2 in the first case, and of $\sigma_{n,j}^2$ to s_n^2 in the second, is negligible as $n\to\infty$.

Proposition 7.8 (Lindeberg and $\max_{j\leq n}\frac{\sigma_j^2}{s_n^2}$, $\max_{j\leq r_n}\frac{\sigma_{n,j}^2}{s_n^2}$) *If $\{X_j\}_{j=1}^{\infty}$ are independent random variables defined on the probability space $(\mathcal{S},\mathcal{E},\lambda)$ that satisfy Lindeberg's condition with $\mu_j=E[X_j]$, $\sigma_j^2=Var[X_j]$, and $s_n^2\equiv\sum_{j=1}^{n}\sigma_j^2$, then as $n\to\infty$:*

$$\lim_{n\to\infty} \max\nolimits_{j\leq n} \frac{\sigma_j^2}{s_n^2} = 0. \tag{7.9}$$

If $\left\{\{X_{n,j}\}_{j=1}^{r_n}\right\}_{n=1}^{\infty}$ is a triangular array of random variables defined on $(\mathcal{S},\mathcal{E},\lambda)$ that satisfies Lindeberg's condition with $\mu_{n,j}=E[X_{n,j}]$, $\sigma_{n,j}^2=Var[X_{n,j}]$ and $s_n^2\equiv\sum_{j=1}^{r_n}\sigma_{n,j}^2$, then as $n\to\infty$:

$$\lim_{n\to\infty} \max\nolimits_{j\leq r_n} \frac{\sigma_{n,j}^2}{s_n^2} = 0. \tag{7.10}$$

Proof. *We prove (7.9) and leave (7.10) as an exercise.*

From (1) of Remark 7.7, we obtain that for any $t>0$:

$$\frac{\sigma_j^2}{s_n^2} \leq t^2 + \frac{1}{s_n^2} \int_{|X_j-\mu_j|\geq ts_n} (X_j-\mu_j)^2 \, d\lambda.$$

Thus:

$$\max_{j\leq n} \frac{\sigma_j^2}{s_n^2} \leq t^2 + \frac{1}{s_n^2} \sum\nolimits_{j=1}^{n} \int_{|X_j-\mu_j|\geq ts_n} (X_j-\mu_j)^2 \, d\lambda.$$

If the Lindeberg condition is satisfied, this integral sum converges to 0 with n, and so:

$$\lim_{n\to\infty} \max_{j\leq n} \frac{\sigma_j^2}{s_n^2} \leq t^2.$$

Since t is arbitrary, this proves:

$$\lim_{n\to\infty} \max_{j\leq n} \frac{\sigma_j^2}{s_n^2} = 0.$$

■

Exercise 7.9 (i.i.d.$\{X_{n,j}\}_{j=1}^{r_n}$ for each n) *Assume that for each n, $\{X_{n,j}\}_{j=1}^{r_n}$ are independent and also identically distributed. In other words, assume that in each row of the triangular array that $X_{n,j} = X_n$ has a fixed distribution with mean μ_n and variance σ_n^2. Show that (7.7) is equivalent to the assumption that for all $t > 0$:*

$$\lim_{n\to\infty} \int_{|Z_n|\geq t\sqrt{r_n}} Z_n^2 d\lambda = 0, \tag{7.11}$$

where $Z_n \equiv (X_n - \mu_n)/\sigma_n$ denotes the normalized version of X_n as in (7.4).

Example 7.10 (Classical central limit theorem) *When $X_{n,j}$ are defined on the probability space $(\mathcal{S}, \mathcal{E}, \lambda)$ and identically distributed for all n and j, as assumed for classical central limit theorem, the Lindeberg condition is always satisfied.*

With fixed mean μ and variance σ^2, the requirement of (7.11) can be written with $Z \equiv \frac{X-\mu}{\sigma}$:

$$\lim_{n\to\infty} \int_{|Z|\geq t\sqrt{r_n}} Z^2 d\lambda = 0.$$

Since $r_n \to \infty$ by assumption, and Z^2 is integrable by the existence of the variance of X, this limit is satisfied, as is Lindeberg's condition by Exercise 7.9.

The following statement and proof of the **Lindeberg central limit theorem** are given in the more general context of triangular arrays. Simply assuming that $X_{n,j} = X_j$, $\mu_{n,j} = \mu_j$, $\sigma_{n,j}^2 = \sigma_j^2$ and $r_n = n$ for all n produces the version applicable to $S_n \equiv \sum_{j=1}^{n} X_j$, where the collection $\{X_j\}_{j=1}^{\infty}$ is assumed independent but not necessarily identically distributed. We state this version as Proposition 7.13.

Exercise 7.11 (On Y_n) *Confirm that Y_n below is of the form in (7.4) with $W = \sum_{j=1}^{r_n} X_{n,j}$.*

Proposition 7.12 (Lindeberg's central limit theorem: Triangular arrays) *For each n, let $\{X_{n,j}\}_{j=1}^{r_n}$ be independent random variables where $r_n \to \infty$ as $n \to \infty$. Let $\mu_{n,j} = E[X_{n,j}]$, $\sigma_{n,j}^2 = Var[X_{n,j}]$ and $s_n^2 = \sum_{j=1}^{r_n} \sigma_{n,j}^2$, and define:*

$$Y_n = \sum_{j=1}^{r_n} \left(\frac{X_{n,j} - \mu_{n,j}}{s_n} \right).$$

If (7.7) is satisfied for all $t > 0$, then as $n \to \infty$:

$$F_{Y_n} \Rightarrow \Phi, \tag{7.12}$$

where $\Phi(x)$ is the distribution function of the standard normal.

Proof. *Defining $X'_{n,j} \equiv (X_{n,j} - \mu_{n,j})/s_n$, then $Y_n = \sum_{j=1}^{r_n} X'_{n,j}$ with $\mu'_{n,j} \equiv E[X'_{n,j}] = 0$, $\left(\sigma'_{n,j}\right)^2 \equiv Var[X'_{n,j}] = \sigma_{n,j}^2/s_n^2$, and $\left(s'_n\right)^2 \equiv \sum_{j=1}^{r_n} \left(\sigma'_{n,j}\right)^2 = 1$. Hence proving this result for independent $\{X'_{n,j}\}_{j=1}^{r_n}$ with $\mu'_{n,j} = 0$ and $\left(s'_n\right)^2 = 1$ yields the general result. We drop the prime notation for convenience and denote $X'_{n,j}$ by $X_{n,j}$.*

Repeating the derivation of (2) of Proposition 7.4 but with $y = Yt$, it follows that with $C_{n,j}(t)$ denoting the characteristic function of such $X_{n,j}$, that for any $\epsilon > 0$:

$$\begin{aligned}
\left|C_{n,j}(t) - (1 - t^2\sigma_{n,j}^2/2)\right| &\leq E\left[\min\left(\left|tX_{n,j}\right|^3/6, \left|tX_{n,j}\right|^2\right)\right] \\
&\leq \int_{|X_{n,j}|<\epsilon} \left|tX_{n,j}\right|^3/6 d\lambda + \int_{|X_{n,j}|\geq\epsilon} \left|tX_{n,j}\right|^2 d\lambda \\
&\leq \frac{\epsilon t^3\sigma_{n,j}^2}{6} + t^2 \int_{|X_{n,j}|\geq\epsilon} \left|X_{n,j}\right|^2 d\lambda.
\end{aligned}$$

Thus since $s_n^2 = \sum_{j=1}^{r_n} \sigma_{n,j}^2 = 1$*:*

$$\sum_{j=1}^{r_n} \left|C_{n,j}(t) - (1 - t^2\sigma_{n,j}^2/2)\right| \leq \frac{\epsilon t^3}{6} + t^2 \sum_{j=1}^{r_n} \int_{|X_{n,j}| \geq \epsilon} |X_{n,j}|^2 \, d\lambda.$$

Since ϵ *is arbitrary, it follows from (7.7) that:*

$$\sum_{j=1}^{r_n} \left|C_{n,j}(t) - (1 - t^2\sigma_{n,j}^2/2)\right| \to 0 \text{ as } n \to \infty. \tag{1}$$

To obtain (7.12), we claim that (1) *implies that for all* t*:*

$$C_n(t) \to \exp\left[-t^2/2\right], \tag{2}$$

where $C_n(t) = \prod_{j=1}^{r_n} C_{n,j}(t)$ *is the characteristic function of* Y_n*. Given* (2)*, (7.12) then follows from (5.44) and Lévy's continuity theorem of Proposition 5.30.*

To prove (2)*, recall Proposition 7.8 with* $s_n^2 = 1$*, that the Lindeberg condition assures that* $\max_{j \leq n} \sigma_{n,j}^2 \to 0$ *as* $n \to \infty$*. So given* t *there exists* N *so that for* $n \geq N$ *and all* j*:*

$$0 \leq 1 - t^2\sigma_{n,j}^2/2 \leq 1. \tag{3}$$

For such $n \geq N$*, and recalling that* $|C_{n,j}(t)| \leq 1$ *by Proposition 5.19, apply (7.2) with* $\lambda = 1$ *to produce:*

$$\left|\prod_{j=1}^{r_n} C_{n,j}(t) - \prod_{j=1}^{r_n} (1 - t^2\sigma_{n,j}^2/2)\right| \leq \sum_{j=1}^{r_n} \left|C_{n,j}(t) - (1 - t^2\sigma_{n,j}^2/2)\right|.$$

Then by the estimate in (1)*:*

$$C_n(t) = \prod_{j=1}^{r_n} (1 - t^2\sigma_{n,j}^2/2) + \epsilon_n, \tag{4}$$

where $|\epsilon_n| \to 0$ *as* $n \to \infty$*.*

Now (3) *implies that* $\exp\left(-t^2\sigma_{n,j}^2/2\right) \leq 1$*, so applying (7.2) and then (7.1) obtains for* $n \geq N$*:*

$$\begin{aligned}
&\left|\prod_{j=1}^{r_n} \exp\left(-t^2\sigma_{n,j}^2/2\right) - \prod_{j=1}^{r_n} (1 - t^2\sigma_{n,j}^2/2)\right| \\
\leq\ & \sum_{j=1}^{r_n} \left|\exp\left(-t^2\sigma_{n,j}^2/2\right) - (1 - t^2\sigma_{n,j}^2/2)\right| \\
\leq\ & \sum_{j=1}^{r_n} \left(t^2\sigma_{n,j}^2/2\right)^2 \exp\left(t^2\sigma_{n,j}^2/2\right) \\
\leq\ & \frac{t^4}{4} \max_j \left[\exp\left(t^2\sigma_{n,j}^2/2\right)\right] \sum_{j=1}^{r_n} \sigma_{n,j}^4 \\
\leq\ & \frac{t^4}{4} \max_j \left[\exp\left(t^2\sigma_{n,j}^2/2\right)\right] \max_j \sigma_{n,j}^2 \sum_{j=1}^{r_n} \sigma_{n,j}^2.
\end{aligned}$$

As $\sum_{j=1}^{r_n} \sigma_{n,j}^2 = 1$ *and* $\max \sigma_{n,j}^2 \to 0$*:*

$$\begin{aligned}
\prod_{j=1}^{r_n} (1 - t^2\sigma_{n,j}^2/2) &= \prod_{j=1}^{r_n} \exp\left[-t^2\sigma_{n,j}^2/2\right] + \epsilon_n' \\
&= \exp\left[-t^2/2\right] + \epsilon_n',
\end{aligned} \tag{5}$$

where $|\epsilon_n'| \to 0$ *as* $n \to \infty$*.*

Combining estimates (4) *and* (5) *yields:*

$$C_n(t) = \exp\left[-t^2/2\right] + \epsilon_n + \epsilon_n',$$

and the result in (2) *follows.* ■

Proposition 7.13 (Lindeberg's central limit theorem: independent sums) *Let $\{X_j\}_{j=1}^{\infty}$ be independent random variables with $\mu_j = E[X_j]$, $\sigma_j^2 = Var[X_j]$ and $s_n^2 = \sum_{j=1}^{n}\sigma_j^2$, and define:*

$$Y_n = \sum\nolimits_{j=1}^{n}\left(\frac{X_j-\mu_j}{s_n}\right).$$

If (7.6) is satisfied for all t, then as $n\to\infty$:

$$F_{Y_n} \Rightarrow \Phi, \tag{7.13}$$

where $\Phi(x)$ is the distribution function of the standard normal.
Proof. *This follows from Proposition 7.12 with $X_{n,j} = X_j$, $\mu_{n,j} = \mu_j$, $\sigma_{n,j}^2 = \sigma_j^2$ and $r_n = n$ for all n.* ■

The following corollary addresses the special case of Proposition 7.12 when $s_n^2 \to \sigma^2$ and $\mu_n \to \mu$.

Corollary 7.14 (When $s_n^2 \to \sigma^2$ and $\mu_n \to \mu$.) *For each n, let $\{X_{n,j}\}_{j=1}^{r_n}$ be independent random variables with $\mu_{n,j} = E[X_{n,j}]$ and $\sigma_{n,j}^2 = Var[X_{n,j}]$. Assume that as $n\to\infty$ that $r_n\to\infty$, and:*

$$s_n^2 \equiv \sum\nolimits_{j=1}^{r_n}\sigma_{n,j}^2 \to \sigma^2, \quad \mu_n \equiv \sum\nolimits_{j=1}^{r_n}\mu_{n,j} \to \mu.$$

If (7.7) is satisfied for all t, then as $n\to\infty$:

$$F_{Z_n} \Rightarrow \Phi, \tag{7.14}$$

where:

$$Z_n = \frac{\sum_{j=1}^{r_n} X_{n,j} - \mu}{\sigma},$$

and $\Phi(x)$ is the distribution of the standard normal.

In other words, the distribution function of $\sum_{j=1}^{r_n} X_{n,j}$ converges in distribution to the normal distribution with mean μ and variance σ^2.
Proof. *With Y_n defined as in Proposition 7.12:*

$$Y_n = \sum\nolimits_{j=1}^{r_n}\left(\frac{X_{n,j}-\mu_{n,j}}{s_n}\right),$$

it follows that:

$$Z_n = c_nY_n + d_n$$

where $c_n \equiv s_n/\sigma$ and $d_n \equiv (\mu_n - \mu)/\sigma$. Since $c_n \to 1$ and $d_n \to 0$, Proposition II.9.16 obtains that Z_n converges in distribution to the same distribution function as Y_n, which is Φ by Proposition 7.12. ■

Remark 7.15 (Lindeberg converse) *The Lindeberg condition is nearly best possible for a central limit theorem. It was proved in 1935 by* ***William Feller*** *(1906–1970) that if:*

$$\lim_{n\to\infty} \max\nolimits_{j\le r_n}\frac{\sigma_{n,j}^2}{s_n^2} = 0,$$

then the central limit theorem conclusion in (7.12) implies the Lindeberg condition in (7.7).

When the above theorem is stated in terms of Lindeberg's sufficient and Feller's necessary conditions, it is often referred to as the ***Lindeberg-Feller central limit theorem.***

Despite its theoretical strength, the Lindeberg condition can be difficult to verify, even in relatively simple cases. The following is a relatively simple example with $\{X_j\}_{j=1}^{\infty}$ binomial variates.

Example 7.16 ($\{X_j\}_{j=1}^{\infty}$ binomial variates) *Let $S_n = \sum_{j=1}^{n} X_j$ where $\{X_j\}_{j=1}^{\infty}$ are independent with X_j binomially distributed with parameter $p_j \equiv X_j^{-1}(1)$, $0 < p_j < 1$, and $1 - p_j \equiv X_j^{-1}(0)$, so $\mu_j = p_j$, $\sigma_j^2 = p_j(1-p_j)$ and $s_n^2 = \sum_{j=1}^{n} p_j(1-p_j)$. For any j define $m_j = \min[p_j, 1-p_j]$ and $M_j = \max[p_j, 1-p_j]$.*

Then with $A_n \equiv \{|X_j - p_j| \geq ts_n\}$:

$$\int_{A_n} (X_j - p_j)^2 \, d\lambda = \begin{cases} p_j\,(1-p_j)\,, & ts_n \leq m_j, \\ p_j\,(1-p_j)^2\,, & m_j < ts_n \leq M_j, \quad p_j = m_j, \\ p_j^2\,(1-p_j)\,, & m_j < ts_n \leq M_j, \quad p_j = M_j, \\ 0, & ts_n > M_j. \end{cases}$$

Since all $M_j < 1$ by definition, it follows that if $ts_n > 1$ then:

$$\sum\nolimits_{j=1}^{n} \int_{|X_j - p_j| \geq ts_n} (X_j - p_j)^2 \, d\lambda = 0.$$

- *If $s_n^2 \to \infty$, this assures that Lindeberg's condition is satisfied. The condition on $s_n^2 \to \infty$ is obtained for example if $0 < a \leq p_j \leq b < 1$ for all j, but this is not necessary as $p_j = 1/j$ or $p_j = 1 - 1/j$ demonstrate.*
- *If $s_n^2 \to s^2$, then $\liminf p_j = 0$ and/or $\limsup p_j = 1$, so $\liminf m_j = 0$ and/or $\limsup M_j = 1$. Hence only the limit in the middle rows of the above table needs investigation. Assuming for simplicity that $p_j \to 0$ as $j \to \infty$, then for $0 < ts_n < 1$:*

$$\frac{1}{s_n^2} \sum\nolimits_{j=1}^{n} \int_{A_n} (X_j - p_j)^2 \, d\lambda = \frac{\sum_{j=1}^{n} p_j\,(1-p_j)^2}{\sum_{j=1}^{n} p_j(1-p_j)} \geq \min_{j \leq n} (1-p_j)\,.$$

Similarly, if $p_j \to 1$ this ratio exceeds $\min_{j \leq n} p_j$. So in these cases of convergent variance, $s_n^2 \to s^2$, the Lindeberg condition is not satisfied.

7.1.3 Lyapunov's Central Limit Theorem

An important alternative sufficient condition to ensure (7.12) was discovered by **Aleksandr Lyapunov** (1857–1918), and called **Lyapunov's condition.** The proof of the associated central limit theorem is relatively easy and amounts to demonstrating that Lyapunov's condition implies Lindeberg's condition. In theory this implies that Lyapunov's condition is weaker than Lindeberg's. In applications, it is often far easier to verify, and thus can be more useful in practice.

As was the case for the **Lindeberg central limit theorem,** the following statement and proof of **Lyapunov's central limit theorem** is given in the more general context of triangular arrays. Simply assuming that $X_{n,j} = X_j$, $\mu_{n,j} = \mu_j$, $\sigma_{n,j}^2 = \sigma_j^2$, and $r_n = n$ for all n produces the version applicable to $S_n \equiv \sum_{j=1}^{n} X_j$, where the collection $\{X_j\}_{j=1}^{\infty}$ is assumed independent but not necessarily identically distributed.

Definition 7.17 (Lyapunov's condition) *With notation of Definition 7.6, a triangular array $\left\{\{X_{n,j}\}_{j=1}^{r_n}\right\}_{n=1}^{\infty}$ is said to satisfy **Lyapunov's condition** if for some $\delta > 0$, $\left|X_{n,j} - \mu_{n,j}\right|^{2+\delta}$ is integrable for all n, j, and:*

$$\lim_{n\to\infty} \frac{1}{s_n^{2+\delta}} \sum\nolimits_{j=1}^{r_n} E\left[\left|X_{n,j} - \mu_{n,j}\right|^{2+\delta}\right] = 0. \tag{7.15}$$

Exercise 7.18 (i.i.d.$\{X_{n,j}\}_{j=1}^{r_n}$ for each n) *Assume that for each n, $\{X_{n,j}\}_{j=1}^{r_n}$ are independent and also identically distributed. In other words, assume that in each row of the triangular array that $X_{n,j} = X_n$ has a fixed distribution with mean μ_n and variance σ_n^2. Show that (7.15) is equivalent to the assumption that for some $\delta > 0$:*

$$\lim_{n\to\infty} \frac{1}{r_n^{\delta/2}} \int |Z_n|^{2+\delta} d\lambda = 0, \tag{7.16}$$

where $Z_n = (X_n - \mu_n)/\sigma_n$ is the normalized version of X_n as in (7.4).

Then prove that this special case of Lyapunov's condition implies the special case of Lindeberg's condition in (7.11).

Example 7.19 (Classical central limit theorem) *When $X_{n,j}$ are defined on $(\mathcal{S}, \mathcal{E}, \lambda)$ and independent and identically distributed for all n, j, as is assumed for classical central limit theorem, Lyapunov's condition need not be satisfied simply because $X_{n,j}$ may only have two moments and thus $\left|X_{n,j} - \mu_{n,j}\right|^{2+\delta}$ need not be integrable for any $\delta > 0$.*

In this case, the classical central limit theorem is thus a little more general than is the result from Lyapunov's central limit theorem below. However, if $|X_n - \mu_n|^{2+\delta}$ is integrable for some $\delta > 0$, then Lyapunov's condition is indeed satisfied.

With fixed mean μ and variance σ^2, the requirement of (7.16) can be written:

$$\lim_{n\to\infty} \frac{1}{r_n^{\delta/2}} \int |Z|^{2+\delta} d\lambda = 0,$$

where $Z = \frac{X-\mu}{\sigma}$. Since $r_n \to \infty$ by assumption, and $|X - \mu|^{2+\delta}$ is assumed integrable, this limit is satisfied, as is Lyapunov's condition by Exercise 7.18.

Lyapunov's central limit theorem is next. The proof generalizes the last demonstration of Exercise 7.18, that Lyapunov's condition implies Lindeberg's condition.

Proposition 7.20 (Lyapunov's central limit theorem) *For each n let $\{X_{n,j}\}_{j=1}^{r_n}$ be independent random variables where $r_n \to \infty$ as $n \to \infty$. Denote $\mu_{n,j} = E[X_{n,j}]$, $\sigma_{n,j}^2 = Var[X_{n,j}]$ and $s_n^2 = \sum_{j=1}^{r_n} \sigma_{n,j}^2$, and define*

$$Y_n = \sum_{j=1}^{r_n} \left(\frac{X_{n,j} - \mu_{n,j}}{s_n}\right).$$

If (7.15) is satisfied for some $\delta > 0$, then as $n \to \infty$:

$$F_{Y_n} \Rightarrow \Phi, \tag{7.17}$$

where $\Phi(x)$ is the distribution of the standard normal.

Proof. We show that (7.15) implies the Lindeberg condition in (7.7) and then apply Proposition 7.12.

To this end, given $t > 0$:

$$\begin{aligned}
&\frac{1}{s_n^2} \sum_{j=1}^{r_n} \int_{|X_{n,j}-\mu_{n,j}|\geq ts_n} \left(X_{n,j} - \mu_{n,j}\right)^2 d\lambda \\
\leq\ &\frac{1}{s_n^2} \sum_{j=1}^{r_n} \int_{|X_{n,j}-\mu_{n,j}|\geq ts_n} \frac{\left|X_{n,j} - \mu_{n,j}\right|^{2+\delta}}{(ts_n)^\delta} d\lambda \\
\leq\ &\frac{1}{t^\delta} \frac{1}{s_n^{2+\delta}} \sum_{j=1}^{r_n} E\left[\left|X_{n,j} - \mu_{n,j}\right|^{2+\delta}\right].
\end{aligned}$$

This last expression converges to 0 as $n \to \infty$ by (7.15), and the proof is complete. ■

Example 7.21 ($\{X_j\}_{j=1}^{\infty}$ binomial variates) *Recall Example 7.16 on the sum of independent binomial random variables in the case where $s_n^2 \to \infty$.*

When

$$0 < a \leq p_j \leq b < 1,$$

it is an exercise to verify that (7.15) is satisfied with $\delta = 1$, and hence the central limit theorem applies as noted previously.

Now $s_n^2 \to \infty$ does not imply that p_j is bounded away from 0 and 1 as the example $p_j = 1/j$ confirms. But in the general case of $s_n^2 \to \infty$, note that $\{X_j\}_{j=1}^{\infty}$ are uniformly bounded with $|X_j - p_j| \leq 1$, and hence since $s_n^2 \to \infty$,

$$\frac{1}{s_n^3} \sum\nolimits_{j=1}^{n} E\left[|X_j - p_j|^3\right] \leq \frac{s_n^2}{s_n^3} \to 0.$$

Thus 7.15 is always satisfied in this case with $\delta = 1$.

Exercise 7.22 *Show that the above example generalizes. If $\{X_j\}_{j=1}^{\infty}$ are uniformly bounded and $s_n^2 \to \infty$, then the central limit theorem applies.*

7.1.4 A Central Limit Theorem on $\mathbb{R}^n$

Recall Definition 6.17 that a random vector $Y \equiv (Y_1, Y_2, ..., Y_n)$ is said to have a **multivariate normal distribution,** or a **multivariate Gaussian distribution,** if there exists a nondegenerate, positive semidefinite $n \times n$ matrix C and an n-vector μ so that the characteristic function of Y has the form (6.11):

$$C_Y(t) = \exp\left[i\mu \cdot t - \frac{1}{2}t^T C t\right].$$

It is also common to say that the probability measure induced by this distribution is multivariate normal. As noted in Exercise 6.16, C is the covariance matrix of the component variates and μ is the mean vector.

The purpose of this section is to derive a multivariate version of the central limit theorem, and we do so using the **Cramér-Wold device** or the **Cramér-Wold theorem** of Proposition 5.42. This very short proof provides a good example of the power and utility of this earlier result.

To this end, let $\{X_m\}_{m=1}^{\infty}$ be a collection of independent random vectors with range in $\mathbb{R}^n$ and with a common distribution function. Thus $\{X_m\}_{m=1}^{\infty}$ are independent and identically distributed, or in short, **i.i.d.**

Let $X_m \equiv (X_{m1}, X_{m2}, ..., X_{mn})$, and note that we need only assume that $E\left[X_{mj}^2\right] < \infty$ for all j. Then $\nu_j \equiv E\left[X_{mj}\right] < \infty$ for all j by the triangle inequality of Proposition V.3.42 and **Lyapunov's inequality** of Corollary IV.4.56. In addition, $\sigma_{jj} \equiv E\left[(X_{mj} - \nu_j)^2\right] < \infty$ for all j by linearity of the integral, and $\sigma_{ij} \equiv E\left[(X_{mi} - \nu_i)(X_{mj} - \nu_j)\right] < \infty$ for all $i \neq j$ by the **Cauchy-Schwarz inequality** of Proposition IV.4.47. This notational change for variance, σ_{jj} instead of σ_j^2, is common and convenient when we also use the covariances of the component variates, denoted σ_{ij}.

Proposition 7.23 (Central Limit Theorem on $\mathbb{R}^n$) *Let $\{X_m\}_{m=1}^{\infty}$ be a collection of independent random vectors with range in $\mathbb{R}^n$, $X_m \equiv (X_{m1}, X_{m2}, ..., X_{mn})$, $C \equiv (\sigma_{ij})$ the $n \times n$ covariance matrix of component variates, and $\nu = (\nu_1, \nu_2, ..., \nu_n)$, the vector of means. Define $S_m \equiv \sum_{j=1}^{m} X_j$, and:*

$$Y_m \equiv \left(\frac{S_m - m\nu}{\sqrt{m}}\right).$$

Then as $m \to \infty$:

$$F_{Y_m} \Rightarrow \Phi_C, \tag{7.18}$$

where Φ_C is the distribution function of the multivariate normal in (6.11) with covariance matrix C, mean vector $\mu = 0$, and characteristic function:

$$C_\Phi(t) = \exp\left[-\frac{1}{2}t^T Ct\right].$$

Proof. *Let $Z \equiv (Z_1, Z_2, ..., Z_n)$ denote a multivariate normal random vector with covariance matrix C and $\mu = 0$. By the Cramér-Wold theorem of Proposition 5.42, weak convergence $Y_m \Rightarrow Z$ will follow if it can be shown that $t \cdot Y_m \Rightarrow t \cdot Z$ for all $t \in \mathbb{R}^n$, where $x \cdot y \equiv \sum_{j=1}^m x_j y_j$ is the dot product of these vectors.*

To this end, define $X'_j = t \cdot X_j$. By linearity of the integral,

$$E\left[X'_j\right] = t \cdot \nu \equiv \nu_t,$$

and:

$$\begin{aligned} Var\left[X'_j\right] &= E\left[\left(\sum\nolimits_{j=1}^m (X_j - \nu_j)\, t_j\right)^2\right] \\ &= \sum\nolimits_{j=1}^m \sum\nolimits_{i=1}^m t_i t_j \sigma_{ij} \\ &= t^T Ct. \end{aligned}$$

With $\sigma_t \equiv \sqrt{t^T Ct}$:

$$\frac{t \cdot Y_m}{\sigma_t} = \frac{\sum_{j=1}^m X'_j - m\nu_t}{\sigma_t \sqrt{m}},$$

and the classical central limit theorem of Proposition 7.4 states that:

$$\frac{t \cdot Y_m}{\sigma_t} \Rightarrow W,$$

where W is a standard normal variate. This obtains that $t \cdot Y_m \Rightarrow \sigma_t W$, a normal variate with mean 0 and variance σ_t^2, by Slutsky's theorem of Proposition II.5.29.

By the same calculations as above, $Z' \equiv t \cdot Z$ has mean 0 and variance σ_t^2, and is a normal variate by Proposition 6.23.

Thus $t \cdot Y_m \Rightarrow t \cdot Z$ for all $t \in \mathbb{R}^n$, and the Cramér-Wold theorem completes the proof. ■

Remark 7.24 *As is the case for the standard versions of the central limit theorem above, this theorem also provides a weak convergence result for the average random vector, $\overline{S}_m \equiv \frac{1}{m}\sum_{j=1}^m X_j$, since:*

$$\frac{S_m - m\nu}{\sqrt{m}} = \left(\overline{S}_m - \nu\right)\sqrt{m}.$$

7.2 Distribution Families Related under Addition

It is common in applications to seek the distribution function of a **sum of independent random variables** given the individual distributions. In many applications, these random variables will also be identically distributed. This investigation was initiated in Section IV.2.2 and extended in Section 2.2.

A direct though cumbersome way to determine this distribution function is by evaluating the Lebesgue-Stieltjes convolution in (2.12), or more commonly, the convolution of the associated densities as in (2.15) when these exist. This approach can quickly become tedious, even for a very simple example like the continuous uniform distribution on $[0, 1]$.

A far more elegant approach is to use moment generating functions or characteristic functions. For sums of independent random variables $\{X_i\}_{i=1}^m$, if $M_{X_i}(t)$ exists for all i for $t \in (-t_0, t_0)$ and $t_0 > 0$, then $M_X(t)$ exists for $X = \sum_{i=1}^m X_i$ on this same interval by Proposition 4.20, and:

$$M_X(t) = \prod_{i=1}^m M_{X_i}(t).$$

Similarly, since $C_{X_i}(t)$ always exists for all t, Proposition 5.23 obtains:

$$C_X(t) = \prod_{i=1}^m C_{X_i}(t).$$

Consequently, if we know $\{M_{X_i}(t)\}_{i=1}^m$, respectively $\{C_{X_i}(t)\}_{i=1}^m$, then we know $M_X(t)$, respectively $C_X(t)$. Thus the distribution function for X can potentially be found by inspection if this moment generating function or characteristic function is recognizable.

Recognizing $M_X(t)$, respectively $C_X(t)$, uniquely identifies the distribution function by the uniqueness results of Proposition 5.53 for $M_X(t)$, respectively Proposition 5.29 for $C_X(t)$. Specifically, if $M_X(t) = M_Y(t)$ on $(-t_0, t_0)$, or $C_X(t) = C_Y(t)$ for all t, then X and Y have the same distribution function.

Of course this approach does not always work. If $\{X_i\}_{i=1}^m$ are independent Student T random variables, $M_{X_i}(t)$ does not exist. Further, the known formula for $C_{X_i}(t)$ is defined in terms of a power series, making $\prod_{i=1}^m C_{X_i}(t)$ essentially impossible to identify. But in this case, the convolution approach is also unwieldy, and there appears to be no accessible approach to working with sums of independent Student T variables.

Even when $M_{X_i}(t)$ and/or $C_{X_i}(t)$ are readily calculated, the resultant $M_X(t)$ or $C_X(t)$ may be unrecognizable. In the special case where $C_X(t)$ is integrable, we can in theory recover the continuous density function for X by the inversion formula in (5.56), but this may or may not be successful.

We begin by recalling the definition of "equal in distribution."

Definition 7.25 ($=_d$) *Given random variables or vectors X and Y on a probability space $(\mathcal{S}, \mathcal{E}, \lambda)$ with associated distribution functions F_X and F_Y, we say that X* ***and*** *Y* ***are equal in distribution, denoted:***

$$X =_d Y,$$

if for all x:

$$F_X(x) = F_Y(x). \tag{7.19}$$

Remark 7.26 (On $=_d$ and characteristic functions) *It follows from the uniqueness theorems of Propositions 5.29 and 5.39 that $X =_d Y$ if and only of $C_X(t) = C_Y(t)$ for all t.*

Definition 7.27 (Distribution families related under addition) *By a* ***distribution family*** *is meant the collection of all distribution functions formed by all valid choices of the defining numerical parameters.*

Given independent and identically distributed random variables $\{X_i\}_{i=1}^m$, we will say that the distribution family of X_i is ***related under addition to the distribution family of*** *X if $X =_d \sum_{i=1}^m X_i$. We use this same terminology when $\{X_i\}_{i=1}^m$ are independent and from the same distribution family but with different parameters.*

If X and X_i have the same distribution family, we will then say that the given distribution family is ***closed under addition.***

Exercise 7.28 *Prove that the distribution family of X_i is related under addition to the distribution family of X if and only if $C_X(t) = \prod_{i=1}^m C_{X_i}(t)$ for all t.*

In this section, we identify some useful relationships. Examples of closed and related distribution families follow, as well as families for which these methods fail to identify the distribution family.

We split the section into discrete and continuous distributions. All sums of random variables below are assumed to be of **independent random variables.**

7.2.1 Discrete Distributions

1. **Discrete Rectangular Distribution on** $[0,1]$

   If R_j has a discrete rectangular distribution on $[0,1]$ with parameter n, then from (5.32):

   $$C_{R_j}(t) = \frac{\exp[i(1+1/n)t] - \exp[it/n]}{n(\exp[it/n]-1)}.$$

   Hence $R = \sum_{j=1}^m R_j$ has characteristic function:

   $$C_R(t) = \left(\frac{\exp[i(1+1/n)t] - \exp[it/n]}{n(\exp[it/n]-1)}\right)^m,$$

   which does not lend itself to ready identification for $n > 1$.

   When $n = 1$, this distribution reduces to an example the **delta function** of (5.34) with $x_0 = 1$ and $C_\delta(t) = e^{itx_0}$. The **delta distribution family is closed under addition.** If R_j has parameter x_j then R has a delta distribution with parameter $x_0 = \sum_{j=1}^m x_j$, but one does not need characteristic functions to elucidate this fact.

2. **Binomial Distribution**

   If B_j has a binomial distribution with parameters p and $n \in \mathbb{N}$, then from (5.35):

   $$C_{B_j}(t) = \left(1 + p(e^{it}-1)\right)^n.$$

   Hence $B = \sum_{j=1}^m B_j$ has characteristic function:

   $$C_B(t) = \left(1 + p(e^{it}-1)\right)^{nm},$$

   and B is binomial with parameters p and nm.

   When the B_j have parameters p and n_j, then B is again binomial with parameters p and $\sum_{j=1}^m n_j$. However, if B_j have parameters p_j and n_j, then B is no longer binomial and its distribution is not readily identifiable.

   The **binomial distribution family is thus closed under addition** when the parameter p is fixed.

3. **Geometric Distribution**

   If G_j has a geometric distribution with parameter p, where $0 < p < 1$, then from (5.36):

   $$C_{G_i}(t) = \frac{p}{1-(1-p)e^{it}},$$

   and $G = \sum_{j=1}^m G_j$ has characteristic function:

   $$C_G(t) = \left(\frac{p}{1-(1-p)e^{it}}\right)^m.$$

This is recognizable from (5.37) as the characteristic function of the negative binomial with parameters p and m. When the p parameter is not fixed, the distribution of G is not readily identifiable.

Thus the **geometric distribution family is related under addition to the negative binomial distribution family** when the parameter p is fixed.

4. **Negative Binomial Distribution**

   As noted in 3 above, if NB_j has a negative binomial distribution with parameters p and positive integer k_j, then by (5.37):

$$C_{NB_j}(t) = \left(\frac{p}{1-(1-p)e^{it}}\right)^{k_j},$$

   and $NB = \sum_{j=1}^{m} NB_j$ is again negative binomial with parameters p and $\sum_{j=1}^{m} k_j$. The **negative binomial family is thus closed under addition** when p is fixed.

   When the p parameter is not fixed, the distribution of NB is not readily identifiable.

5. **Poisson Distribution**

   If P_j is Poisson with parameter λ_j, then from (5.38):

$$C_{P_j}(t) = \exp[\lambda_j(e^{it} - 1)],$$

   and $P = \sum_{j=1}^{m} P_j$ has characteristic function $C_P(t) = \exp[\lambda(e^{it} - 1)]$ with $\lambda = \sum_{j=1}^{m} \lambda_j$. Hence, the **Poisson family is closed under addition.**

7.2.2 Continuous Distributions

1. **Continuous Uniform Distribution**

   If U_j is uniform on $[a, b]$, then from (5.39):

$$C_{U_j}(t) = \frac{e^{ibt} - e^{iat}}{t(b-a)},$$

   and like the discrete version, $U = \sum_{j=1}^{m} U_j$ has an unrecognizable characteristic function.

2. **Exponential Distribution**

   If E_j is exponential with parameter λ, then from (5.40):

$$C_{E_j}(t) = (1 - it/\lambda)^{-1}.$$

   Thus $E = \sum_{j=1}^{m} E_j$ has characteristic function $C_E(t) = (1 - it/\lambda)^{-m}$, which is gamma with parameters λ, m.

   So for fixed λ, the **exponential distribution family is related under addition to the gamma distribution family.**

3. **Gamma Distribution**

   If Γ_j is gamma with parameters λ, α_j, then from (5.40):

$$C_{\Gamma_j}(t) = (1 - it/\lambda)^{-\alpha_j},$$

and $\Gamma = \sum_{j=1}^{m} \Gamma_j$ has the characteristic function of a gamma with parameters λ and $\alpha = \sum_{j=1}^{m} \alpha_j$.

Hence the **gamma distribution family is closed under addition** when λ is fixed.

As noted in Remark IV.1.30, when $\lambda = 1/2$ and $\alpha = n/2$, a gamma variate is called a **chi-squared random variable with n degrees of freedom,** denoted $\chi^2_{n\ d.f.}$. Thus the **chi-squared distribution family is closed under addition** since λ is fixed by definition. Specifically, the sum of m chi-squared random variables with n_j degrees of freedom is chi-squared with $n = \sum_{j=1}^{m} n_j$ degrees of freedom.

4. **Beta Distribution**

If β has a beta distribution with parameters $v > 0$ and $w > 0$, it has the somewhat unwieldy characteristic function of (5.42):

$$C_\beta(t) = 1 + \sum_{n=1}^{\infty} \prod_{j=0}^{n-1} \left(\frac{v+j}{v+w+j} \right) \frac{(it)^n}{n!},$$

and it is apparent that products of such functions will not be readily identifiable.

5. **Cauchy Distribution**

If C_j is Cauchy with parameters γ_j and x_j, then from (5.43):

$$C_{C_j}(t) = \exp\left(ix_j t - \gamma_j |t|\right),$$

and so $C = \sum_{j=1}^{m} C_j$ has characteristic function $C_C(t) = \exp\left(ix_0 t - \gamma |t|\right)$ with $x_0 = \sum_{j=1}^{m} x_j$ and $\gamma = \sum_{j=1}^{m} \gamma_j$.

Hence, the **Cauchy family is closed under addition.**

From this observation, we have the somewhat surprising corollary below that also provides a limitation on the conclusion of the central limit theorem when a random variable, like the Cauchy, has no finite moments.

Corollary 7.29 *If $\{C_j\}_{j=1}^{m}$ are independent, identically distributed Cauchy random variables with parameters γ and x_0, then $\overline{C} = \sum_{j=1}^{m} C_j/m$ is Cauchy with the same parameters.*

Proof. *Apply the above calculation and (5.47).* ■

Hence, averaging i.i.d. Cauchy variates produces the same distribution, and there is no hope that a central limit theorem could ever apply.

6. **Normal Distribution**

If N_j is normal with parameters μ_j, σ_j^2, then by (5.44):

$$C_{N_j}(t) = \exp\left(i\mu_j t - \frac{1}{2}\sigma_j^2 t^2\right),$$

and hence $N = \sum_{j=1}^{m} N_j$ has characteristic function $C_N(t) = \exp\left(i\mu t - \frac{1}{2}\sigma^2 t^2\right)$ with $\mu = \sum_{j=1}^{m} \mu_j$ and $\sigma^2 = \sum_{j=1}^{m} \sigma_j^2$.

Thus the **normal family is closed under addition.**

7.3 Infinitely Divisible Distributions

The process of the previous section, of adding independent, identically distributed (i.i.d.) variates from one distribution family to obtain a variate from the same or a different distribution family, can sometimes be reversed. Namely, given X and n, it is sometimes possible to find independent and identically distributed $\{X_i\}_{i=1}^n$ so that $X =_d \sum_{i=1}^n X_i$.

But first, a comment on the notion of equal in distribution that will be useful below.

Remark 7.30 (On "$=_d$") *If $X = (X_1, ..., X_m)$ and $Y = (Y_1, ..., Y_m)$ with $m > 1$, let $I = \{i_1, ..., i_k\} \subset \{1, 2, ..., m\}$, and $x_J \equiv (x_{j_1}, x_{j_2}, ..., x_{j_{m-k}})$ for $j_k \in J \equiv \widetilde{I}$. Recalling Definition 1.45 on marginal distribution functions, if $X =_d Y$ it follows from (7.19) that for all $x_I \equiv (x_{i_1}, x_{i_2}, ..., x_{i_k})$:*

$$F_{X,I}(x_I) \equiv \lim_{x_J \to \infty} F_X(x_1, x_2, ..., x_m) = F_{Y,I}(x_I).$$

In other words, all marginal distribution functions are also equal, and thus

$$X_I =_d Y_I.$$

In particular, for all j:

$$X_j =_d Y_j \tag{1}$$

The converse is not true. If (1) holds for all j, this does not imply that $X =_d Y$. Indeed by Sklar's theorem of Proposition II.7.16, there are infinitely many joint distribution functions with the same marginal distribution functions $\{F_j(x_j)\}_{j=1}^m$.

We are ready to introduce infinitely divisible distributions.

Definition 7.31 (Infinitely divisible random variable/vector) *A random variable (or vector) X is said to **infinitely divisible** or **I.D.**, and sometimes the associated distribution (or joint distribution) function F is said to **infinitely divisible** or **I.D.**, if for any $n \in \mathbb{N}$, there exists independent, identically distributed random variables (or vectors) $\{X_j^{(n)}\}_{j=1}^n$, so that:*

$$X =_d \sum_{j=1}^n X_j^{(n)}. \tag{7.20}$$

Here "$=_d$" means "equal in distribution."

*A characteristic function $C(t)$ of an infinitely divisible random variable (or vector) is said to be an **infinitely divisible** characteristic function.*

The criterion in (7.20) has implications for marginal random vectors, and can be equivalently expressed in terms of characteristic functions.

Proposition 7.32 (On infinite divisibility) *If the random vector $X = (X_1, ..., X_m)$ is infinitely divisible, then so too are all component random variables $\{X_j\}_{j=1}^m$, and all marginal random vectors $\{X_I\}_{I \subset \{1,2,...,m\}}$.*

The random variable/vector X is infinitely divisible if and only if for every n, there exists a random variable/vector $X^{(n)}$ so that:

$$C_X(t) = [C_{X^{(n)}}(t)]^n. \tag{7.21}$$

Thus, X is infinitely divisible if and only if $[C_X(t)]^{1/n}$ is a characteristic function for all n.

Proof. *The result on component random variables and marginal random vectors follows from Remark 7.30.*

Given (7.21), let $\{X_j^{(n)}\}_{j=1}^n$ *be independent random variables/vectors with the distribution of such* $X^{(n)}$ *and define:*

$$X' \equiv \sum\nolimits_{j=1}^n X_j^{(n)}.$$

Then $C_{X'}(t) = [C_{X^{(n)}}(t)]^n$ *by (5.48) or (5.62), and thus* $C_X(t) = C_{X'}(t)$. *The uniqueness results of Propositions 5.29 and 5.39 assure that* X' *has the distribution function of* X, *and thus* $X' =_d X$, *which is (7.20).*

Conversely, given (7.20), (7.21) follows by (5.48) or (5.62) as above.

If X *is infinitely divisible, then* $[C_X(t)]^{1/n}$ *is the characteristic function of* $X^{(n)}$ *by (7.21). Conversely, if* $[C_X(t)]^{1/n}$ *is a characteristic function, then it is the characteristic function of some distribution function* F. *By Propositions 1.20 and 1.24, there exists a random variable/vector* $X^{(n)}$ *on a probability space with this distribution function, and thus* $[C_X(t)]^{1/n} = C_{X^{(n)}}(t)$. *This is (7.21) and thus* X *is infinitely divisible.* ■

An important subset of the infinitely divisible distributions is the class of **stable distributions.** We will define them here and apply them in the discussion of discrete time asset models, but will otherwise not develop many of their special properties. See **Sato** (1999) for more details.

Definition 7.33 (Stable random variable) *A random vector* X *with range in* $\mathbb{R}^m$ *(or random variable with* $m = 1$*) is said to* ***stable,*** *and sometimes the associated distribution (or joint distribution) function* F *is said to* ***stable,*** *if for any* $n \in \mathbb{N}$ *there exists real* $c_n > 0$ *and* $d_n \in \mathbb{R}^m$ *so that:*

$$\sum\nolimits_{j=1}^n X_j =_d c_n X + d_n, \tag{7.22}$$

where $\{X_j\}_{j=1}^n$ *are independent, identically distributed random vectors (or variables) with distribution* F.

The variate X *is* ***strictly stable*** *if (7.22) is satisfied with* $d_n = 0$.

Exercise 7.34 *Prove that if* $X = (X_1, ..., X_m)$ *is stable with* $m > 1$, *then so too are all component random variables* $\{X_j\}_{j=1}^m$, *and all marginal random vectors* $\{X_I\}_{I \subset \{1,2,...,m\}}$. *Hint: Remark 7.30.*

Exercise 7.35 *Prove that* X *is stable if and only if for every* n, *there exists* $c_n > 0$ *and* $d_n \in \mathbb{R}^m$ *so that:*

$$[C_X(t)]^n = e^{t \cdot d_n} C_X(c_n t). \tag{7.23}$$

Hint: Recall (5.47) and (5.61) for one direction, and uniqueness of characteristic functions for the other.

Exercise 7.36 *By considering representations of sums in (7.22) to* n, m, *and* mn, *show that:*

$$c_{mn} = c_n c_m, \qquad d_{mn} = c_m d_n + n d_m. \tag{7.24}$$

Note: *For a stable variate, it is actually the case that* $c_n = n^{1/\alpha}$ *for some* $0 < \alpha \leq 2$. *See section VI.1 of* ***Feller*** *(1971) for the proof, and the next exercise for examples.*

Exercise 7.37 *Show that if* X *is stable and has two moments, then:*

$$c_n = \sqrt{n}, \qquad d_n = (n - \sqrt{n})\mu_X.$$

This then provides an explicit example of (7.24).

If X *is strictly stable with two moments, why does this result assure that* $d_n = 0$?

It was noted above that the stable distributions are a subset of the infinitely divisible distributions, and this subset is proper as will be seen in Exercise 7.40.

Proposition 7.38 (Stable $\subset$ ID) *If X is stable, then X is infinitely divisible.*
Proof. *If X is stable, then from (7.22):*

$$X =_d \frac{1}{c_n}\left(\sum\nolimits_{j=1}^n X_j - d_n\right) = \sum\nolimits_{j=1}^n \left(\frac{X_j - d_n/n}{c_n}\right).$$

Thus (7.20) is satisfied with $X_j^{(n)} = [X_j - d_n/n]/c_n$. ■

Example 7.39 (Infinitely divisible distributions) *While the general question related to identifying all infinitely divisible distributions is a difficult one, the prior section provides many examples, some of which are also stable. See Exercise 7.40.*

1. ***Degenerate (or Delta) Distribution:*** *Defined in (5.34) to assign measure 1 to a single point x_0, the associated characteristic function is $C_D(t) = e^{ix_0t}$ in one dimension, $C_D(t) = e^{ix_0\cdot t}$ in multivariate notation. This simple distribution is infinitely divisible since $[C_D(t)]^{1/n}$ is the characteristic function of the degenerate distribution with parameter x_0/n.*

   When $x_0 = 0$, $C_D(t) = 1$ and for all n, $X_j^{(n)} = X$ in (7.20). By uniqueness of Propositions 5.29 and 5.39, this is the only distribution function for which $C_D(t) = 1$ for all t.

2. ***Poisson Distribution:*** *By (5.38), the characteristic function of the Poisson with parameter $\lambda > 0$ is:*
$$C_P(t) = \exp[\lambda(e^{it} - 1)].$$
   Thus the Poisson is infinitely divisible since $[C_P(t)]^{1/n}$ is the characteristic function of the Poisson distribution with parameter λ/n.

3. ***Gamma Distribution:*** *The characteristic function of the gamma with parameters λ and α is given in (5.40):*
$$C_\Gamma(t) = (1 - it/\lambda)^{-\alpha},$$
   and thus the Gamma is infinitely divisible since $[C_\Gamma(t)]^{1/n}$ is the characteristic function of the gamma distribution with parameters λ and α/n.

   As a corollary to this, the ***exponential distribution*** *with parameter λ is also infinitely divisible, but with $[C_E(t)]^{1/n}$ the characteristic function of the gamma distribution with parameters λ and $1/n$.*

4. ***Cauchy Distribution:*** *By (5.43), the characteristic function of the Cauchy with parameters γ and x is:*
$$C_C(t) = \exp(ixt - \gamma|t|).$$
   Thus the Cauchy is infinitely divisible since $[C_C(t)]^{1/n}$ is the characteristic function of the Cauchy distribution with parameters γ/n and x/n.

5. ***Normal Distribution:*** *The characteristic function of the normal distribution with parameters $\sigma > 0$ and μ and is given in (5.44):*
$$C_N(t) = \exp\left(i\mu t - \frac{1}{2}\sigma^2 t^2\right),$$
   and thus the normal is infinitely divisible since $[C_N(t)]^{1/n}$ is the characteristic function of the normal distribution with parameters μ/n and $\sigma/\sqrt{n}$.

6. ***Multivariate Normal Distribution:*** *By (5.45), the characteristic function of the multivariate normal random vector* $Y \equiv (Y_1, Y_2, ..., Y_n)$ *is given by:*

$$C_{MN}(t) = \exp\left[i\mu \cdot t - \frac{1}{2}t^T Ct\right],$$

where C *is the (positive semidefinite) covariance matrix of the random vector* $(Y_1, Y_2, ..., Y_n)$ *and defined by* $C_{ij} = E[(Y_i - \mu_i)(Y_j - \mu_j)]$*, and* μ *is the vector of first moments,* $\mu_j = E[Y_j]$*. The multivariate normal is thus infinitely divisible because* $[C_{MN}(t)]^{1/n} = \exp\left[i(\mu/n) \cdot t - \frac{1}{2}t^T (C/n) t\right]$ *is the characteristic function of a multivariate normal, noting that* C/n *is positive semidefinite.*

7. ***Negative Binomial Distribution:*** *The characteristic function of the negative binomial with parameters p and positive integer k is given in (5.37):*

$$C_{NB}(t) = \left(\frac{p}{1-(1-p)e^{it}}\right)^k.$$

Thus $[C_{NB}(t)]^{1/n}$ *looks a lot like a negative binomial, but with parameters p and* k/n*, where the latter value is in general a non-integer.*

For the ***geometric distribution,*** *where* $C_G(t)$ *equals* $C_{NB}(t)$ *with* $k = 1$*, again* $[C_G(t)]^{1/n}$ *looks a lot like a negative binomial, but with parameters p and* $1/n$*. As it turns out, both the geometric and negative binomial are infinitely divisible distributions.*

Recall that k is an integer in the classical definition of the negative binomial, with density function:

$$f_{NB}(j) = \binom{j+k-1}{k-1} p^k (1-p)^j, \quad j = 0, 1, 2, ..$$

This density can be interpreted as the probability of j tails before the kth head in a series of coin flips where the probability of a head is p. But mathematically, $f_{NB}(j)$ *is a perfectly valid density function for any real* $k > 0$*.*

This follows from ***Newton's generalized binomial theorem,*** *named for* ***Isaac Newton*** *(1642–1727):*

$$(1-x)^{-k} = \sum\nolimits_{j=0}^{\infty} \binom{j+k-1}{k-1} x^j, \qquad |x| < 1, \tag{1}$$

where

$$\binom{j+k-1}{k-1} \equiv \binom{j+k-1}{j} \equiv \frac{(j+k-1)(j+k-2)\cdots k}{j!}.$$

Letting $x = 1 - p$ *obtains that* $f_{NB}(j)$ *is a probability density for all* $k > 0$*, and* $C_{NB}(t)$ *given above is the associated characteristic function. Details are left as an exercise in Newton's theorem in* (1).

Thus $[C_{NB}(t)]^{1/n}$ *is the characteristic function of a* ***generalized negative binomial*** *with parameters p and* k/n*, and the same is true for* $[C_G(t)]^{1/n}$*, but with parameters p and* $1/n$*.*

8. ***Compound Poisson Distribution:*** *The Poisson random variable* X_P *with parameter* λ *can be formally defined as a "random" sum:*

$$X_P = \sum\nolimits_{j=1}^{N} D_j, \tag{7.25}$$

where N has the given Poisson distribution, and $\{D_j\}_{j=1}^{\infty}$ are independent degenerate random variables with $x_0 = 1$ and thus $C_D(t) = e^{it}$.

This is intuitively plausible, but should be formally checked by calculating the characteristic function of this random variable. For this we need the law of total expectation in (4.53), conditioning on N.

First:

$$E\left[e^{iX_Pt}|N\right] = \left[C_D(t)\right]^N = e^{itN},$$

since conditional on N, X_P is a fixed summation of i.i.d. variates with characteristic function equal to the product of the component characteristic functions by Proposition 5.23. Thus by (4.53):

$$C_{X_P}(t) = E\left[e^{itN}\right] = e^{-\lambda}\sum\nolimits_{n=0}^{\infty} e^{itn}\frac{\lambda^n}{n!} = \exp[\lambda(e^{it} - 1)].$$

By (5.38) and uniqueness of Proposition 5.29, X_P is Poisson with parameter λ.

A **compound Poisson random variable** *or* **random vector** *X_{CP} is defined exactly as in (7.25), but where the $\{Y_j\}_{j=1}^{\infty}$ is any independent, identically distributed collection of random variables or vectors:*

$$X_{CP} = \sum\nolimits_{j=1}^{N} Y_j. \tag{7.26}$$

An example of this was seen with the aggregate loss model of Example 4.35.

Repeating the above derivation using general vector notation obtains:

$$E\left[e^{it\cdot X_{CP}}|N\right] = \left[C_Y(t)\right]^N,$$

and:

$$C_{CP}(t) = \exp[\lambda(C_Y(t) - 1)]. \tag{7.27}$$

From this it is clear that X_{CP} is infinitely divisible, and $\left[C_{CP}(t)\right]^{1/n}$ is the characteristic function of a compound Poisson with parameters λ/n and the same component variates/vectors $\{Y_j\}_{j=1}^{\infty}$.

For reasons that will be clear below, $C_{CP}(t)$ is sometimes written in terms of the $C_Y(t)$-defining Lebesgue-Stieltjes integral:

$$C_{CP}(t) = \exp\left[\lambda\int_{\mathbb{R}^m}\left(e^{it\cdot x} - 1\right)d\lambda_Y\right], \tag{7.28}$$

where λ_Y is the induced probability measure of Proposition 1.30 associated with Y. See Proposition 7.49 and Remark 7.50.

9. **Lognormal Distribution:** *As noted in Section 5.4.2, although $C_L(t)$ exists in theory, there is no known closed form version of this function. Thus there is no direct way to evaluate if $C_L^{1/n}(t)$ is a characteristic function, or equivalently, if the lognormal is infinitely divisible.*

   But in 1977, it was proved by **Olof Thorin** *that the lognormal is infinitely divisible. Thorin proved that this distribution is the weak limit of generalized gamma convolutions, which he introduced and proved infinitely divisible. The final result then follows from item 3 of Proposition 7.42.*

Exercise 7.40 *Identify which of the above distributions are stable, and determine the parameters $c_n > 0$ and d_n in (7.22).*

With the exception of items 7 and 9 of Example 7.39, demonstrating that the above random variables were infinitely divisible was generally relatively easy. Looking at other examples proves how hard this identification can sometimes be.

Example 7.41 (Infinitely divisible distributions?) *The* ***binomial distribution*** *with parameters p and m has characteristic function given in (5.35) by $C_B(t) = (1+p(e^{it}-1))^m$, and thus $[C_B(t)]^{1/n} = (1+p(e^{it}-1))^{m/n}$. While not the characteristic function of a standard binomial, one wonders if there is a definitional extension of the binomial as there was for the negative binomial in Example 7.39, with this characteristic function.*

As it turns out, there is no such extension because the random variable B is bounded. Thus by item 5 of Proposition 7.42, the binomial is not infinitely divisible.

As another example, the ***continuous uniform distribution on*** *$[a-r, a+r]$ has characteristic function given in (5.39) and (5.20) by:*

$$C_U(t) = e^{iat}\frac{e^{irt} - e^{-irt}}{2rt} = e^{iat}\frac{\cos rt}{rt}.$$

It is again not obvious if $[C_U(t)]^{1/n}$ is a characteristic function. But we can again conclude that it is not.

Item 1 of Proposition 7.42 states that the characteristic function of an infinitely divisible distribution can have no real zeros, so $C(t) \neq 0$ for all $t \in \mathbb{R}$. Certainly $C_U(t)$ has real zeros, and so the continuous uniform distribution is not infinitely divisible. This random variable is also bounded, so item 5 of Proposition 7.42 also applies to obtain this conclusion.

Since the binomial $C_B(t)$ has no real zeros, but is still not infinitely divisible, the property in item 1, that $C(t) \neq 0$ for all $t \in \mathbb{R}$, is necessary for infinite divisibility but is not sufficient.

Proposition 7.42 (Properties of infinitely divisible X) *Some properties of infinitely divisible (I.D.) random variables/vectors follow:*

1. *If X is I.D., then $C_X(t) \neq 0$ for all $t \in \mathbb{R}^m$.*

2. *If X_1 and X_2 are I.D., random vectors, then $X \equiv X_1 + X_2$ is I.D.*

    (a) *Equivalently, if $C_1(t)$ and $C_2(t)$ are I.D. characteristic functions, then $C(t) \equiv C_1(t)C_2(t)$ is I.D.*

    (b) *Equivalently, if F_1 and F_2 are I.D., then F defined as in (2.12) is I.D.*

    These results are then true for all finite combinations (sums, products, convolutions).

3. *If $\{X_j\}_{j=1}^{\infty}$ are I.D. random vectors and $X_j \to_d X$, then X is I.D.*

    (a) *Equivalently, if $\{C_j(t)\}_{j=1}^{\infty}$ are I.D. characteristic functions and there exists a characteristic function $C(t)$ so that $C_j(t) \to C(t)$ for all t, then $C(t)$ is I.D.*

    (b) *Equivalently, if $\{F_j\}_{j=1}^{\infty}$ are I.D. and $F_j \Rightarrow F$, then F is I.D.*

4. *$C(t)$ is I.D. if and only if $[C(t)]^{\alpha}$ is I.D. for all real $\alpha > 0$.*

5. *If X is I.D. and bounded, then X is degenerate and $C_X(t) = C_D(t) \equiv e^{ix_0 \cdot t}$ for some x_0.*

Proof. *For items 1-4, the proof often is independent of whether we are addressing random variables or random vectors, other than which proposition on characteristic functions is referenced. For these results, we provide a proof in the context of random vectors with range $\mathbb{R}^m$, and then setting $m = 1$ obtains the result for random variables. Details are left as an exercise. For item 5, we provide two proofs.*

1. *If X is I.D. then $[C_X(t)]^{1/n}$ is a characteristic function for all n by (7.21). Defining $g(t) = \lim_{n\to\infty}[C_X(t)]^{1/n}$, it is apparent that $g(t) = 0$ if $C_X(t) = 0$, and we claim that $g(t) = 1$ otherwise. To prove this, if the complex number $C_X(t) \neq 0$, then $C_X(t) = e^{a+bi}$ is well-defined by specifying this as the principal value, meaning $b \in (-\pi, \pi]$. Then by Euler's formula in (5.17), as $n \to \infty$:*

$$[C_X(t)]^{1/n} = e^{a/n}(\cos(b/n) + i\sin(b/n)) \to 1.$$

   Now $C_F(0) = 1$, and so by continuity there is a real rectangle $R \equiv \prod_{j=1}^m(-\epsilon, \epsilon)_j$ so that $C_X(t) \neq 0$ on R. Thus $g(t) = 1$ on R and $g(t)$ is continuous at $t = 0$. By Corollary 5.41 to Lévy's continuity theorem, $g(t)$ is a characteristic function and hence is continuous. It now follows that $g(t) \equiv 1$ on $\mathbb{R}^m$. Thus $C_X(t) \neq 0$ on $\mathbb{R}^m$.

2. *If X_1 and X_2 are I.D., then $[C_{X_1}(t)]^{1/n}$ and $[C_{X_2}(t)]^{1/n}$ are characteristic functions for all n. Specifically, $[C_{X_1}(t)]^{1/n} = C_{X_1^{(n)}}(t)$ and similarly $[C_{X_2}(t)]^{1/n} = C_{X_2^{(n)}}(t)$. But $C_X(t) = C_{X_1}(t)C_{X_2}(t)$ by Proposition 5.33, and $C_X(t)$ is infinitely divisible since $[C_X(t)]^{1/n} = C_{X_1^{(n)}}(t)C_{X_2^{(n)}}(t)$ is the characteristic of $X_1^{(n)} + X_2^{(n)}$. This prove items 2 and 2.a. by Proposition 7.32. Item 2.b follows from Proposition 2.10 since F defined as in (2.12) is the distribution function of X.*

3. *$X_j \to_d X$ if and only if $C_{X_j}(t) \to C_X(t)$ for all t by Lévy's continuity theorem of Proposition 5.40. Thus for given n, $\left[C_{X_j}(t)\right]^{1/n} \to [C_X(t)]^{1/n}$ for all t. Since $\left[C_{X_j}(t)\right]^{1/n}$ is a characteristic function by infinite divisibility and $[C_X(t)]^{1/n}$ is continuous at $t = 0$, Lévy's continuity theorem of Corollary 5.41 assures that $[C_X(t)]^{1/n}$ is a characteristic function. This is true for all n and hence X is I.D. by Proposition 7.32. This proves items 3 and 3.a, while item 3.b follows from Definition 3.8.*

4. *If $[C(t)]^\alpha$ is I.D. for all real $\alpha > 0$, then $C(t)$ is I.D. by definition. Conversely, if $C(t)$ is infinitely divisible and $\alpha = p/q$ is rational, then $[C^\alpha(t)]^{1/n} = [C^p(t)]^{1/nq}$. Now $C^p(t)$ in infinitely divisible by item 2, and thus $[C^p(t)]^{1/nq}$ is a characteristic function for all n by definition. So $C^\alpha(t)$ is infinitely divisible for all positive rationals α. For general real α, if $p_j/q_j \to \alpha$ then $C^{p_j/q_j}(t) \to C^\alpha(t)$ for all t and $C^\alpha(t)$ is infinitely divisible by item 3.*

5. *If infinitely divisible X is defined and bounded on $(\mathcal{S}, \mathcal{E}, \lambda)$, then $\Pr[|X| > M] = 0$ for some $M < \infty$, where by definition, $\Pr[|X| > M] \equiv \lambda\left[X^{-1}\left[(-\infty, M) \bigcup (M, \infty)\right]\right]$. We claim that this implies that $\Pr[\left|X^{(n)}\right| > M/n] = 0$ for $X^{(n)}$ in (7.20). To prove this, assume that $\Pr[\left|X^{(n)}\right| > M/n] = p > 0$. Then since $X = \sum_{j=1}^n X_j^{(n)}$ is an independent sum, and leaving the first inequality as an exercise:*

$$\Pr[|X| > M] \geq \Pr\left(\bigcap\nolimits_{j=1}^n \left[\left|X_j^{(n)}\right| > M/n\right]\right) = p^n > 0,$$

   a contradiction.

(a) ***Random variable:*** *By boundedness, all moments of* $X_j^{(n)}$ *are well-defined, and:*

$$Var[X_j^{(n)}] \equiv E\left[\left(X_j^{(n)}\right)^2\right] - \left[E\left(X_j^{(n)}\right)\right]^2 \leq E\left[\left(X_j^{(n)}\right)^2\right] \leq \left(\frac{M}{n}\right)^2.$$

Thus by independence and (4.36):

$$Var[X] = \sum\nolimits_{j=1}^{n} Var[X_j^{(n)}] \leq n\left(\frac{M}{n}\right)^2.$$

Letting $n \to \infty$ *obtains that* $Var[X] = 0$, *so* X *is degenerate by Chebyshev's inequality of Proposition IV.4.34. Specifically, if* $\alpha > 0$:

$$\Pr\left[|X - E[X]| \geq \alpha\right] \leq \frac{Var[X]}{\alpha^2} = 0. \tag{1}$$

Thus $X = E[X] \equiv x_0 \in \mathbb{R}$ *with probability* 1.

(b) ***Random vector:*** *Let* $X = (X_1, ..., X_m)$ *be defined on a probability space* $(\mathcal{S}, \mathcal{E}, \lambda)$. *With* $E[X] = (E[X_1], ..., E[X_m])$ *and* $|X|$ *defined as in (2.14), for* $\alpha > 0$:

$$\{|X - E[X]| \geq \alpha\} \subset \bigcup\nolimits_{j=1}^{m} \{|X_j - E[X_j]| \geq \alpha/\sqrt{n}\}.$$

Since $\{X_j\}_{j=1}^m$ *are infinitely divisible by Remark 7.30, and bounded by assumption, it follows from finite subadditivity of measures and* (1) *that:*

$$\Pr\left[|X - E[X]| \geq \alpha\right] \leq \sum\nolimits_{j=1}^{m} \Pr\{|X_j - E[X_j]| \geq \alpha/\sqrt{n}\} = 0.$$

Thus $X = E[X] \equiv x_0 \in \mathbb{R}^m$ *with probability* 1.

■

Item 5 states that while there exists bounded, infinitely divisible random variables/vectors, these must be degenerate with $X \equiv x_0$ with probability 1. A corollary to this is that the associated random variables $\{X^{(n)}\}_{n=1}^{\infty}$ in Definition 7.31 are also degenerate with $X^{(n)} \equiv x_0/n$ with probability 1. Thus $C_{X^{(n)}}(t) = e^{ix_0t/n}$ or $C_{X^{(n)}}(t) = e^{ix_0 \cdot t/n}$, and so $C_{X^{(n)}}(t) \to 1$ as $n \to \infty$.

Now $1 = C_{D_0}(t)$, the characteristic function of the degenerate random variable D_0 with $x_0 = 0$ in $\mathbb{R}$ or $\mathbb{R}^m$. Thus for a bounded infinitely divisible X, the random variables $\{X^{(n)}\}_{n=1}^{\infty}$ in Definition 7.31 satisfy $X^{(n)} \to_d D_0$ by Lévy's continuity theorem.

Interestingly, the proof of item 1 of Proposition 7.42 shows that this conclusion is true in general.

Corollary 7.43 (If X is I.D., then $X^{(n)} \to_d D_0$) *If* X *is an infinitely divisible random variable/vector on* $\mathbb{R}/\mathbb{R}^m$, *then with* $\{X^{(n)}\}_{n=1}^{\infty}$ *in Definition 7.31:*

$$X^{(n)} \to_d D_0, \tag{7.29}$$

where $D_0 = 0 \in \mathbb{R}/\mathbb{R}^m$ *with probability* 1, *a degenerate random variable/vector.*

Thus if F_n *is the distribution function of* $X^{(n)}$, *then:*

$$\lim_{n\to\infty} F_n(x) = \begin{cases} 1, & x_j > 0 \text{ for all } j, \\ 0, & x_j < 0 \text{ for some } j. \end{cases}$$

Proof. *By definition,* $[C_F(t)]^{1/n} \equiv C_{X^{(n)}}(t)$ *is the characteristic function of* $X^{(n)}$*, and by the proof of item 1 of Proposition 7.42,* $C_{X^{(n)}}(t)$ *converges to a characteristic function* $g(t) = 1$ *on* $\mathbb{R}/\mathbb{R}^m$*. Since this is the characteristic function of* X_{D_0}*,* $g(t) = C_{D_0}(t)$ *and* $C_{X^{(n)}}(t) \to C_{D_0}(t)$ *for all* t *by uniqueness of Propositions 5.29 and 5.39. Lévy's continuity theorem now obtains that* $X^{(n)} \to_d D_0$*.*

The conclusion on $\lim_{n\to\infty} F_n(x)$ *is the definition of weak convergence, that* $\lim_{n\to\infty} F_n(x) = F_{D_0}(x)$ *at all continuity points of* $F_{D_0}(x)$*. These continuity points are in the* 2^m *sectors defined in terms of* $x_j > 0$ *or* $x_j < 0$*.* ∎

7.3.1 De Finetti's Theorem

In this section, we investigate two characterizations of infinitely divisible characteristic functions. The first result provides an early characterization of infinite divisibility called **de Finetti's theorem**. A second more recent result, which we summarize without proof, is known as the **Lévy-Khintchine representation theorem.**

De Finetti's theorem is named for **Bruno de Finetti** (1906–1985), who is credited with introducing the study of infinite divisibility in 1929. This result states that infinitely divisible distributions can be characterized as the weak limit of compound Poisson distributions, recalling item 8 of Example 7.39. Stated another way using Lévy's continuity theorem, infinitely divisible characteristic functions are characterized as the pointwise limits of compound Poisson characteristic functions.

For this proof we need a simple but important result.

Lemma 7.44 **(**$g(t) \equiv \exp[\lambda(C(t)-1)]$ **is I.D.)** *If* $C(t)$ *is a characteristic function of any distribution function* F *on* $\mathbb{R}^m$ *and* $\lambda > 0$*, then* $g(t) \equiv \exp[\lambda(C(t)-1)]$ *is an infinitely divisible characteristic function.*

Proof. *Let* F *be the distribution function for which* $C_F(t) = C(t)$*, and* X *a random vector with distribution* F *as given in Proposition 1.24. Let* $\{X_j\}_{j=1}^{\infty}$ *be independent and identically distributed random vectors with distribution* F *as given by Proposition II.4.4.*

Define the compound Poisson random vector:

$$X_{CP} = \sum\nolimits_{j=1}^{N} X_j,$$

where N *has a Poisson distribution with parameter* λ*. Then by (7.27),* $g(t)$ *is the characteristic function of* X_{CP}*.*

By Proposition 7.32, $g(t)$ *so defined is infinitely divisible since* $[g(t)]^{1/n} = \exp[\lambda_n(C(t)-1)]$ *is the characteristic function with* $\lambda_n = \lambda/n$ *and* X*.* ∎

We state de Finetti's theorem in the context of convergence of characteristic functions. By **Lévy's continuity theorem** of Proposition 5.40, this can be equivalently stated in terms of weak convergence of the associated joint distribution functions.

Proposition 7.45 (de Finetti's theorem) *A characteristic function* $C(t)$ *on* $\mathbb{R}^m$ *is infinite divisible if and only if there exists a sequence of positive real numbers* $\lambda_n > 0$*, and characteristic functions* $C_n(t)$*, so that:*

$$C(t) = \lim_{n\to\infty} \exp[\lambda_n(C_n(t)-1)]. \tag{7.30}$$

Equivalently, a characteristic function $C(t)$ *on* $\mathbb{R}^m$ *is infinite divisible if and only if* $C(t)$ *is the limit of characteristic functions of compound Poisson random vectors.*

Proof. *If the characteristic function $C(t)$ is given as the limit in (7.30), then since each* $\exp\left[\lambda_n\left(C_n(t)-1\right)\right]$ *is infinitely divisible by Lemma 7.44, item 3 of Proposition 7.42 assures that $C(t)$ is infinitely divisible.*

Conversely, assume that $C(t)$ is infinitely divisible. Then $C^{1/n}(t)$ is a characteristic function for all integers $n>0$ by Proposition 7.32, and thus $g_n(t) \equiv \exp\left[n\left(C^{1/n}(t)-1\right)\right]$ *is infinitely divisible by Lemma 7.44. Left to prove is that for all t:*

$$\lim_{n\to\infty} \exp\left[n\left(C^{1/n}(t)-1\right)\right] = C(t). \tag{1}$$

Once proved, (7.30) then holds with $\lambda_n = n$ and $C_n(t) = C^{1/n}(t)$.

For (1), *since $C(t) \neq 0$ for all $t \in \mathbb{R}^m$ by item 1 of Proposition 7.42, fix t and let $C(t) = e^{a+bi}$. To be well-defined let $b \in (-\pi, \pi]$. Using a Taylor series in (5.16):*

$$\begin{aligned} n\left(C^{1/n}(t)-1\right) &= \sum\nolimits_{k=1}^{\infty} \frac{(a+bi)^k}{k!n^{k-1}} \\ &= a+bi+\sum\nolimits_{k=2}^{\infty} \frac{(a+bi)^k}{k!n^{k-1}}. \end{aligned} \tag{2}$$

Recalling (5.18):

$$\left|\sum\nolimits_{k=2}^{\infty} \frac{(a+bi)^k}{k!n^{k-1}}\right| \leq \frac{1}{n}\left|e^{a+bi}-1-a-bi\right|.$$

Thus the summation in (2) *converges to* 0 *as $n \to \infty$, and so for all t:*

$$\exp\left[n\left(C^{1/n}(t)-1\right)\right] \to C(t),$$

and (1) *is proved.*

The proof of Lemma 7.44 obtains that each $\exp\left[\lambda_n\left(C_n(t)-1\right)\right]$ *in (7.30) is the characteristic function of a compound Poisson random vector, and thus the equivalence of the first and second statement of the proposition.* ∎

Remark 7.46 (On de Finetti's theorem) *We make a few observations on De Finetti's theorem:*

1. ***Compound Poisson construction:*** *By Lévy's continuity theorem of Proposition 5.40 and the above proof, any infinitely divisible distribution function is the weak limit of compound Poisson distribution functions, with $\lambda_n = n$ and $C_n(t) = C^{1/n}(t)$, where $C(t)$ is the characteristic function of this distribution. Thus this $C_n(t)$ is the characteristic function of the random vectors in the random summation in (7.26). It is worth a moment to better understand this construction.*

   Let X be an infinitely divisible random vector with joint distribution function F and characteristic function $C_F(t)$. By definition of I.D., for every n there exists i.i.d. random vectors $\{X_j^{(n)}\}_{j=1}^{n}$ so that as in (7.20):

   $$X =_d \sum\nolimits_{j=1}^{n} X_j^{(n)}.$$

   The characteristic function for $X_j^{(n)}$ is then $C_F^{1/n}(t)$.

   De Finetti's theorem states that we can construct a sequence of compound Poisson random vectors, $\{X_n\}_{n=1}^{\infty}$ as follows. Define:

   $$X_n = \sum\nolimits_{j=1}^{N_n} X_j^{(n)},$$

where N_n is Poisson with parameter $\lambda_n = n$, and where $X_j^{(n)}$ here has the same distribution as $X_j^{(n)}$ in (7.20).

It is not the case that $X_n =_d X$, since the compound Poisson summation replicates X in distribution only when $N_n = n$ the mean of λ_n by IV.(4.69). However, de Finetti's result states that:

$$X_n \to_d X.$$

2. ***Uniqueness:** De Finetti's theorem does not assert uniqueness in this representation of infinitely divisible $C(t)$ with $\lambda_n = n$ and $C_n(t) = C^{1/n}(t)$.*

   Looking at the second half of the proof, it is also the case that $C^{\alpha}(t)$ is a characteristic function for all real $\alpha > 0$ by item 4 of Proposition 7.42, and thus $g_{\alpha,\beta}(t) \equiv \exp\left[\beta\left(C^{\alpha}(t) - 1\right)\right]$ is infinitely divisible by Lemma 7.44 for $\beta > 0$. Now if $C(t) = e^{a+bi}$ as in the above proof:

$$\begin{aligned}\beta_n\left(C^{\alpha_n}(t) - 1\right) &= \sum\nolimits_{k=1}^{\infty} \frac{(a+bi)^k}{k!}\beta_n\alpha_n^k \\ &= (a+bi)\,\beta_n\alpha_n + \sum\nolimits_{k=2}^{\infty} \frac{(a+bi)^k}{k!}\beta_n\alpha_n^k.\end{aligned}$$

   The conclusion that $\exp\left[\beta_n\left(C^{\alpha_n}(t) - 1\right)\right] \to C(t)$ then only requires that $\beta_n\alpha_n \to 1$ and $\alpha_n \to 0$. Thus we have many such parametrizations, though all have the structure as that above.

3. ***A question:** Why does the proof of this representation in item 2 not apply to arbitrary characteristic functions $C(t)$? Why must $C(t)$ be infinitely divisible?*

   First given that $\beta_n\alpha_n \to 1$ and $\alpha_n \to 0$, the proof that $\exp\left[\beta_n\left(z^{\alpha_n} - 1\right)\right] \to z$ for $z \in \mathbb{C}$ requires only that $z \neq 0$. So when $z = C(t)$, the characteristic function of an I.D. distribution, this is assured by item 1 of Proposition 7.42. But other characteristic functions which are not infinitely divisible can also have this property. For example the binomial distribution's characteristic function has no real zeros, yet is not infinitely divisible as noted in Example 7.41.

   *The resolution is that in addition to $C(t) \neq 0$, this proof also requires that $C^{\alpha_n}(t)$ be a characteristic function for some sequence $\alpha_n \to 0$. The implication of de Finetti's theorem is that any such characteristic function **must** be infinitely divisible.*

Corollary 7.47 (de Finetti's theorem) *A characteristic function $C(t)$ is infinitely divisible if and only if $C^{\alpha_n}(t)$ is a characteristic function for some sequence $\alpha_n \to 0$.*

Proof. *Infinitely divisible characteristic functions have this property by definition with $\alpha_n = 1/n$.*

Conversely, if such a sequence exists, then the proof of item 1 of Proposition 7.42 applies, replacing $1/n$ there with α_n, and concluding that $C(t) \neq 0$ for all t. The construction in the second half of de Finetti's proof then obtains that:

$$\lim_{n\to\infty} \exp\left[\alpha_n^{-1}\left(C^{\alpha_n}(t) - 1\right)\right] = C(t),$$

and thus $C(t)$ is infinitely divisible by de Finetti's theorem. ■

Remark 7.48 *Each of the distribution functions in Example 7.39 can be rewritten as the weak limit of compound Poisson variates with the same underlying distribution functions for the component variates derived there, and $\lambda_n = n$. It is worth a few moments to think through the details of such constructions, especially when X is compound Poisson. What does the compound Poisson approximation look like, and why does it work?*

It is natural to wonder if there is a way to characterize all characteristic functions that are possible as limits of the characteristic functions of compound Poisson variates as in (7.30). Put another way and without reference to de Finetti's theorem, given an infinitely divisible distribution function F on $\mathbb{R}^m$, is there a natural way to characterize the mathematical form of $C(t)$, the associated characteristic function?

As it turns out there have been several developments from papers in the late 1920s and 1930s by **Bruno de Finetti** (1906–1985), **Andrey Kolmogorov** (1903–1987), **Paul Lévy** (1886–1971) and **Alexandre Khintchine** (a.k.a. **Aleksandr Khinchin**, 1894–1959). A common version of these results is known as the **Lévy-Khintchine representation theorem** which we state next, discussing only parts of the proof. Full details on this result can be found in **Sato** (1999).

Proposition 7.49 (Lévy-Khintchine representation theorem)

1. *Let F be an infinitely divisible distribution function on $\mathbb{R}^m$ and $C(t)$ the associated characteristic function. Then with $t \in \mathbb{R}^m$:*

$$C(t) = \exp\left[i\gamma \cdot t - \frac{1}{2}t^T At + \int_{\mathbb{R}^m} \left(e^{ix\cdot t} - 1 - i\,(x \cdot t)\,\chi_B(x)\right) d\nu\right], \tag{7.31}$$

where $\gamma \in \mathbb{R}^m$, A is a symmetric, positive semi-definite $m \times m$ matrix, $B = \{x \in \mathbb{R}^m |\, |x| \leq 1\}$ and $\chi_B(x)$ the characteristic function of B, and ν is a ***Lévy measure.*** *A Lévy measure is not-necessarily finite, and satisfies:*

$$\nu(\{0\}) = 0, \text{ and, } \int_{\mathbb{R}^m} \min(|x|^2, 1) d\nu < \infty. \tag{7.32}$$

2. *The representation in (7.31) is unique in terms of γ, A, and ν with the given properties.*

3. *Given γ, A, and ν with the given properties, there exists an infinitely divisible distribution function F on $\mathbb{R}^m$ with associated characteristic function $C(t)$ given as in (7.31).*

Remark 7.50 *We make a few observations on this result:*

1. ***Well-definedness:*** *Given a Lévy measure ν, the integral in (7.31) is well-defined since:*

$$\int_{\mathbb{R}^m} \left(e^{ix\cdot t} - 1 - i\,(x \cdot t)\,\chi_B(x)\right) d\nu = \int_B \left(e^{ix\cdot t} - 1 - ix \cdot t\right) d\nu + \int_{\mathbb{R}^m - B} \left(e^{ix\cdot t} - 1\right) d\nu.$$

The first integrand is bounded on B by (5.16) and the Cauchy-Schwarz inequality:

$$\left|e^{ix\cdot t} - 1 - ix \cdot t\right| \leq \frac{1}{2}|x|^2 |t|^2 + O\left(|x|^3 |t|^3\right).$$

The second integrand satisfies $\left|e^{ix\cdot t} - 1\right| \leq 2$ on $\mathbb{R}^m - B$ by (5.18). Thus each integral is finite by (7.32).

2. ***Special Case:*** *If the Lévy measure ν is in fact finite on B, meaning $\int_B d\nu < \infty$, then the Lévy-Khintchine representation simplifies to:*

$$C(t) = \exp\left[i\widetilde{\gamma} \cdot t - \frac{1}{2}t^T At + \int_{\mathbb{R}^m} \left(e^{ix\cdot t} - 1\right) d\nu\right], \tag{7.33}$$

where $\widetilde{\gamma}$ is defined componentwise:

$$\widetilde{\gamma}_j = \gamma_j - \int_V x_j d\nu.$$

Thus:

$$C(t) = C_1(t)C_2(t),$$

and by Proposition 5.33, the underlying random vector X is the **independent sum** *of a* **multivariate normal vector** *X_1 with parameters $\widetilde{\gamma}$ and A as in (6.11), and a* **compound Poisson vector** *X_2 as in (7.28) with $\lambda_Y = \nu/c$ and $\lambda = c$, where $c = \int_{\mathbb{R}^m} d\nu$.*

$$\int_{\mathbb{R}^m} (e^{ix\cdot t} - 1)\, d\nu = c\int_{\mathbb{R}^m} (e^{ix\cdot t} - 1)\, d\left(\frac{\nu}{c}\right).$$

Note that such X_2 exists by Proposition 1.24.

For this representation and result it is enough that $\int_B |x|\, d\nu < \infty$ to make $\widetilde{\gamma}$ well-defined.

3. **Comment on Proof of part 1 and de Finetti's theorem:** *The proof of part 1 of the Lévy-Khintchine representation theorem is the most intricate, and begins with de Finetti's theorem.*

   Given infinitely divisible F, de Finetti's theorem assures the existence of compound Poisson distributions $\{F_n\}$ with $F_n \Rightarrow F$. By (7.28), each such F_n has an associated characteristic function $C_n(t)$ that has the structure of (7.31) with $A_n = 0$, $\gamma_n = 0$, and $\nu_n \equiv \lambda_n \lambda_{X_n}$. As above, here λ_{X_n} is the measure associated with F_{X_n}, the distribution function of the X_n-variates in the summation in (7.26).

   Thus the essence to the proof of part 1 of the Lévy-Khintchine representation theorem is to prove that if each $C_n(t)$ has the structure of (7.31), then $F_n \Rightarrow F$ if and only if F is infinitely divisible and $C(t)$ has the structure of (7.31). Further, the parameters in the structure for $C(t)$ are inherited from the associated structures of the $C_n(t)$.

4. **Comment on Proof of part 3:** *The proof of part 3 is accessible with the tools already developed. We provide an outline.*

   Proof. *Given γ, A, and ν with the given properties, define:*

$$C_n(t) = \exp\left[i\gamma\cdot t - \frac{1}{2}t^T At + \int_{|x|>1/n} \left(e^{ix\cdot t} - 1 - i\,(x\cdot t)\,\chi_B(x)\right) d\nu\right].$$

   By (7.32), ν is a finite measure on $\{|x| > 1/n\}$, and thus as in the special case of item 2, $C_n(t)$ is the characteristic function of an infinitely divisible distribution function.

   Now $C_n(t) \to C(t)$ in (7.31) for all t since $\nu(\{0\}) = 0$, so by Lévy's continuity theorem of Corollary 5.41, such $C(t)$ is a characteristic function if it is continuous at $t = 0$. For this investigation only the integral in the expression for $C(t)$ need be considered. As noted in item 1, this integrand is bounded by a ν-integrable function. Since the integrand is continuous in t, continuity of the integral at $t = 0$ follows from Lebesgue's dominated convergence theorem of Proposition V.7.4.

   Since $C(t)$ is a characteristic function, it now follows from Proposition 7.42 that as the limit of infinitely divisible characteristic functions, $C(t)$ is infinitely divisible. ■

7.4 Distribution Families Related under Multiplication

Interestingly, some distribution families are related under multiplication. Analogous to Definition 7.27:

Definition 7.51 (Distribution families related under multiplication) *By a* ***distribution family*** *is meant the collection of all distribution functions formed by all valid choices of the defining numerical parameters.*

Given independent and identically distributed random variables $\{X_i\}_{i=1}^m$, we will say that the distribution family of X_i is ***related under multiplication to the distribution family of*** X *if $X =_d \prod_{i=1}^m X_i$. We use this same terminology when $\{X_i\}_{i=1}^m$ are independent and from the same distribution family but with different parameters.*

If X and X_i have the same distribution family, we will then say that the given distribution family is ***closed under multiplication****.*

1. **Lognormal Distribution**:

   The **lognormal distribution** is defined in (5.46) and has moments of all orders in IV.(4.80). This distribution was shown to not have a moment generating function in Example IV.4.32, and no closed form version of its characteristic function is currently known.

   While sums of independent lognormal variates are difficult to characterize, this distribution family has the interesting property that it is **closed under multiplication.** Specifically, if L_i is lognormal with parameters μ_i and σ_i^2, then by definition $L_i = \exp(N_i)$ where N_i is normal with parameters μ_i, σ_i^2. By Proposition II.3.56, $\{L_i\}_{i=1}^m$ are independent lognormal if and only if $\{N_i\}_{i=1}^m$ are independent normal.

   Defining:

   $$L \equiv \prod\nolimits_{i=1}^m L_i,$$

   then since $L = \exp\left[\sum_{i=1}^m N_i\right]$, it follows that L has a lognormal distribution with parameters $\mu = \sum_{i=1}^m \mu_i$ and $\sigma^2 = \sum_{i=1}^m \sigma_i^2$.

2. **Log-X Distribution**

   If X is a random variable defined on $(\mathcal{S}, \mathcal{E}, \lambda)$ with distribution function $F_X(x)$, we say that a random variable Y on $(0, \infty)$ is a log-X random variable if $Y \equiv \exp(X)$.

   Then for $y > 0$:

   $$F_Y(y) \equiv \lambda\left[Y^{-1}(-\infty, y]\right] = \lambda\left[X^{-1}(-\infty, \ln y]\right],$$

   and thus:

   $$F_Y(y) = F_X(\ln y). \tag{7.34}$$

   If X has a Lebesgue measurable density function $f_X(x)$, then Y has a density function. To see this, since:

   $$F_Y(y) = \int_{-\infty}^{\ln y} f_X(z)dz,$$

   Proposition III.3.39 and the chain rule obtain for $y > 0$:

   $$f_Y(y) = \frac{f_X(\ln y)}{y}, \quad m\text{-a.e.} \tag{7.35}$$

   While such f_Y need not be finite at $y = 0$, for any $\epsilon > 0$:

   $$\int_\epsilon^\infty f_Y(y)dy = \int_{\ln \epsilon}^\infty f_X(x)dx.$$

   Letting $\epsilon \to 0$, it follows that f_Y is integrable on $[0, \infty)$ and is well-defined as an improper integral.

Analogous to the lognormal example, if the distribution family of X is closed under addition, then the distribution function family of Y is closed under multiplication. For example, the log-gamma distribution is closed under multiplication if all gamma variates are defined with the same parameter λ.

More generally, if the distribution family of X is related under addition to the distribution family of X', then the distribution family of log-X is related under multiplication to the distribution family of log-X'. For example, the exponential family is related under addition to the gamma family, all with common parameter λ, and so the log-exponential family is related under multiplication to the log-gamma family.

If X has a discrete density function $f_X(x)$ defined on $\{x_i\}_{i=1}^N$, where N can be finite or infinite, then $Y = \exp(X)$ had a discrete density function defined on $\{y_i\}_{i=1}^N \equiv \{\exp x_i\}_{i=1}^N$ by

$$f_Y(y_i) = f_X(\ln y_i). \tag{7.36}$$

8

Discrete Time Asset Models in Finance

In this chapter we investigate asset price models in discrete time which are envisioned as two "dimensional" in the sense of having both spatial and a temporal distributional specifications. Often it is the temporal distribution that is explicitly specified, at least approximately, and then the spatial distribution is inferred. The tools of the prior chapters will be applied to investigate models that have both useful and mathematically tractable characteristics.

In the first section, asset price models are introduced in terms of both additive and multiplicative temporal specifications, and the induced spatial distributions are derived. These model specifications are refined in the next section by introducing and illustrating various notions of scalability within a model, where this term identifies how models are related as the time-step $\Delta t \to 0$. The final section investigates limiting distributions associated with scalable models, and unsurprisingly, the earlier central limit theorems play a prominent role.

In this chapter we use the terminology of "asset prices" to simplify the language. The results discussed will apply equally well to the modeling of other financial variables when such frameworks are appropriate.

8.1 Models of Asset Prices

By a **temporal model of asset prices** is meant a collection of distributional assumptions on how asset prices change between various points in time. These distributions are typically parametrized to explicitly reflect the model time-step Δt which is often defined relative to some terminal time T by $\Delta t = T/n$.

By a **spatial model of asset prices** is meant a collection of distributional assumptions on asset prices at various future points in time, denoted $j\Delta t$, for integer $1 \le j \le n$. In most cases, temporal distributional assumptions are "inputs" to a given model, while the spatial distributional assumptions implied by these inputs are the "outputs" of the model.

While what follows also applies to models of random vectors, we restrict our attention to models of random variables for notational simplicity. We investigate two structures for temporal asset models, with particular interest in connections between temporal and spatial distributions.

Remark 8.1 (On Δt) *As noted above, the time-step Δt is commonly defined in terms of some terminal time T by $\Delta t = T/n$. There is nothing in this chapter that requires this other than a bias to tractability, reinforced by the observation that this model is almost always adequate for the purpose.*

The additive temporal random variables $\{Y_i\}_{i=1}^{\infty}$ in the next section and multiplicative variables $\{Z_i\}_{i=1}^{\infty}$ following are defined under the assumption that they are applicable to the interval $J_i \equiv [(i-1)\Delta t, i\Delta t]$, and are therefore parametrized in practice to reflect a constant time-step Δt.

DOI: 10.1201/9781003275770-8

If preferred, these variates could also be parametrized in terms of time-step Δt_i, with J_i defined accordingly, with little change in results other than more complexity in notation and in the parametrization of the associated distribution functions. This complexity would then be compounded later in the section on scalability of models, whereby the various intervals are further subdivided.

Thus we maintain the simpler model and notation and leave it to the reader to generalize as needed.

8.1.1 Additive Temporal Models

Let X_i denote the asset price at time $i\Delta t$ for some given time-step Δt, with $X_0 = x_0$ given and Δt typically defined in reference to a fixed interval $[0, T]$, commonly by $\Delta t = T/n$. The **additive temporal model** specifies that:

$$X_i = X_{i-1} + Y_i, \tag{8.1}$$

where $\{Y_i\}_{i=1}^{\infty}$ are independent random variables with a given distribution function $F_Y(y)$, which logically is parametrized to depend on Δt.

To simplify notation we will sometimes notationally suppress X_0 by assuming that $X_0 = 0$. Then X_n can be understood as the n-period change in the X-variate, and the asset price at time $n\Delta t$ is then given by $X_0 + X_n$.

This model could in theory be defined without independence, or reflect different distributional assumptions for the various time periods. But in addition to adding complexity, this generalization will complicate assumptions for interval subdivisions contemplated for "scalability" in a later section. It can also be argued that independence is a reasonable assumption for most financial variables, and indeed an assumption compelled by anticipated investor activities in an efficient market.

Remark 8.2 (On market efficiency and independence of $\{Y_i\}_{i=1}^{\infty}$) *As a simple example of dependent $\{Y_i\}_{i=1}^{\infty}$, assume that it was observed in the market that the sign of Y_{i-1} was a predictor of the sign of Y_i. In other words, assume that a positive return on assets in period $i-1$ was noted to be a predictor of a likely positive return during period i, and similarly for negative returns.*

In an efficient market, investors would be expected to attempt to monetize this information by increasing positions toward the end of period $i-1$ when Y_{i-1} was emerging positively, and analogously decreasing positions at the end of period $i-1$ when Y_{i-1} was emerging negatively.

This extra buying and selling would put pressure on asset prices, thereby increasing or decreasing Y_{i-1} at what could be argued would be at the expense of Y_i. Thus positive Y_{i-1} would be increased to Y'_{i-1}, and Y_i correspondingly decreased to Y'_i, while negative Y_{i-1} would be decreased to Y'_{i-1}, and Y_i increased to Y'_i. Logically, this trading would continue as long as future returns were somewhat predictable, meaning at least predictable enough to offset trading costs, and thus would cease when returns were essentially independent.

A similar argument can be made in scenarios where the sign of Y_{i-1} was a predictor of the opposite sign for Y_i, or in other scenarios of return dependence.

Remark 8.3 (On identically distributed $\{Y_i\}_{i=1}^{\infty}$) *Admittedly, assuming that $\{Y_i\}_{i=1}^{\infty}$ are identically distributed is less defensible for asset prices and many financial variables. Logically, in many applications, Y_i ought to depend on X_{i-1}. While sometimes applicable, the principal motivation for introducing such models is to investigate various distributional results, including limiting distributions, and contrast these with the oftentimes more realistic multiplicative temporal models introduced below.*

Example 8.4 (Normal $\{Y_i\}_{i=1}^{\infty}$) *Let $X_0 = 0$ and assume that each Y_i in (8.1) has the normal density function given in (6.1) with mean $\mu = 0$ and variance $\sigma^2\Delta t$. By independence, the joint density function of $\bar{Y} \equiv (Y_1, Y_2, ..., Y_n)$ is given in Proposition 1.58 by $f_{\bar{Y}}(y_1, y_2, ..., y_n) = \prod_{i=1}^{n} f_i(y_i)$, and so:*

$$f_{\bar{Y}}(y_1, y_2, ..., y_n) = \left(2\pi\sigma^2\Delta t\right)^{-n/2} \prod\nolimits_{i=1}^{n} \exp\left(-\frac{y_i^2}{2\sigma^2\Delta t}\right).$$

Given $\bar{Y}$, the joint density function of $\bar{X} \equiv (X_1, X_2, ..., X_n)$ can then be derived from this by the change of variables result in (2.29) of Proposition 2.28. To do so, define the transformation T_n on $\mathbb{R}^n$ by:

$$\begin{gathered} T_n : (y_1, y_2, ..., y_n) \to (x_1, x_2, ..., x_n), \\ x_i = \sum\nolimits_{j=1}^{i} y_j. \end{gathered} \tag{8.2}$$

As a sum of independent normals, $X_i = \sum_{j=1}^{i} Y_j$ is normal by a characteristic function argument using (5.48) and the uniqueness result of Proposition 5.29, and X_i has mean 0 and variance $i\sigma^2\Delta t$.

In addition, T_n is a linear transformation (Definition 6.1) on $\mathbb{R}^n$ with associated lower triangular matrix:

$$A_n = \begin{pmatrix} 1 & 0 & 0 & \cdots & 0 \\ 1 & 1 & 0 & \cdots & 0 \\ 1 & 1 & 1 & \cdots & 0 \\ \vdots & \vdots & \vdots & \cdots & \vdots \\ 1 & 1 & 1 & \cdots & 1 \end{pmatrix}. \tag{8.3}$$

That is, $T_n(y_1, y_2, ..., y_n) = A_n y$, where consistent with convention, the n-vector $y = (y_1, y_2, ..., y_n)$ is identified as a column matrix for this matrix multiplication. Further A_n has determinant $\det A_n = 1$, and is thus nonsingular, so A_n^{-1} is well-defined.

It then follows from (2.29) of Proposition 2.28 that with x and y denoting the respective n-vectors, the density function of $\bar{X} = (X_1, X_2, ..., X_n)$ is given:

$$f_{\bar{X}}(x) = f_{\bar{Y}}\left(A_n^{-1}x\right)\left|\det\left(A_n^{-1}\right)\right|.$$

Now $A_n^{-1} : (x_1, x_2, ..., x_n) \to (y_1, y_2, ..., y_n)$ is defined by $y_1 = x_1$, and $y_k = x_k - x_{k-1}$ for $k > 1$, so the matrix of A_n^{-1} is:

$$A_n^{-1} = \begin{pmatrix} 1 & 0 & 0 & \cdots & 0 & 0 \\ -1 & 1 & 0 & \cdots & 0 & 0 \\ 0 & -1 & 1 & \cdots & 0 & 0 \\ \vdots & \vdots & \vdots & \cdots & \vdots & \vdots \\ 0 & 0 & 0 & \cdots & -1 & 1 \end{pmatrix}.$$

This matrix has determinant equal to 1, since $\det A_n^{-1} = 1/\det A_n$ as noted in Remark V.4.29.

Hence:

$$\begin{aligned} & f_{\bar{X}}(x_1, x_2, ..., x_n) \\ & = f_1(x_1) \prod\nolimits_{i=2}^{n} f_i(x_i - x_{i-1}) \\ & = \left(2\pi\sigma^2\Delta t\right)^{-n/2} \exp\left(-\frac{x_1^2}{2\sigma^2\Delta t}\right) \prod\nolimits_{i=2}^{n} \exp\left(-\frac{(x_i - x_{i-1})^2}{2\sigma^2\Delta t}\right). \end{aligned} \tag{1}$$

As anticipated, we can see from this expression that the $\{X_i\}_{i=1}^n$ *are not independent random variables, recalling Corollary 1.57.*

Comparing with (6.6) of Proposition 6.5, $\bar{X} \equiv (X_1, X_2, ..., X_n)$ *so defined in* (1) *has a multivariate normal distribution with covariance matrix* $C = A_n A_n^T \sigma^2 \Delta t$*, where this notation implies that all components of* $A_n A_n^T$ *are to be multiplied by* $\sigma^2 \Delta t$*. This extra factor of* $\sigma^2 \Delta t$ *adjusts for* Y_i *here having variance* $\sigma^2 \Delta t$ *in contrast to a variance of* 1 *in Proposition 6.5. Hence* $\det C = (\sigma^2 \Delta t)^n$*, and with* $C^{-1} = (A_n^T)^{-1} A_n^{-1}/\sigma^2 \Delta t$*, the density function in* (1) *can be expressed:*

$$f_{\bar{X}}(x_1, x_2, ..., x_n) = (2\pi)^{-n/2} [\det C]^{-1/2} \exp\left[-\frac{1}{2} x^T C^{-1} x\right].$$

More generally we have the following.

Proposition 8.5 (On $F_{\bar{X}}$ and $\lambda_{\bar{X}}$ given $\lambda_{\bar{Y}}$) *Assume that the vector of temporal variates* $\bar{Y} = (Y_1, Y_2, ..., Y_n)$ *has independent components, joint distribution function* $F_{\bar{Y}}(y_1, y_2, ..., y_n) = \prod_{i=1}^n F_Y(y_i)$*, and associated Borel measure* $\lambda_{\bar{Y}} = \prod_{i=1}^n \lambda_{Y_i}$ *by Proposition 1.55.*

Then with transformation T_n *defined in (8.2) and vector of spatial variates* $\bar{X} = (X_1, X_2, ..., X_n)$*,* $\lambda_{\bar{X}}$ *is well-defined on* $B \in \mathcal{B}(\mathbb{R}^n)$ *by:*

$$\lambda_{\bar{X}}(B) \equiv \lambda_{\bar{Y}}(T_n^{-1} B), \tag{8.4}$$

and thus:

$$F_{\bar{X}}(x_1, x_2, ..., x_n) \equiv \lambda_{\bar{Y}}\left(T_n^{-1}\left[\prod\nolimits_{i=1}^n (-\infty, x_i]\right]\right).$$

Proof. *Defining the transformation* T_n *as above, it follows that* $\lambda_{\bar{X}} \equiv (\lambda_{\bar{Y}})_{T_n}$*, the measure of Definition 1.28 induced by* T_n*. Thus (8.4) follows by (1.20), and then letting* $B = \prod_{i=1}^n (-\infty, x_i]$ *obtains* $F_{\bar{X}}$*.* ■

When the distribution function $F_{\bar{Y}}(y)$ has an associated density function $f_{\bar{Y}}(y)$, the density function $f_{\bar{X}}$ for $(X_1, X_2, ..., X_n)$ generalizes the result in Example 8.4. Similarly, if the joint distribution function $F_{\bar{X}}$ has a joint density function $f_{\bar{X}}$, this induces a joint density function $f_{\bar{Y}}$, though the component variates will in general not be independent.

Proposition 8.6 (On $f_{\bar{Y}} \to f_{\bar{X}}$ and $f_{\bar{X}} \to f_{\bar{Y}}$) *Let* A_n *be the matrix defined in (8.3) so that* $A_n \bar{Y} = \bar{X}$*.*

1. *If the vector of temporal variates* $\bar{Y} = (Y_1, Y_2, ..., Y_n)$ *has a joint density function* $f_{\bar{Y}}$*, then the joint density function of the vector of spatial variates* $\bar{X} = A_n \bar{Y}$ *is given by:*

$$f_{\bar{X}}(x) = f_{\bar{Y}}(A_n^{-1} x).$$

Hence if $\{Y_i\}_{i=1}^n$ *are independent:*

$$f_{\bar{X}}(x_1, x_2, ..., x_n) = f_{Y_1}(x_1) \prod\nolimits_{j=2}^n f_{Y_j}(x_j - x_{j-1}). \tag{8.5}$$

2. *If* $\bar{X} = (X_1, X_2, ..., X_n)$ *has a joint density function* $f_{\bar{X}}$*, then the joint density function of* $\bar{Y} = A_n^{-1} \bar{X}$ *is given by:*

$$f_{\bar{Y}}(y) = f_{\bar{X}}(A_n y).$$

Thus:

$$f_{\bar{Y}}(y_1, y_2, ..., y_n) = f_{\bar{X}}\left(y_1, \sum\nolimits_{i=1}^2 y_i, ..., \sum\nolimits_{i=1}^n y_i\right). \tag{8.6}$$

Proof. *Since* $\det(A_n^{-1}) = 1/\det(A_n) = 1$, *the general results for* $f_{\bar{X}}(x)$ *and* $f_{\bar{Y}}(y)$ *are (2.29) of Proposition 2.28, while (8.5) and (8.6) reflect the definition of* A_n^{-1} *and* A_n. ■

Example 8.7 ($\bar{X}$ multivariate normal) *Let* $f_{\bar{X}}(x)$ *denote the joint density function of* $\bar{X} = (X_1, X_2, ..., X_n)$, *assumed to be multivariate normally distributed as in (6.6) with mean vector* μ *and covariance matrix* C:

$$f_{\bar{X}}(x) = (2\pi)^{-n/2} [\det C]^{-1/2} \exp\left[-\frac{1}{2}(x-\mu)^T C^{-1}(x-\mu)\right].$$

Proposition 6.23 with $\bar{Y} = A_n^{-1}\bar{X}$ *obtains that* $\bar{Y}$ *is multivariate normally distributed with mean vector* $\tilde{\mu} = A_n^{-1}\mu$ *and covariance* $\tilde{C} = A_n^{-1}C\left(A_n^{-1}\right)^T$:

$$f_{\bar{Y}}(y) = (2\pi)^{-n/2} \left[\det \tilde{C}\right]^{-1/2} \exp\left[-\frac{1}{2}(y-\tilde{\mu})^T \tilde{C}^{-1}(y-\tilde{\mu})\right].$$

Since $\det(A_n^{-1}) = 1$, *it follows that* $\det \tilde{C} = \det C$ *by Remark V.4.29, while* $\tilde{C}^{-1} = A_n^T C^{-1} A_n$ *by Exercise 6.4. Thus:*

$$\begin{aligned}(y-\tilde{\mu})^T \tilde{C}^{-1}(y-\tilde{\mu}) &= [A_n(y-\tilde{\mu})]^T C^{-1}[A_n(y-\tilde{\mu})] \\ &= [A_n y - \mu]^T C^{-1}[A_n y - \mu],\end{aligned}$$

and:

$$f_{\bar{Y}}(y) = (2\pi)^{-n/2} [\det C]^{-1/2} \exp\left[(A_n y - \mu)^T C^{-1}(A_n y - \mu)\right].$$

It is an exercise to verify that this expression agrees with (8.6).

8.1.2 Multiplicative Temporal Models

Let X_i again denote the asset price at time $i\Delta t$ for some given time-step Δt, with $X_0 = x_0$ given and Δt defined in reference to a fixed interval $[0, T]$ by $\Delta t = T/n$. The **multiplicative temporal model** specifies that with given $X_0 > 0$:

$$X_i = X_{i-1} \exp Z_i, \tag{8.7}$$

where $\{Z_i\}_{i=1}^{\infty}$ are independent random variables with a given distribution function $F_Z(x)$, which again logically, is parametrized to depend on Δt.

By Proposition II.3.56, $\{Z_i\}_{i=1}^{\infty}$ are independent random variables if and only if $\{Z_i'\}_{i=1}^{\infty} \equiv \{\exp Z_i\}_{i=1}^{\infty}$ are independent random variables, but the exponential specification in (8.7) is convenient in applications where we do not want $\{X_i\}_{i=1}^{\infty}$ to change sign. In this formulation, the nonnegative multiplicative factor $\exp Z_i$ reflects the **continuously compounded period return,** $Z_i \equiv \ln[X_i/X_{i-1}]$.

The assumption of independence is justified as in Remark 8.2. In this case, the assumption that returns are identically distributed is at least somewhat defensible, reflecting stability of investor risk preferences, and this assumption is common in applications.

To simplify notation, we sometimes suppress X_0 by assuming $X_0 = 1$, and hence X_i can be understood as the i-period continuously compounded total return for the X-variate. The value of the asset price at time $i\Delta t$ is then $X_0 X_i$.

Example 8.8 (Binomial/normal $\{Z_i\}_{i=1}^{\infty}$) *With* $X_0 > 0$ *given, define* $\{Z_i\}_{i=1}^{\infty}$ *with binomial variates:*

$$Z_i = \mu\Delta t + \sigma\sqrt{\Delta t}b_i,$$

where $\{b_i\}_{i=1}^{\infty}$ are independent and binomially distributed to equal ± 1 with probability $1/2$. Since this binomial variate has respective mean and variance of 0 and 1, the first two moments of Z_i are:

$$E[Z_i] = \mu \Delta t, \qquad Var[Z_i] = \sigma^2 \Delta t. \tag{1}$$

Alternatively, define normally distributed Z_i by:

$$Z_i = \mu \Delta t + \sigma \sqrt{\Delta t} N_i,$$

where $\{N_i\}_{i=1}^{\infty}$ are independent standard normal variates with respective mean and variance of 0 and 1. Thus the first two moments of Z_i equal those in (1) for the binomial model.

1. ***Binomial** $\{Z_i\}_{i=1}^{\infty}$: It follows that:*

$$X_i = X_0 \exp\left(\mu i \Delta t + \sigma \sqrt{\Delta t} \sum\nolimits_{j=1}^{i} b_j\right), \tag{8.8}$$

and hence each X_i is log-binomial. Recalling Section 7.4, this means that $\ln(X_i)$ is binomial.

For $k < i$:

$$X_i = X_k \exp\left(\mu(i-k)\Delta t + \sigma \sqrt{\Delta t} \sum\nolimits_{j=k+1}^{i} b_j\right),$$

and X_i and X_k are apparently not independent.

To formalize this, assume X_k is defined with given binomial variates $\sum_{j=1}^{k} b'_j$. Then given X_k, the conditional variate $X_i|X_k$ is defined for $i > k$ by:

$$X_i|X_k = X_0 \exp\left(\mu i \Delta t + \sigma \sqrt{\Delta t}\left[\sum\nolimits_{j=1}^{k} b'_j + \sum\nolimits_{j=k+1}^{i} b_j\right]\right). \tag{2}$$

Since $-k \leq \sum_{j=1}^{k} b'_j \leq k$ is fixed, the joint density function $f(X_k, X_i)$ equals 0 for any X_i-variate with $\sum_{j=1}^{i} b_j > \sum_{j=1}^{k} b'_j + (i-k)$ or $\sum_{j=1}^{i} b_j < \sum_{j=1}^{k} b'_j - (i-k)$.

There will always be at least one value of X_i with this property, and hence:

$$f(X_k, X_i) = 0 \neq f(X_k) f(X_i),$$

and so X_i and X_k are not independent by Corollary 1.57.

Alternatively, in terms of conditional density functions, $f(X_i|X_k)$ is log-binomial, and by (2):

$$\begin{aligned} E[\ln(X_i|X_k)] &= \ln X_0 + i\mu \Delta t + \sigma \sqrt{\Delta t} \sum\nolimits_{j=1}^{k} b'_j, \\ Var[\ln(X_i|X_k)] &= (i-k)\sigma^2 \Delta t. \end{aligned}$$

Similarly, $f(X_i)$ is log-binomial with respective moments $\ln X_0 + i\mu\Delta t$ and $\sigma^2 i \Delta t$. Hence:

$$f(X_i|X_k) \neq f(X_i),$$

and X_i and X_k are not independent by Proposition II.1.34.

2. ***Normal** $\{Z_i\}_{i=1}^{\infty}$: Each X_i is lognormal:*

$$X_i = X_0 \exp\left(\mu i \Delta t + \sigma \sqrt{\Delta t} \sum\nolimits_{j=1}^{i} N_j\right), \tag{8.9}$$

and it follows that for $k < i$:

$$X_i = X_k \exp\left(\mu(i-k)\Delta t + \sigma\sqrt{\Delta t}\sum\nolimits_{j=k+1}^{i} N_j\right).$$

As above it follows that $f(X_i|X_k) \neq f(X_i)$, and thus X_i and X_k are not independent.

More generally, given $\bar{Z} \equiv (Z_1, Z_2, ..., Z_n)$, the joint density function of $\bar{X} \equiv (X_1, X_2, ..., X_n)$ can again be derived from the change of variables result in (2.29) of Proposition 2.28. To do so, define the continuously differentiable and one-to-one transformation g_n on $\mathbb{R}^n$ by:

$$\begin{aligned} g_n &: (z_1, z_2, ..., z_n) \to (x_1, x_2, ..., x_n) \\ x_j &= x_0 \exp\left[\sum\nolimits_{i=1}^{j} z_i\right]. \end{aligned} \tag{8.10}$$

Proposition 8.9 (On $F_{\bar{X}}$ and $\lambda_{\bar{X}}$ given $\lambda_{\bar{Z}}$) *Assume that the vector of continuously compounded returns $\bar{Z} = (Z_1, Z_2, ..., Z_n)$ has independent components, a joint distribution function $F_{\bar{Z}}(z_1, z_2, ..., z_n) = \prod_{i=1}^{n} F_{Z_i}(z_i)$, and associated Borel measure $\lambda_{\bar{Z}} = \prod_{i=1}^{n} \lambda_{Z_i}$ by Proposition 1.55.*

Then with g_n defined in (8.10) and vector of spatial variates $\bar{X} = (X_1, X_2, ..., X_n)$, $\lambda_{\bar{X}}$ is well-defined on $B \in \mathcal{B}(\mathbb{R}^n)$by:

$$\lambda_{\bar{X}}(A) \equiv \lambda_{\bar{Z}}(g_n^{-1}A), \tag{8.11}$$

and thus:

$$F_{\bar{X}}(x_1, x_2, ..., x_n) \equiv \lambda_{\bar{Z}}\left(g_n^{-1}\left[\prod\nolimits_{i=1}^{n}(-\infty, x_i]\right]\right).$$

Proof. *Defining the transformation g_n as above, it follows that $\lambda_{\bar{X}} \equiv (\lambda_{\bar{Z}})_{g_n}$, the measure of Definition 1.28 induced by g_n. Thus (8.11) follows by (1.20), and then letting $B = \prod_{i=1}^{n}(-\infty, x_i]$ obtains $F_{\bar{X}}$.* ■

When the distribution function $F_{\bar{Z}}(z)$ has an associated density function $f_{\bar{Z}}(z)$, the density function $f_{\bar{X}}$ for $\bar{X} = (X_1, X_2, ..., X_n)$ generalizes the result in Example 8.8. Similarly, if the joint distribution function $F_{\bar{X}}$ has a joint density function $f_{\bar{X}}$, this induces a joint density function $f_{\bar{Z}}$, though the component variates will in general not be independent.

To derive the relationship between the density functions of $\bar{X}$ and $\bar{Z}$ by (2.29) of Proposition 2.28, we must determine the associated Jacobian matrix and determinant of the transformation $g_n(z_1, z_2, ..., z_n)$ defined in (8.10). First, $\partial x_j/\partial z_k = x_j$ for $k \leq j$ and $\partial x_j/\partial z_k = 0$ otherwise, so the Jacobian matrix of g_n is lower triangular. Since $x_j > 0$ by construction, the absolute value of the Jacobian determinant is given by:

$$\begin{aligned} \left|\det\left(\frac{\partial g_n(z)}{\partial z}\right)\right| &= \prod\nolimits_{j=1}^{n} x_j \\ &= x_0^n \prod\nolimits_{j=1}^{n} \exp\left[(n-j+1)\, z_j\right]. \end{aligned} \tag{1}$$

Because g_n is invertible, the inverse function theorem (Proposition V.4.37) obtains that the Jacobian matrix of g_n^{-1} satisfies:

$$\left(\frac{\partial g_n^{-1}(x)}{\partial x}\right) = \left(\left.\frac{\partial g_n(z)}{\partial z}\right|_{g_n^{-1}(x)}\right)^{-1},$$

and thus by (1):

$$\left|\det\left(\frac{\partial g_n^{-1}(x)}{\partial x}\right)\right| = \left(\prod\nolimits_{j=1}^{n} x_j\right)^{-1}.$$

This determinant can also be calculated explicitly by noting that $g_n^{-1} : (x_1, x_2, ..., x_n) \to (z_1, z_2, ..., z_n)$ is defined by:

$$z_j = \ln\left[\frac{x_j}{x_{j-1}}\right],$$

with $x_0 > 0$ given, and noting that the Jacobian matrix of g_n^{-1} is lower triangular.

Proposition 8.10 (On $f_{\bar{Z}} \to f_{\bar{X}}$ and $f_{\bar{X}} \to f_{\bar{Z}}$) *Let g_n be the transformation defined in (8.10) so that $g_n(\bar{Z}) = \bar{X}$.*

1. *If the vector of continuously compounded returns $\bar{Z} = (Z_1, Z_2, ..., Z_n)$ has a joint density function $f_{\bar{Z}}$, then the joint density function of the vector of spatial variates $\bar{X} = g_n(\bar{Z})$ is given by:*

$$f_{\bar{X}}(x) = \left(\prod\nolimits_{j=1}^n x_j\right)^{-1} f_{\bar{Z}}\left(\ln\left[\frac{x_1}{x_0}\right], \ln\left[\frac{x_2}{x_1}\right], ..., \ln\left[\frac{x_n}{x_{n-1}}\right]\right).$$

Hence if $\{Z_i\}_{i=1}^n$ are independent:

$$f_{\bar{X}}(x_1, x_2, ..., x_n) = \prod\nolimits_{j=1}^n \frac{1}{x_j} f_{Z_j}\left(\ln\left[\frac{x_j}{x_{j-1}}\right]\right). \tag{8.12}$$

2. *If $\bar{X} = (X_1, X_2, ..., X_n)$ has a joint density function $f_{\bar{X}}$, then the joint density function of $\bar{Z} = g_n^{-1}(\bar{X})$ is given by:*

$$f_{\bar{Z}}(z) = x_0^n \prod\nolimits_{j=1}^n \left[\exp(n-j+1) z_j\right] f_{\bar{X}}\left(\exp z_1, \prod\nolimits_{j=1}^2 \exp z_j, ..., \prod\nolimits_{j=1}^n \exp z_j\right). \tag{8.13}$$

Proof. *By (2.29):*

$$f_{\bar{X}}(x) = f_{\bar{Z}}\left(g_n^{-1}(x)\right)\left|\det\left(\frac{\partial g_n^{-1}(x)}{\partial x}\right)\right|,$$

and the result in item 1 reflects $g_n^{-1}(x)$ and its Jacobian determinant above, while (8.12) is Corollary 1.57.

The result of item 2 similarly follows from:

$$f_{\bar{Z}}(z) = f_{\bar{X}}(g_n(z))\left|\det\left(\frac{\partial g_n(z)}{\partial z}\right)\right|.$$

■

Example 8.11 (Binomial/normal $\{Z_i\}_{i=1}^\infty$) *Recall Example 8.8:*

1. ***Normal returns:*** *If continuously compounded returns $Z_i = \mu\Delta t + \sigma\sqrt{\Delta t} N_i$ with $\{N_i\}_{i=1}^n$ independent standard normals with mean 0 and variance 1, we have from (8.7), that spatial variates are given by:*

$$\begin{aligned} X_i &= \prod\nolimits_{j=1}^i \exp\left(\mu\Delta t + \sigma\sqrt{\Delta t} N_j\right) \\ &= \exp\left[\mu i\Delta t + \sigma\sqrt{\Delta t}\sum\nolimits_{i=1}^i N_j\right]. \end{aligned}$$

For such n-tuples $(x_1, x_2, ..., x_n)$, *(8.12) obtains:*

$$\begin{aligned} & f_{\bar{X}}(x_1, x_2, ..., x_n) \\ = & \prod_{i=1}^{n} \frac{1}{x_i \sigma \sqrt{\Delta t}\sqrt{2\pi}} \exp\left[-\frac{1}{2\sigma^2 \Delta t}\left(\ln\left[\frac{x_i}{x_{i-1}}\right] - \mu\Delta t \right)^2 \right] \\ = & (2\pi\sigma^2\Delta t)^{-n/2}\left(\prod_{i=1}^{n} x_i\right)^{-1} \exp\left[-\frac{1}{2\sigma^2\Delta t}\sum_{i=1}^{n}\left(\ln\left[\frac{x_i}{x_{i-1}}\right] - \mu\Delta t\right)^2 \right]. \end{aligned}$$

2. ***Binomial returns:*** *If* $Z_i = \mu\Delta t + \sigma\sqrt{\Delta t} b_i$ *with* $\{b_i\}_{i=1}^n$ *independent and binomially distributed to equal* 1 *or* -1 *with probabilities* p *and* $1-p$ *respectively, we cannot apply (8.12) to discrete density functions as this followed from a result in Lebesgue integration theory.*

   However, $\ln\left[\frac{x_i}{x_{i-1}}\right]$ *equals* $\mu\Delta t \pm \sigma\sqrt{\Delta t}$ *for all* i, *and hence* $f_X(x_1, x_2, ..., x_n) = 0$ *unless there is an n-tuple* $(b_1, b_2, ..., b_n)$ *of binomial variates so that for each* $i \leq n$:

$$\begin{aligned} x_i &= \prod_{j=1}^{i} \exp\left(\mu\Delta t + \sigma\sqrt{\Delta t} b_j\right) \\ &= \exp\left[\mu i \Delta t + \sigma\sqrt{\Delta t}\sum_{j=1}^{i} b_j\right]. \end{aligned}$$

   If $(x_1, x_2, ..., x_n)$ *is such an n-tuple:*

$$f_{\bar{X}}(x_1, x_2, ..., x_n) = \prod_{i=1}^{n} f_{Z_i}\left(\ln\left[\frac{x_i}{x_{i-1}}\right]\right).$$

   Since $\ln\left[\frac{x_i}{x_{i-1}}\right] = Z_i$, *this obtains:*

$$f_{\bar{X}}(x_1, x_2, ..., x_n) = p^k(1-p)^{n-k},$$

   where k *denotes the number of indexes in* $(b_1, b_2, ..., b_n)$ *for which* $b_i = 1$. *For any* k *there will be* $\binom{n}{k}$ *such n-tuples* $(x_1, x_2, ..., x_n)$, *all of which result in the same value of* X_n.

8.1.3 Simulating Asset Price Paths

In either temporal model it is of interest to simulate asset price paths $(X_1, X_2, ..., X_n)$ in the case of independent Y-variates for the additive model, or independent Z-variates for the multiplicative model. Recalling Chapter IV.5, in both cases it is often sufficient to only generate independent n-tuples of continuous, uniformly distributed variates $\{U_j\}_{j=1}^n$ defined on $[0, 1]$, and to apply Proposition II.4.9.

- In the case where the Y_j variates of the additive temporal model are independent and identically distributed with distribution function F_Y, it follows from this result that:

$$\{Y_j\}_{j=1}^n \equiv \{F_Y^*(U_j)\}_{j=1}^n,$$

  defines a collection of independent variates with the given distribution function. For this statement, recall that F_Y^* denotes the **left continuous inverse** of F_Y of Definition II.3.12.

- For independent and identically distributed Z_j variates:

$$\{Z_j\}_{j=1}^n \equiv \{F_Z^*(U_j)\}_{j=1}^n,$$

similarly defines an independent sample of returns in the multiplicative model.

When a distribution function F is continuous and strictly monotonically increasing, then $F^* = F^{-1}$ by Proposition II.3.22.

In either case, each such sample $\{U_j\}_{j=1}^n$ provides a sample "path" $\bar{X} \equiv (X_1, X_2, ..., X_n)$ of asset variates given X_0 in the respective models.

8.2 Scalable Asset Models

As noted in this chapter's introduction, in most cases, temporal distributional assumptions are "inputs" to a given model, while the spatial distributional assumptions implied by these inputs are the "outputs" to the model. Thus it makes sense to contemplate desirable properties of such input assumptions that will make the model more useful.

While the identification and justification of appropriate distributional assumptions is the responsibility of the model builder, another significant assumption is the dependence of these distributional assumptions on the time-step Δt. Below we define a temporal model of asset prices or other financial variables to be **scalable** when assumptions exist for all Δt, and then refine this notion when such assumptions possess one of three useful properties which we identify. While the third general notion of scalability will appear in the limiting results of Section 8.3, an actual model with this property will not be seen until Section 9.2.

Another useful property of a given model is the **identifiability** of the induced spatial models. This notion is necessarily subjective, and reflects the scope of the portfolio of distributions in the model builder's tool kit. In many instances this can be achieved with certain temporal assumptions, but this then requires that such assumptions are independently defensible.

When satisfied, the property of identifiability allows the modeler to evaluate the reasonableness of induced "output" spatial distributions given the "input" temporal distributional assumptions and the application in hand. This can then lead to an evaluation of the reasonableness of the temporal assumptions.

To define "scalability" of the temporal variates Y_i and Z_i introduced above, we refine this notation to explicitly link the temporal variates to the time-step $\Delta t \equiv T/n$, where $[0, T]$ denotes the time interval over which the model is applied. Given time-step $\Delta t \equiv T/n$, we denote the independent, identically distributed temporal variates of either model by $\{V_i^{(n)}\}_{i=1}^n$, and define for $j = 1, ..., n$:

$$U_j^{(n)} \equiv \sum\nolimits_{i=1}^j V_i^{(n)}.$$

Thus $V_i^{(n)}$ is the temporal variate for the period $[(i-1)\Delta t, i\Delta t]$, while $U_j^{(n)}$ is the spatial variate at time $j\Delta t$. In other words, $V_i^{(n)} = Y_i$ and $U_j^{(n)} = X_j$ in the notation of additive temporal model, while $V_i^{(n)} = Z_i$ and $U_j^{(n)} = \ln[X_j/X_0]$ in the multiplicative temporal model.

This notation also identifies the Δt-refinements, in that what will be denoted $U_j^{(nm)}$, $V_i^{(nm)}$ etc., will be these variates associated with $\Delta t' \equiv \Delta t/m = T/nm$.

Definition 8.12 (Scalable asset models) *A model of asset prices or other financial variables over the time period $[0, T]$ is said to be* **scalable** *if for any n, there exists distributional assumptions on independent, identically distributed $\{V_i^{(n)}\}_{i=1}^n$, defined relative to time-step $\Delta t \equiv T/n$, which define associated variates $\{U_j^{(n)}\}_{j=1}^n$.*

1. *A model is said to be* **perfectly scalable** *or* **exactly scalable** *if given independent, identically distributed $\{V_i^{(n)}\}_{i=1}^n$ associated with $\Delta t \equiv T/n$ and any positive integer m, the independent, identically distributed variates $\{V_k^{(mn)}\}_{k=1}^{mn}$ obtain for $1 \leq j \leq n$:*
$$U_{mj}^{(mn)} =_d U_j^{(n)}, \tag{8.14}$$
where $U_{mj}^{(mn)} \equiv \sum_{i=1}^{mj} V_i^{(mn)}$, and $=_d$means **equal in distribution.**
In other words, for $1 \leq j \leq n$:
$$\sum\nolimits_{i=1}^{mj} V_i^{(mn)} =_d \sum\nolimits_{i=1}^{j} V_i^{(n)}.$$

2. *A model is said to be* **approximately scalable to order** *$k_0 \geq 1$ if all $V^{(n)}$ and $V^{(mn)}$ variates have k_0 moments, and with $U_{mj}^{(mn)}$ defined above:*
$$E\left[\left(U_{mj}^{(mn)}\right)^k\right] = E\left[\left(U_j^{(n)}\right)^k\right], \tag{8.15}$$
for $1 \leq j \leq n$ and $k \leq k_0$.
Equivalently by Proposition IV.4.24, this property can be stated in terms of the central moments. Specifically, that (8.15) is satisfied for $k = 1$, and for $2 \leq k \leq k_0$:
$$E\left[\left(U_{mj}^{(mn)} - E\left(U_{mj}^{(mn)}\right)\right)^k\right] = E\left[\left(U_j^{(n)} - E\left(U_j^{(n)}\right)\right)^k\right].$$

3. *A model is said to be* **asymptotically scalable to order** *$k_0 \geq 1$ if all $V^{(n)}$ and $V^{(mn)}$ variates have k_0 moments, and for $k \leq k_0$ and $1 \leq j \leq n$ there exist finite $\mu'_k(t)$ with $t \equiv j\Delta t$ so that:*
$$E\left[\left(U_{mj}^{(mn)}\right)^k\right] \to \mu'_k(t), \text{ as } m \to \infty. \tag{8.16}$$
As in item 2, this can also be specified in terms of central moments, that (8.16) is satisfied for $k = 1$, and for $2 \leq k \leq k_0$ there exist finite $\mu_k(t)$ so that:
$$E\left[\left(U_{mj}^{(mn)} - E\left(U_{mj}^{(mn)}\right)\right)^k\right] \to \mu_k(t), \text{ as } m \to \infty.$$

As noted above, temporal distributions are typically the inputs to the model because empirical data in finance is almost always temporal, and hence provides the greatest insights into this dimension. We can observe how financial variables change over given Δt time-steps, but cannot directly observe two outcomes of a given variate at any point in time. The associated spatial distribution of the model is then an output of these temporal assumptions and not an independent assumption. However, if the model provides unrealistic spatial results, this may then provide a cautionary note on the reasonableness of the temporal assumptions.

The value of a scalable model is that it allows an evaluation of the properties and implications of the model as $\Delta t \to 0$. This then provides a linkage between the more tractability model in discrete time, and the continuous time models of finance, the subject of later books.

8.2.1 Properties of Scalable Models

For the criteria of scalability, we begin with a characterization of perfect scalability that may well have been anticipated. For this result, recall Definition 7.31.

Proposition 8.13 (Perfect scalability and I.D. temporal variates) *A model is perfectly scalable if and only if for any n, the temporal random variable $V^{(n)}$ is infinitely divisible.*

Proof. *If $V^{(n)}$ is infinitely divisible for given n, then for any m there exists independent and identically distributed variates $\{V_i^{(mn)}\}_{i=1}^m$ so that:*

$$V^{(n)} =_d \sum\nolimits_{i=1}^{m} V_i^{(mn)}, \tag{1}$$

and thus by Proposition 5.23:

$$C_{V^{(n)}}(t) = C^m_{V^{(mn)}}(t). \tag{2}$$

Taking a jth power for $1 \leq j \leq n$, it follows from (2) and Proposition 5.23 again that:

$$C_{U_j^{(n)}}(t) = C_{U_{mj}^{(mn)}}(t),$$

which is (8.14) by Proposition 5.29.

If the model is perfectly scalable, then for any n and m, Definition 8.12 obtains the existence of $\{V_i^{(mn)}\}_{i=1}^m$ with:

$$\sum\nolimits_{i=1}^{m} V_i^{(mn)} =_d V^{(n)},$$

and thus $V^{(n)}$ is infinitely divisible by Definition 7.31. ■

Corollary 8.14 (Perfect scalability and I.D. spatial variates) *A model is perfectly scalable if and only if for any n, the spatial random variables $\{U_j^{(n)}\}_{j=1}^n$ are infinitely divisible.*

Proof. *By Proposition 8.13, a temporal model is perfectly scalable if and only if $V^{(n)}$ is infinitely divisible for any n. But then $\{U_j^{(n)}\}_{j=1}^n$ are infinitely divisible by definition. Conversely, if $\{U_j^{(n)}\}_{j=1}^n$ are infinitely divisible then so too is $U_1^{(n)} \equiv V_1^{(n)}$, and the model is perfectly scalable by item 2 of Proposition 7.42.* ■

The next result proves that scalability on any basis is preserved under a well-defined affine transformation of the $V^{(n)}$ temporal variates. In other words, if a model is scalable on any basis with $\{V^{(n)}\}_{n=1}^\infty$, then so too is the asset model with $\{a^{(n)}V^{(n)} + b^{(n)}\}_{n=1}^\infty$ if parameters are appropriately defined.

Note that while the representation in (8.17) defines $a^{(n)}$ and $b^{(n)}$ in terms of $a^{(1)}$ and $b^{(1)}$, this also implies the subdivision algorithm:

$$a^{(nm)} \equiv a^{(n)}, \qquad b^{(nm)} \equiv b^{(n)}/m.$$

Proposition 8.15 (Affine transformations and scalability) *Given arbitrary $a^{(1)} \neq 0$ and $b^{(1)}$, if a model defined by $\{V^{(n)}\}_{n=1}^\infty$ is perfectly, approximately, or asymptotically scalable, then the model defined by $\{a^{(n)}V^{(n)} + b^{(n)}\}_{n=1}^\infty$ is analogously scalable if for all n:*

$$a^{(n)} \equiv a^{(1)}, \qquad b^{(n)} \equiv \frac{b^{(1)}}{n}. \tag{8.17}$$

In other words, the model $\{\widetilde{V}^{(n)}\}_{n=1}^\infty$ defined by:

$$\widetilde{V}^{(n)} = a^{(1)}V^{(n)} + \frac{b^{(1)}}{n}, \tag{8.18}$$

is similarly scalable.

Proof. *Given n and independent, identically distributed $\{V_i^{(n)}\}_{i=1}^n$ associated with $\Delta t \equiv T/n$, and any m, there exists i.i.d. $\{V_i^{(mn)}\}_{i=1}^{mn}$ so that for $1 \leq j \leq n$, either (8.14) is satisfied, or (8.15) or (8.16) is satisfied for $k \leq k_0$.*

For a model defined with $\{a^{(n)}V_j^{(n)} + b^{(n)}\}_{j=1}^n$ to be analogously scalable, we require the existence of i.i.d. $\{\widetilde{V}_i^{(nm)}\}_{i=1}^{mn}$ so that for $1 \leq j \leq n$:

- *Perfectly scalable:*

$$\sum\nolimits_{i=1}^{mj} \widetilde{V}_i^{(mn)} =_d a^{(n)} \sum\nolimits_{i=1}^{j} V_i^{(n)} + jb^{(n)}. \tag{1}$$

- *Approximately scalable:*

$$E\left[\left(\sum\nolimits_{i=1}^{mj} \widetilde{V}_i^{(mn)}\right)^k\right] = E\left[\left(a^{(n)} \sum\nolimits_{i=1}^{j} V_i^{(n)} + jb^{(n)}\right)^k\right], \; k \leq k_0. \tag{2}$$

- *Asymptotically scalable:*

$$E\left[\left(\sum\nolimits_{i=1}^{mj} \widetilde{V}_i^{(mn)}\right)^k\right] \to \tilde{\mu}'_k(t), \text{ as } m \to \infty. \tag{3}$$

With $V_i^{(nm)}$ as in Definition 8.12, define:

$$\widetilde{V}_i^{(nm)} \equiv a^{(mn)} V_i^{(nm)} + b^{(mn)}.$$

The scalability requirements in (1) – (3) reduce to:

- *Perfectly scalable:*

$$a^{(mn)} \sum\nolimits_{i=1}^{mj} V_i^{(mn)} + mjb^{(mn)} =_d a^{(n)} \sum\nolimits_{i=1}^{j} V_i^{(n)} + jb^{(n)},$$

- *Approximately scalable:*

$$E\left[\left(a^{(mn)} \sum\nolimits_{i=1}^{mj} V_i^{(mn)} + mjb^{(mn)}\right)^k\right] = E\left[\left(a^{(n)} \sum\nolimits_{i=1}^{j} V_i^{(n)} + jb^{(n)}\right)^k\right], \; k \leq k_0.$$

- *Asymptotically scalable:*

$$E\left[\left(a^{(mn)} \sum\nolimits_{i=1}^{mj} V_i^{(mn)} + mjb^{(mn)}\right)^k\right] \to \tilde{\mu}'_k(t), \text{ as } m \to \infty,$$

for $t = j\Delta t$ and $\tilde{\mu}'_k(t)$ to be specified below.

The first result follows given (8.17) and the respective assumption on $\{V_j^{(n)}\}_{j=1}^n$. For the second and third, given (8.17) and the binomial theorem:

$$E\left[\left(a^{(mn)} \sum\nolimits_{i=1}^{mj} V_i^{(mn)} + mjb^{(mn)}\right)^k\right]$$
$$= \sum\nolimits_{i=0}^{k} \binom{k}{i} \left(a^{(1)}\right)^i \left(\frac{j}{n} b^{(1)}\right)^{k-i} E\left[\left(\sum\nolimits_{i=1}^{mj} V_i^{(mn)}\right)^i\right]. \tag{4}$$

In the approximately scalable case, by (8.15) for $k \leq k_0$, another application of the binomial theorem and (8.17):

$$\sum_{i=0}^{k} \binom{k}{i} \left(a^{(1)}\right)^i \left(\frac{j}{n} b^{(1)}\right)^{k-i} E\left[\left(\sum_{i=1}^{mj} V_i^{(mn)}\right)^i\right] = E\left[\left(a^{(n)} \sum_{i=1}^{j} V_i^{(n)} + j b^{(n)}\right)^k\right].$$

For asymptotic scalability, by (8.16) with $t = j\Delta t$ and $k \leq k_0$:

$$\sum_{i=0}^{k} \binom{k}{i} \left(a^{(1)}\right)^i \left(\frac{j}{n} b^{(1)}\right)^{k-i} E\left[\left(\sum_{i=1}^{mj} V_i^{(mn)}\right)^i\right] \to \sum_{i=0}^{k} \binom{k}{i} \left(a^{(1)}\right)^i \left(\frac{j}{n} b^{(1)}\right)^{k-i} \mu_i'(t),$$

which obtains the definition of $\tilde{\mu}_k'(t)$. ■

Example 8.16 (Targeting model moments when $V^{(n)}$ has two moments) *Assume that all $V^{(n)}$ temporal variates have two moments.*

If the model is exactly scalable or approximately scalable to order $k_0 \geq 2$, this obtains a well-defined mean μ_{U_T} and variance $\sigma^2_{U_T}$ for the terminal spatial financial variable $U_n^{(n)} \equiv U_T$, recalling that $T = n\Delta t$. One can then define a transformed asset model with $\left(a^{(1)},\ b^{(1)}\right)$ to ***target a desired mean $\mu_{\widetilde{U}_T}$ and variance $\sigma^2_{\widetilde{U}_T}$***.

Given $\{V^{(n)}\}_{n=1}^{\infty}$, it follows from independence that for $1 \leq j \leq n$:

$$E\left[U_j^{(n)}\right] = j\mu_{V^{(n)}}, \quad Var\left[U_j^{(n)}\right] = j\sigma^2_{V^{(n)}}. \tag{1}$$

With $\widetilde{V}^{(n)} = a^{(1)}V^{(n)} + \frac{b^{(1)}}{n}$ and $\widetilde{U}_j^{(n)}$ analogously defined:

$$E\left[\widetilde{U}_j^{(n)}\right] = j\left(a^{(1)}\mu_{V^{(n)}} + \frac{b^{(1)}}{n}\right), \quad Var\left[\widetilde{U}_j^{(n)}\right] = j\left(a^{(1)}\sigma_{V^{(n)}}\right)^2. \tag{2}$$

Setting $j = n$ in (2), it follows that $E\left[\widetilde{U}_n^{(n)}\right] = \mu_{\widetilde{U}_T}$ and $Var\left[\widetilde{U}_n^{(n)}\right] = \sigma^2_{\widetilde{U}_T}$ when:

$$\begin{aligned} na^{(1)}\mu_{V^{(n)}} + b^{(1)} &= \mu_{\widetilde{U}_T}, \\ n\left(a^{(1)}\right)^2 \sigma^2_{V^{(n)}} &= \sigma^2_{\widetilde{U}_T}. \end{aligned} \tag{3}$$

Since $n\mu_{V^{(n)}} = \mu_{U_T}$ and $n\sigma^2_{V^{(n)}} = \sigma^2_{U_T}$ by (1), these moments are obtained with:

$$a^{(1)} = \frac{\sigma_{\widetilde{U}_T}}{\sigma_{U_T}}, \quad b^{(1)} = \mu_{\widetilde{U}_T} - \frac{\sigma_{\widetilde{U}_T}}{\sigma_{U_T}}\mu_{U_T}. \tag{8.19}$$

In other words, the transformed temporal model is defined by (8.18):

$$\widetilde{V}^{(n)} = \frac{\sigma_{\widetilde{U}_T}}{\sigma_{U_T}} V^{(n)} + \frac{1}{n}\left(\mu_{\widetilde{U}_T} - \frac{\sigma_{\widetilde{U}_T}}{\sigma_{U_T}}\mu_{U_T}\right).$$

This can be expressed in the more intuitive formula:

$$\widetilde{V}^{(n)} = \frac{\sigma_{\widetilde{U}_T}}{\sqrt{n}}\left(\frac{V^{(n)} - \mu_{V^{(n)}}}{\sigma_{V^{(n)}}}\right) + \frac{\mu_{\widetilde{U}_T}}{n}. \tag{8.20}$$

In this representation, all $V^{(n)}$ are ***normalized,*** *meaning shifted to have mean zero and scaled to have variance one.*

Exercise 8.17 (Targeting model moments when $V^{(n)}$ has two moments) *Determine* $(a^{(1)},\ b^{(1)})$ *to target a desired mean* $\mu_{\tilde{U}_T}$ *and variance* $\sigma^2_{\tilde{U}_T}$ *in the case where the asset model is asymptotically scalable. Hint: You now have to express* (3) *in terms of a limit as* $n \to \infty$.

The next result addresses the following question in the affirmative:

Can we build variously scalable models using affine transformations of a fixed random variable V? In other words, can we build such a model by defining:

$$V^{(n)} \equiv a^{(n)}V + b^{(n)}, \tag{8.21}$$

for appropriate $\{a^{(n)}, b^{(n)}\}$?

As seen above for (8.17), the specifications identified in (8.22) and (8.24) below define $a^{(n)}$ and $b^{(n)}$ in terms of arbitrary $a^{(1)} \neq 0$ and $b^{(1)}$. In the latter case of (8.24), this implies the subdivision algorithm:

$$a^{(nm)} \equiv a^{(n)}/\sqrt{m}, \qquad b^{(nm)} \equiv b^{(n)}/m.$$

For the subdivision algorithm for (8.22), see Exercise 8.19.

Proposition 8.18 (Scalable models with fixed V) *Given arbitrary* $a^{(1)} \neq 0$ *and* $b^{(1)}$:

1. *If V is stable, the model defined in (8.21) by* $\{a^{(n)}V + b^{(n)}\}_{n=1}^{\infty}$ *is perfectly scalable if* $a^{(n)}$ *and* $b^{(n)}$ *are defined by:*

$$a^{(n)} \equiv \frac{a^{(1)}}{c_n}, \qquad b^{(n)} \equiv \frac{1}{n}\left(b^{(1)} - a^{(1)}\frac{d_n}{c_n}\right), \tag{8.22}$$

with $c_n > 0$ *and* d_n *given in (7.22).*

2. *If V is strictly stable,* $\{a^{(n)}V + b^{(n)}\}_{n=1}^{\infty}$ *is perfectly scalable if* $a^{(n)}$ *and* $b^{(n)}$ *are defined by:*

$$a^{(n)} \equiv \frac{a^{(1)}}{c_n}, \qquad b^{(n)} \equiv \frac{b^{(1)}}{n}. \tag{8.23}$$

3. *If V has two moments with* $E[V] = 0$, *the model defined in (8.21) by* $\{a^{(n)}V + b^{(n)}\}_{n=1}^{\infty}$ *is approximately scalable to order* $k_0 = 2$ *if* $a^{(n)}$ *and* $b^{(n)}$ *are defined by:*

$$a^{(n)} \equiv \frac{a^{(1)}}{\sqrt{n}}, \qquad b^{(n)} \equiv \frac{b^{(1)}}{n}. \tag{8.24}$$

Proof. *Assume that V is stable. Given n, let* $\{V_i\}_{i=1}^{n}$ *be independent with the distribution of V. Since V is stable, (7.22) obtains that for every m there exists real* $c_m > 0$ *and* d_m *so that if* $\{V_{i_k}\}_{k=1}^{m}$ *are independent and have the distribution of V:*

$$\sum\nolimits_{k=1}^{m} V_{i_k} =_d c_m V_i + d_m.$$

Since $c_m > 0$:

$$V_i =_d \frac{1}{c_m}\sum\nolimits_{k=1}^{m}\left(V_{i_k} - \frac{d_m}{m}\right).$$

Thus for $1 \leq j \leq n$, *given (8.22):*

$$\begin{aligned}
\sum\nolimits_{i=1}^{j}\left(a^{(n)}V_i + b^{(n)}\right) &=_d \sum\nolimits_{i=1}^{j}\left[\frac{a^{(n)}}{c_m}\sum\nolimits_{k=1}^{m}\left(V_{i_k} - \frac{d_m}{m}\right) + b^{(n)}\right] \\
&= \sum\nolimits_{k=1}^{m}\sum\nolimits_{i=1}^{j}\left[\frac{a^{(n)}}{c_m}V_{i_k} + \frac{1}{m}\left(b^{(n)} - a^{(n)}\frac{d_m}{c_m}\right)\right] \\
&= \sum\nolimits_{k=1}^{mj}\left[a^{(mn)}V_k + b^{(mn)}\right].
\end{aligned}$$

This last step requires Exercise 8.19, and this implies that $\{a^{(n)}V + b^{(n)}\}_{n=1}^{\infty}$ *is perfectly scalable.*

If V *is strictly stable then* $d_m = 0$ *by definition, and hence (8.23) follows from (8.22).*

For item 3, let $\{a^{(n)}V_i + b^{(n)}\}_{i=1}^{n}$ *and* m *be given, where* $\{V_i\}_{i=1}^{n}$ *are independent with the distribution of* V*, and where* V *has two moments with* $E[V] = 0$*. Given independent such variates* $\{\{V_{i_k}\}_{k=1}^{m}\}_{i=1}^{n}$*, it follows from* $b^{(mn)} = b^{(n)}/m$ *that for* $1 \leq j \leq n$*:*

$$\begin{aligned} E\left[\sum\nolimits_{i=1}^{j}\sum\nolimits_{k=1}^{m}\left(a^{(mn)}V_{i_k} + b^{(mn)}\right)\right] &= \sum\nolimits_{i=1}^{mj} b^{(mn)} \\ &= \sum\nolimits_{i=1}^{j} b^{(n)} \\ &= E\left[\sum\nolimits_{i=1}^{j}\left(a^{(n)}V_i + b^{(n)}\right)\right]. \end{aligned}$$

Equating second moments is equivalent to equating variances. Ignoring the additive constants, if $a^{(mn)} = a^{(n)}/\sqrt{m}$*:*

$$Var\left[a^{(mn)}\sum\nolimits_{i=1}^{j}\sum\nolimits_{k=1}^{m} V_{i_k}\right] = Var\left[a^{(n)}\sum\nolimits_{i=1}^{j} V_i\right].$$

■

Exercise 8.19 ($a^{(mn)}$, $b^{(mn)}$ in (8.22)) *Prove that from (8.22), we obtain the subdivision algorithm:*

$$a^{(nm)} \equiv \frac{a^{(n)}}{c_m}, \qquad b^{(n)} \equiv \frac{1}{m}\left(b^{(n)} - a^{(n)}\frac{d_m}{c_m}\right).$$

Hint: Exercise 7.36.

Example 8.20 (Targeting model moments 1 with Fixed V) *If* V *has two moments with* $E[V] = 0$ *as in item 3 of Proposition 8.18, it then follows from independence and (8.24) that with* $V^{(n)} \equiv a^{(n)}V + b^{(n)}$*:*

$$E\left[U_n^{(n)}\right] = b^{(1)}, \quad Var\left[U_n^{(n)}\right] = \left(a^{(1)}\right)^2 \sigma_V^2.$$

Thus given V*, one can* ***target the desired mean*** $\mu_{\tilde{U}_T}$ ***and variance*** $\sigma^2_{\tilde{U}_T}$ *of the terminal spatial financial variable* $U_n^{(n)} \equiv U_T$ *with* $\left(a^{(1)}, b^{(1)}\right)$ *defined by:*

$$a^{(1)} = \frac{\sigma_{\tilde{U}_T}}{\sigma_V}, \quad b^{(1)} = \mu_{\tilde{U}_T}. \tag{8.25}$$

In other words, with $V^{(n)} \equiv a^{(n)}V + b^{(n)}$*, the model* $\{V^{(n)}\}_{n=1}^{\infty}$ *is defined by:*

$$V^{(n)} = \frac{\sigma_{\tilde{U}_T}}{\sqrt{n}}\frac{V}{\sigma_V} + \frac{\mu_{\tilde{U}_T}}{n}. \tag{8.26}$$

As in Example 8.16, note that here, V *is normalized to have variance* 1*, recalling that it was defined to have mean* 0*.*

Exercise 8.21 (Targeting model moments 2 with Fixed V) *Prove that the same results of Example 8.20 follow from (8.22) for strictly stable* V *of case 2 of Proposition 8.18, if* V *has two moments with* $E[V] = 0$*. Generalize this result for stable* V*. Hint: Exercise 8.19.*

Example 8.22 (When V is binomial B or normal N) *An example of this approximately scalable result was seen in Example 8.8. For the multiplicative model with $\Delta t \equiv T/n$, the return variate $Z_j \equiv Z_j^{(n)}$ was seen to be:*

$$Z_j^{(n)} \equiv \mu\Delta t + \sigma\sqrt{\Delta t}B_j,$$

with $\{B_j\}_{j=1}^n$ independent and binomially distributed to equal ± 1 with probability $1/2$. Hence $E[B] = 0$, and the parameterization there can be expressed as $Z^{(n)} \equiv a^{(n)}B + b^{(n)}$:

$$a^{(n)} = \frac{\sigma\sqrt{T}}{\sqrt{n}}, \qquad b^{(n)} = \frac{\mu T}{n}. \tag{1}$$

This is consistent with (8.24) with $a^{(1)} = \sigma\sqrt{T}$ and $b^{(1)} = \mu T$, and with (8.26), since $\mu_{U_T} = \mu T$, $\sigma_{U_T} = \sigma\sqrt{T}$ and $\sigma_B = 1$ in this model.

Example 8.8 also defined:

$$Z_j \equiv \mu\Delta t + \sigma\sqrt{\Delta t}N_j,$$

with $\{N_j\}_{j=1}^n$ independent and standard normally distributed. Such N is strictly stable with $c_n = \sqrt{n}$ and $d_n = 0$ from Exercise 7.37. Then (1) *is again satisfied, and this is consistent with (8.23), as well as (8.25) as noted in Exercise 8.21.*

8.2.2 Scalable Additive Models

In this section we identify a variety of distributional results that create scalable additive models, and also address identifiability of the induced spatial random variables.

Assume that we are given a spatial asset price vector $X = (X_1, X_2, ..., X_n)$, and independent, identically distributed additive temporal variates $Y = (Y_1, Y_2, ..., Y_n)$ related as in (8.1), but suppressing X_0:

$$X_j = \sum\nolimits_{i=1}^{j} Y_i.$$

When needed for clarity in the scalability discussion below, we superscript variates as in the previous section:

$$X_j^{(n)} = \sum\nolimits_{i=1}^{j} Y_i^{(n)}.$$

Recall that the significance of n is in defining the size of the time-step $\Delta t \equiv T/n$, where $[0, T]$ is the modeling period. We continue with the convention that constant X_0 is often suppressed, and hence X_j equals the j-period change in this variate, with the actual value of the asset equal to $X_0 + X_j$ at step j, or equivalently, at time $j\Delta t$.

Exercise 8.23 (On independence of $\{X_i\}_{i=1}^n$) *While $\{Y_i\}_{i=1}^n$ are assumed to be independent, $\{X_i\}_{i=1}^n$ cannot be independent, with one exception.*

1. *Show that if the variance of Y_j exists, $\sigma_Y^2 > 0$, then the covariance of the variates X_j and X_k is given by:*
$$Cov(X_j, X_k) = \min(j,k)\sigma_Y^2 > 0. \tag{8.27}$$
Verify that this proves that X_j and X_k are not independent. Hint: Use proof by contradiction and Exercise 6.49.

2. *Without assuming the existence of σ_Y^2, use characteristic functions and (5.48) to show that if X_j and X_k are independent for $j \neq k$, then the distribution function of Y must be the delta function of (5.34) and so $Y = y_0$ with probability 1. This requires the uniqueness result of Proposition 5.29. Thus $\{Y_i\}_{i=1}^n$ cannot be independent otherwise.*

3. *If the variance of Y_j exists, $\sigma_Y^2 = 0$, show that Y is degenerate, meaning that $Y = y_0$ with probability 1. Thus X_j and X_k are independent for $j \neq k$ by item 2. Hint: Chebyshev's inequality of Proposition IV.4.34.*

The criterion of identifiability of the spatial variates is satisfied when, as discussed in Section 7.2, the distribution family of the Y-variates is related under addition to the distribution family of the X-variates. Recalling such examples, we summarize results below.

In practice, it is uncommon to abandon the assumption that the Y-variates are identically distributed, since this complicates the scalability discussion. However, for completeness we present the more general results.

Example 8.24 (Identifiability) *Examples of temporal variates that will produce identifiable spatial variates in the additive temporal model follow.*

1. *If $\{Y_i\}_{i=1}^n$ are independent* ***standard binomials with parameters*** *p and $\{n_i\}_{i=1}^n$, then each X_j is binomial with parameters p and $\sum_{i=1}^j n_i$. If all $n_i = 1$, then each X_j* ***is binomial with parameters*** *p and j.*

2. *If $\{Y_i\}_{i=1}^n$ are independent* ***negative binomials with parameters*** *p and $\{k_i\}_{i=1}^n$, then each X_j* ***is negative binomial with parameters*** *p and $\sum_{i=1}^j k_i$. If all $k_i = k$, then each X_j* ***is negative binomial with parameters*** *p and kj. If all $k_i = 1$, then Y_i has a* ***geometric distribution.***

3. *If $\{Y_i\}_{i=1}^n$ are independent* ***Poisson with parameters*** *$\{\lambda_i\}_{i=1}^n$, then each X_j* ***is Poisson with parameter*** *$\sum_{i=1}^j \lambda_i$, which simplifies to $j\lambda$ when all $\lambda_i = \lambda$.*

4. *If $\{Y_i\}_{i=1}^n$ are independent* ***gamma with parameters*** *λ and $\{\alpha_i\}_{i=1}^n$, then each X_j* ***is gamma with parameters*** *λ and $\sum_{i=1}^j \alpha_i$. If all $\alpha_i = \alpha$, then each X_j* ***is gamma with parameters*** *λ and $j\alpha$. If all $\alpha_i = 1$, then Y_i has an* ***exponential distribution.***

5. *If $\{Y_i\}_{i=1}^n$ are independent* ***Cauchy variates with parameters*** *$\{\gamma_i\}_{i=1}^n$ and $\{x_i\}_{i=1}^n$, then each X_j* ***is Cauchy with parameters*** *$\sum_{i=1}^j \gamma_i$* ***and*** *$\sum_{i=1}^j x_i$. This simplifies as above when $\{Y_i\}_{i=1}^n$ are identically distributed.*

6. *If $\{Y_i\}_{i=1}^n$ are independent* ***normal variates with parameters*** *$\{\mu_i\}_{i=1}^n$ and $\{\sigma_i^2\}_{i=1}^n$, then each X_j* ***is normal with parameters*** *$\sum_{i=1}^j \mu_i$* ***and*** *$\sum_{i=1}^j \sigma_i^2$. This simplifies as above when $\{Y_i\}_{i=1}^n$ are identically distributed.*

7. *If $\{Y_i\}_{i=1}^n$ are independent* ***compound Poisson variates with parameters*** *$\{\lambda_i\}_{i=1}^n$* ***and driving variate in (7.26) denoted*** *W (rather than X to avoid notational confusion), then each X_j* ***is compound Poisson with parameters*** *$\sum_{i=1}^j \lambda_i$* ***and driving variate*** *W. Again, this simplifies as above when $\{Y_i\}_{i=1}^n$ are identically distributed.*

For scalability, we now return to the assumption of identically distributed $\{Y_i\}_{i=1}^n$. To assume otherwise is to have a model for which n, and the associated time-step $\Delta t \equiv T/n$, are "special" in the sense that the various $[(i-1)\Delta t, i\Delta t]$-subintervals of $[0,T]$ require different distributional assumptions. While there may well be a rationale for doing so in special cases, it would now introduce significant complexity in terms of scalability.

Specifically, when each Y_i is further refined to variates defined with respect to $\Delta t \equiv T/nm$, does one then return to the assumption of i.i.d. variates $\{Y_{i_k}^{(m)}\}_{k=1}^m$ with $\sum_{k=1}^m Y_{i_k}^{(m)} = Y_i$ in distribution or moments, or again assume that these variates are independent but not identically distributed?

Such generalizations do not seem to have apparent applications and we thus return to i.i.d. temporal variates.

Example 8.25 (Scalability) *Reviewing the above distributions in order, but now restricting $\{Y_i\}_{i=1}^n$ to be identically distributed, we have the following.*

1. *If X_j is* **standard binomial with parameters** *p and j, then the moment generating function is $M_{X_j}(t) = (1 + p(e^t - 1))^j$ and the distribution function of X_j is not infinitely divisible by item 5 of Proposition 7.42. Hence the binomial temporal model is* **not perfectly scalable** *by Corollary 8.14. However:*

   (a) *Given m, we can approximately represent each $X_j = \sum_{i=1}^{mj} Y_i^{(mn)}$ as an independent sum since:*

   $$\begin{aligned}\left[M_{X_j}(t)\right]^{1/mj} &= (1 + p(e^t - 1))^{1/m} \\ &= 1 + \frac{p}{m}(e^t - 1) + O\left(m^{-1}\right).\end{aligned}$$

   Thus the binomial model appears to be nearly approximately scalable with $Y_i^{(mn)}$ binomial with parameters $\frac{p}{m}$ and 1.

   With the variate $Y_i^{(mn)}$ defined this way, we achieve **approximate scalability to order** *$k_0 = 1$* **but not** *$k_0 = 2$:*

   $$E\left[\sum\nolimits_{i=1}^{mj} Y_i^{(mn)}\right] = jp = E\left[\sum\nolimits_{i=1}^{j} Y_i\right],$$

   $$Var\left[\sum\nolimits_{i=1}^{mj} Y_i^{(mn)}\right] = jp\left(1 - \frac{p}{m}\right) \neq jp(1-p) = Var\left[\sum\nolimits_{i=1}^{j} Y_i\right],$$

   and the discrepancy in the variance estimates worsens as $m \to \infty$.

   However, this model is **asymptotically scalable to order** *$k_0 = 2$, noting that $Var\left[\sum_{i=1}^{mj} Y_i^{(mn)}\right] \to jp = E\left[\sum_{i=1}^{mj} Y_i^{(mn)}\right]$. Thus in the limit, with mean equal to variance, this looks like a Poisson model for asset prices. Recalling the "weak law of small numbers" in Proposition IV.6.8, it is indeed the case that the limiting spatial distributions are Poisson.*

   (b) *Using an affine transformation of a fixed standard binomial for $\{Y_i\}_{i=1}^n$, a better result is produced. Analogous to the classical binomial model for equity prices noted in Example 8.8, define:*

   $$X_j = \sum\nolimits_{i=1}^{j} \left(\mu\Delta t + \sigma\sqrt{\Delta t} b_i\right),$$

   with $\{b_i\}_{i=1}^n$ independent and binomially distributed to equal 1 or -1 with probability $p = 1/2$. As above, $\Delta t = T/n$ for some fixed horizon time T, so X_j denotes the change in price up to time $j\Delta t$. Rewriting with superscripts to reflect this subdivision:

   $$X_j^{(n)} = \sum\nolimits_{i=1}^{j} \left(\mu T/n + \sigma\sqrt{T/n} b_i\right).$$

   Thus $Y_i^{(n)} = c^{(n)} b_i + d^{(n)}$ with $c^{(n)} = \sigma\sqrt{T/n}$ and $d^{(n)} = \mu T/n$.

   Given m, we now define binomial $X_{mj}^{(mn)}$ analogously by: $X_{mj}^{(mn)} \equiv \sum_{i=1}^{mj} (\mu T/mn + \sigma\sqrt{T/mn} b_i)$. In effect, we are simply redefining $\Delta t' = T/mn$, so now $Y_i^{(mn)} = c^{(mn)} b_i + d^{(mn)}$ with $c^{(mn)} = \sigma\sqrt{T/mn}$ and $d^{(mn)} = \mu T/mn$. With this parametrization we have **approximate scalability to order** *$k_0 = 2$, meaning* *that with $\Delta t = T/n$:*

$$\begin{aligned} E\left[X_j^{(n)}\right] &= \mu j \Delta t = E\left[X_{mj}^{(mn)}\right], \\ Var\left[X_j^{(n)}\right] &= \sigma^2 j \Delta t = Var\left[X_{mj}^{(mn)}\right]. \end{aligned}$$

That $X_j^{(n)}$ and $X_{mj}^{(mn)}$ are actually different binomial random variables can be appreciated by evaluating the respective moment generating functions.
This model can be generalized for $p \neq 1/2$, but this requires redefining $c^{(n)}$ and $d^{(n)}$ above in order to maintain the original calibrated values of $E\left[X_j^{(n)}\right]$ and $Var\left[X_j^{(n)}\right]$. See Chapter 8 of ***Reitano (2010)*** *for more on this calibration.*

2. *If X_j is* ***negative binomial with parameters*** *p and kj, then the distribution function of X_j is infinitely divisible and hence this model is* ***perfectly scalable*** *with $Y_i^{(mn)}$ independent negative binomial with parameters p and k/m.*

3. *If X_j is* ***Poisson with parameter*** *$j\lambda$, then the distribution function of X_j is infinitely divisible and thus* ***perfectly scalable*** *with $Y_i^{(mn)}$ independent Poisson with parameter λ/m.*

4. *If X_j is* ***gamma with parameters*** *λ and $j\alpha$, then the distribution function of X_j is infinitely divisible and thus* ***perfectly scalable*** *with $Y_i^{(mn)}$ independent gamma with parameters λ and α/m.*

5. *If X_j is* ***Cauchy with parameters*** *jx_0 and $j\gamma$, then the distribution function of X_j is infinitely divisible and thus* ***perfectly scalable*** *with $Y_i^{(mn)}$ independent Cauchy with parameters x_0/m and γ/m.*

6. *If X_j is* ***normal with parameters*** *$j\mu$ and $j\sigma^2$, then the distribution function of X_j is infinitely divisible and thus* ***perfectly scalable*** *with $Y_i^{(mn)}$ independent normal with parameters μ/m and σ^2/m.*

7. *If X_j is* ***compound Poisson with parameters*** *$j\lambda$* ***and driving variate*** *W, then the distribution function of X_j is infinitely divisible and thus* ***perfectly scalable*** *with $Y_i^{(mn)}$ independent compound Poisson with parameters λ/m and driving variate W.*

8.2.3 Scalable Multiplicative Models

In this section we investigate scalability and identifiability of multiplicative models. As expected, the prior sections' results are applicable by definition, though we investigate in greater detail the implications of approximate scalability in this context.

Assume we are given temporal returns $\{Z_i\}_{i=1}^n$ and spatial $\{X_i\}_{i=1}^n$, where these variates are related as in (8.7), but suppressing X_0:

$$X_j = \prod\nolimits_{i=1}^{j} \exp Z_i = \exp\left[W_j\right],$$

with:

$$W_j \equiv \sum\nolimits_{i=1}^{j} Z_i.$$

As in the additive temporal model above, when needed for notational clarity in the scalability discussion, we superscript variates as:

$$X_j^{(n)} = \exp\left[\sum\nolimits_{i=1}^{j} Z_i^{(n)}\right] = \exp\left[W_j^{(n)}\right].$$

As above, the significance of n is to define the size of the time step, $\Delta t \equiv T/n$, where $[0, T]$ is the modeling period.

Before turning to our investigation, we make some general observations:

1. **Identifiability:** Assume that the distribution family of the Zs is related under addition to the distribution family of Ws. Then if the spatial variate $W_j \equiv \sum_{i=1}^{j} Z_i$ has distribution function F_{w_j}, then $X_j = \exp\left[\sum_{i=1}^{j} Z_i\right]$ has a log-F_{w_j} distribution as defined in (7.34). Thus the models in Example 8.24 apply equally well in this context, substituting Z for Y and W for X.

2. **Perfect Scalability:** If Z_j is infinitely divisible, then so too is W_j for all j, and hence this multiplicative model is perfectly scalable by Proposition 8.13. That is, for any m there are independent, identically distributed variates $\{Z_i^{(mn)}\}_{i=1}^{mn}$ so that with $W_{mj}^{(nm)} \equiv \sum_{i=1}^{mj} Z_i^{(mn)}$:

$$W_j^{(n)} =_d W_{mj}^{(nm)}.$$

   It then follows that for all j:

$$X_j^{(n)} =_d X_{mj}^{(mn)}, \tag{1}$$

   where:

$$X_{mj}^{(mn)} \equiv \exp W_{mj}^{(nm)}.$$

   Thus the perfectly scalable models of Example 8.25 work within the multiplicative framework, and provide distributional results on the spatial variates $\{X_j^{(n)}\}_{j=1}^{n}$ by (1).

3. **Approximate Scalability:** If the W-distributions are only approximately scalable, meaning that the first k_0 moments of $\sum_{i=1}^{j} Z_i^{(n)}$ match those of $\sum_{i=1}^{mj} Z_i^{(mn)}$, then by definition, the resulting multiplicative temporal models are approximately scalable. Thus approximately scalable additive temporal models give rise to approximately scalable multiplicative temporal models, **by definition.** However, this observation does not address properties of the spatial variates.

We next investigate results related to the existence of moments of the spatial variates $\{X_j^{(n)}\}_{j=1}^{n}$. The first result states that in perfectly scalable multiplicative models, if moments exist, they scale appropriately.

Proposition 8.26 (Perfect scalability and moments of $\{X_j^{(n)}\}_{j=1}^{n}$) *If a multiplicative model is perfectly scalable, then $\{X_j^{(n)}\}_{j=1}^{n}$ have moments up to order $k_0 \geq 1$ if and only if $\{X_{mj}^{(mn)}\}_{j=1}^{n}$ have moments up to order k_0 for all m. Further, these moments are consistent, in that for $1 \leq j \leq n$, $k \leq k_0$ and all m:*

$$E\left[\left(X_j^{(n)}\right)^k\right] = E\left[\left(X_{mj}^{(mn)}\right)^k\right].$$

Proof. *To simplify notation, let* $Y = X_j^{(n)}$ *and* $Z = X_{mj}^{(mn)}$ *be defined on* $(\mathcal{S}, \mathcal{E}, \lambda)$, *as is justified by Proposition 1.20. Then* $F_Y(y) = F_Z(y)$ *for all* y *by* (1) *above, and thus* $\lambda_Y = \lambda_Z$ *for the induced Borel measures in (1.21).*

For any k *for which the integrals exist, Proposition V.4.18 obtains:*

$$\begin{aligned} E\left[Y^k\right] &\equiv \int_{\mathcal{S}} Y^k d\lambda = \int_{\mathbb{R}} y^k d\lambda_Y, \\ E\left[Z^k\right] &\equiv \int_{\mathcal{S}} Z^k d\lambda = \int_{\mathbb{R}} z^k d\lambda_Z. \end{aligned}$$

Since $\lambda_Y = \lambda_Z$, $E\left[Y^k\right]$ *exists if and only if* $E\left[Z^k\right]$ *exists, and when they exist, these moments agree.* ■

The next result relates existence of moments of $X \equiv \exp W$ with existence of the moment generating function for W.

Proposition 8.27 (Moments of X_j, multiplicative model) *Let* $X \equiv \exp W$ *where* $W \equiv \sum_{i=1}^{j} Z_i$ *and* $\{Z_i\}_{i=1}^{j}$ *are independent and identically distributed. If the moment generating function* $M_Z(t)$ *exists for* $0 \leq |t| \leq k_0$ *with* $k_0 \geq 1$ *an integer, then* X *has moments up to order* k_0. *Further, for* $k \leq k_0$:

$$E\left[X^k\right] = M_W(k), \tag{8.28}$$

where $M_W(k) = M_Z^j(t)$.

In particular, if $M_Z(t)$ *exists for all* t, *then (8.28) is satisfied for all* k.

Proof. *Assume, as is justified by Proposition 1.20, that* W *and* X *are defined on the probability space* $(\mathcal{S}, \mathcal{E}, \lambda)$:

$$W,\ X : (\mathcal{S}, \mathcal{E}, \lambda) \rightarrow (\mathbb{R}, \mathcal{B}(\mathbb{R})).$$

Since $M_Z(t)$ *exists for* $0 \leq |t| \leq k_0$, *it follows from Proposition 4.20 that* $M_W(t)$ *exists for* $0 \leq |t| \leq k_0$ *and* $M_W(k) = M_Z^j(t)$.

Existence of $M_W(k)$ *and Proposition V.4.18 obtain:*

$$M_W(k) \equiv \int_{\mathcal{S}} e^{Wk} d\lambda = \int_{\mathbb{R}} e^{yk} d\lambda_W, \tag{1}$$

where λ_W *is the probability measure on* $(\mathbb{R}, \mathcal{B}(\mathbb{R}))$ *induced by* W *of (1.21).*

If $E\left[X^k\right]$ *exists for given* k, *then by the same change of variable:*

$$E\left[X^k\right] \equiv \int_{\mathcal{S}} X^k d\lambda = \int_{\mathbb{R}} x^k d\lambda_X, \tag{2}$$

where λ_X *is the probability measure on* $(\mathbb{R}^+, \mathcal{B}(\mathbb{R}^+))$ *induced by* X.

Define a transformation:

$$T : \left(\mathbb{R}^+, \mathcal{B}\left(\mathbb{R}^+\right), \lambda_X\right) \rightarrow \left(\mathbb{R}, \mathcal{B}(\mathbb{R}), (\lambda_X)_T\right),$$

by:

$$Tx = \ln x.$$

Since T *is continuous on* $\mathbb{R}^+$, T^{-1} *maps open sets to open sets by Proposition I.3.12, so* $T^{-1}(\mathcal{B}(\mathbb{R})) \subset \mathcal{B}(\mathbb{R}^+)$. *Thus* T *is measurable and* $(\lambda_X)_T$ *is well-defined as in (1.20) of Definition 1.28:*

$$(\lambda_X)_T(A) \equiv \lambda_X\left(T^{-1}(A)\right),\ A \in \mathcal{B}(\mathbb{R}). \tag{3}$$

Further, by (3) *and (1.20):*

$$\lambda_X\left(T^{-1}(A)\right) \equiv \lambda\left(X^{-1}T^{-1}(A)\right) = \lambda\left((TX)^{-1}(A)\right),$$

and since $TX = W$, *it follows that* $(\lambda_X)_T = \lambda_W$.

Another application of Proposition V.4.18 then obtains:

$$\int_{\mathbb{R}} x^k d\lambda_X = \int_{\mathbb{R}} e^{yk} d\left(\lambda_X\right)_T = \int_{\mathbb{R}} e^{yk} d\lambda_W.$$

Hence from (1) *it follows that* $E\left[X^k\right]$ *exists for* $k \leq k_0$ *and* $E\left[X^k\right] = M_W(k)$. ■

The prior result provides an insight into when moments of $X_j^{(n)}$ and $X_{mj}^{(mn)}$ can agree in a general scalable multiplicative model. Recall that by Definition 8.12, to be scalable means that for any n, there exists distributional assumptions on independent, identically distributed $\{Z_i^{(n)}\}_{i=1}^n$, defined relative to time-step $\Delta t \equiv T/n$, which define associated variates $\{W_i^{(n)}\}_{i=1}^n$, where $W_j^{(n)} \equiv \sum_{i=1}^j Z_i^{(n)}$.

Proposition 8.28 (Equality of moments of $\{X_j^{(n)}\}_{j=1}^n$ and $\{X_{mj}^{(mn)}\}_{j=1}^n$) *Given a scalable multiplicative model, assume that there exists n and $k_0 \geq 1$ so that $\{X_j^{(n)}\}_{j=1}^n$ have moments up to order k_0 where $X_j^{(n)} \equiv \exp W_j^{(n)}$. Assume further that for given m, that $\{X_{mj}^{(mn)}\}_{j=1}^n$ have moments up to order k_0, where $X_{mj}^{(mn)} \equiv \exp W_{mj}^{(mn)}$, and that $M_{Z^{(n)}}(t)$ and $M_{Z^{(mn)}}(t)$ exist for $0 \leq |t| \leq k_0$.*

Then for $k \leq k_0$:

$$E\left[\left(X_j^{(n)}\right)^k\right] = E\left[\left(X_{mj}^{(mn)}\right)^k\right], \tag{8.29}$$

if and only if:

$$M_{W_j^{(n)}}(k) = M_{W_{mj}^{(mn)}}(k).$$

Proof. *This follows directly from Proposition 8.27, that for given j, m, and $k \leq k_0$, $E\left[\left(X_j^{(n)}\right)^k\right]$ and $E\left[\left(X_{mj}^{(mn)}\right)^k\right]$ exist with:*

$$\begin{aligned} E\left[\left(X_j^{(n)}\right)^k\right] &= M_{W_j^{(n)}}(k), \\ E\left[\left(X_{mj}^{(mn)}\right)^k\right] &= M_{W_{mj}^{(mn)}}(k). \end{aligned}$$

■

A corollary to Proposition 8.28 is that an approximately scalable multiplicative model will not in general obtain equality of moments of the spatial variates as in (8.29), for any k.

Proposition 8.29 $\left(\textbf{Approximate scalability and } E\left[\left(X_j^{(n)}\right)^k\right], E\left[\left(X_{mj}^{(mn)}\right)^k\right]\right)$ *If a multiplicative model is approximately scalable to order k_0, and $M_{Z^{(n)}}(t)$ and $M_{Z^{(mn)}}(t)$ exist for $0 \leq |t| \leq k_0$, then $E\left[\left(X_j^{(n)}\right)^k\right]$ and $E\left[\left(X_j^{(mn)}\right)^k\right]$ exist for any $k \leq k_0$, but in general:*

$$E\left[\left(X_j^{(n)}\right)^k\right] \neq E\left[\left(X_{mj}^{(mn)}\right)^k\right]. \tag{8.30}$$

However, if $M_{Z^{(n)}}(t)$ and $M_{Z^{(mn)}}(t)$ exist for all t, and thus $k_0 = \infty$, then (8.29) is satisfied for all j and k.

Proof. *Fixing j and m, it follows from approximate scalability that for $k \leq k_0$:*

$$E\left[\left(W_j^{(n)}\right)^k\right] = E\left[\left(W_{mj}^{(mn)}\right)^k\right] \equiv \mu_k'.$$

Applying Proposition IV.4.27 on the power series for the moment generating function, for $0 \leq |t| \leq k_0$:

$$\begin{aligned} M_{W_j^{(n)}}(t) &= \sum\nolimits_{i=0}^{k_0} t^i \mu_i'/i! + \sum\nolimits_{i=k_0+1}^{\infty} t^i \mu_i'\left(W_j^{(n)}\right)/i!, \\ M_{W_{mj}^{(mn)}}(t) &= \sum\nolimits_{i=0}^{k_0} t^i \mu_i'/i! + \sum\nolimits_{i=k_0+1}^{\infty} t^i \mu_i'(W_{mj}^{(mn)})/i!, \end{aligned}$$

where $\mu_i'\left(W_j^{(n)}\right)$ and $\mu_i'(W_{mj}^{(mn)})$ are the higher moments for these variates.

By Proposition 8.27, both $E\left[\left(X_j^{(n)}\right)^k\right]$ and $E\left[\left(X_{mj}^{(mn)}\right)^k\right]$ exist for $k \leq k_0$, and by (8.28):

$$E\left[\left(X_j^{(n)}\right)^k\right] - E\left[\left(X_{mj}^{(mn)}\right)^k\right] = \sum\nolimits_{i=k_0+1}^{\infty} k^i \left[\mu_i'\left(W_j^{(n)}\right) - \mu_i'(W_{mj}^{(mn)})\right] \Big/ i!.$$

Thus (8.30) is satisfied, unless $k_0 = \infty$. ■

Hence, if a multiplicative model is approximately scalable to order $k_0 < \infty$, with temporal variates that have moment generating functions, even lower order moments of $X_j^{(n)}$ and $X_{mj}^{(mn)}$ will in general not agree. However, the next example illustrates that such a model can still be useful.

Example 8.30 (Example 8.8, Binomial $\{Z_i\}_{i=1}^n$) *Recall Example 8.8 of log-binomial $X_j^{(n)} \equiv \exp \sum_{i=1}^j Z_j^{(n)}$ for $1 \leq j \leq n$ with binomial $Z_j^{(n)} \equiv a^{(n)} b_i + b^{(n)}$. The parameters $a^{(n)}$ and $b^{(n)}$ of this example are defined relative to a fixed horizon time T and initial time step $\Delta t = T/n$, so the price $X_j^{(n)}$ represents the asset price at time $j\Delta t = jT/n$ assuming $X_0 = 1$. Recalling that b_i is binomially distributed to equal 1 or -1 with probability $p = 1/2$, while $a^{(n)} = \sigma\sqrt{T/n}$ and $b^{(n)} = \mu T/n$, obtains:*

$$X_j^{(n)} = \exp\left[\sum\nolimits_{i=1}^{j} \left(b_i \sigma\sqrt{T/n} + \mu T/n\right)\right].$$

This model is approximately scalable to order $k_0 = 2$. Specifically, with $a^{(mn)} = \sigma\sqrt{T/mn}$ and $b^{(mn)} = \mu T/mn$, the distribution functions of $W_j^{(n)} = \sum_{i=1}^j \left(a^{(n)} b_i + b^{(n)}\right)$ and $W_{mj}^{(mn)} \equiv \sum_{i=1}^{mj} \left(a^{(mn)} b_i + b^{(mn)}\right)$ agree to two moments:

$$\begin{aligned} E\left[W_j^{(n)}\right] &= E\left[W_{mj}^{(mn)}\right] = \mu jT/n, \\ Var\left[W_j^{(n)}\right] &= Var\left[W_{mj}^{(mn)}\right] = \sigma^2 jT/n. \end{aligned} \tag{1}$$

By Exercise 8.31:

$$E\left[\left(W_j^{(n)}\right)^p\right] \neq E\left[\left(W_{mj}^{(mn)}\right)^p\right], \; p > 2,$$

and thus this model is approximately scalable exactly to order $k_0 = 2$.

By Proposition 8.27, we can compare the moments of $X_j^{(n)} \equiv \exp W_j^{(n)}$ *and* $X_{mj}^{(mn)} \equiv \exp W_j^{(mn)}$ *by evaluating the moment generating functions of* $W_j^{(n)}$ *and* $W_{mj}^{(mn)}$*. As a sum of independent random variables:*

$$\begin{aligned} M_{W_j^{(n)}}(k) &= \frac{1}{2^j}\left(\exp\left[\left(\mu T/n + \sigma\sqrt{T/n}\right)k\right] + \exp\left[\left(\mu T/n - \sigma\sqrt{T/n}\right)k\right]\right)^j \\ &= \frac{1}{2^j}\exp\left[\mu kjT/n\right]\left(\exp\left[\sigma k\sqrt{T/n}\right] + \exp\left[-\sigma k\sqrt{T/n}\right]\right)^j, \end{aligned}$$

and similarly:

$$M_{W_{mj}^{(mn)}}(k) = \frac{1}{2^{mj}}\exp\left[\mu kjT/n\right]\left(\exp\left[\sigma k\sqrt{T/mn}\right] + \exp\left[-\sigma k\sqrt{T/mn}\right]\right)^{mj}.$$

To have $E\left[\left(X_j^{(n)}\right)^k\right] = E\left[\left(X_{mj}^{(mn)}\right)^k\right]$ *for some* k *requires that* $M_{W_j}(k) = M_{W_{mj}^{(mn)}}(k)$ *by Proposition 8.28. This would require that with* $c \equiv \exp\left[\sigma k\sqrt{T/n}\right] > 1$ *and* $m > 1$*:*

$$\frac{c + c^{-1}}{2} = \left(\frac{c^{1/\sqrt{m}} + c^{-1/\sqrt{m}}}{2}\right)^m.$$

However, this is not possible since using some calculus, the function on the right is increasing for real $m \geq 1$*, and so equality cannot occur for* $c > 1$*. Hence* $M_{W_j}(k) \neq M_{W_{mj}^{(mn)}}(k)$ *for all* $k > 0$*, and so by the above result,* $E\left[\left(X_j^{(n)}\right)^k\right] \neq E\left[\left(X_{mj}^{(mn)}\right)^k\right]$ *for all* k*, illustrating the general result above.*

However, it is an exercise to show that for any $t \in \mathbb{R}$*,* $M_{W_{mj}^{(mn)}}(t)$ *has a well-defined limit as* $m \to \infty$*, and specifically:*

$$M_{W_{mj}^{(mn)}}(t) \to \exp\left[\mu\left(jT/n\right)t + \frac{1}{2}\sigma^2\left(jT/n\right)t^2\right]. \tag{2}$$

This implies that for m *large, that* $M_{W_{mj}^{(mn)}}(t)$ *is nearly independent of* m*, and hence, so too is* $E\left[\left(X_{mj}^{(mn)}\right)^k\right] = M_{W_{mj}^{(mn)}}(k)$*. See Remark 8.32.*

Exercise 8.31 $\left(E\left[\left(W_j^{(n)}\right)^p\right] \neq E\left[\left(W_{mj}^{(mn)}\right)^p\right], p > 2\right)$ *Using the notation of Example 8.30, prove that for odd* p*:*

$$E\left[\left(W_j^{(n)} - E\left[W_j^{(n)}\right]\right)^p\right] = E\left[\left(W_{mj}^{(mn)} - E\left[W_{mj}^{(mn)}\right]\right)^p\right],$$

but for even $p > 2$*:*

$$E\left[\left(W_j^{(n)} - E\left[W_j^{(n)}\right]\right)^p\right] \neq E\left[\left(W_{mj}^{(mn)} - E\left[W_{mj}^{(mn)}\right]\right)^p\right],$$

From this and Proposition IV.4.24, derive that:

$$E\left[\left(W_j^{(n)}\right)^p\right] \neq E\left[\left(W_{mj}^{(mn)}\right)^p\right], \ p > 2.$$

Remark 8.32 (On $\lim M_{W_{mj}^{(mn)}}(t)$**)** *As an introduction to the topic of the next section, note that* (2) *of Example 8.30 implies that for all* j, n *with* $1 \leq j \leq n$, *the limiting distribution of* $W_{mj}^{(mn)}$ *as* $m \to \infty$ *is normal with mean* $\mu jT/n$ *and variance* $\sigma^2 jT/n$. *This follows because by IV.(4.78), the function in the limit is the moment generating function of this variate, while Proposition 5.53 assures that the normal variate is uniquely determined by this moment generating function.*

It thus follows by (1) of Example 8.30 that for $t = k$ *for any* k, *the expression on the right in* (2) *is the value of the limit of* $E\left[\left(X_{mj}^{(mn)}\right)^k\right]$ *as* $m \to \infty$:

$$E\left[\left(X_{mj}^{(mn)}\right)^k\right] \to \exp\left[\mu\left(jT/n\right)k + \frac{1}{2}\sigma^2\left(jT/n\right)k^2\right].$$

The lognormal distribution with parameters $\mu jT/n$ *and* $\sigma^2 jT/n$ *also has these same moments by IV.(4.80). However, Example IV.4.58 illustrates that the lognormal distribution is not uniquely determined by these moments. Thus based just on* (2), *we can only say that as* $m \to \infty$, *the limiting distribution of the price* X_t *for* $t = jT/n$ *is consistent with that of the lognormal distribution with these parameters.*

But we can say more by noting that:

$$\Pr\left[X_{mj}^{(mn)} \leq x\right] = \Pr\left[W_{mj}^{(mn)} \leq \ln x\right]. \tag{1}$$

Thus by the central limit theorem of Proposition IV.6.13, as $m \to \infty$:

$$\Pr\left[X_{mj}^{(mn)} \leq x\right] \to \Phi\left(\frac{\ln x - \mu jT/n}{\sqrt{\sigma^2 jT/n}}\right),$$

where Φ *is the distribution function of the standard normal in (6.2).*

This then confirms that the limiting distribution of the price X_t *for* $t = j\Delta t$ *is indeed lognormal with parameters* $\mu jT/n$ *and* $\sigma^2 jT/n$.

Exercise 8.33 *Provide details for the application of the central limit theorem in Remark 8.32. Hint: Express:*

$$\frac{W_{mj}^{(mn)} - \mu jT/n}{\sqrt{\sigma^2 jT/n}}$$

as a random sum, and show that this sum is equivalent to:

$$\sum\nolimits_{k=1}^{mj} b_k/\sqrt{mj}.$$

Justify the application of Proposition IV.6.13, and then apply to (1) *of Remark 8.32.*

8.3 Limiting Distributions of Scalable Models

In this section, we discuss limiting distributions of scalable asset models as $m \to \infty$. As above and as typical in applications, we assume a fixed future time horizon denoted T, time-steps initially defined by $\Delta t = T/n$ for some $n \geq 1$, and refinements of these time-steps obtained by dividing Δt by m. Thus the distributional properties of the model as $m \to \infty$ can be equivalently framed in terms of the distributional properties as $\Delta t' \equiv T/mn \to 0$.

We can dispense with the perfectly scalable class of models quickly because by definition, these models are independent of m, and hence there is no limit to consider. Specifically, by definition $X_{mj}^{(mn)} =_d X_j^{(n)}$ for all m in the additive model. In the multiplicative model, $W_{mj}^{(mn)} =_d W_j^{(n)}$ for all m where $W_j^{(n)} = \sum_{i=1}^{j} Z_i^{(n)}$, and thus since $X = \exp W$, it follows again that $X_{mj}^{(mn)} =_d X_j^{(n)}$ for all m. Recalling that $X_j^{(n)}$ denotes the asset price at time $t = j\Delta t$, this implies that in a perfectly scalable model the distribution of X_t is independent of partitioning of the interval $[0, T]$ for all $t = jT/n$, for all j, n.

The investigation in this section will derive the following result. Under the hypothesis that the model is approximately or asymptotically scalable to order $k_0 \geq 2$, and subject to a regularity condition on the temporal variates, there are only two possible limiting distributions for scalable asset models:

1. Additive temporal models $\Rightarrow$ Normal distribution
2. Multiplicative temporal models $\Rightarrow$ Lognormal distribution

These results are independent of the subinterval distributions of the $Y_i^{(n)}$ and $Z_i^{(n)}$ variates, and not surprisingly reflect an application of the central limit theorems of Section 7.1.

The needed regularity condition is not assured by approximate or asymptotic scalability to order $k_0 \geq 2$, as examples illustrate, but this assumption is always satisfied for the models of Proposition 8.18.

Thus in order to achieve other spatial distributions, a more general approach is needed for the specification of the associated temporal distributions. Specifically, one must either abandon independence of these variates, or abandon the assumption that these variates are identically distributed, or both. This then introduces the question of how such distributions are specified, and what are their properties, and this general investigation is the topic of the next three books.

8.3.1 Scalable Additive Models

For approximately and asymptotically scalable models, the question of limiting distributions as $m \to \infty$, or equivalently, as $\Delta t' \equiv T/mn \to 0$ is of apparent interest. We first note, however, that limiting distributions need not exist for such models, nor be useful.

Example 8.34 (Another binomial model) *Modifying the approximately scalable binomial model in Example 8.25 somewhat, we begin with* $X_j^{(n)} = \sum_{i=1}^{j} \left(\mu\Delta t + \sigma\sqrt{\Delta t} b_i\right)$, *with* $\sigma > 0$ *and* $\{b_i\}_{i=1}^{n}$ *independent and binomially distributed to equal* 1 *or* -1 *with probability* $p = 1/2$. *As always,* $\Delta t = T/n$ *for some fixed horizon time* T, *and hence* $X_j^{(n)}$ *is the value of the change in the spatial variate to time* $j\Delta t = jT/n$. *Given* $m \geq 1$, *we now approximate* $X_j^{(n)}$ *with binomial* $X_{mj}^{(mn)}$ *defined by:*

$$X_{mj}^{(mn)} \equiv \sum_{i=1}^{mj} \left(\mu T/mn + \sigma^m \sqrt{T/mn} b_i\right).$$

With this parametrization, the model is approximately scalable to order $k_0 = 1$, *since:*

$$E\left[X_j^{(n)}\right] = \mu jT/n = E\left[X_{mj}^{(mn)}\right].$$

However, since $Var\left[X_j^{(n)}\right] = \sigma^2 jT/n$, *and:*

$$Var\left[X_{mj}^{(mn)}\right] = \sigma^{2m} jT/n,$$

approximate scalability to order 2 *occurs only if* $\sigma = 1$.

Further:

1. *If* $0 < \sigma < 1$, *then* $Var\left[X_{mj}^{(mn)}\right] \to 0$ *as* $m \to \infty$, *and hence by Chebyshev's inequality (Proposition IV.4.34),* $X_{mj}^{(mn)} \to_d \mu jT/n$. *In other words,* $X_{mj}^{(mn)}$ *converges in distribution to the degenerate distribution with value* $\mu jT/n = E\left[X_j^{(n)}\right]$.

2. *If* $\sigma > 1$, *then* $Var\left[X_{mj}^{(mn)}\right] \to \infty$ *as* $m \to \infty$, *and the same is true for all even central moments:*
$$E\left[\left(X_{mj}^{(mn)} - \mu jT/n\right)^{2k}\right] \to \infty,$$
while all odd central moments are 0 *by symmetry.*

Alternatively for $\sigma > 0$, *recall Lyapunov's condition in (7.15). If* $\delta > 0$, *then as* $m \to \infty$:
$$\frac{1}{(\sigma^{2m} jT/n)^{1+\delta/2}} \sum\nolimits_{i=1}^{mj} E\left[\left|\sigma^m \sqrt{T/mn} b_i\right|^{2+\delta}\right] = \frac{1}{(mj)^{\delta/2}} \to 0.$$
So by Lyapunov's central limit theorem of Proposition 7.20, as $m \to \infty$:
$$\frac{X_{mj}^{(mn)} - \mu jT/n}{\sqrt{\sigma^{2m} jT/n}} \Rightarrow Z,$$
the standard normal variate Z.

Thus for any $a > 0$ *and* m *large:*
$$\Pr\left[\left|X_{mj}^{(mn)} - \mu jT/n\right| < a\right] \approx \Phi\left(a \Big/ \sqrt{\sigma^{2m} jT/n}\right) - \Phi\left(-a \Big/ \sqrt{\sigma^{2m} jT/n}\right),$$
and this probability converges to 0 *for* $\sigma > 1$, *and to* 1 *for* $0 < \sigma < 1$.

To set the stage for the following result, assume that we are given time horizon T, additive temporal $Y^{(n)} = (Y_1^{(n)}, Y_2^{(n)}, ..., Y_n^{(n)})$ and spatial $X^{(n)} = (X_1^{(n)}, X_2^{(n)}, ..., X_n^{(n)})$ related by:
$$X_j^{(n)} = X_0 + \sum\nolimits_{i=1}^{j} Y_i^{(n)}.$$
As above, X_0 denotes the initial asset value, and $\{Y_i^{(n)}\}_{i=1}^n$ are independent and identically distributed on some space $(\mathcal{S}, \mathcal{E}, \lambda)$, and parametrized for time-step $\Delta t \equiv T/n$ for some $n \geq 1$. Thus $X_j^{(n)}$ is the asset price variate for time $t = j\Delta t$.

If this temporal model is **scalable**, then given m, there are independent, identically distributed $\{Y_i^{(mn)}\}_{i=1}^{mn}$, parametrized for time-step $\Delta t' \equiv T/mn$, so that:
$$X_{mj}^{(mn)} \equiv X_0 + \sum\nolimits_{i=1}^{mj} Y_i^{(mn)}$$
is the refined asset price at time $t = j\Delta t$. Such a model is approximately scalable or asymptotically scalable to order k_0 when $E\left[\left(X_{mj}^{(mn)}\right)^k\right]$ satisfies (8.15) or (8.16), respectively, for $k \leq k_0$ and $1 \leq j \leq n$.

For fixed n and j with $1 \leq j \leq n$, the collections of random variables $\{Y_i^{(mn)}\}_{i=1}^{mj}$ for $m = 1, 2, ...$, form a triangular array of Definition 7.6. Thus the question of the limiting distribution of $X_{mj}^{(mn)} \equiv \sum_{i=1}^{mj} Y_i^{(mn)}$ is precisely the question addressed by the Lindeberg and Lyapunov central limit theorems of Propositions 7.12 and 7.20, so the following results will not surprise.

Proposition 8.35 (Limits of scalable additive models) *With the above notation, define:*

$$\widetilde{Y}^{(mn)} \equiv \frac{Y^{(mn)} - \mu^{(mn)}}{\sigma^{(mn)}},$$

where $\mu^{(mn)} \equiv E\left[Y^{(mn)}\right]$ *and* $\left(\sigma^{(mn)}\right)^2 \equiv Var\left[Y^{(mn)}\right]$, *and assume that for all* $t > 0:$

$$\lim_{m\to\infty} \int_{\left|\widetilde{Y}^{(mn)}\right| > t\sqrt{mn}} \left(\widetilde{Y}^{(mn)}\right)^2 d\lambda = 0. \tag{8.31}$$

1. *If the additive temporal model is approximately scalable to order* $k_0 \geq 2$, *then for all* j *with* $1 \leq j \leq n$:

$$\frac{X_{mj}^{(mn)} - E\left[X_j^{(n)}\right]}{\sqrt{Var\left[X_j^{(n)}\right]}} \to_d N, \tag{8.32}$$

   as $m \to \infty$, *where* N *is standard normal. In other words, the distribution function* $F_{X_{mj}^{(mn)}}$ *converges weakly to the normal distribution with mean* $E\left[X_j^{(n)}\right]$ *and variance* $Var\left[X_j^{(n)}\right]$.

2. *If the additive temporal model is asymptotically scalable to order* $k_0 \geq 2$, *then for all* j *with* $1 \leq j \leq n$ *and* $t \equiv j\Delta t$:

$$\frac{X_{mj}^{(mn)} - \mu_1'(t)}{\sqrt{\mu_2(t)}} \to_d N, \tag{8.33}$$

   as $m \to \infty$. *In other words, the distribution function* $F_{X_{mj}^{(mn)}}$ *converges weakly to the normal distribution with mean* $\mu_1'(t)$ *and variance* $\mu_2(t)$.

Proof. *First,* $\{Y_i^{(mn)}\}_{i=1}^{mn}$ *are i.i.d. for each* m, *and it follows from Exercise 7.9 that* $\left\{\{Y_i^{(mn)}\}_{i=1}^{mj}\right\}_{m=1}^{\infty}$ *satisfy Lindeberg's condition if for all* $t > 0$, *and with* $\widetilde{Y}^{(mn)}$ *defined above:*

$$\lim_{m\to\infty} \int_{\left|\widetilde{Y}^{(mn)}\right| > t\sqrt{mj}} \left(\widetilde{Y}^{(mn)}\right)^2 d\lambda = 0.$$

This is equivalent to (8.31), and thus Lindeberg's central limit theorem of Proposition 7.12 and Corollary 7.14 apply.

In item 1, approximate scalability obtains by (8.15):

$$E\left[X_{mj}^{(mn)}\right] = E\left[X_j^{(n)}\right], \qquad Var\left[X_{mj}^{(mn)}\right] = Var\left[X_j^{(n)}\right],$$

and so:

$$\frac{\sum_{i=1}^{mj}\left(Y_i^{(mn)} - E\left[Y_i^{(mn)}\right]\right)}{\sqrt{\sum_{i=1}^{mj} Var\left[Y_i^{(mn)}\right]}} = \frac{X_{mj}^{(mn)} - E\left[X_j^{(n)}\right]}{\sqrt{Var\left[X_j^{(n)}\right]}}.$$

Thus (8.32) is a restatement of (7.12).

For item 2, asymptotic scalability obtains by (8.16):

$$\sum\nolimits_{i=1}^{mj} E\left[Y_i^{(mn)}\right] \to \mu_1'(t), \qquad \sum\nolimits_{i=1}^{mj} Var\left[Y_i^{(mn)}\right] \to \mu_2(t),$$

and (8.33) is a restatement of (7.14). ■

Remark 8.36 (On (8.31)) *As logic demands, Lindeberg's condition in (8.31) cannot in general be satisfied for a perfectly scalable model for which the temporal variates have two moments. By Corollary 8.14, $X_j^{(n)}$ is then infinitely divisible so the distribution functions of $X_{mj}^{(mn)}$ and $X_j^{(n)}$ agree. Thus the limiting result in (8.32) would be impossible unless $X_j^{(n)}$ is normally distributed.*

In addition, Lindeberg's condition cannot be satisfied for the asymptotically scalable binomial model of Example 8.25, where $X_{mj}^{(mn)}$ was seen to converge weakly to the Poisson distribution with parameter $\lambda = jp$. See Exercise 8.37.

In the general case, $Var(\widetilde{Y}^{(mn)}) = 1$ obtains that:

$$1 \geq \int_{\left|\widetilde{Y}^{(mn)}\right|>t\sqrt{mn}} \left(\widetilde{Y}^{(mn)}\right)^2 d\lambda \geq mnt^2 \Pr\left[\left|\widetilde{Y}^{(mn)}\right| > t\sqrt{mn}\right]. \tag{1}$$

Thus a necessary but not sufficient condition for (8.32) is that:

$$mnt^2 \Pr\left[\left|\widetilde{Y}^{(mn)}\right| > t\sqrt{mn}\right] \to 0, \text{ as } m \to \infty. \tag{2}$$

This can be expressed in the **little-***o notation of Definition 7.2:*

$$\Pr\left[\left|\widetilde{Y}^{(mn)}\right| > t\sqrt{mn}\right] = o\left(\frac{1}{m}\right), \text{ as } m \to \infty.$$

Exercise 8.37 (Example 8.25 and Lindeberg's condition) *As claimed in Remark 8.36, prove that the asymptotically scalable binomial model of Example 8.25, where binomial $Y_i^{(mn)} = 1$ with probability $\frac{p}{m}$ and $Y_i^{(mn)} = 0$ with probability $1 - \frac{p}{m}$, does not satisfy Lindeberg's condition. Hint: Convert the domain of integration in (8.31) from $\left|\widetilde{Y}^{(mn)}\right| > t\sqrt{mn}$ to a statement about $Y^{(mn)}$. Show that you can choose t small enough so that the upper bound of 1 is achieved in (1) of Remark 8.36.*

The next example identifies an approximately scalable model where the upper bound of 1 is again achieved in (1) of Remark 8.36, and thus Lindeberg's condition fails.

Example 8.38 $\left(\int_{\left|\widetilde{Y}^{(mn)}\right|>t\sqrt{mn}} \left(\widetilde{Y}^{(mn)}\right)^2 d\lambda \to 1\right)$ *Denoting $E[Y^{(n)}] = \mu^{(n)}$ and $\sqrt{Var[Y^{(n)}]} = \sigma^{(n)}$, define:*

$$Y^{(mn)} = \begin{cases} -\sqrt{m}\sigma^{(n)} + \mu^{(n)}/m, & p = 1/2m^2, \\ \mu^{(n)}/m, & p = 1 - 1/m^2, \\ \sqrt{m}\sigma^{(n)} + \mu^{(n)}/m, & p = 1/2m^2. \end{cases}$$

A calculation obtains that $E[Y^{(mn)}] = \mu^{(n)}/m$ and $\sqrt{Var[Y^{(mn)}]} = \sigma^{(n)}/\sqrt{m}$, and thus this model is approximately scalability to order $k_0 = 2$.

Also, in the notation of Proposition 8.35:

$$\widetilde{Y}^{(mn)} = \begin{cases} -m, & p = 1/2m^2, \\ 0, & p = 1 - 1/m^2, \\ m, & p = 1/2m^2, \end{cases}$$

and letting $t = 1$ we see that for all $m > n$:

$$\int_{\left|\widetilde{Y}^{(mn)}\right|>\sqrt{mn}} \left(\widetilde{Y}^{(mn)}\right)^2 d\lambda = 1.$$

In addition, the limit in (8.31) is 1 for all t.

Corollary 8.39 (Limits of scalable additive models) *With the notation of Proposition 8.35, assume that for some* $\delta > 0$ *that* $\left(\widetilde{Y}^{(mn)}\right)^{2+\delta}$ *is integrable for all* m, *and that:*

$$\lim_{m\to\infty} \frac{1}{(mn)^{\delta/2}} \int \left|\widetilde{Y}^{(mn)}\right|^{2+\delta} d\lambda = 0. \tag{8.34}$$

Then items 1 and 2 of Proposition 8.35 remain valid.

Proof. *This is Lyapunov's condition restated as in Exercise 7.18.*

By direct calculation:

$$\int_{\left|\widetilde{Y}^{(mn)}\right|>t\sqrt{mn}} \left(\widetilde{Y}^{(mn)}\right)^2 d\lambda \leq \int_{\left|\widetilde{Y}^{(mn)}\right|>t\sqrt{mn}} \frac{\left|\widetilde{Y}^{(mn)}\right|^{2+\delta}}{\left(t\sqrt{mn}\right)^{\delta}} d\lambda \leq \frac{1}{t^{\delta}(mn)^{\delta/2}} \int \left|\widetilde{Y}^{(mn)}\right|^{2+\delta} d\lambda,$$

and thus (8.34) implies (8.31). ■

Exercise 8.40 $\left(\lim_{m\to\infty} \frac{1}{(mn)^{\delta/2}} \int \left|\widetilde{Y}^{(mn)}\right|^{2+\delta} d\lambda = \infty\right)$ *Since Lyapunov's condition implies Lindeberg's condition, the Lindeberg condition failures of Exercise 8.37 and Example 8.38 must imply Lyapunov condition failures. Verify this. Hint: For Example 8.38, verify that:*

$$\frac{1}{(mn)^{\delta/2}} \int \left|\widetilde{Y}^{(mn)}\right|^{2+\delta} d\lambda = (m/n)^{\delta/2},$$

and thus this limit is unbounded for all $\delta > 0$. *Derive the comparable result for Exercise 8.37.*

The next result identifies a class of approximately scalable models for which the conclusion of Proposition 8.35 always holds.

Proposition 8.41 (Limits of scalable models: $Y^{(n)} \equiv a^{(n)}V + b^{(n)}$**)** *Given arbitrary* $a^{(1)} \neq 0$ *and* $b^{(1)}$, *let* V *have two moments with* $E[V] = 0$.

Then the approximately scalable to order $k_0 = 2$ *model of Proposition 8.18, defined in (8.21) by:*

$$Y^{(n)} \equiv a^{(n)}V + b^{(n)}$$

with:

$$a^{(n)} \equiv \frac{a^{(1)}}{\sqrt{n}}, \qquad b^{(n)} \equiv \frac{b^{(1)}}{n},$$

satisfies the requirements of Proposition 8.35.

Thus by (8.32), for all j *with* $1 \leq j \leq n$:

$$\frac{X_{mj}^{(mn)} - E\left[X_j^{(n)}\right]}{\sqrt{Var\left[X_j^{(n)}\right]}} \to_d N, \text{ as } m \to \infty,$$

with N *the standard normal variate, and where:*

$$\begin{aligned} E\left[X_j^{(n)}\right] &= X_0 + \frac{j}{n} b^{(1)}, \\ Var\left[X_j^{(n)}\right] &= \frac{j}{n}\left(a^{(1)}\sigma_V\right)^2. \end{aligned}$$

In other words, as $m \to \infty$:

$$X_{mj}^{(mn)} \to_d N\left(X_0 + \frac{j}{n}b^{(1)}, \frac{j}{n}\left(a^{(1)}\sigma_V\right)^2\right).$$

Proof. *Recall item 3 of Proposition 8.18, that if V is a random variable with two moments and $E[V] = 0$, the parametrization given by $\{Y_i^{(n)}\}_{i=1}^n \equiv \{a^{(n)}V_i + b^{(n)}\}_{i=1}^n$ obtains an approximately scalable model to order $k_0 = 2$ if as implied by (8.24):*

$$a^{(mn)} \equiv a^{(n)}/\sqrt{m}, \qquad b^{(mn)} \equiv b^{(n)}/m.$$

Hence $\mu^{(mn)} = b^{(mn)}$ and $\sigma^{(mn)} = a^{(mn)}\sigma_V$ in the notation of Proposition 8.35, and:

$$\widetilde{Y}^{(mn)} = \frac{V}{\sigma_V}.$$

Thus $\left(\widetilde{Y}^{(mn)}\right)^2$ is integrable by existence of σ_V^2, and (8.31) is then satisfied since $\widetilde{Y}^{(mn)}$ is independent of m.

Since $X_j^{(n)} = X_0 + \sum_{i=1}^j Y_i^{(n)}$, the moments follow from independence of $\{Y_i^{(n)}\}_{i=1}^n$. ■

Example 8.42 (Additive binomial temporal model) *An example of a Proposition 8.41 model was seen in Example 8.22, where:*

$$Y_i^{(n)} \equiv \mu\Delta t + \sigma\sqrt{\Delta t}b_i,$$

with $\{b_i\}_{i=1}^n$ independent and binomially distributed to equal ± 1 with probability $1/2$. Hence $E[b_i] = 0$, and with $\Delta t \equiv T/n$, the parameterization for $\{Y_i^{(n)}\}_{i=1}^n$ is:

$$a^{(n)} = \sigma\sqrt{T/n}, \qquad b^{(n)} = \mu T/n.$$

This approximately scalable additive temporal model with $\sigma_b^2 = 1$ thus has the property that for all n and $1 \leq j \leq n$:

$$X_{mj}^{(mn)} \to_d N(X_0 + j\Delta t\mu, j\Delta t\sigma^2), \ m \to \infty.$$

8.3.2 Scalable Multiplicative Models

We next turn to **multiplicative temporal models**, the conclusions of which are simplified by the Mann-Wald theorem of Proposition 3.26. For the following result, assume that we are given time horizon T, $n \geq 1$, temporal $Z^{(n)} = (Z_1^{(n)}, Z_2^{(n)}, ..., Z_n^{(n)})$ and spatial $X^{(n)} = (X_1^{(n)}, X_2^{(n)}, ..., X_n^{(n)})$ related by:

$$X_j^{(n)} = X_0 \exp\left(\sum\nolimits_{i=1}^j Z_i^{(n)}\right).$$

Recall X_0 denotes the initial asset value, and $\{Z_i^{(n)}\}_{i=1}^n$ are independent and identically distributed on some probability space $(\mathcal{S}, \mathcal{E}, \lambda)$, with distribution function appropriate for time-step $\Delta t \equiv T/n$.

If this temporal model is **scalable**, then given m, there are independent, identically distributed $\{Z_i^{(mn)}\}_{i=1}^{mn}$, parametrized for time-step $\Delta t' \equiv T/mn$, so that:

$$X_{mj}^{(mn)} \equiv X_0 \exp\left(\sum\nolimits_{i=1}^{mj} Z_i^{(mn)}\right),$$

is the refined asset price at time $t = j\Delta t$. Denoting:

$$W_j^{(n)} \equiv \sum\nolimits_{i=1}^{j} Z_i^{(n)},$$

and similarly for $W_{mj}^{(mn)}$, such a model is approximately scalable or asymptotically scalable to order k_0 when $E\left[\left(W_{mj}^{(mn)}\right)^k\right]$ satisfies (8.15) or (8.16), respectively, for $k \leq k_0$ and $1 \leq j \leq n$.

For fixed n and j with $1 \leq j \leq n$, the collections of random variables $\{Z_i^{(mn)}\}_{i=1}^{mj}$ for $m = 1, 2, ...,$ form a triangular array of Definition 7.6. Thus the question of the limiting distribution of $W_{mj}^{(mn)} \equiv \sum_{i=1}^{mj} Z_i^{(mn)}$ is precisely the question addressed by the Lindeberg and Lyapunov central limit theorems of Propositions 7.12 and 7.20, so the following results will again not surprise. The Mann-Wald theorem then obtains limiting results for $X_{mj}^{(mn)}$.

Proposition 8.43 (Limits of scalable multiplicative models) *With the above notation, define:*

$$\widetilde{Z}^{(mn)} \equiv \frac{Z^{(mn)} - \mu^{(mn)}}{\sigma^{(mn)}},$$

where $\mu^{(mn)} \equiv E\left[Z^{(mn)}\right]$ *and* $\left(\sigma^{(mn)}\right)^2 \equiv Var\left[Z^{(mn)}\right]$, *and assume that for all* $t > 0$:

$$\lim_{m\to\infty} \int_{\left|\widetilde{Z}^{(mn)}\right| > t\sqrt{mn}} \left(\widetilde{Z}^{(mn)}\right)^2 d\lambda = 0. \tag{8.35}$$

1. *If the multiplicative temporal model is approximately scalable to order* $k_0 \geq 2$, *then for all* j *with* $1 \leq j \leq n$:

$$\frac{\ln\left[X_{mj}^{(mn)}/X_0\right] - E\left[W_j^{(n)}\right]}{\sqrt{Var\left[W_j^{(n)}\right]}} \to_d N, \tag{8.36}$$

with N *a standard normal variate. Thus:*

$$X_{mj}^{(mn)} \to_d X_0 \exp\left(E\left[W_j^{(n)}\right] + N\sqrt{Var\left[W_j^{(n)}\right]}\right), \tag{8.37}$$

and $F_{X_{mj}^{(mn)}}$ *converges weakly to the lognormal distribution with parameters* $\mu = \ln X_0 + E\left[W_j^{(n)}\right]$ *and* $\sigma^2 = Var\left[W_j^{(n)}\right]$.

2. *If the multiplicative temporal model is asymptotically scalable to order* $k_0 \geq 2$, *then for all* j *with* $1 \leq j \leq n$ *and* $t \equiv j\Delta t$:

$$\frac{\ln\left[X_{mj}^{(mn)}/X_0\right] - \mu_1'(t)}{\sqrt{\mu_2(t)}} \to_d N, \tag{8.38}$$

with N *a standard normal variate. Thus:*

$$X_{mj}^{(mn)} \to_d X_0 \exp\left(\mu_1'(t) + N\sqrt{\mu_2(t)}\right), \tag{8.39}$$

and $F_{X_{mj}^{(mn)}}$ *converges weakly to the lognormal distribution with parameters* $\mu = \ln X_0 + \mu_1'(t)$ *and* $\sigma^2 = \mu_2(t)$.

Proof. *First, $\{Z_i^{(mn)}\}_{i=1}^{mn}$ are i.i.d. for each m, and it follows as in the proof of Proposition 8.35 that $\left\{\{Z_i^{(mn)}\}_{i=1}^{mj}\right\}_{m=1}^{\infty}$ satisfy Lindeberg's condition and thus Lindeberg's central limit theorem of Proposition 7.12 and Corollary 7.14 apply.*

For item 1, it follows from (8.32) of Proposition 8.35 that:

$$\frac{W_{mj}^{(mn)} - E\left[W_j^{(n)}\right]}{\sqrt{Var\left[W_j^{(n)}\right]}} \to_d N, \tag{1}$$

which is (8.36) by definition of $W_{mj}^{(mn)}$. The Mann-Wald theorem of Proposition 3.26 and $h(x) = \exp\left(x\sqrt{Var\left[W_j^{(n)}\right]} + E\left[W_j^{(n)}\right] + \ln X_0\right)$ then obtains (8.37).

Item 2 is derived analogously and left as an exercise. ■

Corollary 8.44 (Limits of scalable multiplicative models) *With the notation of Proposition 8.43, assume that for some $\delta > 0$ that $\left(\widetilde{Z}^{(mn)}\right)^{2+\delta}$ is integrable for all m, and that:*

$$\lim_{m\to\infty} \frac{1}{(mn)^{\delta/2}} \int \left|\widetilde{Z}^{(mn)}\right|^{2+\delta} d\lambda = 0. \tag{8.40}$$

Then items 1 and 2 of Proposition 8.43 remain valid.

Proof. *Exactly as in Corollary 8.39, (8.40) implies (8.35), and the proof is complete.* ■

Exercise 8.37 and Example 8.38 provide examples of asymptotically scalable and approximately scalable models, respectively, which do not satisfy the regularity conditions of (8.40) or (8.35). Thus these also provide examples within the multiplicative framework for which a limiting lognormal distribution is not obtained.

The following exercise reformulates Proposition 8.41.

Exercise 8.45 (Scalable models: $Z^{(n)} \equiv a^{(n)}V + b^{(n)}$) *Formulate and prove a restatement of Proposition 8.41 for multiplicative models, where $Z^{(n)} \equiv a^{(n)}V + b^{(n)}$. Here $a^{(n)} \equiv a^{(1)}/\sqrt{n}$ and $b^{(n)} \equiv b^{(1)}/n$ for arbitrary $a^{(1)} \neq 0$ and $b^{(1)}$, and V is a random variable with two moments and $E[V] = 0$. Hint: Investigate $\widetilde{Z}^{(mn)}$ in this model, and show that $\left(\widetilde{Z}^{(mn)}\right)^2$ is integrable, and thus (8.35) is satisfied.*

While "only" a special case of Exercise 8.45, with V defined as a binomial b, we state the following result as a proposition because of the importance of this model in finance.

Proposition 8.46 (Multiplicative binomial model) *Let:*

$$X_j^{(n)} \equiv X_0 \exp\left[\sum\nolimits_{i=1}^{j} Z_i^{(n)}\right],$$

where:

$$Z_i^{(n)} = \mu\Delta t + \sigma\sqrt{\Delta t} b_i,$$

with $\{b_i\}_{i=1}^n$ independent and binomially distributed to equal 1 or -1 with probability $p = 1/2$. As above, $\Delta t \equiv T/n$ for some fixed horizon time T, and thus $X_j^{(n)}$ is the price variate at time $t = j\Delta t$ for any integer $1 \leq j \leq n$.

Given m and $\{b_i\}_{i=1}^{mn}$, define the refined price variate at time $t = j\Delta t$:

$$X_{mj}^{(mn)} \equiv X_0 \exp\left[\sum\nolimits_{i=1}^{mj} \left(\mu\Delta t/n + \sigma\sqrt{\Delta t/m}b_i\right)\right].$$

Then as $m \to \infty$:

$$\frac{\ln[X_{mj}^{(mn)}/X_0] - \mu j\Delta t}{\sigma\sqrt{j\Delta t}} \to_d N, \tag{8.41}$$

where N denotes a standard normal variate.

Equivalently, $X_{mj}^{(mn)}$ converges in distribution to a lognormal distribution with parameters $\mu j\Delta t + \ln X_0$ and $\sigma^2 j\Delta t$, and denoting this limiting lognormal variate by X_t with $t = j\Delta t$:

$$X_t = X_0 \exp\left[\mu t + \sigma\sqrt{t}N\right]. \tag{8.42}$$

Proof. *This is a special case of Exercise 8.45, with $V = b$.* ■

9

Pricing of Financial Derivatives

In this chapter, we investigate the discrete-time pricing of financial derivatives on various underlying assets where it is assumed that such assets can be modeled in the real world according to the multiplicative binomial temporal models of Chapter 8.

The first section develops the binomial lattice framework for pricing European and American versions of such derivatives based on the notion of a replicating portfolio. The pricing formulas for European derivatives are seen to reduce to the expected value of discounted settlements, with expectations defined relative to new "risk-neutral" probabilities for the binomial lattice.

The next section investigates limiting distributions of various asset prices as $\Delta t \to 0$ under this risk-neutral probability structure, as well as under an alternative probability structure for real-world modeling. Then limits of the binomial lattice prices for general European derivatives are obtained as $\Delta t \to 0$, and for European puts and calls, these limiting results are the famous Black-Scholes-Merton pricing formulas.

Various results related to these limiting prices are then derived, such as derivative price convergence to settlement values, and that European derivative prices satisfy the Black-Scholes-Merton partial differential equation.

Path-dependent financial derivatives are then investigated, and initially modeled within a non-recombining binomial lattice framework to obtain replicating portfolio pricing formulas. Monte Carlo pricing, meaning pricing based on samples of asset price paths, is considered next, as are convergence results to lattice prices as the sample size increases.

The final section investigates lognormal model pricing of financial derivatives, exact and by Monte Carlo methods, and extends the Black-Scholes-Merton framework to path-dependent European derivatives.

9.1 Binomial Lattice Pricing

The multiplicative binomial temporal model of Proposition 8.46 is sometimes referred to as the **binomial return model** of equities, or other assets, in the **real-world measure.** This means that the implied distributions of future asset prices, as determined by μ, σ, and Δt, represent a model for the actual, observable, future outcomes.

It will be assumed that such assets trade without costs or other "frictions" in the market, and that both "long" and "short" positions are implementable.

Given this binomial return model, the payoffs at maturity of a European-style financial derivative on this asset, such as an option, can be achieved by an initial portfolio of the underlying asset and the risk-free asset. To accomplish this, one needs to appropriately rebalance this initial portfolio at the end of each time-step, buying and selling between the underlying asset and the risk-free asset positions.

Such a financial derivative is then said to be **replicated** by this initial portfolio, and it will follow from the so-called **law of one price** that the price of the option must agree

DOI: 10.1201/9781003275770-9

with the price of this initial **replicating portfolio**. It is this last conclusion that unites the mathematics of financial derivative replication and the realities of financial derivative pricing in the financial markets.

For the law of one price, the critical assumption is that both long and short positions in the modeled asset trade freely. The reason for this is that market enforces this law by **risk-free arbitrage,** whereby a market trader is able to take long and short positions in portfolios with identical payoffs, but different prices. Logically going short the higher priced portfolio and going long the lower priced, the trader makes a risk-free profit of the price differential, since a risk-free exit is assured by the assumed equality of the portfolios' payoffs.

As most financial derivatives and all risk-free assets trade freely, long and short, the critical assumption needed for the enforcement of the law of one price is then the assumption that the underlying asset of the financial derivative itself trades freely. In most models one ignores the actual trading costs and other frictions like taxes that exist in the real-world markets.

It will then turn out that the price of this financial derivative, which is the price of this replicating portfolio, equals the risk-free present value of an expectation of the derivative settlement or payoff. For this calculation however, this expectation is not taken with respect to the original real-world probabilities of the model, but instead, with respect to a new set of probabilities which we investigate shortly.

9.1.1 European Derivatives

A **European financial derivative on an underlying asset** X is a financial contract with a specified term T, which provides a settlement value or payoff at time T which is a function of the value of X at that time. European put and call options are specific examples of such derivative securities with payoff functions defined in (9.41) and (9.40).

Options contracts conventionally create nonnegative settlement values to the buyer of the option, who is said to be **long the option**. However, there is nothing in the mathematics of derivatives pricing below that requires nonnegative values, and indeed, the pricing approach below applies equally well to forward or other derivative contracts for which the payoff, now better called a "settlement," may be positive or negative.

When the payoff is strictly nonnegative, the seller or "writer" of the contract is said to be **short the contract**, and makes payments to the long at time T as contractually required. When the settlement can be positive or negative, the **long** is the party that profits when the price of X increases, while the short profits when X decreases.

Remark 9.1 (On derivative "settlements") *For derivatives pricing, it is the settlement "values" that are required, in some currency units, but this is not to imply that all financial derivatives are settled in cash. Many such derivatives are* ***physical settle,*** *whereby the long and short exchange X for cash, while some financial derivatives are* ***cash settle,*** *by which is meant that one party pays the settlement value to the other party in cash.*

We ignore these details for pricing, and reference ***Hull*** *(2022) and* ***McDonald*** *(2006) for more details on derivatives contracts, markets and trading, as well as pricing.*

To simplify language, we will sometimes use the term "option" in the development below, but it will be clear from the development that other than the explicit form of the Black-Scholes-Merton option pricing formulas for European put and call options, that all of the development below applies equally well to general European derivative securities on X.

Remark 9.2 (Law of one price) *As noted above, the* ***law of one price*** *is predicated on the idea that if the prices of assets or portfolios with identical payoffs do not agree, investors* *can create a **risk-free arbitrage** by going long the cheaper alternative, and shorting the more expensive alternative. The positions are then settled at maturity with perfectly offsetting contractual payoffs, and the investor keeps the initial profit without having taken any risk.*

These buy and sell pressures then adjust prices back to an equilibrium level which ultimately obeys this law. In the real world, "back to equilibrium" need not mean "exactly equal," but only to the point at which the arbitrage is no longer profitable, which is to say, that the price discrepancy is within trading costs.

The law of one price is a reasonable assumption in any market that **allows both long and short positions** in derivatives contracts and underlying assets. Since many derivative contracts are exchange traded and thus freely traded, it is usually the underlying asset that determines the applicability of the law of one price. Examples of such freely traded underlying assets are equities, tradable indexes, currencies, and futures contracts on various assets. Collectively, these are often called **investment assets,** meaning assets held by investors for potential appreciation, and thus these assets are also available for borrowing (for a fee) and shorting.

Exercise 9.3 (The risk-free asset) *A risk-free asset over a period $[0, \Delta t]$ is defined as a security for which one unit can be acquired in a given currency for $B(0) > 0$ at $t = 0$, and which has a certain maturity value in that currency at time $t = \Delta t$ of $B(\Delta t)$. By "certain" is that the investor has no risk that such repayment will not be made, or that the maturity value will change.*

Define the implied risk-free interest rate in continuous units by $B(\Delta t) = B(0)e^{r_c \Delta t}$, or nominal units by $B(\Delta t) = B(0)(1 + r_a \Delta t)$.

Prove by an arbitrage argument that if risk-free assets can be acquired or shorted without friction costs, that the risk-free rate is uniquely defined for any period Δt. Hint: Assume that two such assets exist for given Δt, with $r_c \neq r'_c$ or $r_a \neq r'_a$, and consider a risk-free arbitrage. How does this trade change prices to assure ultimate equality of risk-free rates?

As introduced above, we now investigate the existence of an initial replicating portfolio for a given financial derivative, and identify the price of this portfolio. We will then prove that the price of this portfolio equals the present value of this derivative's expected payoff, where this expectation is calculated within the given binomial model, but with a new probability structure. In essence, we will replace the **real-world probabilities defined with** p by the **risk-neutral probabilities** defined by $q(\Delta t)$, to be identified.

To this end, let X_0 denote the **price of one unit of an asset**, such as the price of one share of a common stock, or one unit of a foreign currency, or one unit of an equity index portfolio. Given T, let the parameters μ and σ be identified for this asset by the assumption that:

$$E\left[\ln[X_T/X_0]\right] = \mu T, \qquad Var\left[\ln[X_T/X_0]\right] = \sigma^2 T. \tag{9.1}$$

Given n and $\Delta t \equiv T/n$, the multiplicative binomial model of Proposition 8.46 obtains asset prices $X_j^{(n)} \equiv X_{j\Delta t}$ at time $t = j\Delta t$ for $1 \leq j \leq n$:

$$X_j^{(n)} = X_0 \exp\left[\sum\nolimits_{i=1}^{j} Z_i^{(n)}\right].$$

Let:

$$Z_i^{(n)} = \mu \Delta t + \sigma\sqrt{\Delta t} b_i,$$

and $\{b_i\}_{i=1}^n$ are independent binomial variates which equal ± 1 with probability $p = \frac{1}{2}$.

The next result shows that this binomial model satisfies the moment assumptions in (9.1), and we can then say that the binomial model is **calibrated to two moments of** X.

Proposition 9.4 (Binomial model X_T) *The above binomial model obtains $n+1$ asset prices X_T at time $T = n\Delta t$:*

$$\left\{X_0 e^{ku(\Delta t)} e^{(n-k)d(\Delta t)}\right\}_{k=0}^{n},$$

with:

$$u(\Delta t) = \mu\Delta t + \sigma\sqrt{\Delta t}, \qquad d(\Delta t) = \mu\Delta t - \sigma\sqrt{\Delta t}, \tag{9.2}$$

and respective probabilities defined with $p = 1/2$ by:

$$\left\{\binom{n}{k} p^k (1-p)^{n-k}\right\}_{k=0}^{n}.$$

Further, for $1 \leq j \leq n$:

$$\begin{aligned} E\left[\ln(X_j^{(n)}/X_0)\right] &= \mu j\Delta t; \\ Var\left[\ln(X_j^{(n)}/X_0)\right] &= \sigma^2 j\Delta t. \end{aligned} \tag{9.3}$$

Proof. *Since each b_i equals ± 1 with probability $1/2$, let k denote the number of variates equal to 1 in a given sample $\{b_i\}_{i=1}^n$. Then $0 \leq k \leq n$, and:*

$$\sum\nolimits_{i=1}^{n} Z_i^{(n)} = ku(\Delta t) + (n-k)\, d(\Delta t), \tag{1}$$

where the "up return" $u(\Delta t)$ and "down return" $d(\Delta t)$ are given in (9.2).

There are 2^n distinct such samples of $\{b_i\}_{i=1}^n$ where we distinguish order, any one of which has probability $p^k(1-p)^{n-k}$, where k is the number of variates equal to 1. This reduces to $1/2^n$ independent of k when $p = 1/2$, but we maintain general notation. For given k, there are $\binom{n}{k}$ such samples in various orders, and thus the probability of (1) is $\binom{n}{k}p^k(1-p)^{n-k}$ as claimed.

For moments:

$$\begin{aligned} \ln(X_j^{(n)}/X_0) &= \sum\nolimits_{i=1}^{j} Z_i^{(n)} \\ &= \mu j\Delta t + \sigma\sqrt{\Delta t}\sum\nolimits_{i=1}^{j} b_i, \end{aligned}$$

and (9.3) follows by independence of $\{b_i\}_{i=1}^n$, that $E\left[\sum_{i=1}^j b_i\right] = 0$ and $Var\left[\sum_{i=1}^j b_i\right] = j$. ■

It is an exercise to check that the same calibration of moments can be achieved with other real-world p with $0 < p < 1$ by properly redefining $u(\Delta t)$ and $d(\Delta t)$. See **Reitano** (2010) for this and other details noted below.

Exercise 9.5 ($0 < p < 1$) *Given p with $0 < p < 1$, show that (9.1) is satisfied with:*

$$u(\Delta t) = \mu\Delta t + a\sigma\sqrt{\Delta t}, \qquad d(\Delta t) = \mu\Delta t - a^{-1}\sigma\sqrt{\Delta t}, \tag{9.4}$$

where $a = \sqrt{(1-p)/p}$.

Exercise 9.6 ($d(\Delta t) < r_c(\Delta t) < u(\Delta t)$) *Prove by an arbitrage argument that for an asset that pays no income:*

$$d(\Delta t) < r_c(\Delta t) < u(\Delta t), \tag{9.5}$$

where $r_c \equiv r_c(\Delta t)$ is given in Exercise 9.3. Hint: Consider proof by contradiction. Why is the absence of income important for this result?

Notation 9.7 (Binomial lattice) *Asset prices will often need to be identified by "time" and "state." In this binomial model, future asset prices are defined at times $t = j\Delta t$ for $j = 1, ..., n$, but then at any time j there are also $j+1$ possible "states" for this price.*

A common notational scheme is to denote this time-state price as $X_{k,j}$, defined by:

$$X_{k,j} = X_0 e^{ku(\Delta t)} e^{(j-k)d(\Delta t)}, \tag{9.6}$$

where $0 \leq j \leq n$ denotes the time and $0 \leq k \leq j$ denotes the state. Of course $X_{0,0} = X_0$, while the $j+1$ state prices at time j are $\{X_0 e^{ku(\Delta t)} e^{(j-k)d(\Delta t)}\}_{k=0}^{j}$.

The notation $X_{k,j}$ is reminiscent of the notation for matrix components under the usual convention of $X_{\text{row,col}}$, where the rows of the array relate to states, and the columns identify times. But this is not a rectangular matrix of course, but a triangular matrix. In this notational scheme, each $X_{k,j}$ is connected by the $u(\Delta t)$ and $d(\Delta t)$ return variates to two, time $j+1$ prices:

$$\begin{array}{ccc} & & X_{k+1,j+1} \\ & \nearrow^{u(\Delta t)} & \\ X_{k,j} & & \\ & \searrow_{d(\Delta t)} & \\ & & X_{k,j+1} \end{array} \tag{9.7}$$

Thus k denotes the number of $u(\Delta t)$ returns over all prior periods.

If the above diagram is completed from $t = 0$ to $t = T$, the grid created is called a ***binomial lattice of asset prices,*** *and provides a handy visual aid in the current development. The above visual representation is compelling since often $d(\Delta t) < 0$ and $u(\Delta t) > 0$ in a model, though this is not compelled by (9.5).*

It is often easier to program this lattice as an upper or lower triangular matrix. For example, an upper triangular matrix will have top row in order of $\{X_{k,k}\}_{k=0}^{n} \equiv \{X_0 e^{ku(\Delta t)}\}_{k=0}^{n}$:

$$\begin{array}{ccc} X_{k,j} & \rightarrow^{u(\Delta t)} & X_{k+1,j+1} \\ & \searrow_{d(\Delta t)} & X_{k,j+1} \end{array}$$

The next result derives the binomial lattice price of a European financial derivative on an underlying asset by first identifying the associated replicating portfolio. As noted above, by **European** is meant that this financial derivative has a formulaic settlement based on the underlying asset price, but only at maturity T. Of course, an investor in such a contract can always obtain an early settlement by selling the contract in the market at the then trading price, but the contract itself allows no early settlement.

Also, by **replicating portfolio** is meant that this portfolio replicates these formulaic settlement values at all lattice prices at maturity.

In order to construct the replicating portfolio, which will contain the underlying asset and risk-free asset, we must assume that both long and short positions in this underlying asset trade freely, and without transaction costs such as bid-ask spreads, or other market frictions such as taxes. We also initially assume that this asset pays no income during the modeling period.

In order to derive that the price of this financial derivative equals the price of this portfolio, we will assume that the law of one price prevails, as discussed in Remark 9.2. We also assume that the contractual settlements are certain to be made, and thus neither counterparty to the contract has credit risk to the other counterparty. In practice, credit risk is essentially eliminated on exchange-traded options by the requirement of **margin accounts** with adequate collateral to guarantee settlement. For over-the-counter financial derivatives markets, similar collateral structures are commonly used.

Remark 9.8 (Binomial lattice price and Market price) *Below we will state that the price of the replicating portfolio must equal the* **market price** *of a financial derivative contract with the given replicated payoffs. The logic will always be based on a risk-free arbitrage argument.*

It should be emphasized that by "market price" is meant, the price in a market where only binomial lattice outcomes are possible! It would be illogical to suggest that these would be the correct prices in the real-world financial markets, in which prices other than those found on the binomial lattice regularly occur.

This then raises the question: Why are binomial lattice prices useful?

The answer is that such prices are only an approximation, but become more useful as $\Delta t \to 0$ and the binomial lattice prices become better and better approximations of the distribution of prices in the real world. Such limiting prices will be addressed below in certain special cases, and are otherwise estimated numerically.

We introduce the replication construction next, assuming that the underlying asset pays no income. This asset is thus often thought of as a common stock which pays no dividends. But it should be noted that this construction only requires that if dividends are payable, that there is no **ex dividend date** during the modeling period of $[0, T]$, and thus no dividend effect on price.

Proposition 9.9 (Binomial lattice price of a European derivative) *Let $O_T[X_T]$ denote the settlement value of a European financial derivative at time $T = n\Delta t$ expressed as a function of the then prevailing price of an asset X. Assume that investors can take long or short positions in X with no transaction costs, that X pays no income, and that the risk-free interest rate with continuous compounding is fixed and equal to r for long and short positions.*

Then the binomial lattice price at time 0 for a replicating portfolio for this financial derivative security is given by:

$$O_0^{(n)}[X_0] = e^{-rT} \sum\nolimits_{j=0}^{n} \binom{n}{j} q^j (1-q)^{n-j} O_T\left[X_0 e^{ju(\Delta t)} e^{(n-j)d(\Delta t)}\right], \tag{9.8}$$

with $u(\Delta t)$ and $d(\Delta t)$ given in (9.2).

Thus, $O_0^{(n)}[X_0]$ equals the risk-free present value of the expected settlements at time $T = n\Delta t$, where this expectation reflects the original binomial distribution of returns, but with ***risk-neutral probability*** *$q \equiv q(\Delta t)$ defined by:*

$$q(\Delta t) \equiv \frac{e^{r\Delta t} - e^{d(\Delta t)}}{e^{u(\Delta t)} - e^{d(\Delta t)}}. \tag{9.9}$$

Proof. *Some of the details of the derivation of this option pricing result are left as an exercise with the following steps.*

1. *Given any of the $j+1$ asset prices at time $j\Delta t$ for $0 \leq j \leq n-1$:*

$$X_{k,j} \in \left\{X_0 e^{ku(\Delta t)} e^{(j-k)d(\Delta t)}\right\}_{k=0}^{j},$$

assume that we are given a ***value function*** *$f_{k,j}(X_{\cdot,\cdot})$, defined on the associated asset prices at time $(j+1)\Delta t$. In other words, assume that given price $X_{k,j}$, that there is a value function that identifies $f_{k,j}(e^{u(\Delta t)} X_{k,j})$ and $f_{k,j}(e^{d(\Delta t)} X_{k,j})$. Show that there are time-state constants $a_{k,j}$ and $b_{k,j}$, so that an initial portfolio of $a_{k,j}$ units of $X_{k,j}$, and $b_{k,j}$ units invested risk-free at continuous rate r, exactly replicates these two values. Hint:*

Noting that the initial price of this portfolio is $P_{j\Delta t}(X_{k,j}) \equiv a_{k,j}X_{k,j} + b_{k,j}$, *identify two equations in these two unknowns, recalling that the asset pays no income. Derive from this:*

$$a_{k,j} = \frac{f_{k,j}(e^{u(\Delta t)}X_{k,j}) - f_{k,j}(e^{d(\Delta t)}X_{k,j})}{\left(e^{u(\Delta t)} - e^{d(\Delta t)}\right)X_{k,j}}, \tag{9.10}$$

and:

$$b_{k,j} = \left(\frac{e^{u(\Delta t)}f_{k,j}(e^{d(\Delta t)}X_{k,j}) - e^{d(\Delta t)}f_{k,j}(e^{u(\Delta t)}X_{k,j})}{e^{u(\Delta t)} - e^{d(\Delta t)}}\right)e^{-r\Delta t}. \tag{9.11}$$

2. *Confirm that the price of this portfolio,* $P_{j\Delta t}(X_{k,j}) \equiv a_{k,j}X_{k,j} + b_{k,j}$, *must equal the market price at time* $j\Delta t$ *in state* k *of a financial contract with time* $(j+1)\Delta t$ *payoffs of* $f_{k,j}(X_{k+1,j+1})$ *if* $u(\Delta t)$ *prevails, and* $f_{k,j}(X_{k,j+1})$ *if* $d(\Delta t)$ *prevails. Hint: Construct risk-free arbitrages if the market price exceeds, or is below, this calculated price, which in turn requires the assumption that you can go long or short* $X_{k,j}$.

3. *Show that the market price* $P_{j\Delta t}(X_{k,j}) \equiv a_{k,j}X_{k,j} + b_{k,j}$ *at time* $j\Delta t$ *of this portfolio can be algebraically manipulated into:*

$$P_{j\Delta t}(X_{k,j}) = e^{-r\Delta t}\left[qf_{k,j}(e^{u(\Delta t)}X_{k,j}) + (1-q)f_{k,j}(e^{d(\Delta t)}X_{k,j})\right],$$

with $q \equiv q(\Delta t)$ *defined in (9.9).*

4. *If* $j = n-1$, *define this value function to be the derivative settlement values:*

$$\begin{aligned} f_{k,n-1}(e^{u(\Delta t)}X_{k,n-1}) &= O_T\left[X_{k+1,n}\right], \\ f_{k,n-1}(e^{d(\Delta t)}X_{k,n-1}) &= O_T\left[X_{k,n}\right]. \end{aligned}$$

Then by item 2, $P_{(n-1)\Delta t}(X_{k,n-1})$ *in item 3 must equal the market price of this financial derivative at time* $(n-1)\Delta t$, *so we label these prices as* $O^{(n)}_{(n-1)\Delta t}\left[X_{k,n-1}\right]$, *and thus for* $0 \leq k \leq n-1$:

$$\begin{aligned} O^{(n)}_{(n-1)\Delta t}\left[X_{k,n-1}\right] &= e^{-r\Delta t}\left[qO_T\left[e^{u(\Delta t)}X_{k,n-1}\right] + (1-q)O_T\left[e^{d(\Delta t)}X_{k,n-1}\right]\right] \\ &= e^{-r\Delta t}\left[qO_T\left[X_{k+1,n}\right] + (1-q)O_T\left[X_{k,n}\right]\right] \end{aligned} \tag{1}$$

5. *The logic and the formula in item 4 can be applied when* $j = n-2$, *defining the value function* $f_{k,n-2}$ *in terms of* $O^{(n)}_{(n-1)\Delta t}$, *deriving for* $0 \leq k \leq n-2$:

$$\begin{aligned} O^{(n)}_{(n-2)\Delta t}\left[X_{k,n-2}\right] &= e^{-r\Delta t}qO^{(n)}_{(n-1)\Delta t}\left[e^{u(\Delta t)}X_{k,n-2}\right] + (1-q)O^{(n)}_{(n-1)\Delta t}\left[e^{d(\Delta t)}X_{k,n-2}\right] \\ &= e^{-r\Delta t}qO^{(n)}_{(n-1)\Delta t}\left[X_{k+1,n-1}\right] + (1-q)O^{(n)}_{(n-1)\Delta t}\left[X_{k,n-1}\right]. \end{aligned}$$

Using (1), rewrite this formula in terms of the settlement function O_T.

6. *By induction, show that* $O^{(n)}_{j\Delta t}\left[X_{k,j}\right]$, *the market price for this option at time* $t = j\Delta t$ *when the asset price is* $X_{k,j}$, *is given for* $0 \leq k \leq j$ *by:*

$$O^{(n)}_{j\Delta t}\left[X_{k,j}\right] = e^{-r(n-j)\Delta t}\sum\nolimits_{i=0}^{n-j}\binom{n-j}{i}q^i(1-q)^{n-j-i}O_T\left[X_{k,j}e^{iu(\Delta t)}e^{(n-j-i)d(\Delta t)}\right]. \tag{2}$$

Hint: Assume that $O^{(n)}_{j\Delta t}[X_{k,j}]$ *is given by this formula. Then by item 3, the replicating portfolios at time* $(j-1)\Delta t$ *satisfy:*

$$P_{(j-1)\Delta t}(X_{k,j-1}) = e^{-r\Delta t}\left[qO^{(n)}_{j\Delta t}\left[e^{u(\Delta t)}X_{k,j-1}\right] + (1-q)O^{(n)}_{j\Delta t}\left[e^{d(\Delta t)}X_{k,j-1}\right]\right].$$

Show that $P_{(j-1)\Delta t}(X_{k,j-1}) = O^{(n)}_{(j-1)\Delta t}[X_{k,j-1}]$.

7. *The proof is complete by noting that this general formula in* (2) *produces* (9.8) *when* $j = 0$.

■

Remark 9.10 ($0 < p < 1$) *It should be noted that the proof of Proposition 9.9 did not reflect the real-world probability* $p = \frac{1}{2}$, *other than in the form of* $u(\Delta t)$ *and* $d(\Delta t)$ *in (9.2). Thus for general* p *of Exercise 9.5, the same derivatives pricing formula is obtained, as is the risk-neutral probability formula in (9.9), but where the definitions of* $u(\Delta t)$ *and* $d(\Delta t)$ *now reflect* p *as noted in that exercise.*

Exercise 9.11 (Rebalancing the replicating portfolio) *By the above derivation, at any node on the binomial lattice with price* $X_{k,j}$ *say,* $O^{(n)}_{j\Delta t}[X_{k,j}]$ *is the price of the replicating portfolio that replicates the next period's option price of* $O^{(n)}_{(j+1)\Delta t}[X_{k+1,j+1}]$ *when* $u(\Delta t)$ *prevails, and* $O^{(n)}_{(j+1)\Delta t}[X_{k,j+1}]$ *when* $d(\Delta t)$ *prevails. In other words,* $O^{(n)}_{j\Delta t}[X_{k,j}] = a_{k,j}X_{k,j} + b_{k,j}$ *where* $a_{k,j}$ *and* $b_{k,j}$ *satisfy (9.10) and (9.11) defined relative to these next period prices.*

Verify that any such replicating portfolio can be rebalanced at the end of the period to produce the next period's needed replicating portfolio. By rebalanced is meant, that one can trade the current portfolio into the needed portfolio with no cost or profit. Thus the initial portfolio, $a_{0,0}X_{0,0} + b_{0,0}$ *can be rebalanced all the way to time* T *to produce the desired settlement values of* $O_T(X_{k,n})$. *Hint: Given* $a_{k,j}X_{k,j} + b_{k,j}$, *show that this has the same market value as* $a_{k+1,j+1}X_{k+1,j+1}+b_{k+1,j+1}$ *when* $u(\Delta t)$ *prevails, and* $a_{k,j+1}X_{k,j+1}+b_{k,j+1}$, *when* $d(\Delta t)$ *prevails.*

The derivation of Proposition 9.9 required that the underlying asset of the financial derivative was governed by the multiplicative binomial model of Proposition 8.46, paid no income, and that long and short positions in this asset traded freely. The prototypical such asset is a common stock with no ex dividend date during the modeling period.

The next exercise generalizes to other common underlying assets.

Exercise 9.12 (Binomial lattice prices: Other underlying assets) *Derive the modification to Proposition 9.9 for three alternative underlying assets, and verify how Proposition 9.9 potentially fails for a fourth.*

1. *Assume that* X ***is the price of one unit of foreign currency*** *in units of the domestic currency, and that Proposition 9.4 and (9.7) are used to model the future states of this price. Assume that the risk-free interest rate on this foreign currency is fixed and equal to* r_f *with continuous compounding, for long and short positions. Repeat the steps in the above proof to show that (9.8) is still valid, but now:*

$$q^{Curr}(\Delta t) = \frac{e^{(r-r_f)\Delta t} - e^{d(\Delta t)}}{e^{u(\Delta t)} - e^{d(\Delta t)}}. \tag{9.12}$$

Hint: Investigate how (9.10) and (9.11) of item 1 *will change if you hold foreign currency in the replicating portfolio, noting that you would invest/borrow it at risk-free* r_f*. Thus both the price of one unit of currency changes, as does the number of units held. Now derive the change in item* 3.

2. *The adaptation of item 1 is frequently used when the underlying asset X is an* ***equity index.*** *When the individual stock dividends of the index are paid somewhat uniformly over the modeling period, it is a reasonable and common approximation to model dividends as if paid continuously at dividend rate of* δ*. If these dividends are then continuously reinvested back into the index, verify that (9.8) is still valid, but now:*

$$q^{Ind}(\Delta t) = \frac{e^{(r-\delta)\Delta t} - e^{d(\Delta t)}}{e^{u(\Delta t)} - e^{d(\Delta t)}}. \tag{9.13}$$

Hint: Mechanically, X_t *changes in price over the period and pays cash dividends that are converted back into the index:*

$$X_t \rightarrow X_{t+\Delta t} + (e^{\delta\Delta t} - 1)X_{t+\Delta t} \rightarrow e^{\delta\Delta t}X_{t+\Delta t}.$$

Admittedly, one might assume dividends are paid on X_t*, but again this is an approximation.*

3. *Assume that X* ***is a futures price,*** *and that Proposition 9.4 and (9.7) are used to model the states of this price in the future. Logically, the derivatives settlement date must be well before the futures delivery date, and we assume this. Repeat the steps in the above proof to show that (9.8) is still valid, but now:*

$$q^{Fut}(\Delta t) = \frac{1 - e^{d(\Delta t)}}{e^{u(\Delta t)} - e^{d(\Delta t)}}. \tag{9.14}$$

Hint: Investigate how (9.10) and (9.11) of item 1 *will change if you hold futures contracts in the replicating portfolio. Recall that if the futures price changes from* $X_{k,j}$ *to* $X_{k+1,j+1}$*, say, the value of one long futures position changes by* $X_{k+1,j+1} - X_{k,j}$*, and this is then multiplied by* $a_{k,j}$*. For item 3 of this proposition's proof, justify why the price of the replicating portfolio is* $b_{k,j}$*.*

4. *Assume that X* ***is the price of a commodity,*** *and that Proposition 9.4 and (9.7) are used to model the future states of this price. Verify how the above derivation may fail:*

   (a) *Item 1 of the proof works mathematically if* $a_{k,j} > 0$*, but note that holding commodities long often requires payment of storage costs. Thus this cost is equivalent to negative income and can be accommodated as for currency but with negative* r_c*.*

   (b) *If the calculated* $a_{k,j} < 0$ *in item 1, this requires a short position in X which is generally not possible (gold is the exception). Thus step 1 can be pushed through for derivatives with* $a_{k,j} > 0$*, meaning derivatives with payoffs that are positively correlated with asset prices (calls for example).*

   (c) *Item 2 of the proof always fails, since the law of one price requires that you can take long or short positions in X, and shorting is generally not possible. For derivatives with* $a_{k,j} > 0$*, step 2 will provide only that* $a_{k,j}X_{k,j} + b_{k,j}$ *is an upper bound to price of the end of period payoffs.*

   One exception to the conclusion in item c is for gold, which though a commodity, has the somewhat unique distinction of being a commodity that investors often hold for its appreciation potential. Thus it is possible to short gold, and the original setup is valid with the small adjustment that there is a cost to borrowing gold, called the "lease rate," which the lender earns, and the borrower pays, much like a currency.

We memorialize the derivations in items 1, 2, and 3 of Exercise 9.11 as a corollary.

Corollary 9.13 (Binomial lattice price: Other assets) *Let $O_T[X_T]$ denote the settlement value of a European financial derivative at time $T = n\Delta t$ expressed as a function of the then prevailing price of an asset X. Assume that investors can take long or short positions in X with no transaction costs.*

Then the binomial lattice price $O_0^{(n)}[X_0]$ is given as in (9.8), with a redefinition of q:

- *If X is the price of a foreign currency with associated risk-free rate r_f, then q is given in (9.12).*
- *If X is the price of a stock index with assumed continuous dividend rate δ, then q is given in (9.13).*
- *If X is the futures price for a contract with delivery date after time T, then q is given in (9.14).*

Proof. *Exercise 9.11.* ■

Notation 9.14 (Expected value pricing) *The European derivatives price in (9.8) can also be written in the simpler and more suggestive notation as:*

$$O_0^{(n)}[X_0] = e^{-rT}E_q[O_T[X_T]], \tag{9.15}$$

where E_q denotes the expectation of the option payoffs $O_T[X_T]$, relative to the risk-neutral probability $q \equiv q(\Delta t)$.

Specifically:

$$\Pr_q[X_{k,n}] = \binom{n}{k} q^k(1-q)^{n-k},$$

where $n\Delta t = T$ and $q(\Delta t)$ is defined in (9.9), or as modified in Exercise 9.11.

This notation is suggestive because it raises the possibility that the limiting price as $n \to \infty$ of this European derivative will be given by the same formula. Then, the E_q-expectation makes sense when taken with respect to the limiting distribution of these binomial distributions of asset prices as $n \to \infty$. We will see below that this possibility is indeed realized.

Example 9.15 (Long forward contract) *As an example, let the European derivative security be a long forward contract on X with forward price F, and settlement at time T. We assume that X pays no income over the contract period.*

The settlement value is given by $F_T[X_T] = X_T - F$, and denoting the price in (9.8) by $F_0^{(n)}[X_0]$:

$$\begin{aligned} F_0^{(n)}[X_0] &= e^{-rT}\sum\nolimits_{j=0}^{n}\binom{n}{j}q^j(1-q)^{n-j}\left[X_0e^{ju(\Delta t)}e^{(n-j)d(\Delta t)} - F\right] \\ &= e^{-rT}\left[X_0\sum\nolimits_{j=0}^{n}\binom{n}{j}\left(qe^{u(\Delta t)}\right)^j\left[(1-q)e^{d(\Delta t)}\right]^{n-j} - F\right], \end{aligned}$$

since $\sum_{j=0}^{n}\binom{n}{j}q^j(1-q)^{n-j} = 1$.

Now $qe^{u(\Delta t)} + (1-q)e^{d(\Delta t)} = e^{r\Delta t}$ obtains:

$$\sum\nolimits_{j=0}^{n}\binom{n}{j}\left(qe^{u(\Delta t)}\right)^j\left[(1-q)e^{d(\Delta t)}\right]^{n-j} = \left[qe^{u(\Delta t)} + (1-q)e^{d(\Delta t)}\right]^n = e^{rn\Delta t},$$

and thus since $n\Delta t = T$:

$$F_0^{(n)}[X_0] = X_0 - Fe^{-rT}. \tag{9.16}$$

Exercise 9.16 (Long forward contract) *Derive (9.16) by constructing a replicating portfolio for this contract at time* 0, *and justifying by arbitrage that the price of this portfolio must agree with the price of this contract. Hint: What position in X and risk-free assets has value $X_T - F$ at time T?*

How does (9.16) change if X is the price of a foreign currency in units of the domestic currency? Verify this price using the approach of Example 9.15, as well as by constructing a replicating portfolio.

Consider next the price of a financial derivative security on X with the identity settlement function, $O_T(x) = x$ at time T. For example, if X is a common stock with no ex dividend date during the modeling period and $q \equiv q(\Delta t)$ in (9.9):

$$O_0^{(n)}[X_0] = e^{-rT} \sum\nolimits_{j=0}^{n} \binom{n}{j} q^j (1-q)^{n-j} \left[X_0 e^{ju(\Delta t)} e^{(n-j)d(\Delta t)} \right] = X_0. \tag{9.17}$$

Stated another way:

$$X_0 = e^{-rT} E_q [X_T]. \tag{9.18}$$

It is compellingly logical that this financial derivative, which delivers either the common stock with price X_T or the equivalent in currency units, would be priced at X_0 today. This contract can be readily replicated by buying one share of stock today, and arbitrage will enforce the identity in (9.17) assuming only that both the stock and derivative contract can be freely traded both long and short.

For the risk-free asset, with constant settlement function $O_T(x) = X_0 e^{rT}$, it is an exercise to check that (9.17) is again satisfied.

For currencies, stock indexes and futures contracts, one obtains a different, yet equally compelling result, as will be explained below.

Exercise 9.17 (On (9.17) and generalizations) *Derive (9.17), and then show that using (9.8) and the respective q in (9.12), (9.12), or (9.14), that:*

1. ***Currencies:*** *A contract with settlement value of one unit of currency has current price:*

$$O_0^{(n)}[X_0] = e^{-rT} \sum\nolimits_{j=0}^{n} \binom{n}{j} q^j (1-q)^{n-j} \left[X_0 e^{ju(\Delta t)} e^{(n-j)d(\Delta t)} \right] = e^{-r_f T} X_0. \tag{9.19}$$

This financial derivative can be replicated by purchasing $e^{-r_f T}$ units of foreign currency today at price $e^{-r_f T} X_0$ and investing at rate r_f.

2. ***Stock index:*** *A contract with settlement value of one unit of the index has current price:*

$$O_0^{(n)}[X_0] = e^{-rT} \sum\nolimits_{j=0}^{n} \binom{n}{j} q^j (1-q)^{n-j} \left[X_0 e^{ju(\Delta t)} e^{(n-j)d(\Delta t)} \right] = e^{-\delta T} X_0. \tag{9.20}$$

This financial derivative can be replicated by purchasing $e^{-\delta T}$ units of the index today at price $e^{-\delta T} X_0$, and reinvesting dividends in the index.

3. ***Futures:*** *A contract with settlement value of the then futures price:*

$$O_0^{(n)}[X_0] = e^{-rT} \sum\nolimits_{j=0}^{n} \binom{n}{j} q^j (1-q)^{n-j} \left[X_0 e^{ju(\Delta t)} e^{(n-j)d(\Delta t)} \right] = e^{-rT} X_0. \tag{9.21}$$

This financial derivative can be replicated by investing $e^{-rT} X_0$ of cash at r and going long one futures contract on X_0 with delivery date past time T. Why does this work?

The probability identified as q in (9.9), (9.12), (9.13), or (9.14) is called the **risk-neutral probability** for the respective securities. One may wonder why.

Remark 9.18 (On risk-neutral probabilities) *In a* ***risk-neutral world****, which is characterized by a linear utility function (see* **Reitano** *(2010), Section 9.8.8), every risky asset is priced as the risk-free present value of the expected cash flows. Since the riskiness of these cash flows is not reflected in the expectation, and discounting is performed with the risk-free rate, this is indeed a risk-neutral world.*

By (9.17), the current price X_0 of some assets, such as a common stock with no dividends, equals the risk-free present value of its expected price at time T, where this expectation is calculated using the risk-neutral probability as identified in (9.9). Thus q as defined in (9.9), equals the unique probability that would be used in a risk-neutral world to obtain the current price of X_0 for such a stock.

The three pricing equations in Exercise 9.17 reflect this same idea, but need to be restructured. With settlement function $O_T(x) = xe^{cT}$ and c defined by r_f or δ, the identities in (9.19) and (9.20) can be expressed:

$$O_0^{(n)}[X_0] = X_0.$$

Thus the risk-neutral price now for $X_T e^{r_f T}$ in currency at time T is X_0, and similarly for $X_T e^{\delta T}$ of the index. In each case, this agrees with the current price of the currency/index, noting that we can then let interest accumulate and reinvest dividends.

For (9.19), consider the settlement function $O_T(x) = x - X_0$, which is the settlement value on a long futures contract. It then follows that:

$$O_0^{(n)}[X_0] = 0,$$

the current price of this contract.

In summary, the name ***risk-neutral probabilities*** *implies that in each case, these are the probabilities that would be used in a risk-neutral world to obtain current market prices.*

The real world is not risk-neutral certainly. But many financial derivatives can be priced as if we lived in a risk-neutral world, using the risk-neutral probabilities derived above, due to the replicating portfolio constructions of Proposition 9.9 and Exercise 9.12. If a financial derivative can be replicated, then the derivative's price equals the price of this portfolio by the law of one price, and thus risk-neutral pricing of underlying assets obtains risk-neutral pricing of financial derivatives.

Remark 9.19 (On risk-neutral pricing) *It is tempting to think that replication and the risk-neutral pricing approach above is an inessential convenience. Indeed, within the context of the real-world binomial models using any p, it may well seem possible to represent X_0:*

$$X_0 = e^{-r_{RW}T} E_p[X_T],$$

where r_{RW} is now the return demanded by investors in the real world, and reflecting risk.

With $p = 1/2$ for the definitions of $u(\Delta t)$ and $d(\Delta t)$ given in (9.2), the above formula produces:

$$X_0 = e^{-r_{RW}T} \left(\frac{e^{u(\Delta t)} + e^{d(\Delta t)}}{2} \right)^n X_0.$$

Thus r_{RW} can be calculated:

$$e^{r_{RW}} = \left(\frac{e^{u(\Delta t)} + e^{d(\Delta t)}}{2} \right)^{1/\Delta t}. \tag{1}$$

But this is then a dead end for pricing financial derivatives on X. While such r_{RW} may be the appropriate rate to discount for the risk of X_T in the real world, the risk of $O_T[X_T]$ will in general be quite different, and indeed may be negatively correlated with the risk of X_T.

Put another way, let r'_{RW} be defined relative to a given derivative by:

$$O_0^{(n)}[X_0] = e^{-r'_{RW}T} E_p[O_T[X_T]].$$

Given market pricing of $O_0^{(n)}[X_0]$, this rate can be reverse-engineered as above, will in theory differ for every derivative, and need not even be positive. However, in the absence of market prices, this rate is unobservable, and thus this approach cannot be used to determine financial derivative prices.

The power of risk-neutral pricing is that this approach allows expected present value pricing of derivatives with risk-free rates, which are independently observable.

Exercise 9.20 *Check that as $\Delta t \to 0$:*

$$r_{RW} \to \mu + \frac{1}{2}\sigma^2.$$

Hint: Take a logarithm of the above identity in (1)*, then approximate the exponentials and logarithm with Taylor series. Focus on the terms up to $O(\Delta t)$, and show that the higher order remainder terms converge to* 0.

9.1.2 American Options

For European options, there is a well-defined long position, the buyer of the option, and short position, the seller of the option, and the long controls the exercise of the option in the sense of deciding whether or not to exercise. But because the settlement value is formulaic, one could also argue that it is well-defined when exercise **should** occur. Then on such exercise, the long and short have complementary settlement functions.

General European financial derivatives such as forward contracts also make sense. Here there need be no "option" component *per se*, since exercise is mandatory. Again, both parties have well-defined and complementary settlement functions, where by complementary is meant that the values sum to zero.

An **American financial derivative** provides for a formulaic settlement at time T as does a European derivative, but also provides a formulaic **early settlement option**, meaning at times $0 < t < T$ and sometimes $0 < t_0 \leq t < T$. This early settlement then cancels the remainder of the contract, and thus early settlement requires a real decision. Since the counterparties to such a contract have opposing interests, this structure makes little sense outside the options market.

As in the European case, there is then a well-defined long and short position, and it is the long position that chooses early exercise, or not. For this reason, we abandon more general references to financial derivatives in this section.

By an arbitrage argument, the market price for an American option must equal or exceed its then current exercise value. Also by definition, one would expect the market price for an American option to equal or exceed the market price for an otherwise identical European option, and this is usually the case. A potential anomaly here is an American call option on a dividend paying stock, for which future dividends depress the stock price and thus lower the value of the call. We will not pursue such examples further.

For pricing of American options, we require as above that long and short positions in these derivatives and the underlying assets trade freely, and that there are no trading costs or other market frictions. We also assume that the contractual payments to the long position are certain to be made, and thus the long has no credit risk to the short. In practice, credit risk is essentially eliminated on exchange-traded options by the requirement that the short position establish a **margin account** with adequate collateral to guarantee settlement. For over-the-counter options markets, similar collateral structures are commonly used.

The pricing of European options can be adapted to American options by appropriately adjusting the one period iterative pricing formula in item 4 of Proposition 9.9. For the proof, all references to "item xx" will be to Proposition 9.9.

For this result, we recall the notation of (9.7).

Proposition 9.21 (Binomial lattice price of an American derivative) *Let $O_t[X_t]$ denote the settlement value of an American option at time t expressed as a function of the then prevailing price of an asset X. Assume that investors can take long or short positions in X with no transaction costs, that X pays no income, and that the risk-free interest rate with continuous compounding is fixed and equal to r for both long and short positions.*

Then the binomial lattice price $O_0^{(n)}[X_0]$ at time $t = 0$ for a replicating portfolio for this financial derivative is defined iteratively by:

$$O_{j\Delta t}^{(n)}[X_{k,j}] = \max\left(e^{-r\Delta t}\left[qO_{(j+1)\Delta t}^{(n)}[X_{k+1,j+1}] + (1-q)O_{(j+1)\Delta t}^{(n)}[X_{k+1,j}]\right], O_{j\Delta t}[X_{k,j}]\right), \tag{9.22}$$

where $q \equiv q(\Delta t)$ as in (9.9).

Defining $O_{n\Delta t}^{(n)}[X_{k,n}] \equiv O_T[X_{k,n}]$ for all k, this iteration to $t = 0$ is in order:

- $j = n-1, n-2, ..., 0;$
- *For each j, $k = 0, 1, ..., j$.*

Proof. *At time $t = n\Delta t$, the market price of this derivative is unambiguously equal to the settlement values, and so $O_{n\Delta t}^{(n)}[X_{k,j}] = O_T[X_{k,n}]$ for all $k = 0, ..., n$.*

At time $t = (n-1)\Delta t$, the prices of replicating portfolios are given as in item 1 of the proof of Proposition 9.9 by $P_{(n-1)\Delta t}(X_{k,n-1}) \equiv a_{k,j}X_{k,n-1} + b_{k,n-1}$ for $0 \leq k \leq n-1$, where each such portfolio replicates $O_T^{(n)}[X_{k+1,n}]$ if $u(\Delta t)$ prevails, and $O_T^{(n)}[X_{k,n}]$ if $d(\Delta t)$ prevails. If early exercise value $O_{(n-1)\Delta t}[X_{k,n-1}] < P_{(n-1)\Delta t}(X_{k,n-1})$, then as in item 2 it follows that $P_{(n-1)\Delta t}(X_{k,n-1}) = O_{(n-1)\Delta t}^{(n)}(X_{k,n-1})$ is the market price; if $O_{(n-1)\Delta t}[X_{k,n-1}] \geq P_{(n-1)\Delta t}(X_{k,n-1})$, then $O_{(n-1)\Delta t}[X_{k,n-1}]$ is the market price. Thus (9.22) is proved at time $(n-1)\Delta t$.

This identical argument can then be repeated at each successive j, and the proof is complete. ■

Remark 9.22 (Computational cost) *In the case of a European financial derivative, the first step iterative formula in* (1) *of the proof of Proposition 9.9 is exactly the iterative formula in (9.22), but without reference to an early settlement value $O_{j\Delta t}[X_{k,j}]$. Indeed, the proof of this result reflected the fact that a European derivative can also be evaluated with the above iterative formula, modified to omit the early settlement values.*

Fortunately, these calculations collapse into the simplified formula in (9.8) for a European financial derivative.

For an American financial derivative, there is no such simplification. The above formula must be iterated n-times to obtain $\{O_{(n-1)\Delta t}^{(n)}[X_{k,n-1}]\}_{k=0}^{n-1}$, then $(n-1)$-times to obtain $\{O_{(n-2)\Delta t}^{(n)}[X_{k,n-2}]\}_{k=0}^{n-2}$, and so forth, ultimately requiring $\sum_{j=1}^{n} j = \frac{1}{2}n(n+1)$ iterations.

Corollary 9.23 (Binomial lattice price: Other assets) *Let $O_t[X_t]$ denote the settlement value of an American option at time t expressed as a function of the then prevailing price of an asset X. Assume that investors can take long or short positions in X with no transaction costs.*

Then in the following cases, the binomial lattice price $O_0^{(n)}[X_0]$ is given iteratively as in (9.22) with a redefinition of q:

- *If X is the price of a foreign currency with associated risk-free rate r_f, then q is given in (9.12).*
- *If X is the price of a stock index with continuous dividend rate δ, then q is given in (9.13).*
- *If X is the futures price for a contract with delivery date after time T, then q is given in (9.14).*

Proof. *Left as an exercise in generalizing Proposition 9.21 as in Exercise 9.12.* ∎

Thus for American options, at each node of the binomial lattice, the market price (recall Remark 9.8) of the financial derivative is the greater of:

- The current exercise value, $O_{j\Delta t}[X_{k,j}]$, and,
- The value of all future exercise opportunities as reflected in the expected present value.

In contrast to the European option formula, this pricing formula is necessarily iterative and does not reduce to a simple formula at $t = 0$ as does the European price in (9.8).

It should be noted that $O_0^{(n)}[X_0]$ is indeed the price of a replicating portfolio for this option, but not necessarily the price of a replicating portfolio for the maturity settlements defined by $O_T[X_{k,n}]$. Instead, an American option price provides a portfolio at time 0 that can be rebalanced to the optimal exercise date, which might be that of an early settlement or one at final maturity. Thus, this replicating portfolio replicates the prices of this option throughout its existence.

9.2 Limiting Risk-Neutral Asset Distribution

Given the above derivative pricing results, a logical next question now becomes:

What is the limiting distribution of X_T, or $X_T^{(n)}$ in the notation of the prior chapter, as $n \to \infty$?

We cannot simply apply the result of Proposition 8.46 on multiplicative binomial temporal models because here the binomial probability is a function of Δt, $q \equiv q(\Delta t)$, or equivalently, q is a function of n since $\Delta t = T/n$.

However, an immediate corollary of the next result is that:

$$\lim_{\Delta t \to 0} q(\Delta t) = \frac{1}{2}, \tag{9.23}$$

and this generalizes to $\lim_{\Delta t \to 0} q(\Delta t) = p$ when $u(\Delta t)$ and $d(\Delta t)$ are redefined to reflect $p \neq 1/2$. So it is natural to expect that the limiting asset distribution under the risk-neutral q probabilities will agree with the limiting distribution under the real-world p probabilities of Proposition 8.46.

But this expectation is soon to be dashed. It turns out that $q(\Delta t)$ approaches $1/2$ (or p) relatively slowly. Indeed, this convergence is slow enough to shift the mean of the limiting distribution from that in Proposition 8.46, though it will not change the variance.

9.2.1 Analysis of the Probability $q(\Delta t)$

To investigate the limiting distribution of assets under $q(\Delta t)$ will require a more informative representation of this probability as a function of Δt. To this end, we prove the following expansion for $q(\Delta t)$ to $O(\Delta t^{3/2})$. For a generalization of these results to $O(\Delta t^2)$, the reader can as an exercise extend the below proof an additional step, or see Proposition 9.155 in **Reitano** (2010), which also generalizes the result to $p \neq 1/2$.

To simplify notation from this point forward, we first consolidate the various formulas for $q(\Delta t)$ in (9.9), (9.12), (9.13), and (9.14).

Notation 9.24 (On $q^{Gen}(\Delta t)$) *With $u(\Delta t)$ and $d(\Delta t)$ defined in (9.2), the general formulas for $q(\Delta t)$ for the various underlying assets can be expressed:*

$$q^{Gen}(\Delta t) = \frac{e^{R(r)\Delta t} - e^{d(\Delta t)}}{e^{u(\Delta t)} - e^{d(\Delta t)}}, \tag{9.24}$$

where $R(r)$ is defined by:

$$R(r) = \begin{cases} r, & q^{Gen} = q(\Delta t), \\ r - r_f, & q^{Gen} = q^{Curr}(\Delta t), \\ r - \delta, & q^{Gen} = q^{Ind}(\Delta t), \\ 0, & q^{Gen} = q^{Fut}(\Delta t). \end{cases} \tag{9.25}$$

Proposition 9.25 ($q^{Gen}(\Delta t)$ to $O(\Delta t^{3/2})$) *With $q^{Gen}(\Delta t)$ as in (9.24):*

$$q^{Gen}(\Delta t) = \frac{1}{2} + \frac{1}{2\sigma}\left(R(r) - \mu - \sigma^2/2\right)\sqrt{\Delta t} + O(\Delta t^{3/2}), \tag{9.26}$$

*where the **big-O** error term $O(\Delta t^{3/2})$ in (9.26) means that as $\Delta t \to 0$:*

$$\frac{q^{Gen}(\Delta t) - \frac{1}{2} - \frac{R(r)-\mu-\sigma^2/2}{2\sigma}\sqrt{\Delta t}}{\Delta t^{3/2}} \to C < \infty.$$

Proof. *With $u(\Delta t)$ and $d(\Delta t)$ as in (9.2), a Taylor series analysis obtains:*

$$\begin{aligned} q^{Gen}(\Delta t) &= \frac{\exp\left[\sigma\sqrt{\Delta t} + (R(r) - \mu)\Delta t\right] - 1}{\exp\left[2\sigma\sqrt{\Delta t}\right] - 1} \qquad (1) \\ &= \frac{\sigma\sqrt{\Delta t} + \left[(R(r) - \mu) + \sigma^2/2\right]\Delta t + O(\Delta t^{3/2})}{2\sigma\sqrt{\Delta t} + 2\sigma^2\Delta t + O(\Delta t^{3/2})}. \end{aligned}$$

Dividing numerator and denominator by $\sigma\sqrt{\Delta t}$ obtains the limit noted above, that $\lim_{\Delta t \to 0} q(\Delta t) = \frac{1}{2}$.

Rewriting (1) and substituting $\sqrt{\Delta t} \to x$, let:

$$f(x) = \frac{\exp\left[(R(r) - \mu)x^2\right] - \exp\left[-\sigma x\right]}{\exp\left[\sigma x\right] - \exp\left[-\sigma x\right]},$$

which is infinitely differentiable for $x > 0$. Expanding numerator $N_f(x)$ and denominator $D_f(x)$ using Taylor series obtains:

$$\begin{aligned} N_f(x) &= \sum\nolimits_{n=1}^{\infty} \left[\frac{(R(r)-\mu)^{n/2}}{(n/2)!}\delta_n - \frac{(-\sigma)^n}{n!} \right] x^n \\ &\equiv \sum\nolimits_{n=1}^{\infty} a_n x^n, \\ D_f(x) &= \sum\nolimits_{n=1}^{\infty} \frac{2}{n!}\sigma^n (1-\delta_n) x^n \\ &\equiv \sum\nolimits_{n=1}^{\infty} b_n x^n. \end{aligned}$$

where:

$$\delta_n \equiv \begin{cases} 0, & n \text{ odd}, \\ 1, & n \text{ even}. \end{cases}$$

An explicit power series long division is simplified by noting that $b_2 = 0$ and yields:

$$f(x) = \frac{a_1}{b_1} + \frac{a_2}{b_1}x + \left(a_3 - \frac{a_1 b_3}{b_1}\right)x^2 + O(x^3).$$

A substitution then obtains (9.26). ■

Example 9.26 (q-Moments of binomial b_q) *Denote by b_q the above binomial b in the risk-neutral probability q, so $b_q = 1$ with probability q, and $b_q = -1$ with probability $1-q$.*

To simplify notation, express $q^{Gen}(\Delta t)$ in (9.26) by:

$$q^{Gen} = \frac{1}{2} + a\sqrt{\Delta t} + O(\Delta t^{3/2}).$$

A calculation obtains:

$$\begin{aligned} E_q[b_q] &= 2q - 1 = 2a\sqrt{\Delta t} + O(\Delta t^{3/2}), \\ Var_q[b_q] &= 4q(1-q) = 1 - 4a^2\Delta t + O(\Delta t^{3/2}). \end{aligned}$$

While (9.23) is satisfied for all risk-neutral probability expressions in (9.26), the presence of the $\sqrt{\Delta t}$ term in these expansions causes this convergence to be very slow as $\Delta t \to 0$. The most important consequence of this slow convergence of $q(\Delta t)$ to $1/2$, is that the mean of $\ln(X_j^{(n)}/X_0)$ is shifted significantly relative to (9.3), while the effect on variance is minimal and goes to zero with Δt.

Proposition 9.27 (q-Moments of $\ln(X_j^{(n)}/X_0)$) *Given T and n with $\Delta t = T/n$, if $X_t \equiv X_j^{(n)}$ denotes the random variable of binomial lattice prices at time $t = j\Delta t$, then under the q-probabilities of (9.26):*

$$\begin{aligned} E_{q(\Delta t)}[\ln(X_t/X_0)] &= (R(r) - \sigma^2/2)t + O(\Delta t), \\ Var_{q(\Delta t)}[\ln(X_t/X_0)] &= \sigma^2 t + O(\Delta t), \end{aligned} \tag{9.27}$$

where $R(r)$ is given in (9.25).

Proof. *Since $X_j^{(n)}/X_0 = \prod_{k=1}^{j} X_k^{(n)}/X_{k-1}^{(n)}$, with $X_0^{(n)} = X_0$ by definition, it follows that:*

$$\ln(X_j^{(n)}/X_0) = \sum\nolimits_{k=1}^{j} \ln\left(X_k^{(n)}/X_{k-1}^{(n)}\right),$$

where by definition of $X_k^{(n)}$:

$$\ln\left(X_k^{(n)}/X_{k-1}^{(n)}\right) = \mu\Delta t + \sigma\sqrt{\Delta t}b_{q_k}. \tag{1}$$

Here $\{b_{q_k}\}_{k=1}^n$ are independent binomial variates which equal 1 with probability q, and -1 with probability $1-q$, where in the notation of Example 9.26 and Proposition 9.25:

$$\begin{aligned} q &= \frac{1}{2} + a\sqrt{\Delta t} + O(\Delta t^{3/2}) \\ &\equiv \frac{1}{2} + \frac{1}{2\sigma}\left(R(r) - \mu - \sigma^2/2\right)\sqrt{\Delta t} + O(\Delta t^{3/2}). \end{aligned} \tag{2}$$

By Proposition II.3.56, independence of $\{b_{q_k}\}_{k=1}^n$ assures independence of $\left\{\ln\left(X_k^{(n)}/X_{k-1}^{(n)}\right)\right\}_{k=1}^n$, and thus:

$$\begin{aligned} E_q\left[\ln(X_j^{(n)}/X_0)\right] &= \sum_{k=1}^j E_q\left[\ln(X_k^{(n)}/X_{k-1}^{(n)})\right], \\ Var_q\left[\ln(X_j^{(n)}/X_0)\right] &= \sum_{k=1}^j Var_q\left[\ln(X_k^{(n)}/X_{k-1}^{(n)})\right]. \end{aligned} \tag{3}$$

By (1), *Example 9.26, then* (2)*:*

$$\begin{aligned} E_q\left[\ln(X_k^{(n)}/X_{k-1}^{(n)})\right] &= \mu\Delta t + \left(2a\sqrt{\Delta t} + O(\Delta t^{3/2})\right)\sigma\sqrt{\Delta t} \\ &= (R(r) - \sigma^2/2)\Delta t + O(\Delta t^2), \end{aligned}$$

and similarly:

$$\begin{aligned} Var_q\left[\ln(X_k^{(n)}/X_{k-1}^{(n)})\right] &= \sigma^2\Delta t\left(1 - 4a^2\Delta t + O(\Delta t^{3/2})\right) \\ &= \sigma^2\Delta t + O(\Delta t^2). \end{aligned}$$

The result in (9.27) then follows by addition in (3), *noting that $jO(\Delta t^2)$ is at most $nO(\Delta t^2) = O(\Delta t)$, recalling that $\Delta t = T/n$.* ■

9.2.2 Limiting Asset Distribution under $q(\Delta t)$

Under the real-world probability $p = \frac{1}{2}$, Proposition 8.46 states that as $\Delta t \to 0$, the limiting distribution of $\ln[X_t/X_0]$ is normal for $t = jT/n$ for any n and all $1 \leq j \leq n$, with moments:

$$E_p\left[\ln[X_t/X_0]\right] = \mu t, \qquad Var_p\left[\ln[X_t/X_0]\right] = \sigma^2 t.$$

The expressions in (9.27) make clear that even before deriving the limiting distribution of X_t under the risk-neutral probability $q(\Delta t)$, the mean and variance of $\ln[X_t/X_0]$ for such t must be to $O(\Delta t)$:

$$E_{q(\Delta t)}\left[\ln[X_t/X_0]\right] = (R(r) - \sigma^2/2)t, \qquad Var_{q(\Delta t)}\left[\ln[X_t/X_0]\right] = \sigma^2 t,$$

with $R(r)$ defined in (9.25) for the various asset classes.

Remark 9.28 (On μ) *Focusing on the shift in the mean of $\ln[X_t/X_0]$ when we move from p to $q(\Delta t)$, it becomes clear by inspection of (9.25) and is quite notable, that in the limit under $q(\Delta t)$, the real-world parameter μ has disappeared! The same result occurs with the more general model of (9.4) as seen in Proposition 9.159 of* **Reitano** *(2010).*

The modeling implications of this are discussed in the next section.

Returning to limiting distributions, the real-world asset model is **approximately scalable** to order $k_0 = 2$ with probability $p = \frac{1}{2}$. Under q, this model is not approximately scalable, but is **asymptotically scalable** to order $k_0 = 2$ by (9.27). Thus Proposition 8.43 applies subject to the verification of Lindeberg's condition in (8.35).

This verification is the focus of the next result.

Proposition 9.29 (Limiting distribution of $X_j^{(n)}$ under $q(\Delta t)$ as $\Delta t \to \infty$) *Given a fixed horizon time T and $\Delta t = T/n$, initial price $X_0 > 0$, and parameters μ and $\sigma > 0$, let:*

$$X_j^{(n)} = X_0 \exp\left[\sum\nolimits_{i=1}^{j} Z_i^{(n)}\right],$$

where $Z_i^{(n)} = \mu\Delta t + \sigma\sqrt{\Delta t} b_i^{(n)}$. Here $\{b_i^{(n)}\}_{i=1}^{n}$ are independent and binomially distributed to equal 1 or -1 with respective probabilities q_n and $1 - q_n$, where $q_n \equiv q(\Delta t)$ is defined for the various underlying assets with (9.24). Thus $X_j^{(n)}$ is the price variate at time $t = jT/n$ for integer $0 < j \leq n$ in this multiplicative temporal model.

Given m, define the refined price variate $X_{mj}^{(mn)}$ at time $t = jT/n$:

$$X_{mj}^{(mn)} \equiv X_0 \exp\left[\sum\nolimits_{i=1}^{mj} Z_i^{(mn)}\right],$$

where $Z_i^{(mn)} = \mu\Delta t_m + \sigma\sqrt{\Delta t_m} b_i^{(mn)}$ with $\Delta t_m \equiv \Delta t/m$, and $\{b_i^{(mn)}\}_{i=1}^{mn}$ defined as above relative to $q_{mn} \equiv q(\Delta t_m)$.

Then for all $t \equiv j\Delta t$ for $1 \leq j \leq n$, as $m \to \infty$:

$$\frac{\ln[X_{mj}^{(mn)}/X_0] - (R(r) - \sigma^2/2)t}{\sigma\sqrt{t}} \to_d N, \tag{9.28}$$

where N denotes the standard normal variate, and $R(r)$ is defined in (9.25).

Equivalently, $X_{mj}^{(mn)}$ converges in distribution for all $1 \leq j \leq n$ to the lognormally distributed variate X_t for $t = j\Delta t$, with parameters $\ln X_0 + (R(r) - \sigma^2/2)t$ and $\sigma^2 t$:

$$X_{mj}^{(mn)} \to_d X_t \equiv X_0 \exp\left[(R(r) - \sigma^2/2)t + \sigma\sqrt{t}N\right]. \tag{9.29}$$

Proof. *For given $t = j\Delta t$, define a triangular array indexed by $m \geq 1$ by:*

$$\{Z_i^{(mn)}\}_{i=1}^{mj} \equiv \{\mu\Delta t_m + \sigma\sqrt{\Delta t_m} b_i^{(mn)}\}_{i=1}^{mj},$$

where as above, $\Delta t_m \equiv \Delta t/m$. Then $\ln\left[X_{mj}^{(mn)}/X_0\right] = \sum_{i=1}^{mj} Z_i^{(mn)}$, an independent sum for each m by Proposition II.3.56, and the mean $\mu^{(mn)}$ and variance $\left(\sigma^{(mn)}\right)^2$ of $Z^{(mn)}$ are given in the proof of Proposition 9.27:

$$\mu^{(mn)} = (R(r) - \sigma^2/2)\Delta t_m + O(\Delta t_m), \qquad \left(\sigma^{(mn)}\right)^2 = \sigma^2\Delta t_m + O(\Delta t_m). \tag{1}$$

Define $\widetilde{Z^{(mn)}} \equiv (Z^{(mn)} - \mu^{(mn)})/\sigma^{(mn)}$:

$$\begin{aligned} \widetilde{Z^{(mn)}} &= \frac{\left(\mu - (R(r) - \sigma^2/2)\right)\Delta t_m + \sigma\sqrt{\Delta t_m} b_i^{(mn)} - O(\Delta t_m)}{\sqrt{\sigma^2\Delta t_m + O(\Delta t_m)}} \\ &= \frac{\left(\mu - (R(r) - \sigma^2/2)\right)\sqrt{\Delta t_m} + \sigma b_i^{(mn)} - O(\sqrt{\Delta t_m})}{\sigma\sqrt{1 + O(1)}}, \end{aligned}$$

where by $O(1)$ is meant that this is a bounded constant as $\Delta t_m \to 0$

To apply Proposition 8.43, it is enough to show (8.35), that for all $t > 0$*:*

$$\lim_{m\to\infty}\int_{\left|\widetilde{Z^{(mn)}}\right|>t\sqrt{mn}}\left(\widetilde{Z^{(mn)}}\right)^2 d\lambda = 0.$$

But as $m \to \infty$*:*

$$\left|\widetilde{Z^{(mn)}}\right| \to c,$$

and thus $\left|\widetilde{Z^{(mn)}}\right|$ *is bounded for* m *large and (8.35) is satisfied.*

The results in (9.28) and (9.29) now follow from item 2 of Proposition 8.43. ■

9.3 A Real-World Model under $p(\Delta t)$

The result of the prior section should give every good model builder pause. We recap.

1. Real-World Model: *To build a real-world model for the various asset classes identified in (9.25), we require appropriate parameters* μ *and* σ*. The model over any interval* $[t, t+\Delta t]$ *then becomes:*

$$X_{t+\Delta t} = X_t \exp\left[\mu\Delta t + \sigma\sqrt{\Delta t}b\right], \tag{1}$$

with b *a binomial random variable, equal to* ± 1 *with probability* $p = \frac{1}{2}$*. This is equivalent to the* $u(\Delta t)$ *and* $d(\Delta t)$ *continuous returns of (9.2):*

$$u(\Delta t) = \mu\Delta t + \sigma\sqrt{\Delta t}, \qquad d(\Delta t) = \mu\Delta t - \sigma\sqrt{\Delta t}.$$

By Proposition 8.46, we obtain the limiting result for the various asset classes:

$$X_{t+\Delta t} = X_t \exp\left[N\left(\mu\Delta t, \sigma^2\Delta t\right)\right]. \tag{2}$$

This result allows one to both validate this model, as well as calibrate it.

- **Validate:** *Given real-world pricing data with* Δt *time-steps, verify that* $\ln[X_{t+\Delta t}/X_t]$ *is approximately normal.*
- **Calibrate:** *Given this real-world data, estimate the model parameters by:*

$$\mu = \frac{1}{\Delta t}E\left[\ln[X_{t+\Delta t}/X_t]\right], \quad \sigma^2 = \frac{1}{\Delta t}Var\left[\ln[X_{t+\Delta t}/X_t]\right].$$

2. Risk-Neutral Model: *For the pricing of various financial derivatives on these asset classes, we again began with the model in* (1)*, but discovered that for derivatives pricing, that the real-world probability* $p = \frac{1}{2}$ *played no role. Instead, this asset model was better understood with binomial* b *equalling* $+1$ *with probability* $q(\Delta t)$*, where this "risk-neutral" probability varied for the various asset classes as in (9.24):*

$$q^{Gen}(\Delta t) = \frac{e^{R(r)\Delta t} - e^{d(\Delta t)}}{e^{u(\Delta t)} - e^{d(\Delta t)}}, \tag{3}$$

with $R(r)$ *defined in (9.25).*

For derivatives pricing we also required the risk-free rate r *for every asset class, the foreign risk-free rate* r_f *for the various currencies, and the approximately continuous dividend rate* δ *for equity indexes.*

Under $q^{Gen}(\Delta t)$, *the limiting result of Proposition 9.29 for the various asset classes becomes:*

$$X_{t+\Delta t} = X_t \exp\left[N\left((R(r) - \sigma^2/2)\Delta t, \sigma^2 \Delta t\right)\right],$$

and in this model, the real-world parameter μ *is nowhere to be found.*

Conclusion 9.30 (Redefine $u(\Delta t)$ **and** $d(\Delta t)$**)** *Since the parameter* μ *is lost in the limiting distribution for financial derivatives pricing, there is no logical reason to reflect this parameter in the binomial lattice in the definition of* $u(\Delta t)$ *and* $d(\Delta t)$. *Indeed, it would be a serendipitous moment for a modeler to learn that such an extraneous parameter in the model improved results. Thus it would make more sense to discard* μ *for derivatives pricing, and in our binomial lattices, define:*

$$\tilde{u}(\Delta t) = \sigma\sqrt{\Delta t}, \qquad \tilde{d}(\Delta t) = -\sigma\sqrt{\Delta t}. \tag{9.30}$$

All the prior results on derivatives pricing would continue to hold, since they did not depend on the value of μ, *and thus the only change would be that* $q^{Gen}(\Delta t)$ *in (3) would use* $\tilde{u}(\Delta t)$ *and* $\tilde{d}(\Delta t)$.

This simplification makes sense for derivatives pricing, but we now will have two binomial lattice constructions, one for the real world with $u(\Delta t)$ *and* $d(\Delta t)$, and one for derivatives pricing with $\tilde{u}(\Delta t)$ and $\tilde{d}(\Delta t)$.

However, there is a standard approach to real-world modeling using the pricing lattice $\tilde{u}(\Delta t)$ and $\tilde{d}(\Delta t)$, and a real-world probability $p \equiv p(\Delta t)$:

$$p(\Delta t) = \frac{\exp\left(\left[\mu + \frac{\sigma^2}{2}\right]\Delta t\right) - e^{-\sigma\sqrt{\Delta t}}}{e^{\sigma\sqrt{\Delta t}} - e^{-\sigma\sqrt{\Delta t}}}. \tag{9.31}$$

Proposition 9.31 (Limiting distribution of $X_j^{(n)}$ **under** $p(\Delta t)$ **as** $\Delta t \to \infty$**)** *Given a fixed horizon time* T *and* $\Delta t = T/n$, *initial price* $X_0 > 0$, *and parameter* $\sigma > 0$, *let:*

$$X_j^{(n)} = X_0 \exp\left[\sum\nolimits_{i=1}^{j} Z_i^{(n)}\right],$$

where $Z_i^{(n)} = \sigma\sqrt{\Delta t} b_i^{(n)}$. *Here* $\{b_i^{(n)}\}_{i=1}^{n}$ *are independent and binomially distributed to equal 1 or −1 with respective probabilities* p_n *and* $1 - p_n$, *where* $p_n \equiv p(\Delta t)$ *of (9.31). Thus* $X_j^{(n)}$ *is the price variate at time* $t = jT/n$ *for integer* $0 < j \leq n$ *in this multiplicative temporal model.*

Given m, *define the refined price variate* $X_{mj}^{(mn)}$ *at time* $t = jT/n$:

$$X_{mj}^{(mn)} \equiv X_0 \exp\left[\sum\nolimits_{i=1}^{mj} Z_i^{(mn)}\right],$$

where $Z_i^{(mn)} = \sigma\sqrt{\Delta t_m} b_i^{(mn)}$ *with* $\Delta t_m \equiv \Delta t/m$, *and* $\{b_i^{(mn)}\}_{i=1}^{mn}$ *defined as above relative to* $p_{mn} \equiv p(\Delta t_m)$.

Then for all $t \equiv j\Delta t$ *for* $1 \leq j \leq n$, *as* $m \to \infty$:

$$\frac{\ln[X_{mj}^{(mn)}/X_0] - \mu t}{\sigma\sqrt{t}} \to_d N, \tag{9.32}$$

where N *denotes the standard normal variate.*

Equivalently, $X_{mj}^{(mn)}$ converges in distribution for all $1 \leq j \leq n$ to the lognormally distributed variate X_t for $t = j\Delta t$, with parameters $\ln X_0 + \mu t$ and $\sigma^2 t$:

$$X_{mj}^{(mn)} \to_d X_t \equiv X_0 \exp\left[\mu t + \sigma\sqrt{t}N\right]. \tag{9.33}$$

Proof. *Define:*

$$p(\Delta t) = \frac{\exp(A\Delta t) - e^{-\sigma\sqrt{\Delta t}}}{e^{\sigma\sqrt{\Delta t}} - e^{-\sigma\sqrt{\Delta t}}},$$

with A to be determined.

Since the proof of Proposition 9.25 did not depend on the definition of $R(r)$, (9.26) obtains:

$$p(\Delta t) = \frac{1}{2} + \frac{1}{2\sigma}\left(A - \sigma^2/2\right)\sqrt{\Delta t} + O(\Delta t^{3/2}),$$

noting that we have set $\mu = 0$ in that result. Similarly, Proposition 9.27 did not depend on the definition of $R(r)$, and (9.27) obtains:

$$\begin{aligned} E_{p(\Delta t)}\left[\ln(X_t/X_0)\right] &= (A - \sigma^2/2)t + O(\Delta t), \\ Var_{p(\Delta t)}\left[\ln(X_t/X_0)\right] &= \sigma^2 t + O(\Delta t). \end{aligned}$$

The limiting distribution of Proposition 9.29 is again independent of the form of $R(r)$, and thus by (9.29):

$$X_{mj}^{(mn)} \to_d X_t \equiv X_0 \exp\left[(A - \sigma^2/2)t + \sigma\sqrt{t}N\right].$$

Comparing with (9.32), this model will obtain the desired real-world distribution of Proposition 8.46 if we define:

$$A = \mu + \frac{\sigma^2}{2}.$$

■

Remark 9.32 (On (9.31)) *The reader may well be surprised at the formulation in (9.31), since in many option pricing references one would see:*

$$\bar{p}(\Delta t) = \frac{\exp(\bar{\mu}\Delta t) - e^{-\sigma\sqrt{\Delta t}}}{e^{\sigma\sqrt{\Delta t}} - e^{-\sigma\sqrt{\Delta t}}}. \tag{1}$$

In this parametrization, $\bar{\mu}$ is the mean ***arithmetic return,*** *and works with the model:*

$$\bar{X}_{t+\Delta t} = \bar{X}_t\left[1 + \bar{\mu}\Delta t + \sigma\sqrt{\Delta t}b\right].$$

In this book's development, μ is the mean ***continuous return,*** *and works with the above model:*

$$X_{t+\Delta t} = X_t \exp\left[\mu\Delta t + \sigma\sqrt{\Delta t}b\right].$$

For arithmetic returns, if $T = n\Delta t$, then:

$$\bar{X}_T = X_0 \prod\nolimits_{j=1}^{n}\left[\bar{\mu}\Delta t + \sigma\sqrt{\Delta t}b_j\right], \tag{2}$$

while in the above continuous return model:

$$X_T = X_0\left[\sum\nolimits_{j=1}^{n}\left(\mu\Delta t + \sigma\sqrt{\Delta t}b_j\right)\right].$$

It then turns out that for the arithmetic return model, as for the continuous return model, the limiting distribution as $\Delta t \to 0$ is lognormal, with parameters:

$$\bar{X}_T \equiv X_0 \exp\left[\left(\bar{\mu} - \frac{\sigma^2}{2}\right)T + \sigma\sqrt{T}N\right], \tag{3}$$

while as seen above:

$$\bar{X}_T \equiv X_0 \exp\left[\mu T + \sigma\sqrt{T}N\right].$$

In summary, the real-world probability in (1) *uses an arithmetic return $\bar{\mu}$, which when compounded in* (2) *obtains a lognormal limiting distribution with mean $\left(\bar{\mu} - \frac{\sigma^2}{2}\right)T$. In the formulation here with real-world probability in (9.31), μ is a continuous return and obtains a lognormal limiting distribution with mean μT.*

This obtains the identity:

$$\mu = \bar{\mu} - \frac{\sigma^2}{2}.$$

The intuition here is that volatile arithmetic returns will compound to a total return that is less than the arithmetic return, and indeed, less by half the variance.

Remark 9.33 (On μ) *As noted in the previous section, it is common to price financial derivatives under the assumption that $\mu = 0$. However, for completeness, we continue to reflect this parameter in the following developments, and leave it to the reader to set $\mu = 0$ if desired.*

9.4 Limiting Price of European Derivatives

We now have all the ingredients for a final limiting formula for the price of a European financial derivative on an underlying security which is actively traded, long or short. To utilize the above results, we also assume that this security can be traded with no costs, and that the future price of this security at any time T has a lognormal distribution with parameters $\mu T + \ln X_0$ and $\sigma^2 T$, and thus by Proposition 8.46 can be approximated by an appropriately parametrized binomial return distribution with $p = 1/2$.

Recalling the previous development, given settlement date T and n, define $\Delta t \equiv T/n$ and binomial lattice returns $e^{u(\Delta t)}$ and $e^{d(\Delta t)}$ as in (9.2):

$$u(\Delta t) = \mu\Delta t + \sigma\sqrt{\Delta t}, \qquad d(\Delta t) = \mu\Delta t - \sigma\sqrt{\Delta t}.$$

While we can assume $\mu = 0$ as noted in the prior section, we maintain this parameter for notational continuity with the earlier development.

The above trading assumptions on the underlying asset allowed the construction of replicating portfolios everywhere on the lattice. These portfolios contained positions in the underlying security and a risk-free security, and replicated derivative values in the next period. Moving forward in time, these replicating portfolios could be rebalanced between security positions at the end of each Δt-period in a self-financing way, meaning without changing the market value of the portfolio. Further, at time T, the final portfolio replicated the financial derivative's payoff in any of the $n + 1$ possible price states. Assuming the law of one price as reinforced by arbitrage, this assured that the price of this financial derivative at time 0 must agree with the price of the initial replicating portfolio.

The price of this initial portfolio could then be formulaically expressed in terms of the financial derivative's settlements in these $n+1$ price states. Indeed, this price equalled the risk-free present value of the expected payoff as seen in (9.8). This expectation is calculated based on the originally assumed binomial distribution of underlying asset prices at time T, but with a new "risk-neutral" binomial probability denoted $q \equiv q(\Delta t)$. The expression for q depended on the underlying asset, as is summarized in (9.24).

Now the binomial distribution of underlying asset prices under q converges weakly to a lognormal distribution as $n \to \infty$ by Proposition 9.29. It thus seems natural to expect that the limiting price of a financial derivative satisfies the expectations formula of (9.15), but with the limiting lognormal distribution used for expectations in place of the binomial.

We now prove this result. The essence of the proof is as follows:

With $X_T^{(n)}$ denoting the time T binomial model prices of the underlying asset given n, Proposition 9.29 obtains that $X_T^{(n)} \to_d X_T$. In other words, $X_T^{(n)}$ converges in distribution as $n \to \infty$ to a lognormal variate X_T with parameters $\ln X_0 + (R(r) - \sigma^2/2)t$ and $\sigma^2 t$, and with $R(r)$ defined in (9.25). The financial derivative payoff $O_T\left[X_T^{(n)}\right]$ is a random variable for any Borel measurable payoff function O_T, and thus we must prove convergence, $O_T\left[X_T^{(n)}\right] \to_d O_T\left[X_T\right]$, as well as convergence of the expected values of these variates, $E\left[O_T\left[X_T^{(n)}\right]\right] \to E\left[O_T\left[X_T\right]\right]$.

Weak convergence of $O_T\left[X_T^{(n)}\right]$ will follow readily from prior results, but to prove convergence of expectations will require an additional assumption.

Before turning to the general result, we provide a simpler result with a much simpler proof.

Remark 9.34 (On the Black-Scholes-Merton pricing formula) *We call the following pricing formula a "Black-Scholes-Merton" result in recognition of the seminal 1973 papers of* ***Fischer Black*** *(1938–1995) and* ***Myron S. Scholes*** *(b. 1941), and,* ***Robert C. Merton*** *(b. 1944). While the general result echoes these earlier results, it is important to note that the approach of starting with replication within a binomial return model was not contemplated in these 1973 papers.*

The approach used by Black-Scholes and Merton was in spirit similar to that above, in the sense that they "replicated" the option with a portfolio of X-assets and risk-free assets, and hence concluded that the option must therefore have a price equal to the price of this replicating portfolio. But mathematically, these seminal results were derived within the theory of stochastic processes, a subject that will be studied and applied in later books.

The binomial lattice or "tree" approach to option pricing followed here was introduced in 1979 in another seminal paper for this theory, by ***John C. Cox*** *(b. 1943),* ***Stephen A. Ross*** *(1944–2017), and* ***Mark Rubinstein*** *(1944–2019). It is often referred to as the* ***Cox-Ross-Rubenstein binomial lattice model for option pricing.***

Proposition 9.35 (A Black-Scholes-Merton pricing formula) *Let $X_0 > 0$, $T > 0$, μ, and $\sigma > 0$ be fixed. Let $\Delta t \equiv T/n$ for integer n, and for $0 < j \leq n$ define:*

$$X_j^{(n)} \equiv X_0 \exp\left[\sum\nolimits_{i=1}^{j} \left(\mu \Delta t + \sigma\sqrt{\Delta t} b_i\right)\right],$$

where $\{b_i\}_{i=1}^n$ are independent and binomially distributed to equal 1 or -1 with respective probabilities q and $1-q$. Here $q \equiv q\left(\Delta t\right)$ is defined for various underlying assets in (9.24).

If $O_0^{(n)}[X_0]$ is given as in (9.8) with a **bounded and continuous** *settlement function $O_T(x)$, then as $n \to \infty$:*

$$O_0^{(n)}[X_0] \to O_0[X_0] \equiv e^{-rT} \int_0^\infty O_T(y) f_L(y)dy,$$

where $f_L(y)$ denotes the lognormal density function with parameters $\ln X_0 + (R(r) - \sigma^2/2)T$ and $\sigma^2 T$, and where $R(r)$ is defined in (9.25).

Proof. *Letting $j = n$, (9.29) states that $X_n^{(n)} \to_d X_T$ as $n \to \infty$, where X_T has the above lognormal distribution. If $F_{(n)}$ and F_L are the associated binomial and lognormal distribution functions, then by Definition 3.2 this can be equivalently stated $F_{(n)} \Rightarrow F_L$, or in terms of the induced probability measures, $\mu_{(n)} \Rightarrow \mu_L$.*

Given bounded and continuous $O_T(x)$, item 2 of Proposition III.4.28 states that $O_0^{(n)}[X_0]$ in (9.8) can be expressed as a Riemann-Stieltjes integral:

$$O_0^{(n)}[X_0] = e^{-rT} \int O_T(x)dF_{(n)}.$$

Proposition V.3.63 then obtains that this Riemann-Stieltjes integral is equal, with a small adjustment, to the associated Lebesgue-Stieltjes integral over any bounded interval. Using Riemann-Stieltjes integrability and a limiting argument where this adjustment converges to zero:

$$O_0^{(n)}[X_0] = e^{-rT} \int O_T(x)d\mu_{(n)}.$$

Given $\mu_{(n)} \Rightarrow \mu_L$, the portmanteau theorem of Proposition 3.6 now assures that for every bounded and continuous $O_T(x)$:

$$e^{-rT} \int O_T(x)d\mu_{(n)} \to e^{-rT} \int O_T(x)d\mu_L.$$

By the change of variables of Proposition V.4.8, this latter integral equals the Lebesgue integral of (9.36):

$$e^{-rT} \int O_T(x)d\mu_L = e^{-rT} \int O_T(x)f_L(x)dx,$$

where $f_L(x)$ is the above lognormal density function. ■

In summary, if restricted to bounded continuous $O_T(x)$, the results in (9.36) of the following proposition are derivable with only the portmanteau theorem and various earlier integration and change of variables results. The need for the longer proof below is to generalize this result to Borel measurable payoff functions with finite second moments.

The first consequence of a generalization to Borel measurable payoff functions is that the integral in (9.36) must initially be understood as a Lebesgue integral. By Proposition III.2.56, if $O_T(y)$ is locally bounded, Riemann and absolutely Riemann integrable, then the integral in (9.36) can also be interpreted as a Riemann integral. By Proposition III.1.22, such locally bounded $O_T(y)$ will be Riemann integrable if $O_T(y)$ is continuous almost everywhere.

Thus subject to the existence of the improper integral, (9.36) can also be interpreted as a Riemann integral for payoff functions that are locally bounded and continuous almost everywhere. Put another way, this result can be interpreted as a Riemann integral for any payoff function encountered in the real world.

Proposition 9.36 (General Black-Scholes-Merton pricing formula) *Let $X_0 > 0$, $T > 0$, μ, and $\sigma > 0$ be fixed. Let $\Delta t \equiv T/n$ for integer n, and for $0 < j \leq n$ define:*

$$X_j^{(n)} \equiv X_0 \exp\left[\sum\nolimits_{i=1}^{j}\left(\mu\Delta t + \sigma\sqrt{\Delta t}b_i\right)\right],$$

where $\{b_i\}_{i=1}^n$ are independent and binomially distributed to equal 1 or -1 with respective probabilities q and $1-q$. Here $q \equiv q(\Delta t)$ is defined for various underlying assets in (9.24).

Let $O_0^{(n)}[X_0]$ be given as in (9.8), representing the price of a European financial derivative on X with Borel measurable payoff function $O_T\left[X_T^{(n)}\right]$ at time T, where $X_T^{(n)} \equiv X_n^{(n)}$. Assume that the discontinuity set of O_T has Lebesgue measure zero, and that:

$$\sup_n E_q\left[\left(O_T\left(X_T^{(n)}\right)\right)^2\right] < \infty, \tag{9.34}$$

where this expectation is defined relative to the binomial distribution of $X_T^{(n)}$.

Then as $n \to \infty$:

$$O_0^{(n)}[X_0] \to O_0[X_0] \equiv e^{-rT}E\left[O_T(X_T)\right], \tag{9.35}$$

where X_T denotes the lognormally distributed limit variate of Proposition 9.29.

Thus defined as a Lebesgue integral:

$$O_0[X_0] = e^{-rT}\int_0^\infty O_T(y)\, f_L(y)dy, \tag{9.36}$$

where $f_L(y)$ denotes the lognormal density function with parameters $\ln X_0 + (R(r) - \sigma^2/2)T$ and $\sigma^2 T$, and where $R(r)$ is defined in (9.25). Equivalently:

$$O_0[X_0] \equiv e^{-rT}\int_{-\infty}^\infty O_T(X_0 e^x)\, f_N(x)dx, \tag{9.37}$$

where $f_N(x)$ is the density function of the normal distribution with mean $(R(r) - \sigma^2/2)T$ and variance $\sigma^2 T$.

Proof. *With expectation defined relative to the binomial distribution of $X_T^{(n)} \equiv X_n^{(n)}$, (9.15) states:*

$$O_0^{(n)}[X_0] \equiv e^{-rT}E_q\left[O_T\left(X_T^{(n)}\right)\right],$$

and thus dropping the discount factor e^{-rT}, (9.35) is equivalent to:

$$E_q\left[O_T\left(X_T^{(n)}\right)\right] \to E\left[O_T(X_T)\right], \tag{1}$$

with the latter expectation defined above.

Now $X_T^{(n)} \to_d X_T$ by Proposition 9.29, which implies by Definition 3.2 that $\mu_{(n)} \Rightarrow \mu_L$ for the associated binomial measure $\mu_{(n)}$ and lognormal measure μ_L. By Skorokhod's representation theorem of Proposition II.8.30, there exists random variables $\{Y_T^{(n)}, Y_T\}$ defined on a common probability space $(\mathcal{S}, \mathcal{E}, \lambda)$, with the given probability measures $\{\mu_{(n)}, \mu_L\}$, and such that $Y_T^{(n)}(s) \to Y_T(s)$ for all $s \in \mathcal{S}$. Then by Proposition II.5.21, pointwise convergence assures that $Y_T^{(n)} \to_d Y_T$. In summary, $\{Y_T^{(n)}, Y_T\}$ are defined on a common probability space and have the same distribution functions as $\{X_T^{(n)}, X_T\}$, and thus this proof proceeds with these random variables as a technical simplification.

The Mann-Wald theorem of Proposition 3.26 states that if $Y_T^{(n)} \to_d Y_T$, *then* $O_T\left(Y_T^{(n)}\right) \to_d O_T(Y_T)$ *for any Borel measurable function:*

$$O_T : (\mathbb{R}, \mathcal{B}(\mathbb{R}), m) \to (\mathbb{R}, \mathcal{B}(\mathbb{R}), m),$$

such that $m(D_{O_T}) = 0$, *where* D_{O_T} *denotes the discontinuity set of the* O_T. *Since* $m(D_{O_T}) = 0$ *by assumption, convergence* $O_T\left(Y_T^{(n)}\right) \to_d O_T(Y_T)$ *follows.*

The next step is to prove convergence of expectations in (9.35):

$$E_q\left[O_T\left(Y_T^{(n)}\right)\right] \to E\left[O_T(Y_T)\right], \tag{2}$$

and for this, we require a regularity assumption on $\left\{O_T\left(Y_T^{(n)}\right)\right\}_{n=1}^{\infty}$.

By the above construction on $(\mathcal{S}, \mathcal{E}, \lambda)$:

$$E_q\left[\left(O_T\left(Y_T^{(n)}\right)\right)^2\right] = \int \left(O_T\left(Y_T^{(n)}\right)\right)^2 d\lambda.$$

To apply Proposition V.4.18 for a change variables, recall that $\lambda_{Y_T^{(n)}}$, *the measure on* $\mathbb{R}$ *induced by* $Y_T^{(n)}$, *satisfies* $\lambda_{Y_T^{(n)}} \equiv \mu_{(n)}$ *as noted in Example V.4.14. Thus by Proposition V.4.18:*

$$\int \left(O_T\left(Y_T^{(n)}\right)\right)^2 d\lambda = \int (O_T(y))^2 \, d\mu_{(n)} \equiv E_q\left(\left(O_T\left(X_T^{(n)}\right)\right)^2\right), \tag{3}$$

where the last step is justified since $Y_T^{(n)}$ *and* $X_T^{(n)}$ *have the same distribution.*

Applying (9.34) to (3) obtains:

$$\sup_n \int \left(O_T\left(Y_T^{(n)}\right)\right)^2 d\lambda < \infty,$$

and thus by Proposition 4.25:

$$\int O_T\left(Y_T^{(n)}\right) d\lambda \to \int O_T(Y_T) \, d\lambda. \tag{4}$$

By the same change of variables as above, (4) proves (2) and (9.35).

For (9.36), since X_T *and* Y_T *have the same lognormal distribution, the change of variables of Propositions V.4.18 and V.4.8 obtain:*

$$\int O_T(Y_T) \, d\lambda = \int O_T(y) \, d\mu_L = \int_0^{\infty} O_T(y) \, f_L(y) dy, \tag{5}$$

where f_L *is the lognormal density function with the above parameters.*

The final expression in (9.37) is obtained by another change of variables with Proposition V.4.18, defining the transformation $T : (\mathbb{R}, \mathcal{B}(\mathbb{R}), \mu_N) \to (\mathbb{R}^+, \mathcal{B}(\mathbb{R}^+))$ *by* $T : x \to X_0 e^x$. *Here,* μ_N *is the measure induced by the normal random variable identified above with mean* $(R(r) - \sigma^2/2)T$ *and variance* $\sigma^2 T$. *By this Book V result:*

$$\int O_T(y) d\left(\mu_N\right)_T \equiv \int O_T(Tx) \, d\mu_N, \tag{6}$$

where $(\mu_N)_T$ *is the measure on* $\mathbb{R}^+$ *induced by* T *and* μ_N.

By Definition 1.28, for any Borel set A:

$$(\mu_N)_T(A) = \mu_N(T^{-1}(A)).$$

Letting $A = (-\infty, x]$ *obtains:*

$$\begin{aligned}(\mu_N)_T((-\infty, x]) &= \mu_N((-\infty, \ln(x/X_0)) \\ &= \mu_L((-\infty, x]).\end{aligned}$$

Thus the distribution function associated with $(\mu_N)_T$ *is* F_L*, the lognormal distribution function.*

This and (6) obtain:

$$\int_0^\infty O_T(y)d\mu_L = \int O_T(X_0 e^x)\, d\mu_N,$$

and the final step for (9.37) is another substitution using Proposition V.4.8, transforming the $d\mu_N$*-integral to a Lebesgue integral:*

$$\int O_T(X_0 e^x)\, d\mu_N = \int O_T(X_0 e^x)\, f_N(x)dx.$$

■

Remark 9.37 (On (9.34)) *To conclude the convergence of* $E_q\left[O_T\left(X_T^{(n)}\right)\right] \to E[O_T[(X_T)]]$ *from* $O_T\left(X_T^{(n)}\right) \to_d O_T(X_T)$*, it is sufficient to require that* $\left\{O_T\left(X_T^{(n)}\right)\right\}_{n=1}^\infty$ *are uniformly integrable by Proposition 4.23. By Proposition 4.25, this conclusion is also obtained if* $\sup_n E\left[\left|O_T\left(X_T^{(n)}\right)\right|^{1+\delta}\right] < \infty$ *for some* $\delta > 0$*, so the assumption in (9.34) with* $\delta = 1$ *is more than needed.*

Alternatively, we can replace (9.34) with the assumption:

$$\sup_n E_q\left[\left|O_T\left(X_T^{(n)}\right)\right|\right] < \infty, \qquad \sup_n Var_q\left[O_T\left(X_T^{(n)}\right)\right] < \infty, \tag{9.38}$$

since:

$$E_q\left[\left(O_T\left(X_T^{(n)}\right)\right)^2\right] \leq Var_q\left(O_T\left[X_T^{(n)}\right]\right) + \left(E_q\left[\left|O_T\left[X_T^{(n)}\right]\right|\right]\right)^2.$$

The advantage of the assumption that $\sup_n E_q\left[\left|O_T\left[X_T^{(n)}\right]\right|^{1+\delta}\right] < \infty$ *for any* $\delta > 0$*, is that one then obtains the uniform absolute mean bound in (9.38) for free by Lyapunov's inequality of Corollary IV.4.56.*

While a restriction as in (9.34) or (9.38) is needed for the theoretical development of Proposition 9.36, in the real world it can be argued that $\left\{O_T\left(X_T^{(n)}\right)\right\}_{n=1}^\infty$ *are always uniformly integrable without explicit additional assumptions. This follows because for any financial derivative, these payoffs are in fact uniformly bounded by global GDP, say.*

Exercise 9.38 ($E\left[(O_T(X_T))^2\right] < \infty$**)** *Show that (9.34) assures that* $E\left[(O_T(X_T))^2\right] < \infty$*, where this expectation is defined relative to the lognormal distribution of* X_T*. In other words, prove that:*

$$\int_0^\infty (O_T(y))^2 f_L(y)dy < \infty. \tag{9.39}$$

Hint: Define $\{Y_T^{(n)}, Y_T\}$ as in the above proof on a common probability space $(\mathcal{S}, \mathcal{E}, \lambda)$, with the same distributions as $\{X_T^{(n)}, X_T\}$, and with $Y_T^{(n)}(s) \to Y_T(s)$ for all $s \in \mathcal{S}$. Recall Fatou's lemma of Proposition V.3.19 to show that:

$$\int_{\mathcal{S}} (O_T [Y_T])^2 \, d\lambda < \infty$$

is implied by:

$$\sup_n \int_{\mathcal{S}} \left(O_T\left(Y_T^{(n)}\right)\right)^2 d\lambda < \infty.$$

Justify the application of this proposition, and then justify all integral change of variables to the desired result.

Show that it is also the case that $Var\left[O_T(X_T)\right] < \infty$.

Example 9.39 (Long forward contract) *The price of a long forward contract on an asset X which pays no income, or on a currency or stock index, is given by:*

$$O_0[X_0] = e^{(R(r)-r)T} X_0 - Fe^{-rT},$$

where F is the forward price. This equals the price of the replicating portfolio, which is a long position in X, or $e^{-r_f T}X$ or $e^{-\delta T}X$, respectively, and a borrowing at the risk-free rate that matures for F.

This price is also produced with the formula above. To see this, note that $O_T(x) = x - F$ and thus by (9.37):

$$\begin{aligned} O_0[X_0] &\equiv e^{-rT} \int_{-\infty}^{\infty} (X_0 e^x - F) f_N(x) dx \\ &= e^{-rT} \left[X_0 \int_{-\infty}^{\infty} e^x f_N(x) dx - F \right]. \end{aligned}$$

The integral is the moment generating function of this normal density evaluated at $t = 1$. Recalling IV.(4.78), this integral has value $e^{R(r)T}$, and the result follows.

9.4.1 Black-Scholes-Merton Option Pricing

Applying Proposition 9.36 to European put and call options, one arrives at the famous **Black-Scholes-Merton formulas for the prices of a European put and call option.** As noted in the previous section, this result is named for the seminal papers in 1973 of **Fischer Black** (1938–1995) and **Myron S. Scholes** (b. 1941), and, **Robert C. Merton** (b. 1944), and for which Merton and Scholes received the 1997 Nobel Prize in Economics. Black was deceased by that time, and this award is not made posthumously.

The settlement functions for these options at time T are defined in reference to a fixed **strike price** K by:

1. **Call Option:** This gives the long the right to buy X at time T for K, so:

$$O_T^C = \max\left[X_T - K, 0\right]. \tag{9.40}$$

2. **Put Option:** This gives the long the right to sell X at time T for K, so:

$$O_T^P = \max\left[K - X_T, 0\right]. \tag{9.41}$$

Notation 9.40 *One often sees these options referred to as* ***vanilla European calls*** *and* ***vanilla European puts*** *to distinguish them from the more exotic versions of such options discussed below, which are then labelled* ***exotic European puts and calls.*** *Analogously, one sees the* ***American-style*** *versions of these options with corresponding labels. That said, the terminology for financial derivatives is ever evolving, and it can be hazardous to make assumptions about contractual obligations.*

To simplify the above payoff expressions, it is common to write these as:

$$O_T^C = (X_T - K)^+, \qquad O_T^P = (K - X_T)^+.$$

In any case, these settlement functions represent the monetary value of exercise, but can disguise the exercise transaction.

Many puts and calls are ***physical settle,*** *meaning that exercise truly requires the buying of X for K for a call, and the selling of X for K for a put. This will then only be logical for the long if $X_T > K$ for the call, or $X < K$ for the put. The put is then automatically monetized for $K - X_T$, since the sold asset was either just acquired for X_T or previously held and had value of X_T. Either way, the wealth of the long increased by $K - X_T$. For a call, physical settle can be monetized by the long by simply selling X at or near the then price of X_T.*

Options on equity indexes are usually ***cash settle*** *for practical reasons, and thus the settlement function portrays the actual transaction.*

In order to justify the application of Proposition 9.36 to these options, we must verify only that (9.34) is satisfied, since continuity is apparent. The easiest way to do this is to note that since $O_T^C\left(X_T^{(n)}\right)O_T^P\left(X_T^{(n)}\right) \equiv 0$:

$$E_q\left[\left(O_T^C\left(X_T^{(n)}\right) - O_T^P\left(X_T^{(n)}\right)\right)^2\right] = E_q\left[\left(O_T^C\left(X_T^{(n)}\right)\right)^2\right] + E_q\left[\left(O_T^P\left(X_T^{(n)}\right)\right)^2\right].$$

Now:

$$O_T^C\left(X_T^{(n)}\right) - O_T^P\left(X_T^{(n)}\right) \equiv X_T^{(n)} - K,$$

the payoff of a long forward, and so:

$$E_q\left[\left(O_T^C\left(X_T^{(n)}\right)\right)^2\right] + E_q\left[\left(O_T^P\left(X_T^{(n)}\right)\right)^2\right] = E_q\left[\left(X_T^{(n)} - K\right)^2\right].$$

To confirm that (9.34) is satisfied for both options, it is enough to prove:

$$\sup_n E_q\left[\left(X_T^{(n)} - K\right)^2\right] < \infty.$$

and this can be done by showing that $\lim_n E_q\left[\left(X_T^{(n)} - K\right)^2\right]$ exists.

To this end, note that $E_q\left[X_T^{(n)}\right] = X_0 e^{R(r)T}$ as is implied by (9.18) and Exercise 9.17, with $R(r)$ defined in (9.25). It thus follows that:

$$\begin{aligned} E_q\left[\left(X_T^{(n)} - K\right)^2\right] &= E_q\left[\left(X_T^{(n)}\right)^2\right] - 2KX_0e^{R(r)T} + K^2 \\ &\to X_0^2 e^{(2R(r)+\sigma^2)T} - 2KX_0e^{R(r)T} + K^2. \end{aligned}$$

The next exercise provides the details on this last step.

Exercise 9.41 $\left(\textbf{On } \lim_{n\to\infty} E_q\left[\left(X_T^{(n)}\right)^2\right]\right)$ *Prove that with $R(r)$ defined in (9.25):*

$$\lim_{n\to\infty} E_q\left[\left(X_T^{(n)}\right)^2\right] = X_0^2 e^{(2R(r)+\sigma^2)T}. \tag{9.42}$$

Hint:

1. *Using the binomial probabilities under q, derive that:*

$$E_q\left[\left(X_T^{(n)}\right)^2\right] = X_0^2\left(qe^{2u(\Delta t)} + (1-q)e^{2d(\Delta t)}\right)^n,$$

with $u(\Delta t)$ and $d(\Delta t)$ defined in (9.2).

2. *Substitute for q for various underlying assets using (9.24), and then derive:*

$$E_q\left[\left(X_T^{(n)}\right)^2\right] = X_0^2\left(e^{(\mu+R(r))\Delta t+\sigma\sqrt{\Delta t}} + e^{(\mu+R(r))\Delta t-\sigma\sqrt{\Delta t}} - e^{2\mu\Delta t}\right)^n.$$

3. *Take a logarithm and since $n = T/\Delta t$:*

$$\begin{aligned}
&\lim_{n\to\infty} \ln\left(e^{(\mu+R(r))\Delta t+\sigma\sqrt{\Delta t}} + e^{(\mu+R(r))\Delta t-\sigma\sqrt{\Delta t}} - e^{2\mu\Delta t}\right)^n \\
&= T\lim_{\Delta t\to 0+}\frac{1}{\Delta t}\ln\left(e^{(\mu+R(r))\Delta t+\sigma\sqrt{\Delta t}} + e^{(\mu+R(r))\Delta t-\sigma\sqrt{\Delta t}} - e^{2\mu\Delta t}\right) \\
&= T\left.\frac{df}{dx}\right|_{x=0}
\end{aligned}$$

where:

$$f(x) = \ln\left(e^{(\mu+R(r))x+\sigma\sqrt{x}} + e^{(\mu+R(r))x-\sigma\sqrt{x}} - e^{2\mu x}\right).$$

Justify that this function is right differentiable at $x = 0$.

4. *Evaluate this derivative, using a Taylor approximation for the evaluation at $x = 0$, and justify the expression for $\lim_{n\to\infty} E_q\left[\left(X_T^{(n)}\right)^2\right]$ in (9.42) from this result.*

Remark 9.42 *The formula in (9.42) agrees with the second moment formula in IV.(4.80) for the lognormal distribution parameterized as in Proposition 9.29. Thus the convergence in distribution in this model, $X_T^{(n)} \to_d X_T$, also obtains convergence of two moments:*

$$E_q\left[\left(X_T^{(n)}\right)^j\right] \to E_q\left[(X_T)^j\right], \qquad j = 1, 2.$$

*In Proposition 9.159 of **Reitano** (2010) is proved that the moment generating function of $\ln\left[X_T^{(n)}/X_0\right]$, which has a binomial distribution under q defined in (9.9), converges to the moment generating function of the associated normal distribution.*

The final result for the **Black-Scholes-Merton formulas** follows.

Proposition 9.43 (Black-Scholes-Merton pricing formulas) *For a **European call option**, the limiting price in (9.36) becomes:*

$$O_0^C(X_0) = e^{-(r-R(r))T}X_0\Phi(d_1) - e^{-rT}K\Phi(d_2), \tag{9.43a}$$

while the result for a ***European put option:***

$$O_0^P(X_0) = e^{-rT}K\Phi(-d_2) - e^{-(r-R(r))T}X_0\Phi(-d_1). \tag{9.44}$$

Here Φ denotes the distribution function of the standard normal,

$$d_1 = \frac{\ln(X_0/K) + (R(r) + \sigma^2/2)T}{\sigma\sqrt{T}}, \qquad d_2 = \frac{\ln(X_0/K) + (R(r) - \sigma^2/2)T}{\sigma\sqrt{T}}. \tag{9.45}$$

and $R(r)$ is defined in (9.25) for the various underlying assets.

Proof. *These formulas are assigned as exercises, and are derived by an explicit evaluation of the associated Riemann integrals using a change of variables in (9.36).* ■

9.5 Properties of Black-Scholes-Merton Prices

9.5.1 Price Convergence to Payoff

Rewriting the price of a European derivative in (9.37) at time $t < T$:

$$O_t[X_t] \equiv e^{-r(T-t)} \int_{-\infty}^{\infty} O_T\left(X_t \exp\left[(R(r) - \sigma^2/2)(T-t) + x\sigma\sqrt{T-t}\right]\right)\phi(x)dx, \tag{9.46}$$

with $R(r)$ defined in (9.25) for the various underlying assets. Here we use a change of variables to put the normal distribution parameters into the payoff function, and thus $\phi(x) \equiv e^{-x^2/2}/\sqrt{2\pi}$, the standard normal density function of (6.2).

A natural question arises from (9.46).

If $X_t \to X_T$ as $t \to T$, meaning convergence of the realized prices, must $O_t[X_t] \to O_T(X_T)$ with $O_t[X_t]$ defined in (9.46)?

Of course this convergence would in theory be enforced in the real world by arbitrage, but this is a statement about the associated mathematics.

Ignoring $\phi(x)$ which defines a measure as seen below, the integrand in (9.46) converges to $O_T(X_T)$, so this is really a question of "integration to the limit" as introduced in Chapter III.2 and summarized in Section V.3.3.5. As seen in that development, there are several assumptions which allow the passing of the limit into the integrand, and the next result provides a general result. See also Remark 9.45.

Proposition 9.44 (On $O_t[X_t] \to O_T[X_T]$) *If $X_t \to X_T$ as $t \to T$, and for all $t \leq T$:*

$$\int_{-\infty}^{\infty} O_T^2\left(X_t \exp\left[(R(r) - \sigma^2/2)(T-t) + x\sigma\sqrt{T-t}\right]\right)\phi(x)dx < C < \infty, \tag{9.47}$$

then:

$$O_t[X_t] \to O_T(X_T). \tag{9.48}$$

Proof. *By Proposition V.4.8, $O_t[X_t]$ in (9.46) can be expressed:*

$$O_t[X_t] \equiv e^{-r(T-t)} \int_{\mathbb{R}} O_T\left(X_t \exp\left[(R(r) - \sigma^2/2)(T-t) + x\sigma\sqrt{T-t}\right]\right)d\mu_\phi(x),$$

where μ_ϕ is the Borel measure induced by $\phi(x)$. Similarly, (9.47) can be restated:

$$\int_{-\infty}^{\infty} O_T^2\left(X_t \exp\left[(R(r) - \sigma^2/2)(T-t) + x\sigma\sqrt{T-t}\right]\right)d\mu_\phi(x) < C < \infty. \tag{1}$$

Let $\{t_n\}_{n=1}^{\infty}$ be given with $t_n \to \infty$. By Proposition 4.25, (1) *assures that $\{f_{t_n}(x)\}_{n=1}^{\infty}$ defined by:*

$$f_{t_n}(x) \equiv O_T\left(X_{t_n} \exp\left[(R(r) - \sigma^2/2)(T - t_n) + x\sigma\sqrt{T - t_n}\right]\right)$$

is uniformly integrable relative to μ_ϕ. Since $f_{t_n}(x) \to f_T(x)$ for all x, the uniform integrability convergence theorem of Proposition V.3.55 now obtains that:

$$\int_{-\infty}^{\infty} f_{t_n}(x) d\mu_\phi(x) \to \int_{-\infty}^{\infty} f_T(x) d\mu_\phi(x). \tag{2}$$

As (2) *is true for all sequences $\{t_n\}_{n=1}^{\infty}$ with $t_n \to \infty$, the proof is complete.* ■

Remark 9.45 (Other convergence criteria) *The condition in (9.47) can be replaced by other criteria:*

1. ***Uniform integrability:** Proposition 4.25 requires only that $|f_{t_n}(x)|^{1+\delta}$ be integrable and uniformly bounded for some $\delta > 0$, so $\delta = 1$ as above is only an example.*

2. ***Bounded** $O_T(x)$: If $|O_T(x)| \leq C$ then (9.47) is apparently satisfied, and the result follows. Alternatively, the bounded convergence theorem of Proposition V.3.50 obtains the result since $\mu_\phi(\mathbb{R}) = 1$.*

3. ***Dominated** $f_{t_n}(x)$: If $|f_{t_n}(x)| \leq g(x)$ for all sequences $\{t_n\}_{n=1}^{\infty}$ with $t_n \to \infty$ with μ_ϕ-integrable $g(x)$, the result follows from Lebesgue's dominated convergence theorem of Proposition V.3.45.*

9.5.2 Put-Call Parity

It was noted in Section 9.4.1 that at time T, the payoff functions for European put and call options with given strike price K satisfy:

$$O_T^C\left[X_T^{(n)}\right] - O_T^P\left[X_T^{(n)}\right] \equiv X_T^{(n)} - K.$$

This identity has nothing to do with the binomial model of prices, and is true by definition for all values of X_T.

The market interpretation of this result is that if an investor has a portfolio with a long call option and a short put option at time $t = 0$, then at time T this investor will have a portfolio with settlement value $X_T - K$, recalling the settlement discussion of Notation 9.40. This latter expression is also the settlement value for a long T-period **forward contract** on X with forward price K.

The actual contractual obligation of this forward contract to the long is that X must be purchased at time T for amount K. Thus $X_T - K$ is the long's profit and loss position at that time if the asset is immediately liquidated at price X_T, or more generally, the mark-to-market profit and loss at that time. Conversely, the contractual obligation for the short is to sell X for K, and thus the short has a position that is equivalent to a long put and a short call.

This long forward contract on an **asset with no income** can be replicated. If an investor creates a portfolio that is long one unit of such X_0, and is short a risk-free security that matures for K at time T, this portfolio will cost $X_0 - Ke^{-rT}$ and also have value $X_T - K$ at time T. By the law of one price and the assumption that X is actively traded long and short, these portfolios must have identically the same value at time 0 since they have exactly the same values at time T. Given the above identity in settlement values of the put/call portfolio and the long forward contract, this assures the price identity:

$$O_0^C(X_0) - O_0^P(X_0) = X_0 - Ke^{-rT}. \tag{9.49}$$

This relationship is known as **put-call parity,** and often expressed with conventional notation for European put and call premiums:

$$c - p = X_0 - Ke^{-rT}.$$

It states that independent of the individual pricing of European puts and calls, it must be the case that the difference between these prices equals the price of a long or short forward contract. Specifically, $O_0^C(X_0) - O_0^P(X_0)$ equals the price or market value of a long forward, and conversely, $O_0^P(X_0) - O_0^C(X_0)$ equals the market value of a short forward.

The put-call parity identity has two interpretations:

1. It is an **identity in the prices** between European puts, European calls and forward contracts, or equivalently, an identity in the prices of European puts, European calls, the underlying asset, and risk-free assets.

2. It is a **replication formula,** which translates any algebraically equivalent version of this identity into a contract replication statement. For example:

   $$c = p + X_0 - Ke^{-rT}$$

   can be understood as:

   $$\text{long call} = \{\text{long put, long underlying asset, short RF asset}\}.$$

   Thus for replication, positive prices are long positions, and negative prices are short positions.

Exercise 9.46 (Put-call parity for other underlying assets) *Derive put-call parity for options on currencies, stock indexes, and futures, using the above logic. As above, Ke^{-rT} denotes the price of this risk-free asset for the price equation, and an actual risk-free asset for replication.*

1. ***Currency:*** *Let c and p denote option prices (in domestic currency) per unit of foreign currency, and X_0 denote the price in domestic currency for one unit of foreign currency:*

   $$c - p = X_0 e^{-r_f T} - Ke^{-rT}. \tag{9.50}$$

   As a price equation, X_0 is understood in units of domestic currency, while as a replication formula, X_0 denotes one unit of foreign currency.

2. ***Equity index:*** *Let c and p denote option prices per unit of the equity index, and X_0 denote the price for one unit of the index:*

   $$c - p = X_0 e^{-\delta T} - Ke^{-rT}. \tag{9.51}$$

   As in item 1, X_0 is understood in currency units as a price equation, while for replication, it denotes one unit of the equity index.

3. **Futures:** *Let c and p denote option prices on a futures contract on one unit of underlying asset, and X_0 denote the futures price, which is defined per one unit of the underlying asset:*

$$c - p = X_0 e^{-rT} - Ke^{-rT} + \text{“long futures.”} \tag{9.52}$$

As an identity in prices, the long futures position disappears since initially a futures contract has a price of 0. For replication, $X_0 e^{-rT}$ is a risk-free asset with a maturity of X_0, and "long futures" is a long position in a futures contract on one unit of the underlying asset. Hint: When exercised, a long call delivers cash of $(X_T - X_0)^+$ plus a long futures contract at price X_T to the long, while a long put delivers cash of $(X_0 - X_T)^+$ plus a short futures at price X_T to the long.

9.5.3 Black-Scholes-Merton PDE

Various partial derivatives of the price of a financial derivative are called the "Greeks" of the derivative and are often denoted by Greek letters. This is a natural notation because "delta," whether denoted as upper case Δ or lower case δ, has been codified in mathematical analysis as the abbreviation for "change," whether in variates like Δx, or functions $\Delta f \equiv f(x + \Delta x) - f(x)$, or combined in a Taylor series:

$$\Delta f \approx f'(x)\Delta x + \frac{1}{2} f''(x) (\Delta x)^2 + ...$$

The Greeks of a financial derivative are typically expressed numerically, as are prices. However, also like prices, all Greeks are functions of the various parameters which define the derivative and distribution of X_T. For example, $\Delta \equiv \Delta(X_t, \sigma, r, t)$ where the parameter t is primarily relevant because it identifies the important parameter of $T - t$.

The most common Greeks follow, although it should be noted that there are many more that are defined by mixed and higher partial derivatives. These definitions are general and apply beyond the European structures investigated above, and hence we drop this qualifier for this definition.

Definition 9.47 (Common "Greeks") *Let $O_t[X_t]$ denote the price of financial derivative at time t, with underlying asset X. When these partial derivatives exist, the common Greeks are defined as follows:*

1. **Delta:**

$$\Delta \equiv \frac{\partial O_t[X_t]}{\partial X_t}.$$

2. **Gamma:**

$$\Gamma \equiv \frac{\partial^2 O_t[X_t]}{\partial X_t^2}.$$

3. **Vega:**

$$\nu \equiv \frac{\partial O_t[X_t]}{\partial \sigma}.$$

Note that "vega" is not a Greek letter, but the letter commonly used for it is the Greek lower case "nu," or simply the English lower case "v."

4. **Rho:**

$$\rho \equiv \frac{\partial O_t[X_t]}{\partial r}.$$

5. ***Theta:*** *With the parameter t defining the remaining maturity of the contract as $T-t$:*

$$\theta \equiv \frac{\partial O_t\left[X_t\right]}{\partial t}.$$

Note: As is standard practice, partial derivatives are evaluated holding all other variables constant. Thus for $\partial/\partial t$, $X_t \equiv X$ and there is no contribution from the dependence of X on t.

We next investigate the existence of the common Greeks when $O_t\left[X_t\right]$ denotes the price of a European financial derivative as given by the Lebesgue integral in (9.46). **Leibniz's rule,** named for **Gottfried Leibniz** (1646–1716), addresses this question.

While various versions of this rule were investigated in Section V.3.4, Proposition V.3.61 is most applicable, and we restate it here for completeness, with a small change of notation.

Proposition 9.48 (Leibniz integral rule: Complete Lebesgue-Stieltjes) *Given a complete Borel measure space $(Y, \sigma(Y), \mu)$, such as the Lebesgue measure space $(\mathbb{R}^p, \mathcal{M}_L(\mathbb{R}^p), m)$, let $E \in \sigma(Y)$, and $D \subset \mathbb{R}^n$ an open set. Let $g(x,y)$ be defined on $D \times E$, and assume that for all $x \in D$ that:*

1. *$g(x,y)$ is μ-integrable over E;*
2. *$g_{x_k}(x,y) \equiv \frac{\partial g(x_1,\dots,x_n,y)}{\partial x_k}$ exists μ-a.e.;*
3. *$|g_{x_k}(x,y)| \leq h(y)$ μ-a.e., where $h(y)$ is μ-integrable over E.*

Then defined on D:

$$G(x) \equiv \int_E g(x,y)d\mu$$

is differentiable with respect to x_k with:

$$G_{x_k}(x) = \int_E g_{x_k}(x,y)d\mu. \tag{9.53}$$

Proof. *See Proposition V.3.61.* ■

To apply this result, which includes changing some notation, let $f(t,x) \equiv O_t\left[X_t\right]$, the price of a European derivative at time t when the price of the underlying asset is $x \equiv X_t$. With T the final settlement date, $f(t,x)$ can be expressed as a Lebesgue integral as in (9.46) for $t < T$:

$$f(t,x) \equiv e^{-r(T-t)} \int_{-\infty}^{\infty} O_T\left(x \exp\left[\left(R(r) - \sigma^2/2\right)(T-t) + y\sigma\sqrt{T-t}\right]\right)\phi(y)dy, \tag{1}$$

with $R(r)$ defined in (9.25) for the various underlying assets, and $\phi(x) \equiv e^{-x^2/2}/\sqrt{2\pi}$, the standard normal density function of (6.2).

We now state the main result on the existence of the common Greeks, as well as identify an equation satisfied by a subset of these Greeks, specifically Δ, Γ and θ. This result when stated as in (9.55) is known as the **Black-Scholes-Merton partial differential equation.**

Proposition 9.49 (Black-Scholes-Merton PDE) *Assume that settlement value function $O_T(x)$ is twice differentiable m-a.e., and that $O_T(x)$ and these derivatives are Lebesgue measurable and have at most polynomial growth, m-a.e. That is, for $n \leq 2$:*

$$\left|\frac{\partial^n O_T(x)}{\partial x^n}\right| \leq c_n |x|^{d_n}, \ m\text{-a.e.}, \tag{9.54}$$

where we define $\frac{\partial^n O_T(x)}{\partial x^n} \equiv O_T(x)$ for $n = 0$.

Then for $t < T$, all the common Greeks of Definition 9.47 exist.

Further, with $f(t,x) \equiv O_t[X_t]$ as in (1) above, then for $t < T$:

$$\begin{gathered}\tfrac{\partial f}{\partial t} + R(r)x\tfrac{\partial f}{\partial x} + \tfrac{1}{2}\sigma^2 x^2 \tfrac{\partial^2 f}{\partial x^2} = rf, \\ f(T,x) = O_T(x).\end{gathered} \tag{9.55}$$

Thus for $t < T$, the Greeks Δ, Γ and θ satisfy:

$$\theta + R(r)X_t\Delta + \frac{1}{2}\sigma^2 X_t^2 \Gamma = rO_t[X_t]. \tag{9.56}$$

Proof. *To prove the existence of the common Greeks, we must apply Proposition 9.48, and for this we show that (9.54) is sufficient to ensure the dominated conditions of item 3 of that result. We then derive the identities for Δ, Γ and θ, and show that the partial differential equation in (9.55) is satisfied for $t < T$.*

We leave as Exercise 9.50 the existence of ρ and ν, as well as a few details on the justification of the boundary condition $f(T,x) = O_T(x)$ by Proposition 9.44.

To simplify notation in the application of Proposition 9.48, let:

$$\begin{aligned} f(t,x) &\equiv \textstyle\int_{-\infty}^{\infty} g(t,x,y)dy, \\ g(t,x,y) &\equiv e^{-r(T-t)}O_T(x\exp[k(t,y)])\,\phi(y), \\ k(t,y) &\equiv (R(r) - \sigma^2/2)(T-t) + y\sigma\sqrt{T-t}. \end{aligned} \tag{1}$$

Given (t,x), define the open set D of Proposition 9.48 as $D \equiv (t-\epsilon, t+\epsilon) \times (x-\epsilon, x+\epsilon)$, choosing ϵ so that $t+\epsilon < T$. The Lebesgue measurable set E is defined by $E = \mathbb{R}$.

For $(t,x) \in D$, note that $g(t,x,y)$ is Lebesgue integrable on $\mathbb{R}$. Considering the functional form of the components, for $(t,x) \in \bar{D} \equiv [t-\epsilon, t+\epsilon] \times [x-\epsilon, x+\epsilon]$:

$$|g(t,x,y)| \leq a_0 \exp[b_0 d_0 y] \exp\left[-\frac{1}{2}y^2\right], \tag{2}$$

with d_0 defined in (9.54), so item 1 of Proposition 9.48 is satisfied. Similarly, $g_t(t,x,y)$, $g_x(t,x,y)$, and $g_{xx}(t,x,y)$ exist m-a.e., since $O_T(x)$ has two derivatives m-a.e. and the remaining component functions are continuously differentiable on D, so item 2 is also satisfied.

For item 3 of Proposition 9.48, the growth bounds in (9.54) provide the needed result. For example:

$$g_t(t,x,y) = rg(t,x,y) + e^{-r(T-t)}O_T'(x\exp[k(t,y)])\,x\exp[k(t,y)]\,k_t(t,y)\phi(y). \tag{3}$$

The first term is integrable for $(t,x) \in D$ by (2). Since $|k_t(t,y)| \leq a_1 + b_1 y$ on $\bar{D}$, the second term is bounded on D by:

$$a_0(a_1 + b_1 y)\exp[b_1(d_1+1)y]\exp\left[-\frac{1}{2}y^2\right],$$

and thus integrable.

The same analysis proves item 3 of Proposition 9.48 for:

$$g_x(t,x,y) = e^{-r(T-t)}O_T'\left(x\exp\left[k(t,y)\right]\right)\exp\left[k(t,y)\right]\phi(y), \tag{4}$$
$$g_{xx}(t,x,y) = e^{-r(T-t)}O_T''\left(x\exp\left[k(t,y)\right]\right)\exp\left[2k(t,y)\right]\phi(y).$$

Thus $f_t(t,x)$, $f_x(t,x)$, and $f_{xx}(t,x)$ exists as the integrals of the respective derivatives of $g(t,x,y)$, proving the existence of θ, Δ and Γ, respectively.

For (9.55), from (3) and (4), since $k_t(t,y) = -(R(r)-\sigma^2/2) - y\sigma/2\sqrt{T-t}$:

$$\begin{aligned} g_t(t,x,y) &= rg(t,x,y) + x\left[-(R(r)-\sigma^2/2) - \frac{y\sigma}{2\sqrt{T-t}}\right] \\ &\quad \times e^{-r(T-t)}O_T'\left(x\exp\left[k(t,y)\right]\right)\exp\left[k(t,y)\right]\phi(y) \\ &= rg(t,x,y) - \left[(R(r)-\sigma^2/2)\right]xg_x(t,x,y) \\ &\quad -\frac{yx\sigma}{2\sqrt{T-t}}e^{-r(T-t)}O_T'\left(x\exp\left[k(t,y)\right]\right)\exp\left[k(t,y)\right]\phi(y). \end{aligned}$$

Integrating with respect to y and applying the Leibniz formula obtains:

$$\begin{aligned} f_t(t,x) &= rf(t,x) - \left[(R(r)-\sigma^2/2)\right]xf_x(t,x) \\ &\quad -\frac{x\sigma}{2\sqrt{T-t}}e^{-r(T-t)}\int_{-\infty}^{\infty} yO_T'\left(x\exp\left[k(t,y)\right]\right)\exp\left[k(t,y)\right]\phi(y)dy. \end{aligned} \tag{5}$$

Again by the Leibniz rule, $f_{xx}(t,x)$ is obtained by integrating $g_{xx}(t,x,y)$ in (4):

$$f_{xx}(t,x) = e^{-r(T-t)}\int_{-\infty}^{\infty} O_T''\left(x\exp\left[k(t,y)\right]\right)\exp\left[2k(t,y)\right]\phi(y)dy.$$

Noting that:

$$\frac{\partial O_T'\left(x\exp\left[k(t,y)\right]\right)}{\partial y} = x\sigma\sqrt{T-t}O_T''\left(x\exp\left[k(t,y)\right]\right)\exp\left[k(t,y)\right], \tag{6}$$

$f_{xx}(t,x)$ can be expressed:

$$f_{xx}(t,x) = \frac{e^{-r(T-t)}}{x\sigma\sqrt{T-t}}\int_{-\infty}^{\infty}\frac{\partial O_T'\left(x\exp\left[k(t,y)\right]\right)}{\partial y}\exp\left[k(t,y)\right]\phi(y)dy.$$

Performing integration by parts on the integral by Proposition III.3.64 (see Remark 9.51), and noting that the boundary values are zero due to (9.54) and $\phi(y)$, obtains:

$$\begin{aligned} f_{xx}(t,x) &= -\frac{e^{-r(T-t)}}{x\sigma\sqrt{T-t}}\int_{-\infty}^{\infty}\frac{\partial\left[\exp\left[k(t,y)\right]\phi(y)\right]}{\partial y}O_T'\left(x\exp\left[k(t,y)\right]\right)dy \\ &= \frac{e^{-r(T-t)}}{x\sigma\sqrt{T-t}}\int_{-\infty}^{\infty} yO_T'\left(x\exp\left[k(t,y)\right]\right)\exp\left[k(t,y)\right]\phi(y)dy \\ &\quad -\frac{e^{-r(T-t)}}{x}\int_{-\infty}^{\infty} O_T'\left(x\exp\left[k(t,y)\right]\right)\exp\left[k(t,y)\right]\phi(y)dy. \end{aligned}$$

Finally:

$$\begin{aligned} \frac{1}{2}\sigma^2x^2f_{xx}(t,x) &= \frac{x\sigma}{2\sqrt{T-t}}e^{-r(T-t)}\int_{-\infty}^{\infty} yO_T'\left(x\exp\left[k(t,y)\right]\right)\exp\left[k(t,y)\right]\phi(y)dy \\ &\quad -\frac{1}{2}\sigma^2xf_x(t,x), \end{aligned} \tag{7}$$

and a substitution into (5) yields (9.55).

The boundary condition in (5) of $f(T,x) = O_T(x)$ follows from Proposition 9.44, since using the estimates as above, (9.47) is satisfied by (9.54). Details are left as an exercise. ■

Exercise 9.50 (On ρ and ν) *Given the assumptions of the above proposition, prove that $\nu \equiv \frac{\partial O_t[X_t]}{\partial \sigma}$ and $\rho \equiv \frac{\partial O_t[X_t]}{\partial r}$ exist by the Leibniz rule, and justify the details on the application of Proposition 9.44 for the boundary condition $f(T,x) = O_T(x)$ in (9.55).*

Remark 9.51 (On Lebesgue integration by parts) *Proposition III.3.64 summarizes the needed result on Lebesgue integration by parts over an interval $[a,b]$:*

$$(\mathcal{L})\int_a^b f'(y)g(y)dy = f(b)g(b) - f(a)g(a) - (\mathcal{L})\int_a^b f(y)g'(y)dy,$$

under the requirement that both $f(y)$ and $g(x)$ are absolutely continuous. In the notation above:

$$f(y) = O'_T(x\exp[k(t,y)]), \quad g(y) = \exp[k(t,y)]\,\phi(y). \tag{1}$$

Since $g(y)$ is continuously differentiable, this function is absolutely continuous by item 5 of Remark III.3.56.

Now $h(y) \equiv O''_T(x\exp[k(t,y)])\exp[k(t,y)]$ is Lebesgue measurable by assumption, and bounded over any interval $[a,b]$. The positive and negative parts (Definition III.2.43) $h^{\pm}(y)$ are then also measurable and bounded, so Lebesgue integrable by Proposition III.2.15, and thus $h(y)$ is Lebesgue integrable over this interval .

Defining:

$$H(y) = (\mathcal{L})\int_a^y O''_T(x\exp[k(t,z)])\exp[k(t,z)]\,dz,$$

it follows that $H(y)$ is absolutely continuous by Proposition III.3.58. Then by (6) in the proof:

$$H(y) = \frac{1}{x\sigma\sqrt{T-t}}\int_a^y \frac{\partial O'_T(x\exp[k(t,z)])}{\partial z}dz.$$

Finally, by Proposition III.3.62:

$$H(y) = \frac{1}{x\sigma\sqrt{T-t}}\left[O'_T(x\exp[k(t,y)]) - O'_T(x\exp[k(t,a)])\right],$$

and absolute continuity of $f(y)$ in (1) follows from $f(y) = cH(y) + d$.

Also note that $f(y)g(y)$ is 0 in the limit at $y = \pm\infty$ due to the presence of $\phi(y)$ in (1), so integration by parts can be extended over $\mathbb{R}$ as noted in the proof.

Example 9.52 (Forward contract) *By Example 9.39, the price at time $t \leq T$ of a long forward contract on an asset X which pays no income, or on a currency or stock index, is given by:*

$$O_t[X_t] = e^{(R(r)-r)(T-t)}X_t - Fe^{-r(T-t)},$$

where F is the forward price. Hence:

$$\theta = -(R(r)-r)\,e^{(R(r)-r)(T-t)}X_t - Fre^{-r(T-t)},$$
$$\Delta = e^{(R(r)-r)(T-t)},$$
$$\Gamma = 0,$$

and it follows as in (9.56) that:

$$\theta + R(r)X_t\Delta + \frac{1}{2}\sigma^2X_t^2\Gamma = rO_t[X_t].$$

In the case of European puts and calls, the price functions of Proposition 9.43 as rewritten in Exercise 9.53 are seen to be infinitely differentiable in X, σ, r, and t. These put and call prices also satisfy (9.55) by Proposition 9.49, as can be verified by the interested reader.

Exercise 9.53 (Common Greeks of European puts/calls) *Restating the formulas in Proposition 9.43 to be evaluated at time $t < T$:*

- ***European call option:***

$$O_t^C(X_t) = e^{-(r-R(r))(T-t)} X_t \Phi(d_1) - e^{-r(T-t)} K\Phi(d_2), \tag{9.57}$$

- ***European put option:***

$$O_t^P(X_t) = e^{-r(T-t)} K\Phi(-d_2) - e^{-(r-R(r))(T-t)} X_t \Phi(-d_1), \tag{9.58}$$

where Φ denotes the distribution function of the standard normal, $R(r)$ is defined in (9.25) for the various underlying assets, and:

$$\begin{aligned} d_1 &= \frac{\ln(X_t/K)+(R(r)+\sigma^2/2)(T-t)}{\sigma\sqrt{T-t}}, \\ d_2 &= \frac{\ln(X_t/K)+(R(r)-\sigma^2/2)(T-t)}{\sigma\sqrt{T-t}}. \end{aligned} \tag{9.59}$$

Derive the following:

$$\Delta_t^C = e^{-(r-R(r))(T-t)} \Phi(d_1), \qquad \Delta_t^P = -e^{-(r-R(r))(T-t)} \Phi(-d_1). \tag{9.60}$$

$$\Gamma_t^C = \Gamma_t^P = \frac{e^{-(r-R(r))(T-t)} \phi(d_1)}{X_t \sigma \sqrt{T-t}}, \tag{9.61a}$$

where ϕ denotes the density function of the standard normal in (6.2). Hint: While (9.60) looks elementary, note that d_1 and d_2 are functions of X_t. Also, the relationship between put and call Greeks is simplified by using put-call parity.

If you enjoy differentiation, derive formulas for vega, rho, and theta, and verify that (9.55) is satisfied.

9.5.4 Lattice Approximations for "Greeks"

For European puts and calls, the Black-Scholes-Merton formulas provide exact options prices which can, with a little effort, be differentiated to obtain exact formulas for the various Greeks. For most other financial derivatives, no such conveniently differentiable pricing formula exists, and thus, Greeks must be estimated, often from the binomial lattice prices. These prices will be denoted $O_t^{(n)}[X_t]$ here to simplify notation, noting that in general, $t = j\Delta t$ with $\Delta t = T/n$ and $0 \leq j \leq n$.

1. Delta: Recall that the **forward difference approximation** to $f'(x_0)$ for a differentiable function $f(x)$ is defined:

$$f'(x_0) = \frac{f(x_0 + \Delta x) - f(x_0)}{\Delta x} + \mathcal{O}[\Delta x], \tag{9.62}$$

while the **central difference approximation** is defined:

$$f'(x_0) = \frac{f(x_0 + \Delta x) - f(x_0 - \Delta x)}{2\Delta x} + \mathcal{O}\left[(\Delta x)^2\right]. \tag{9.63}$$

Here we use script $\mathcal{O}$ for the **big-*O*** notation of Definition 7.2 to avoid notational confusion with the settlement function O_t. Also, the above errors require that $f(x)$ has two, or three derivatives, respectively, as is justified with a Taylor series analysis.

Thus given a binomial lattice and derivatives price $O_t^{(n)}[X_t]$, the delta of this price can be approximated with these formulas by repeating the lattice calculation for $O_t^{(n)}[X_t + \Delta X]$, or repeating it twice for $O_t^{(n)}[X_t \pm \Delta X]$. This approach is computationally less difficult for European derivatives for which (9.8) provides a simpler formulaic price. For American options, which must be evaluated iteratively by (9.22), the computational price can be significant.

Seemingly an alternative approach, $O_t^{(n)}[X_t]$ is equal to the price of the replicating portfolio for the financial derivative at time t. That is, simplifying the notation of Proposition 9.9:

$$O_t^{(n)}[X_t] = a_t^{(n)} X_t + b_t^{(n)},$$

where $a_t^{(n)}$ denotes number of units of the underlying asset with market value $a_t^{(n)} X_t$, and $b_t^{(n)}$ is the market value of an investment in the risk-free asset at rate r. When the underlying asset is a futures contract, the market value of $a_t^{(n)}$ units of the underlying is 0, and thus $O_t^{(n)}[X_t] = b_t^{(n)}$. The signs of $a_t^{(n)}$ and $b_t^{(n)}$ determine whether the respective positions are long or short.

While it is tempting to think that this formula implies that the option delta, meaning the X_t-calculus derivative, at time t is:

$$\frac{dO_t^{(n)}[X_t]}{dX_t} = a_t^{(n)},$$

this calculation is not justified.

Such a delta calculation from the replicating portfolio requires that $a_t^{(n)}$ and $b_t^{(n)}$ be independent of X_t, and by (9.10) and (9.11), it is clear that they are not. For example by (9.10):

$$a_t^{(n)} = \frac{O_{t+\Delta t}^{(n)}\left[X_t e^{u(\Delta t)}\right] - O_{t+\Delta t}^{(n)}\left[X_t e^{d(\Delta t)}\right]}{X_t e^{u(\Delta t)} - X_t e^{d(\Delta t)}}. \tag{9.64}$$

However, assuming that $O_{t+\Delta t}^{(n)}(x)$ is indeed a differentiable function of x, the formula in (9.64) implies that $a_t^{(n)}$ is a central difference approximation to $\frac{dO_{t+\Delta t}^{(n)}}{dx}$, evaluated with mean price $x_0 = \bar{X}_{t+\Delta t}$:

$$\bar{X}_{t+\Delta t} = \frac{1}{2}\left(X_t e^{u(\Delta t)} + X_t e^{d(\Delta t)}\right),$$

and $\Delta x \equiv .5\left(X_t e^{u(\Delta t)} - X_t e^{d(\Delta t)}\right)$. In other words:

$$a_t^{(n)} \approx \left.\frac{dO_{t+\Delta t}^{(n)}(x)}{dx}\right|_{x=\bar{X}_{t+\Delta t}},$$

and the error between $a_t^{(n)}$ and $\frac{dO_{t+\Delta t}^{(n)}}{dx}$ is then $\mathcal{O}\left[(\Delta x)^2\right] = \mathcal{O}\left[\left(e^{u(\Delta t)} - e^{d(\Delta t)}\right)^2\right]$ when $O_{t+\Delta t}^{(n)}(x)$ has three derivatives.

Recalling (9.2) obtains:

$$\bar{X}_{t+\Delta t} = X_t e^{\mu \Delta t}\left(1 + \frac{1}{2}\sigma^2 \Delta t + \mathcal{O}\left[(\Delta t)^2\right]\right),$$

and:

$$\mathcal{O}\left[(\Delta x)^2\right] = \mathcal{O}\left[\Delta t\right].$$

Hence,

$$a_t^{(n)} = \left.\frac{dO_{t+\Delta t}^{(n)}(x)}{dx}\right|_{x=\bar{X}_{t+\Delta t}} + \mathcal{O}\left[\Delta t\right], \tag{1}$$

and the binomial model $a_t^{(n)}$-coefficients approximate the delta of the financial derivative at time $t+\Delta t$ when the asset price is $\bar{X}_{t+\Delta t} = X_t e^{\mu\Delta t} + \mathcal{O}\left[\Delta t\right].$

As noted in Section 9.3, it is common to set $\mu = 0$ for derivatives pricing, and thus $\bar{X}_{t+\Delta t} = X_t + \mathcal{O}\left[\Delta t\right].$ When Δt is small, $a_t^{(n)}$ is then a good approximation to $\Delta_{t+\Delta t}$, the delta at time $t+\Delta t$ with price X_t. It is consequently not uncommon in practice to use $a_t^{(n)}$ for Δ_t, the delta of the financial derivative at time t.

2. Gamma: Given a twice differentiable function $f(x)$, $f''(x)$ can also be approximated by the **central difference approximation**:

$$f''(x_0) \approx \frac{f(x_0+\Delta x) + f(x_0-\Delta x) - 2f(x_0)}{(\Delta x)^2}. \tag{9.65}$$

By Taylor series analysis, the error of this approximation is $\mathcal{O}\left[\Delta x\right]$ when $f(x)$ has three derivatives, and $\mathcal{O}\left[(\Delta x)^2\right]$ when $f(x)$ has four derivatives.

Thus the gamma of a financial derivative can be estimated by repeating the lattice calculations for $O_t^{(n)}\left[X_t \pm \Delta X\right]$, and these prices could also be used to provide a good approximation to delta with (9.63). That said, such repeated lattice calculations can be computationally expensive.

Since gamma is the derivative of delta, we can also estimate two deltas at time $t+\Delta t$ and then transform these into a gamma estimate. Given X_t, let:

$$a_{t+\Delta t}^u = \frac{O_{t+2\Delta t}^{(n)}\left[X_t e^{2u(\Delta t)}\right] - O_{t+\Delta t}^{(n)}\left[X_t e^{u(\Delta t)+d(\Delta t)}\right]}{X_t e^{2u(\Delta t)} - X_t e^{u(\Delta t)+d(\Delta t)}},$$

where we drop the superscript (n) to make room for u/d. Thus $a_{t+\Delta t}^u$ is the asset position at that node in the sense that $O_{t+\Delta t}^{(n)}\left[e^{u(\Delta t)}X_t\right] = a_t^u e^{u(\Delta t)}X_t + b_t^u$. By the above analysis:

$$a_{t+\Delta t}^u = \left.\frac{dO_{t+2\Delta t}^{(n)}(x)}{dx}\right|_{x=\bar{X}_{t+2\Delta t}^u} + \mathcal{O}\left[\Delta t\right],$$

where $\bar{X}_{t+2\Delta t}^u \equiv .5(X_t e^{2u(\Delta t)} + X_t e^{u(\Delta t)+d(\Delta t)})$.

Defining $a_{t+\Delta t}^d$ and $\bar{X}_{t+2\Delta t}^d$ analogously, gamma can be estimated from these deltas:

$$\frac{a_{t+\Delta t}^u - a_{t+\Delta t}^d}{.5\left(\bar{X}_{t+2\Delta t}^u - \bar{X}_{t+2\Delta t}^d\right)} \approx \left.\frac{d^2 O_{t+2\Delta t}^{(n)}(x)}{dx^2}\right|_{x=\bar{X}_{t+2\Delta t}} \tag{2}$$

with $\bar{X}_{t+2\Delta t} \equiv .5\left(\bar{X}_{t+2\Delta t}^u + \bar{X}_{t+2\Delta t}^d\right)$. It then follows that:

$$\begin{aligned}\bar{X}_{t+2\Delta t} &= X_t\left(\frac{X_t e^{u(\Delta t)} + X_t e^{d(\Delta t)}}{2}\right)^2 \\ &= X_t e^{2\mu\Delta t}\left(1 + \frac{1}{2}\sigma^2\Delta t + \mathcal{O}\left[(\Delta t)^2\right]\right)^2.\end{aligned}$$

Taking $\mu = 0$ as is common, $\bar{X}_{t+2\Delta t} = X_t + \mathcal{O}[\Delta t]$ and the calculation in (2) is a gamma estimate at time $t + 2\Delta t$ with price X_t. It is consequently not uncommon in practice to use this or similar calculation for Γ_t, the gamma of the financial derivative at time t.

3. Theta: For the theta calculation, which is a t-derivative, the information on the binomial lattice is perfectly suited with $\mu = 0$, as is common. Then $X_t e^{u(\Delta t)} e^{d(\Delta t)} = X_t$, and the formula in (9.62):

$$\theta_t \approx \left(\frac{O^{(n)}_{t+2\Delta t}[X_t] - O^{(n)}_t [X_t]}{2\Delta t} \right),$$

with an error of $\mathcal{O}[\Delta t]$.

4. Vega, Rho: Lattice recalculations are necessary for vega and rho because the dependence of $O^{(n)}_t[X_t]$ on σ, respectively r, is functionally complicated.

The volatility assumption σ appears in both the asset return assumptions, $u(\Delta t)$ and $d(\Delta t)$, as well as in the definitions of the risk-neutral probability q in (9.24), while r discounts cash flows, but also appears in most versions of q.

Once derivative prices are recalculated, these Greeks can be estimated as in (9.62) or (9.63).

9.6 Limiting Price of American Derivatives

Though beyond the scope of the materials developed here, there are a number of research results on the convergence as $\Delta t \to 0$ of American option prices derived from binomial models. We provide three references.

Dietmar P. J. Leisen (1998) evaluates the order of convergence for American put option prices under several variants of binomial models, and proposes a model that accelerates this convergence.

The paper of **Lishang Jiang and Min Dai** (1999) proves convergence of binomial prices of American options using methods of numerical analysis, and generalizes to a class of exotic options.

Using advanced tools of stochastic processes that are studied in Books VII-IX, **K. Amin and A. Khanna** (1994) prove that if the sequence of binomial models converges weakly to the stochastic process (i.e., diffusion) of asset prices, then the corresponding sequence of binomial American option prices also converges to the American option price implied by that diffusion.

9.7 Path-Dependent Derivatives

A **path-dependent European financial derivative** is a financial contract with settlement value at time T that depends not only on the value X_T, but also on the values of X_t for $t < T$. In other words, the settlement at time T not only depends on the terminal value of the underlying asset, but also on the "path" of asset prices from X_0 to X_T.

While this description implies that path-dependent derivative settlements reflect continuous time evolution of asset prices, such a contract design would be impossible to settle. In order for both counterparties to be satisfied with settlement calculations, the asset prices used in the settlement function must be objectively verifiable. This quickly leads to the conclusion that only prices at market close will be used, and not intraday prices. Thus the highest frequency possible for the asset prices in a settlement function is daily.

For pricing such derivatives, it is apparent that the binomial lattice pricing approach of Proposition 9.9 requires some modification, since given any of the $n+1$ terminal values of X_T on that lattice, there are in general multiple asset paths which result in this value.

In detail, recall that in the proof of (9.8), we required the valuation of derivative payoffs at expiry, $O_T\left[X_0 e^{ju(\Delta t)} e^{(n-j)d(\Delta t)}\right]$, for $0 \leq j \leq n$. But the time T asset price $X_0 e^{ju(\Delta t)} e^{(n-j)d(\Delta t)}$ has $\binom{n}{j}$ different asset paths which lead to this value, and hence potentially $\binom{n}{j}$ different payoffs for a path-dependent derivative security. Consequently, X_T only uniquely identifies an asset path in the extreme cases of $j=0$ and $j=n$.

In order to price such options, we first restate the binomial pricing result for "standard" European derivatives within a path-based framework.

9.7.1 Path-Based Pricing of European Derivatives

The binomial lattice pricing formula for a European financial derivative in (9.8) states that the price $O_0^{(n)}[X_0]$ at time 0 equals the risk-free present value of the expected value of time T settlements $\{O_T\left[X_0 e^{ju(\Delta t)} e^{(n-j)d(\Delta t)}\right]\}_{j=1}^n$, where this expectation is calculated using the binomial probabilities $\left\{\binom{n}{j} q^j (1-q)^{n-j}\right\}_{j=0}^n$:

$$O_0^{(n)}[X_0] = e^{-rT} \sum\nolimits_{j=0}^{n} \binom{n}{j} q^j (1-q)^{n-j} O_T\left[X_0 e^{ju(\Delta t)} e^{(n-j)d(\Delta t)}\right]$$

The return parameters $u(\Delta t)$ and $d(\Delta t)$ are given in (9.2), q is defined in (9.24) for the various asset classes, and $n = T/\Delta t$.

We can formally restate this formula in terms of the 2^n paths implied by this model, where a path is defined by an $(n+1)$-tuple of prices:

$$(X_0, X_{\Delta t}, X_{2\Delta t}, ..., X_{n\Delta t}),$$

and where for all j:

$$X_{(j+1)\Delta t} = \begin{cases} e^{u(\Delta t)} X_{j\Delta t}, & \Pr = q, \\ e^{d(\Delta t)} X_{j\Delta t}, & \Pr = 1-q. \end{cases}$$

Any such path can be identified by the n-tuple of $u(\Delta t)$ and $d(\Delta t)$ values which generates it, or, by the n-tuple of binomial variates $b_j = \pm 1$ which generates these $u(\Delta t)$ and $d(\Delta t)$ values. The probability of such a path defined by j-$u(\Delta t)$s and $(n-j)$-$d(\Delta t)$s is then $q^j (1-q)^{n-j}$.

Given a European financial derivative and $X_T = X_0 e^{ju(\Delta t)} e^{(n-j)d(\Delta t)}$, there are $\binom{n}{j}$ paths with this terminal value, and thus the associated term in (9.8) can be expressed:

$$\begin{aligned} &\binom{n}{j} q^j (1-q)^{n-j} O_T\left[X_0 e^{ju(\Delta t)} e^{(n-j)d(\Delta t)}\right] \\ &= \sum\nolimits_{\binom{n}{j}} q^j (1-q)^{n-j} O_T\left[(X_0, X_{\Delta t}, X_{2\Delta t}, ..., X_T)\right]. \end{aligned}$$

The last sum is over all $\binom{n}{j}$ paths with the given terminal X_T, and $O_T[(X_0, X_{\Delta t}, X_{2\Delta t}, ..., X_T)]$ is the settlement value of this derivative based on this path, where here, $O_T[(X_0, X_{\Delta t}, X_{2\Delta t}, ..., X_T)] = O_T[X_T]$.

Thus we can express the formula for $O_0^{(n)}[X_0]$ in (9.8) by:

$$O_0^{(n)}[X_0] = e^{-rT}\sum\nolimits_{2^n} q^j(1-q)^{n-j}O_T\left[(X_0, X_{\Delta t}, X_{2\Delta t}, ..., X_{n\Delta t})\right], \tag{9.66}$$

where this summation is over all 2^n n-tuples defined by all possible sequences of returns or binomial variates. As above, the parameter j then equals the number of $u(\Delta t)$s or $+1$s in the given n-tuples.

9.7.2 Lattice Pricing of European PD Derivatives

The reformulated version of the Proposition 9.9 European pricing formula in (9.66) would seem to now formally provide for the pricing of a path-dependent derivative, by simply substituting the path-dependent settlement function of such a derivative. But we cannot justify this without further analysis.

Indeed, the original formula in (9.8) was derived from a period-by-period replicating portfolio argument. The result of this was that $O_0^{(n)}[X_0]$ equaled the market value of a portfolio of underlying assets and risk-free securities, which if appropriately rebalanced each period, would ultimately replicate the $n+1$ settlements at time T for this European derivative.

In order to support the conclusion that the formula in (9.66) provides the price of a path-dependent European derivative, we must replicate this argument. In other words, we must demonstrate that $O_0^{(n)}[X_0]$ so calculated again equals the price of such a portfolio, which after rebalancing, will replicate each of the 2^n path-based settlements $O_T\left[(X_0, X_{\Delta t}, X_{2\Delta t}, ..., X_{n\Delta t})\right]$.

In this section we demonstrate that $O_0^{(n)}[X_0]$, as given in (9.66), is indeed the price of such a portfolio.

To this end, first note that path-dependent settlement functions do not explicitly reflect the value of X_0 when newly created, since this value is known on the pricing date at $t=0$. When traded in the secondary market, for which history has accumulated since original pricing, this settlement function will reflect X_0 and prior asset prices. To simplify notation, we reflect X_0 in the settlement function both because the rest of the path depends on this value, and as a reminder that this function may indeed require this value, and potentially even prior values.

As will be seen, the demonstration of the following result reflects a relatively minor adaptation of the proof of Proposition 9.9 for standard European derivatives. Figure 9.1 provides a visual of the respective frameworks, where for notational simplicity we denote $u = e^{u(\Delta t)}$ and $d = e^{d(\Delta t)}$ with $u(\Delta t)$ and $d(\Delta t)$ defined in (9.2).

The binomial lattice on the left is known as a **recombining lattice**, and reflects the common visual depiction of asset prices underlying Proposition 9.9. It "recombines" simply because, for example:

$$Sud = Sdu. \tag{1}$$

That is, asset prices are the same after two periods of unequal returns, independent of the order of these returns. This property is observed everywhere on the lattice, because the price only depends on the number of u and d factors, not on the order. But this depiction loses the details of the price paths. For example, at terminal price $Sudd$, labelled "2" on the left, there are $\binom{3}{2}=3$ paths that lead to this price, and these are labelled "2" on the right.

In the **nonrecombining lattice** on the right of Figure 9.1, the emphasis is on price paths. There, although (1) remains valid, the different paths leading to these same prices are isolated.

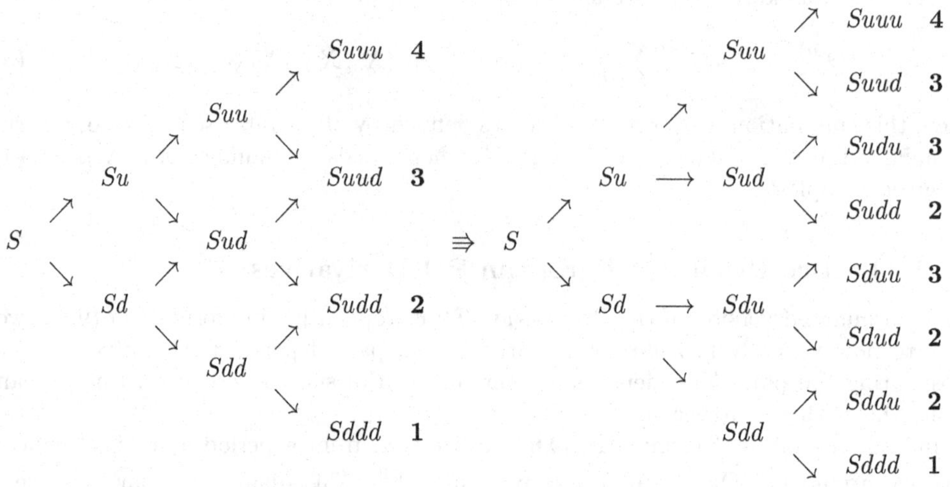

				⇒				
			Suuu 4				*Suu*	↗ *Suuu* 4
		Suu						↘ *Suud* 3
	Su		*Suud* 3			*Su* →	*Sud*	↗ *Sudu* 3
S		*Sud*		⇒	*S*			↘ *Sudd* 2
	Sd		*Sudd* 2			*Sd* →	*Sdu*	↗ *Sduu* 3
		Sdd						↘ *Sdud* 2
			Sddd 1				*Sdd*	↗ *Sddu* 2
								↘ *Sddd* 1

FIGURE 9.1
Binomial Lattices: Recombining and Nonrecombining.

But note that the binomial geometries of these lattices are identical. At any price, there are exactly two prices possible in the next period. Thus the mechanics of replication below will work identically with replication in Proposition 9.9. The primary difference in the two setups is that there are $\sum_{j=1}^{n} j = \frac{1}{2}n(n+1)$ replicating portfolios within the lattice on the left for a standard European derivative, whereas there will be $\sum_{j=1}^{n} 2^{j-1} = 2^n - 1$ replicating portfolios for a path-dependent European derivative.

Proposition 9.54 (Binomial lattice price of a European PD derivative) *Let* $O_T[(X_0, X_{\Delta t}, X_{2\Delta t}, ..., X_{n\Delta t})]$ *denote the time T settlement of a path-dependent European derivative security as a function of the realized asset price path. Assume that investors can take long or short positions in X with no transaction costs, and that the risk-free interest rate with continuous compounding is fixed and equal to r for long and short positions.*

Then the price at time 0 of a replicating portfolio is given by:

$$O_0^{(n)}[X_0] = e^{-rT}\sum\nolimits_{2^n} q^j(1-q)^{n-j} O_T[(X_0, X_{\Delta t}, X_{2\Delta t}, ..., X_{n\Delta t})], \tag{9.67}$$

where the summation is over all 2^n n-tuples of $u(\Delta t)/d(\Delta t)$ return variates, and j denotes the number of $u(\Delta t)$-variates in the given n-tuple. As above, r denotes the risk-free interest rate, $q \equiv q(\Delta t)$ is the risk-neutral probability of $u(\Delta t)$ as given by (9.24) for the various underlying assets, and $T = n\Delta t$.

Proof. *The derivation of this option pricing result requires a small reorganization of the derivation for Proposition 9.9, and in fact the first two steps differ only in notational representation. Again we leave some details as exercises.*

1. *At time $j\Delta t$, given any one of the 2^j asset price paths $(X_0, X_{\Delta t}, X_{2\Delta t}, ..., X_{j\Delta t}) \equiv \bar{X}$ for notational simplicity, assume we are given a value function that identifies $f(\bar{X}, X_{(j+1)\Delta t})$ for the two possible values of $X_{(j+1)\Delta t}$. The time $(j+1)\Delta t$ asset price is given by $X_{(j+1)\Delta t} = X_{j\Delta t}e^{u(\Delta t)}$ with probability q, or $X_{(j+1)\Delta t} = X_{j\Delta t}e^{d(\Delta t)}$ with probability $1-q$, where $X_{j\Delta t}$ is the last component of $\bar{X}$. Then there are real constants a_j, b_j, so that a time $j\Delta t$ portfolio of a_j units of $X_{j\Delta t}$, and b_j units invested at rate r, exactly replicates the 2 values of $f(\bar{X}, X_{(j+1)\Delta t})$ at time $(j+1)\Delta t$.*

For example, if X is an asset with no income:

$$a_j = \frac{f(\bar{X}, e^{u(\Delta t)} X_{j\Delta t}) - f(\bar{X}, e^{d(\Delta t)} X_{j\Delta t})}{\left(e^{u(\Delta t)} - e^{d(\Delta t)}\right) X_{j\Delta t}}, \tag{9.68}$$

and,

$$b_j = \left(\frac{e^{u(\Delta t)} f(\bar{X}, e^{d(\Delta t)} X_{j\Delta t}) - e^{d(\Delta t)} f(\bar{X}, e^{u(\Delta t)} X_{j\Delta t})}{e^{u(\Delta t)} - e^{d(\Delta t)}} \right) e^{-r\Delta t}. \tag{9.69}$$

Analogous results apply for currencies, equity indexes, and futures contracts, as derived in Exercise 9.12.

2. *For any given $(X_0, X_{\Delta t}, X_{2\Delta t}, ..., X_{j\Delta t}) \equiv \bar{X}$, the price of this portfolio at time $j\Delta t$, $a_j X_{j\Delta t} + b_j$, can be algebraically manipulated into the form:*

$$f\left(\bar{X}\right) = e^{-r\Delta t}\left[qf\left(\bar{X}, X_{j\Delta t}e^{u(\Delta t)}\right) + (1-q)f\left(\bar{X}, X_{j\Delta t}e^{d(\Delta t)}\right)\right],$$

where $q \equiv q(\Delta t)$ is given in (9.24) for the various underlying assets.

3. *The formula in item 2 can be applied starting at time $T = (n-1)\Delta t$, at which time $\bar{X} = \left(X_0, X_{\Delta t}, X_{2\Delta t}, ..., X_{(n-1)\Delta t}\right)$, and the above value function $f\left(\bar{X}, X_{n\Delta t}\right)$ is defined by the final payoff values:*

$$O_T\left[(X_0, X_{\Delta t}, X_{2\Delta t}, ..., X_{n\Delta t})\right].$$

Because these time $(n-1)\Delta t$ prices $f\left(\bar{X}\right) \equiv f\left(X_0, X_{\Delta t}, X_{2\Delta t}, ..., X_{(n-1)\Delta t}\right)$ are the prices of replicating portfolios for the derivatives final payoffs, the law of one price obtains that these must equal the 2^{n-1} prices of the derivative at that time:

$$f\left(X_0, X_{\Delta t}, X_{2\Delta t}, ..., X_{(n-1)\Delta t}\right) \equiv O^{(n)}_{(n-1)\Delta t}\left(X_0, X_{\Delta t}, X_{2\Delta t}, ..., X_{(n-1)\Delta t}\right).$$

4. *These 2^{n-1} prices at time $(n-1)\Delta t = T - \Delta t$ then provide the value function $f\left(\bar{X}, X_{(n-1)\Delta t}\right)$ at time $(n-2)\Delta t$, where $\bar{X} = \left(X_0, X_{\Delta t}, X_{2\Delta t}, ..., X_{(n-2)\Delta t}\right)$, producing 2^{n-2} replicating portfolios, and thus 2^{n-2} prices of the derivative at that time:*

$$O^{(n)}_{(n-2)\Delta t}\left(X_0, X_{\Delta t}, X_{2\Delta t}, ..., X_{(n-2)\Delta t}\right).$$

Each such value equals the price of a portfolio at time $T - 2\Delta t$ that replicates the associated time $T - \Delta t$ derivative prices, which in turn provide for replicating portfolios for the time T settlement values.

Continuing in this way, we derive by mathematical induction:

$$\begin{aligned} &O^{(n)}_{j\Delta t}\left(X_0, X_{\Delta t}, X_{2\Delta t}, ..., X_{j\Delta t}\right) \\ &\quad = e^{-r(n-j)\Delta t}\sum\nolimits_{2^{n-j}} q^i(1-q)^{n-j-i} O_T\left(X_0, X_{\Delta t}, ..., X_{j\Delta t}, X_{(j+1)\Delta t}, ..., X_{n\Delta t}\right), \end{aligned} \tag{1}$$

where this summation is over all 2^{n-j}, $(n-j)$-tuples of price paths defined by $\left(X_{(j+1)\Delta t}, ..., X_{n\Delta t}\right)$.

5. *This general formula produces (9.8) when $j = 0$.*

■

Exercise 9.55 *Derive the formula in* (1) *of the above proof by mathematical induction. Hint: The above proof confirms the validity of this formula for* $j = n - 1$, *for which the sum in* (1) *has two terms. Assuming this formula is valid for given* j, *prove it is valid for* $j - 1$ *by replicating the time* j *settlements with the formula in item 2.*

Example 9.56 (Path-dependent financial derivatives) *Examples of path-dependent European financial derivatives are as follows:*

1. ***Asian Options:*** *These put and call options have payoff functions that reflect the average price of the underlying assets. Such averages can be defined as an* ***arithmetic average*** *or* ***geometric average:***

$$X^A \equiv \frac{1}{m}\sum_{j=1}^{m} X_{jT/m}, \qquad X^G \equiv \left(\prod_{j=1}^{m} X_{jT/m}\right)^{1/m},$$

where the averaging frequency implied by m *can be defined in various ways, such as daily, weekly, etc.*

The average price of the underlying assets can then be used in the time T *payoff function as the strike price, or, the reference price.*

   (a) ***Asian strike*** *puts or calls are defined with the standard reference price of* X_T, *but with strike* $K = X^A$ *or* $K = X^G$.

   (b) ***Asian price*** *puts and calls are defined in terms of the reference price of* X^A *or* X^G, *and fixed strike price of* K.

2. ***Lookback Options:*** *These put and call options reflect the* ***maximum*** *or* ***minimum*** *of underlying asset prices:*

$$X^{\max} = \max_j\{X_{jT/m}\}; \qquad X^{\min} = \min_j\{X_{jT/m}\},$$

where the frequency implied by m *can be defined in various ways. Such options are again labelled based on whether the maximum or minimum is used as the reference price, or, strike price.*

   (a) ***Floating lookback call*** *is defined with a reference price of* X_T *and strike* $K = X^{\min}$.

   (b) ***Floating lookback put*** *is defined with reference* X_T *and strike* $K = X^{\max}$.

   (c) ***Fixed lookback call*** *is defined with reference price* $X^{\max}$ *and fixed strike* K.

   (d) ***Fixed lookback put*** *is defined with reference* $X^{\min}$ *and fixed strike* K.

Thus in all cases, the option is defined to maximize the potential payoffs to the long position.

3. ***Barrier Options:*** *In one respect, barrier put and call options have conventional payoff functions defined at time* T *with the usual formulas and a fixed strike* K. *What makes barrier options path-dependent, is that the applicability of this payoff function depends on whether the price* X_t *for* $t < T$ *reaches a given* ***barrier*** B. *These options are classified into* ***knock-in options,*** *for which the payoff function becomes applicable only if the barrier is reached, and* ***knock-out options,*** *for which the payoff function is cancelled if the barrier is reached.*

Knock-in options are then categorized as:

(a) ***Down-and-in,*** *applicable when* $X_0 > B$.

(b) ***Up-and-in,*** *applicable when* $X_0 < B$.

Knock-out options are similarly categorized as:

(a) ***Down-and-out,*** *applicable when* $X_0 > B$.

(b) ***Up-and-out,*** *applicable when* $X_0 < B$.

Each of these four types of barrier options can then be defined as puts or calls, but not all designs produce something useful. For example, if $X_0 < B < K$, *a knock-out call has no value, while a knock-in call has the same value as a vanilla call. It is an exercise to think through other examples, and derive the parity equations:*

$$\begin{aligned} D/I + D/O &= Vanilla, \\ U/I + U/O &= Vanilla. \end{aligned}$$

9.7.3 Lattice Pricing of American PD Derivatives

Though potentially computationally intensive, the approach of Proposition 9.54 for European path-dependent derivatives can be adapted to American path-dependent derivatives in exactly the same way that Proposition 9.21 for American derivatives adapts the European derivatives result of Proposition 9.9. The notation and combinatorics are a bit more complicated however.

At time $j\Delta t$, there are 2^j possible price paths $\bar{X}_j \equiv (X_0, X_{\Delta t}, X_{2\Delta t}, ..., X_{j\Delta t})$, and there is no notationally efficient way to enumerate these paths. Fortunately for the next result, any order works fine, so we simply denote these price paths as $\bar{X}_{k,j} \equiv (X_0, X_{k,\Delta t}, X_{k,2\Delta t}, ..., X_{k,j\Delta t})$, with $k = 1, ..., 2^j$. Given any (k, j), the connected price paths on the lattice are then denoted $(\bar{X}_{k,j}, X_{k,j\Delta t}e^{u(\Delta t)})$ and $(\bar{X}_{k,j}, X_{k,j\Delta t}e^{d(\Delta t)})$, where $u(\Delta t)$ and $d(\Delta t)$ are given as in (9.2).

Proposition 9.57 (Binomial lattice price of an American PD derivative) *Let* $O_T [(X_0, X_{\Delta t}, X_{2\Delta t}, ..., X_{n\Delta t})]$ *denote the time* $T = n\Delta t$ *settlement of a path-dependent American derivative security as a function of the realized asset price path, and similarly for* $0 \leq j < n$ *let* $O_{j\Delta t} [(X_0, X_{\Delta t}, X_{2\Delta t}, ..., X_{j\Delta t})]$ *denote the time* $t = j\Delta t$ *settlement value if exercised.*

Then the binomial lattice price $O_0^{(n)} [X_0]$ *at time* $t = 0$ *for a replicating portfolio for this financial derivative is defined iteratively by:*

$$\begin{aligned} & O_{j\Delta t}^{(n)} (\bar{X}_{k,j}) \\ = \; & \max\left(e^{-r\Delta t}\left[qO_{(j+1)\Delta t}^{(n)}\left(\bar{X}_{k,j}, X_{k,j\Delta t}e^{u(\Delta t)}\right)\right.\right. \\ & \left.\left. +(1-q)O_{(j+1)\Delta t}^{(n)}\left(\bar{X}_{k,j}, X_{k,j\Delta t}e^{d(\Delta t)}\right)\right], O_{j\Delta t}\left[\bar{X}_{k,j}\right]\right), \end{aligned} \tag{9.70}$$

where $\bar{X}_{k,j} \equiv (X_0, X_{k,\Delta t}, X_{k,2\Delta t}, ..., X_{k,j\Delta t})$, *and* $q \equiv q(\Delta t)$ *as in (9.24) for the various assets classes.*

Defining $O_{n\Delta t}^{(n)} [\bar{X}_{n,k}] \equiv O_T [\bar{X}_{n,k}]$ *for all* k, *these* $2^n - 1$ *iterations to* $t = 0$ *are in order:*

- $j = n - 1, n - 2, ..., 0;$
- *For each* j, $k = 1, ..., 2^j$.

Proof. *At time $T = n\Delta t$, the settlement value is unambiguously equal to $O_T\left[\bar{X}_{n,k}\right]$ for any of the 2^n possible price paths.*

For any of the 2^{n-1} possible price paths $\bar{X}_{k,n-1} \equiv \left(X_0, X_{k,\Delta t}, X_{k,2\Delta t}, ..., X_{k,(n-1)\Delta t}\right)$ at time $(n-1)\Delta t$, there are two associated price paths to time $n\Delta t$ defined by $\left(\bar{X}_{k,n-1}, X_{k,(n-1)\Delta t}e^{u(\Delta t)}\right)$ and $\left(\bar{X}_{k,n-1}, X_{k,(n-1)\Delta t}e^{d(\Delta t)}\right)$, with time $n\Delta t$ settlement values of $O_T\left(\bar{X}_{k,n-1}, X_{k,(n-1)\Delta t}e^{u(\Delta t)}\right) \equiv O^{(n)}_{n\Delta t}\left(\bar{X}_{k,n-1}, X_{k,(n-1)\Delta t}e^{u(\Delta t)}\right)$ and $O_T\left(\bar{X}_{k,n-1}, X_{k,(n-1)\Delta t}e^{d(\Delta t)}\right) \equiv O^{(n)}_{n\Delta t}\left(\bar{X}_{k,n-1}, X_{k,(n-1)\Delta t}e^{u(\Delta t)}\right)$.

By Proposition 9.54, the price $R\left(\bar{X}_{k,n-1}\right)$ at time $(n-1)\Delta t$ for a replicating portfolio for these settlements is given by:

$$\begin{aligned} R\left(\bar{X}_{k,n-1}\right) &= e^{-r\Delta t}\left[qO^{(n)}_{n\Delta t}\left(\bar{X}_{k,n-1}, X_{k,(n-1)\Delta t}e^{u(\Delta t)}\right)\right. \\ &\quad \left. +(1-q)O^{(n)}_{n\Delta t}\left(\bar{X}_{k,n-1}, X_{k,(n-1)\Delta t}e^{d(\Delta t)}\right)\right]. \end{aligned}$$

There is also an obtainable settlement value of $O_{(n-1)\Delta t}\left[\bar{X}_{k,n-1}\right]$ at that time. The price of this derivative $O^{(n)}_{(n-1)\Delta t}\left(\bar{X}_{k,n-1}\right)$ at that time is thus the greater value by an arbitrage argument, and is given by (9.70) with $j = n-1$:

$$O^{(n)}_{(n-1)\Delta t}\left(\bar{X}_{k,n-1}\right) = \max\left(R\left(\bar{X}_{k,n-1}\right), O_{(n-1)\Delta t}\left[\bar{X}_{k,n-1}\right]\right).$$

At time $(n-2)\Delta t$, there are 2^{n-2} possible price paths $\bar{X}_{k,n-2} \equiv (X_0, X_{k,\Delta t}, X_{k,2\Delta t}, ..., X_{k,(n-2)\Delta t})$, each with two associated price paths to time $(n-1)\Delta t$, namely $(\bar{X}_{k,n-2}, X_{k,(n-2)\Delta t}e^{u(\Delta t)})$ and $(\bar{X}_{k,n-2}, X_{k,(n-2)\Delta t}e^{d(\Delta t)})$. These two paths have associated time $(n-1)\Delta t$ prices of $O^{(n)}_{(n-1)\Delta t}(\bar{X}_{k,n-2}, X_{k,(n-2)\Delta t}e^{u(\Delta t)})$ and $O^{(n)}_{(n-1)\Delta t}(\bar{X}_{k,n-2}, X_{k,(n-2)\Delta t}e^{d(\Delta t)})$, which can be replicated at price $R(\bar{X}_{k,n-2})$ and compared to settlement value $O_{(n-2)\Delta t}[\bar{X}_{k,n-2}]$, obtaining (9.70) for $j = n-2$.

This iterates to time 0, where there is one price "path" $\bar{X}_{1,0} \equiv X_0$, and two associated price paths $\left(\bar{X}_{1,0}, X_{1,0}e^{u(\Delta t)}\right)$ and $\left(\bar{X}_{1,0}, X_{1,0}e^{d(\Delta t)}\right)$ to time Δt. The price at that time is calculated as above, and the proof is complete. ■

9.7.4 Monte Carlo Pricing of European PD Derivatives

The formula in (9.67) provides an exact valuation of the replicating portfolio for path-dependent European financial derivatives given the binomial model for the price of the underlying asset. It is apparent that this exact approach becomes computationally challenging for n large since the summation requires 2^n terms, and each term requires the generation of a distinct binomial asset path. In contrast, the price of a standard European derivative in (9.8) reflects a summation with only $n+1$ terms, and each term is easily calculated since the needed terminal asset prices are explicitly represented.

But given that the exact price in (9.67) reflects a large universe of 2^n paths, one logically surmises that such a price can be estimated using a sample of paths. Specifically, if a large sample of paths is generated, the distribution of the derivative's payoffs should be close to that predicted in theory, and hence, the price of the derivative based on this sample should be "close" to the theoretical price.

In this section, we investigate path-based pricing and its properties.

While a given path-dependent settlement function may well group paths into far less than 2^n settlement "buckets," we will ignore such groupings in the general development for notational simplicity. The resulting model will then be generally applicable since it will not be assumed in the derivation that different paths produce different payoffs.

Thus we focus on the distribution of paths generated by the model and not on the distribution of payoffs. The distribution of paths is governed by the multinomial distribution, which reflects the multinomial theorem first introduced in IV.(4.40), and applied earlier in this book.

Generalizing the familiar **binomial theorem of** (4.31), the so-called **multinomial theorem** states:

$$\left(\sum_{i=1}^{R} a_i\right)^n = \sum_{n_1,n_2,..n_R} \frac{n!}{n_1!n_2!...n_R!} a_1^{n_1} a_2^{n_2}...a_R^{n_R}, \tag{9.71}$$

where this summation is over all distinct R-tuples $(n_1, n_2, ..n_R)$ with $n_j \geq 0$ and $\sum_{j=1}^{R} n_j = n$.

The combinatorial coefficient $\frac{n!}{n_1!n_2!...n_R!}$ counts the number of times the given factor $a_1^{n_1}a_2^{n_2}...a_R^{n_R}$ arises in this product. This follows because given any n terms $\{a_{i_j}\}_{j=1}^{n}$, where "any" means allowing for repetitions, there are $n!$ possible orderings ignoring the values in this collection. The divisions by the $n_k!$ factors then eliminate the multiple counts associated with orderings of the subsets where $a_{i_j} = a_k$.

Definition 9.58 (Multinomial distribution) *Given $\{p_j\}_{j=1}^{R}$ with $0 < p_j < 1$ and $\sum_{j=1}^{R} p_j = 1$ and fixed $n \in \mathbb{N}$, the **multinomial probability function** or **density function** f_M with parameters $\{p_j\}_{j=1}^{R}$ and n is defined on every integer R-tuple $(n_1, n_2, .., n_R)$ with $n_j \geq 0$ and $\sum n_j = n$ by:*

$$f_M(n_1, n_2, .., n_R) = \frac{n!p_1^{n_1}p_2^{n_2}...p_R^{n_R}}{n_1!n_2!...n_R!}. \tag{9.72}$$

This is indeed a probability function on the collection of such R-tuples $(n_1, n_2, .., n_R)$ since $\sum_{j=1}^{R} p_j = 1$, and thus by the multinomial theorem in (9.71):

$$\sum_{n_1,n_2,...,n_R} f_M(n_1, n_2, .., n_R) = 1,$$

where the summation is over all R-tuples $(n_1, n_2, .., n_R)$ as defined above.

Example 9.59 *Some applications for this distribution follow.*

1. *At a country fair, a girl with a bow and n arrows is shooting at a target with $R-1$ concentric colored rings of different sizes. She has probability p_j of hitting the jth ring, and probability $p_R = 1 - \sum_{j=1}^{R-1} p_j$ of missing the target and hitting the ground. The sample space is then the collection of all R-tuples $(n_1, n_2, .., n_R)$ of results, where each n_j denotes the number of arrows hitting the respective ring, where $j = R$ denotes the number hitting the ground.*

2. *Given a binomial model for asset prices, assume that M sequences of n returns $u(\Delta t)$ and $d(\Delta t)$ in (9.2) are generated with probability of $u(\Delta t)$ equal to $q \equiv q(\Delta t)$, as defined in (9.24) for various asset classes. One question could be: For any nonnegative $(n+1)$-tuple $(M_0, M_1, M_2, .., M_n)$ with $\sum M_j = M$, what is the probability that exactly M_j sequences will have j u-returns for all j?*

   From (9.72) it then follows that with p_j the probability of j u-returns:

$$f_M(M_0, M_1, .., M_n) = \frac{(n+1)!p_0^{M_0}p_1^{M_1}p_2^{M_2}...p_n^{M_n}}{M_0!M_1!M_2!...M_n!}.$$

This is the model for the financial derivatives of Section 9.1. There, settlement value buckets are defined by the final underlying asset price, which in turn is defined by the number of $u(\Delta t)$-returns that occur with probability q. In this application, the probability of an asset path containing exactly j u-returns is thus $p_j = \binom{n}{j} q^j (1-q)^{n-j}$, $j = 0, 1, .., n$.

3. *Generalizing the model in item 2 for the path-dependent derivative valuation, assume that all 2^n possible sequences of n returns are generated and ordered in some manner, where the probability of $u(\Delta t)$ is again equal to $q \equiv q(\Delta t)$. If M sequences of n returns are then generated, and a nonnegative 2^n-tuple $(M_1, M_2, .., M_{2^n})$ given with $\sum M_j = M$, what is the probability that exactly M_j sequences will equal the jth sequence in our listing?*

   By (9.72) it follows that with p_j the probability of the jth sequence on the list, and $N \equiv 2^n$:

$$f_M(M_1, .., M_N) = \frac{N! p_1^{M_1} p_2^{M_2} ... p_N^{M_N}}{M_1! M_2! ... M_N!}.$$

   This is the model for the general path-dependent option because the payoff buckets are defined by the entire sequence of asset prices. This sequence in turn is defined by the exact order of the sequence of $u(\Delta t)$ and $d(\Delta t)$ returns, which occur with respective probabilities q and $1 - q$. In this application, the probability of a path equalling the jth sequence on the list is given by $p_j = q^{k_j}(1 - q)^{n-k_j}$, where k_j denotes the number of $u(\Delta t)$ returns in this kth sequence.

Some properties of the multinomial distribution needed below and assigned as an exercise are summarized in the following proposition. See also Section 7.6.4 of **Reitano** (2010).

Proposition 9.60 (Multinomial moments) *Given the multinomial probability density of (9.72) with parameters $\{p_j\}_{j=1}^R$ and n, and defined on integer R-tuples $(n_1, n_2, .., n_R)$ with $n_j \geq 0$ and $\sum n_j = n$, let $N \equiv (N_1, N_2, .., N_R)$ denote the random vector with range on this collection of R-tuples.*

1. *For any j, the marginal probability function $f(n_j)$ of N_j is binomial with parameters p_j and n:*

$$f(n_j) = \binom{n}{n_j} p_j^{n_j} (1 - p_j)^{n-n_j},$$

   and hence:

$$E[N_j] = np_j, \qquad Var[N_j] = np_j(1 - p_j).$$

2. *For any $j \neq k$, the marginal probability function $f(n_j, n_k)$ of (N_j, N_k) is multinomial with parameters $\{p_j, p_k, 1 - p_j - p_k\}$ and n:*

$$f(n_j, n_k) = \frac{n! p_1^{n_j} p_2^{n_k} (1 - p_j - p_k)^{n-n_j-n_k}}{n_j! n_k! (n - n_j - n_k)!},$$

   and hence:

$$E[N_j N_k] = n(n-1)p_j p_k, \qquad Cov[N_j, N_k] = -np_j p_k.$$

Proof. *Left as an exercise. Hint: These marginal probability functions are obtained by summations over all other variates, and appropriately manipulating such summations and applying (9.71).* ■

With the prepatory work on the multinomial distribution complete, we now address the European derivative pricing result, suppressing the distinction between vanilla and path-dependent European derivatives. As noted in Example 9.59, both path-based models fit within a general multinomial framework by properly defining R and the probabilities $\{p_j\}_{j=1}^R$.

Assume then that M binomial paths have been generated for asset X within the given model, with $\Delta t = T/n$ the time-step used in the definition of period returns $u(\Delta t)$ and $d(\Delta t)$ given in (9.2), and the $u(\Delta t)$-return risk-neutral probability $q(\Delta t)$ given in (9.24) for the various asset classes. As in Example 9.59:

- **Vanilla financial derivatives:** Let $R = n+1$ and $p_j = \binom{n}{j} q^j (1-q)^{n-j}$ for $j = 0, 1, .., n$, where j denotes the number of $u(\Delta t)$-returns. Here, path-buckets are defined in terms of the number of $u(\Delta t)$-returns, so the jth bucket contains $\binom{n}{j}$ paths.

- **Path-dependent derivatives:** Let $R = 2^n$ and $p_j = q^{k_j}(1-q)^{n-k_j}$, where k_j denotes the number of $u(\Delta t)$-returns in the jth path sequence, assuming an arbitrary ordering of all possible 2^n sequences. Here, path-buckets are defined in terms of the listing of all possible 2^n sequences, so each bucket contains one path.

In either model, the generated M paths can be represented as $(M_1, M_2, ..., M_R)$, where M_j denotes the number of paths in the jth bucket, and thus $M = \sum_{j=1}^R M_j$.

Definition 9.61 (Monte Carlo price of a European derivative) *For either model above, let $\{P_k^{(n)}\}_{k=1}^M$ denote an M-sample of price paths for asset X, so:*

$$P_k^{(n)} = (X_0, X_{k,\Delta t}, X_{k,2\Delta t}, ..., X_{k,n\Delta t}),$$

and let $\{M_j\}_{j=1}^R$ denote the number of such paths that fall into each of the R path buckets.

Define the random variable $O_M^{(n)}(X_0)$, called the ***Monte Carlo price estimate:***

$$O_M^{(n)}(X_0) = \frac{e^{-rT}}{M} \sum_{k=1}^M O_T\left(P_k^{(n)}\right). \tag{9.73}$$

Equivalently:

$$O_M^{(n)}(X_0) = \frac{e^{-rT}}{M} \sum_{j=1}^R M_j O_T\left(P_{M_j}^{(n)}\right), \tag{9.74}$$

where $P_{M_j}^{(n)}$ is a representative path in the jth bucket for the vanilla derivative, and is the unique path in this bucket for the path-dependent derivative.

Recalling the notion of independent random vectors from Definition 1.53, we have the following.

Proposition 9.62 ($\{P_k^{(n)}\}_{k=1}^M$ are independent random vectors) *For any M, the price paths $\{P_k^{(n)}\}_{k=1}^M$ are independent random vectors. Thus $\{P_k^{(n)}\}_{k=1}^\infty$ are independent random vectors for any countable collection.*

Proof. *By construction, the collection $\{P_k^{(n)}\}_{k=1}^M$ requires Mn-independent binomial variates $\{b_i^q\}_{i=1}^{Mn}$ defined to equal 1 with probability q, and -1 with probability $1-q$. As above $q \equiv q(\Delta t)$, the risk-neutral probability given in (9.24) for the various asset classes. In detail, the path $P_k^{(n)} = (X_0, X_{k,\Delta t}, X_{k,2\Delta t}, ..., X_{k,n\Delta t})$ is defined in terms of $\{b_i^q\}_{i=(k-1)n+1}^{kn}$ by:*

$$X_{k,j\Delta t} \equiv X_0 \exp\left[\sum_{i=(k-1)n+1}^{(k-1)n+j} \left(\mu\Delta t + \sigma\sqrt{\Delta t} b_i^q\right)\right]. \tag{1}$$

By Exercise II.3.50, independence of the random variables $\{b_i^q\}_{i=1}^{Mn}$ *obtains independence of the random vectors:*

$$\{(b_{(k-1)n+1}^q, b_{(k-1)n+2}^q, ..., b_{kn}^q)\}_{k=1}^M.$$

For each k*, define a transformation* $g_k : \mathbb{R}^n \to \mathbb{R}^n$ *by:*

$$g_k : (b_{(k-1)n+1}^q, b_{(k-1)n+2}^q, ..., b_{kn}^q) \to P_k^{(n)},$$

where each component of $P_k^{(n)}$ *is defined in* (1).

Then since continuous, each g_k *is Borel measurable by Propositions V.2.12. It then follows from Proposition II.3.56 that:*

$$\{g_k(b_{(k-1)n+1}^q, b_{(k-1)n+2}^q, ..., b_{kn}^q)\}_{k=1}^M = \{P_k^{(n)}\}_{k=1}^M,$$

are independent random vectors.

The final statement is Definition 1.53, that an infinite collection of random vectors is deemed independent if any finite subcollection of vectors are independent. ■

To consider $O_M^{(n)}(X_0)$ as a random variable, we must assume that O_T is a Borel measurable function. For vanilla derivatives, since $O_T\left(P_k^{(n)}\right) \equiv O_T\left(X_{k,T}^{(n)}\right)$, we assume that O_T is a Borel measurable function defined on $\mathbb{R}$, while for the path-dependent case, where $O_T\left(P_k^{(n)}\right)$ is defined on $(X_0, X_{k,\Delta t}, X_{k,2\Delta t}, ..., X_{k,n\Delta t})$ or $(X_{k,\Delta t}, X_{k,2\Delta t}, ..., X_{k,n\Delta t})$, we assume O_T is a Borel measurable function defined on $\mathbb{R}^{n+1}$ or $\mathbb{R}^n$.

The general result is stated next. For background on convergence of random variables, see Sections II.5.2 and II.8.5.

Exercise 9.63 ($O_M^{(n)}(X_0) \to_P O_0^{(n)}(X_0)$.) *By Proposition II.5.21, convergence with probability* 1 *in (9.78) implies that* $O_M^{(n)}(X_0)$ ***converges in probability to*** $O_0^{(n)}(X_0)$*:*

$$O_M^{(n)}(X_0) \to_P O_0^{(n)}(X_0).$$

Recall that this means that for any $\epsilon > 0$*:*

$$\Pr\left[\left|O_M^{(n)}(X_0) - O_0^{(n)}(X_0)\right| \geq \epsilon\right] \to 0, \text{ as } M \to \infty. \tag{9.75}$$

Assuming item 1 in Proposition 9.64 and $Var\left[\left\{O_M^{(n)}\left(P_{M_j}^{(n)}\right)\right\}_{j=1}^R\right] < \infty$*, prove convergence in probability directly. Hint: Using Proposition 9.62, prove that:*

$$Var\left[O_M^{(n)}(X_0)\right] = \frac{e^{-2rT}}{M} Var\left[\left\{O_M^{(n)}\left(P_{M_j}^{(n)}\right)\right\}_{j=1}^R\right], \tag{9.76}$$

where the variance of the R settlement values on the right is defined under $\{p_j\}_{j=1}^R$ *defined above. Now apply Chebyshev's inequality of Proposition IV.4.34.*

Proposition 9.64 (Monte Carlo price convergence to binomial price) *With* $O_M^{(n)}(X_0)$ *defined in (9.74) with Borel measurable* O_T*:*

1. *The expected value of* $O_M^{(n)}(X_0)$ *equals the binomial lattice option price:*

$$E[O_M^{(n)}(X_0)] = O_0^{(n)}(X_0), \tag{9.77}$$

where, depending on the model, $O_0^{(n)}(X_0)$ *denotes the vanilla derivative price in (9.8), or the path-dependent derivative price in (9.67).*

2. *If* $Var\left[\left\{O_M^{(n)}\left(P_{M_j}^{(n)}\right)\right\}_{j=1}^{R}\right] < \infty$, *where this variance of the R settlement values is defined under* $\{p_j\}_{j=1}^R$ *defined above, then as* $M \to \infty$:

$$O_M^{(n)}(X_0) \to_1 O_0^{(n)}(X_0). \tag{9.78}$$

In other words, $O_M^{(n)}(X_0)$ ***converges with probability*** 1 ***to*** $O_0^{(n)}(X_0)$.

Proof. *For item 1, Proposition 9.60 states that each* M_j *is binomial. Since expectation is linear:*

$$\begin{aligned} E[O_M^{(n)}(X_0)] &= e^{-rT}\sum\nolimits_{j=1}^{R} E\left[\frac{M_j}{M}\right] O_T\left(P_{M_j}^{(n)}\right) \\ &= e^{-rT}\sum\nolimits_{j=1}^{R} p_j O_T\left(P_{M_j}^{(n)}\right). \end{aligned}$$

Recalling the definition of R and p_j *in the respective models, this expectation equals the price in (9.8) or (9.67), respectively.*

For item 2, assume $\{P_k^{(n)}\}_{k=1}^{\infty}$ *are given and define the random variable* $Y_k = e^{-rT}O_T\left(P_k^{(n)}\right)$, *the present value of the derivatives settlement on the kth path. Then since* O_T *is Borel measurable and* $\{P_k^{(n)}\}_{k=1}^{\infty}$ *are independent random vectors on* $\mathbb{R}^n$ *by Proposition 9.62,* $\{Y_k\}_{k=1}^{\infty}$ *are independent random variables by Proposition II.3.56. In addition,* $\{Y_k\}_{k=1}^{\infty}$ *are identically distributed, equalling* $e^{-rT}O_T\left(P_{M_j}^{(n)}\right)$ *with probability* p_j.

Thus:

$$\begin{aligned} Var\,[Y_k] &= e^{-2rT}Var\left[O_T\left(P_k^{(n)}\right)\right] \\ &\equiv e^{-2rT}Var\left[\left\{O_M^{(n)}\left(P_{M_j}^{(n)}\right)\right\}_{j=1}^{R}\right] < \infty, \end{aligned}$$

where this variance is defined under $\{p_j\}_{j=1}^R$.

Thus:

$$\sum\nolimits_{k=1}^{\infty} Var\,[Y_k]/k^2 < \infty,$$

and since $E\,[Y_k] = O_0^{(n)}(X_0)$ *from item 1, the strong law of large numbers of Proposition IV.6.50 obtains:*

$$\frac{1}{M}\sum\nolimits_{k=1}^{M} Y_k - O_0^{(n)}(X_0) \to_1 0,$$

and the proof is complete. ■

Remark 9.65 (Errors in sample prices) *We make two observations on the errors in path-based pricing of financial derivatives. These apply to both the vanilla financial derivatives of Section 9.1, as well as the path-based financial derivatives of this section. In both cases, we have a well-defined binomial lattice price in Propositions 9.9 and 9.54, though these lattices appear differently as noted in Figure 9.1.*

1. *The price of a European derivative security obtained with sampling in the path-based model above will contain* ***two types of error*** *compared to the theoretically correct price, which we denote by* $O_0(X_0)$. *Denoting by* $O_M^{(n)}(X_0)$ *the price obtained with M paths and time-steps of* $\Delta t \equiv T/n$:

(a) ***Discretization error:*** *This error is defined as:*

$$\varepsilon^D(\Delta t) = O_0(X_0) - O_0^{(n)}(X_0),$$

and represents the error between $O_0(X_0) \equiv O_0^{(\infty)}(X_0)$*, the limiting derivative price as* $n \to \infty$*, and the binomial lattice value* $O_0^{(n)}(X_0)$*.*
This error depends on Δt*, or equivalently* n*, since* $\Delta t = T/n$*. It is caused by discretizing time and asset price movements, which discretizes the ultimate distribution of derivative payoff values.*

(b) ***Estimation error (Sampling error):*** *This is defined as:*

$$\varepsilon^E(\Delta t) = O_0^{(n)}(X_0) - O_M^{(n)}(X_0),$$

and represents the error between the path-based option price estimate, $O_M^{(n)}(X_0)$*, and the binomial lattice value* $O_0^{(n)}(X_0)$*.*
This error is caused by sampling error, that while $E\left[\frac{M_j}{M}\right] = p_j$ *for all* j*, it will "never" occur that* $\frac{M_j}{M} = p_j$ *for every* j*, and hence* $O_M^{(n)}(X_0) \neq O_0^{(n)}(X_0)$*.*

2. *As was noted in Exercise 9.63,* ***estimation error*** *decreases with* $1/M$ *by Chebyshev's inequality:*

$$\Pr\left[\left|O_0^{(n)}(X_0) - O_M^{(n)}(X_0)\right| \geq \epsilon\right] < \frac{e^{-2rT}}{M\epsilon^2} Var\left[\left\{O_M^{(n)}\left(P_{M_j}^{(n)}\right)\right\}_{j=1}^R\right].$$

Consequently, we can in theory choose $\epsilon_M \to 0$ *in such a way that* $M\epsilon_M^2 \to \infty$*, and thereby ensure that estimation error is theoretically eliminated as* $M \to \infty$*.*

In practice, however, this will be a slow and painful process since in order for $M\epsilon_M^2 \to \infty$ *as* $M \to \infty$*, it will be necessary to have either* $\epsilon_M \to 0$ *quite slowly, and/or have* $M\epsilon_M^2 \to \infty$ *quite slowly.*

9.8 Lognormal Pricing Model

In this section we investigate the exact and Monte Carlo pricing of European and path-dependent European financial derivatives, using the lognormal distribution of future asset prices derived in Proposition 9.29.

In the setup for this result, we are given a fixed horizon time T, $\Delta t \equiv T/n$ for some integer n, initial price $X_0 > 0$, and parameters μ and $\sigma > 0$. Asset prices $\{X_j^{(n)}\}_{j=1}^n$ at times $t = j\Delta t$ are then defined by:

$$X_j^{(n)} = X_0 \exp\left[\sum\nolimits_{i=1}^j Z_i^{(n)}\right],$$

where $Z_i^{(n)} = \mu\Delta t + \sigma\sqrt{\Delta t} b_i^{(n)}$. Here $\{b_i^{(n)}\}_{i=1}^n$ are independent and binomially distributed to equal 1 or -1 with respective probabilities q_n and $1 - q_n$, where $q_n \equiv q(\Delta t)$ is defined for the various underlying assets with (9.24). Thus $X_j^{(n)}$ is the price variate at time $t = jT/n$ for integer $0 < j \leq n$ in this multiplicative temporal model.

Given integer m, define the refined price variate $X_{mj}^{(mn)}$ at time $t = jT/n$:

$$X_{mj}^{(mn)} \equiv X_0 \exp\left[\sum\nolimits_{i=1}^{mj} Z_i^{(mn)}\right],$$

where $Z_i^{(mn)} = \mu\Delta t_m + \sigma\sqrt{\Delta t_m}b_i^{(mn)}$ with $\Delta t_m \equiv \Delta t/m$ and $\{b_i^{(mn)}\}_{i=1}^{mn}$ defined as above relative to $q_{mn} \equiv q\left(\Delta t_m\right)$.

Then by Proposition 9.29, $X_{mj}^{(mn)}$ converges in distribution as $m \to \infty$ for $0 < j \leq n$ to the **lognormally distributed variate** X_t with $t \equiv j\Delta t$:

$$X_{mj}^{(mn)} \to_d X_t \equiv X_0 \exp\left[(R(r) - \sigma^2/2)t + \sigma\sqrt{t}N\right], \tag{9.79}$$

where N has a standard normal distribution with mean 0 and variance 1, and $R(r)$ is defined in (9.25) for the various asset classes.

While the result in (9.79) was derived from the binomial model, it is now an exact model for asset prices, and it is natural to investigate derivative pricing within this model.

9.8.1 European Financial Derivatives

The question of pricing financial derivatives within the lognormal framework has already been answered by the general Black-Scholes-Merton pricing formula of Proposition 9.36, where it was presented as the limit of binomial lattice prices as $\Delta t \to 0$.

We begin with a small restatement of the general Black-Scholes-Merton pricing result, which collects some earlier results. The primary objective of this restatement is to highlight that this pricing result reflects the assumption that the real-world distribution of prices is lognormally distributed.

Proposition 9.66 (Price of a European derivative) *Let $X_0 > 0$, $T > 0$, r, μ, and $\sigma > 0$ be fixed, and assume that the asset price X_T at time T has a lognormal distribution:*

$$X_T = X_0 \exp\left[\left(\mu T + \sigma\sqrt{T}N\right)\right],$$

where $N \sim N(0,1)$ is standard normal.

Let $\Delta t \equiv T/n$ for integer n, and for $0 < j \leq n$ define:

$$X_T^{(n)} \equiv X_0 \exp\left[\sum\nolimits_{i=1}^{n}\left(\mu\Delta t + \sigma\sqrt{\Delta t}b_i\right)\right],$$

where $\{b_i\}_{i=1}^{n}$ are independent and binomially distributed to equal 1 or -1 with respective probabilities q and $1-q$, where $q \equiv q(\Delta t)$ is defined for various underlying assets in (9.24).

Assume that a European financial derivative on X has a Borel measurable settlement function $O_T\left[X\right]$ at time T, that the discontinuity set of O_T has Lebesgue measure zero, and that:

$$\sup_n E_q\left(\left(O_T\left[X_T^{(n)}\right]\right)^2\right) < \infty, \tag{9.80}$$

where this expectation is defined relative to the binomial distribution of $X_T^{(n)}$.

Then the price $O_0\left[X_0\right]$ of this financial derivative at time 0 is given by:

$$O_0\left[X_0\right] = e^{-rT}\int_{-\infty}^{\infty} O_T\left(X_0 \exp\left[(R(r) - \sigma^2/2)T + \sigma\sqrt{T}z\right]\right)\varphi\left(z\right)dz, \tag{9.81}$$

where $\varphi\left(z\right)$ is the density of N, and $R(r)$ is defined in (9.25) for the various asset classes.

Proof. *Given lognormally distributed X_T as above, define $X_{T,p}^{(n)}$ identically to $X_T^{(n)}$ but where $\{b_i\}_{i=1}^n$ are independent and binomially distributed to equal 1 or -1 with probability $1/2$. Then by Proposition 8.46, $X_{T,p}^{(n)} \to_d X_T$.*

This binomial model for asset prices then induces a binomial lattice on which this option can be priced to obtain $O_0^{(n)}[X_0]$ of (9.8). Then subject to the above constraints on the settlement function $O_T[X]$, Proposition 9.36 obtains that $O_0^{(n)}[X_0] \to O_0[X_0]$ as $n \to \infty$. ■

While this is a nice mathematical result, one can well imagine that the integral in (9.81) will not be easily evaluated. Indeed, even the relatively simple settlement functions of puts and calls require some effort, as discovered in the exercise of proving Proposition 9.43.

However, (9.81) can be expressed:

$$O_0[X_0] = e^{-rT} E_N \left[O_T \left(X_0 \exp \left[(R(r) - \sigma^2/2)T + \sigma\sqrt{T}N \right] \right) \right], \tag{9.82}$$

where this expectation is defined relative to standard normal N. Defined as mean of this settlement function under N, it makes sense to investigate estimation of this mean with samples.

Recalling Definition 9.61:

Definition 9.67 (Monte Carlo price of a European derivative) *Given a European financial derivative with settlement function $O_T[X]$ at time T, and M-sample of independent standard normals $\{N_k\}_{k=1}^M$, define the random variable $O_M(X_0)$, called the* ***Monte Carlo price estimate:***

$$O_M(X_0) = \frac{e^{-rT}}{M} \sum\nolimits_{k=1}^{M} O_T \left(X_0 \exp \left[(R(r) - \sigma^2/2)T + \sigma\sqrt{T}N_k \right] \right). \tag{9.83}$$

As expected, we have the following result.

Proposition 9.68 (Monte Carlo price) *With $O_M(X_0)$ defined in (9.83), assume that the settlement function $O_T(X)$ satisfies the assumptions of Proposition 9.66. Then:*

1. *The expected value of $O_M(X_0)$ equals the option price in (9.81):*

$$E[O_M(X_0)] = O_0(X_0). \tag{9.84}$$

2. *As $M \to \infty$, $O_M(X_0)$* ***converges with probability*** 1 ***to*** $O_0(X_0)$:

$$O_M(X_0) \to_1 O_0(X_0). \tag{9.85}$$

Proof. *Since expectation is linear, (9.84) follows from (9.83) and (9.82).*

For item 2, assume $\{N_k\}_{k=1}^\infty$ are given and define the random variable:

$$Y_k = e^{-rT} O_T \left(X_0 \exp \left[(R(r) - \sigma^2/2)T + \sigma\sqrt{T}N_k \right] \right).$$

Since O_T is Borel measurable and $\{N_k\}_{k=1}^\infty$ are independent random variables, $\{Y_k\}_{k=1}^\infty$ are independent random variables by Proposition II.3.56. In addition, $\{Y_k\}_{k=1}^\infty$ are identically distributed.

By Exercise 9.38,

$$Var(O_T[X_T]) \equiv Var\left(O_T \left[X_0 \exp \left[(R(r) - \sigma^2/2)T + \sigma\sqrt{T}N \right] \right] \right) < \infty,$$

and so:

$$Var[Y_k] = e^{-2rT}Var[O_T[X_T]] < \infty.$$

Thus:

$$\sum_{k=1}^{\infty} Var[Y_k]/k^2 < \infty,$$

and since $E[Y_k] = O_0(X_0)$ *from item 1, the strong law of large numbers in Proposition IV.6.50 obtains:*

$$\frac{1}{M}\sum_{k=1}^{M} Y_k - O_0(X_0) \to_1 0,$$

and the proof is complete. ■

9.8.2 European PD Financial Derivatives

In this section, we derive an exact pricing formula for path-dependent European derivatives using the above lognormal pricing framework, and then investigate Monte Carlo pricing.

That exact pricing is possible reflects the observation made in the introduction to Section 9.7. While the description of a path-dependent derivative implies that the settlement function reflects the continuous time evolution of asset prices, such a contract design would be impossible to settle. In order for both counterparties to be satisfied with settlement calculations, the asset prices used in the settlement function must be objectively verifiable. This quickly leads to the conclusion that only prices at market close are to be used, and thus the highest pricing frequency possible in a settlement function is daily, and then defined on a trading day basis.

Thus given a European path-dependent option to be settled at time T, define $\Delta t = T/n$ so that the settlement function is exactly calculated with the price vector:

$$(X_0, X_1, ..., X_n) \equiv (X_0, X_{\Delta t}, ..., X_{n\Delta t}).$$

As noted previously, the initial price X_0 will not usually be reflected in the settlement function for a newly traded derivative, but in the secondary market, such derivative settlement functions may not only reflect X_0, but also past prices, which are like X_0, fixed at time 0. Other than the need to redefine the settlement function to include such fixed past prices, the approaches of this section apply without change to these situations. For notational simplicity, we will assume the derivative is newly traded and omit X_0.

We begin with an exact result, where we assume that the Borel measurable settlement function $O_T[(X_{\Delta t}, ..., X_{n\Delta t})]$ at time T is bounded. While not the most general result, it simplifies the verification of (9.80) in the various contexts below, as well as the application of Fubini's theorem.

Proposition 9.69 (Price of a European PD derivative) *Let* $X_0 > 0$, $T > 0$, r, μ, *and* $\sigma > 0$ *be fixed, and for fixed* n *and* $\Delta t \equiv T/n$, *assume that the asset prices* $X_{j\Delta t}$ *for* $1 \leq j \leq n$ *have lognormal distributions:*

$$X_{j\Delta t} = X_{(j-1)\Delta t} \exp\left[\left(\mu\Delta t + \sigma\sqrt{\Delta t}N_j\right)\right],$$

where $\{N_j\}_{j=1}^n$ *are independent standard normal. Assume that a European path-dependent financial derivative on* X *has a bounded Borel measurable settlement function* $O_T(X_{\Delta t}, ..., X_{n\Delta t})$ *at time* T, *and that the discontinuity set of* O_T *has Lebesgue measure zero.*

Then the price $O_0[X_0]$ *of this financial derivative at time* 0 *is given by:*

$$O_0(X_0) = e^{-rT}\int_{\mathbb{R}^n} O_T(X_0\exp(w_1), ..., X_0\exp(w_n))\, f(w)dw, \tag{9.86}$$

where $R(r)$ is defined in (9.25) for the various asset classes, $f(w)$ is a multivariate normal density function of (6.6) with mean vector:

$$\mu \equiv \left(\left(R(r) - \sigma^2/2\right)\Delta t, 2\left(R(r) - \sigma^2/2\right)\Delta t, ..., n\left(R(r) - \sigma^2/2\right)\Delta t\right),$$

and covariance matrix C with elements $\{C_{ij}\}_{i,j=1}^{n}$ given by:

$$C_{ij} = \min(i,j)\sigma^2\Delta t.$$

Proof. *To prove this result, we investigate the iterative pricing of this financial derivative.*

At time $T = n\Delta t$, the price of this derivative is unambiguous, and equal to $O_T(X_1, ..., X_n)$, where as noted above, $(X_1, ..., X_n) \equiv (X_{\Delta t}, ..., X_{n\Delta t})$.

At time $(n-1)\Delta t$, $\bar{X}_{n-1} \equiv (X_1, ..., X_{n-1})$ is known. For integer m, define $\Delta t_m = \Delta t/m$, and let:

$$X_{mn}^{(mn)} \equiv X_{n-1}\exp\left[\sum\nolimits_{i=1}^{m}\left(\mu\Delta t_m + \sigma\sqrt{\Delta t_m}b_i\right)\right],$$

where $\{b_i\}_{i=1}^{m}$ are independent and binomially distributed to equal 1 or -1 with respective probabilities q and $1-q$, and with $q \equiv q(\Delta t)$ is defined for various underlying assets in (9.24).

Then (9.80) is satisfied by boundedness of O_T:

$$\sup_m E_q\left[\left(O_T\left(X_1, ..., X_{n-1}, X_{mn}^{(mn)}\right)\right)^2\right] < \infty, \tag{1}$$

and so the price at time $(n-1)\Delta t$ for any $\bar{X}_{n-1}$ is given by (9.81):

$$O_{(n-1)\Delta t}\left(\bar{X}_{n-1}\right) = e^{-r\Delta t}\int_{-\infty}^{\infty} O_T\left(\bar{X}_{n-1}, X_{n-1}\exp\left[(m + sz_n\right]\right)\varphi\left(z_n\right)dz_n. \tag{2}$$

Here for notational convenience we let $m \equiv \left(R(r) - \sigma^2/2\right)\Delta t$ and $s \equiv \sigma\sqrt{\Delta t}$.

Now consider time $(n-2)\Delta t$, with $\bar{X}_{n-2} \equiv (X_1, ..., X_{n-2})$ known. By (2), the time $(n-1)\Delta t$ price of this derivative depends only on $\bar{X}_{n-1}$, which we can now express in terms of $\bar{X}_{n-2}$ and X_{n-1}:

$$\begin{aligned} O_{(n-1)\Delta t}\left(\bar{X}_{n-1}\right) &= e^{-r\Delta t}\int_{-\infty}^{\infty} O_T\left(\bar{X}_{n-2}, X_{n-1}, X_{n-1}\exp\left[(m + sz_n\right]\right)\varphi\left(z_n\right)dz_n \\ &\equiv O_{(n-1)\Delta t}\left(\bar{X}_{n-2}, X_{n-1}\right). \end{aligned}$$

As this result is true for all $\bar{X}_{n-1}$, it is consequently true for given $\bar{X}_{n-2}$ and all X_{n-1}.

With $X_{m(n-1)}^{(mn)}$ defined as above, again by boundedness:

$$\sup_m E_q\left[\left(O_{(n-1)\Delta t}\left(\bar{X}_{n-2}, X_{m(n-1)}^{(mn)}\right)\right)^2\right] < \infty,$$

and by another application of (9.81), the price at time $(n-2)\Delta t$ for any $\bar{X}_{n-2}$ is given by:

$$\begin{aligned} &O_{(n-2)\Delta t}\left(\bar{X}_{n-2}\right) \\ =\ & e^{-r\Delta t}\int_{-\infty}^{\infty} O_{(n-1)\Delta t}\left(\bar{X}_{n-2}, X_{n-2}\exp\left[(m + sz_{n-1}\right]\right)\varphi\left(z_{n-1}\right)dz_{n-1} \\ =\ & e^{-2r\Delta t}\int_{-\infty}^{\infty}\int_{-\infty}^{\infty} O_T\left(\bar{X}_{n-2}, X_{n-2}\exp\left[(m + sz_{n-1}\right], X_{n-2}\exp\left[(2m + s\left(z_n + z_{n-1}\right)\right]\right) \\ &\times\varphi\left(z_n\right)\varphi\left(z_{n-1}\right)dz_n dz_{n-1}. \end{aligned}$$

This price is valid for all $\bar{X}_{n-2}$, which can then be interpreted with fixed $\bar{X}_{n-3}$, as true for all X_{n-2}. In other words:

$$O_{(n-2)\Delta t}\left(\bar{X}_{n-2}\right) \equiv O_{(n-2)\Delta t}\left(\bar{X}_{n-3}, X_{n-2}\right).$$

Continuing in this way obtains $O_0(X_0)$ as an iterated Lebesgue integral:

$$\begin{aligned} &O_0(X_0) \\ &= e^{-rT}\int_{-\infty}^{\infty}\cdots\int_{-\infty}^{\infty} O_T\left(X_0\exp(m+sz_1),....,X_0\exp\left(nm+s\sum\nolimits_{k=1}^{n} z_k\right)\right) \\ &\quad\times\varphi(z_n)\ldots\varphi(z_1)\,dz_n\cdots dz_1. \end{aligned} \tag{3}$$

By the boundedness assumption of O_T, this integrand is absolutely integrable, and so Fubini's theorem of Proposition V.5.16, as generalized in Remark V.5.19, obtains that this integrand is integrable as an n-dimensional Lebesgue integral.

Thus the iterated integral in (3) equals the n-dimensional Lebesgue integral:

$$O_0(X_0) = e^{-rT}\int_{\mathbb{R}^n} O_T\left(X_0\exp(m+sz_1),....,X_0\exp\left(nm+s\sum\nolimits_{k=1}^{n} z_k\right)\right)\varphi(z)\,dz. \tag{4}$$

Here $z \equiv (z_1,\cdots,z_n)$ and $\varphi(z) \equiv \varphi(z_1)\ldots\varphi(z_n)$ is the multivariate normal density function for independent normals $\{N_j\}_{j=1}^n$ by Corollary 1.57.

For the needed change of variables, let:

$$\mu \equiv \begin{pmatrix} m \\ 2m \\ \vdots \\ nm \end{pmatrix}, \quad A \equiv \begin{pmatrix} s & 0 & 0 & \cdots & 0 \\ s & s & 0 & \cdots & 0 \\ \vdots & \vdots & \vdots & \cdots & \vdots \\ s & s & s & \cdots & 0 \\ s & s & s & s & s \end{pmatrix}, \tag{5}$$

where $m = \left(R(r) - \sigma^2/2\right)\Delta t$, and A is a lower triangular matrix with nonzero elements $s = \sigma\sqrt{\Delta t}$. Note that A has $\det A = s^n \equiv \sigma^n\left(\Delta t\right)^{n/2} \neq 0$, and so A is invertible.

Thus by (4):

$$O_0(X_0) = e^{-rT}\int_{\mathbb{R}^n}\tilde{O}_T\left(Az+\mu\right)\varphi(z)\,dz, \tag{6}$$

where we simplify notation so that given $w \equiv (w_1,\ldots,w_n)$:

$$\tilde{O}_T(w) \equiv O_T\left(X_0\exp(w_1),....,X_0\exp(w_n)\right).$$

Define the transformation $Tz \equiv Az + \mu$, and note that T is one-to-one, continuously differentiable, and with nonzero Jacobian determinant $\det T'(w)$ (recall V.(4.30)):

$$\det T'(w) = \det A = \sigma^n\left(\Delta t\right)^{n/2}.$$

By Proposition V.4.44, a change of variable $w = Tz$ in (6) obtains:

$$\begin{aligned} e^{-rT}\int_{\mathbb{R}^n}\tilde{O}_T\left(Az+\mu\right)\varphi(z)\,dz &\equiv e^{-rT}\int_{\mathbb{R}^n}\tilde{O}_T\left(Tz\right)\varphi\left(T^{-1}[Tz]\right)dz \\ &= \frac{e^{-rT}}{\sigma^n\left(\Delta t\right)^{n/2}}\int_{\mathbb{R}^n}\tilde{O}_T(w)\,\varphi\left(T^{-1}w\right)dw. \end{aligned}$$

Finally, $T^{-1}w = A^{-1}(w-\mu)$, *and so returning to the notation* O_T*:*

$$O_0(X_0) = \frac{e^{-rT}}{\sigma^n(\Delta t)^{n/2}} \int_{\mathbb{R}^n} O_T\left(X_0\exp(w_1),...,X_0\exp(w_n)\right)\varphi\left(A^{-1}(w-\mu)\right)dw, \tag{7}$$

and the proof is complete with the following exercise. ■

Exercise 9.70 *Let* $\varphi(z) \equiv \varphi(z_1)\ldots\varphi(z_n)$ *denote the multivariate normal density function for independent normals* $\{N_i\}_{i=1}^n$*:*

$$\varphi(z) = (2\pi)^{-n/2}\exp\left[z^T I z\right],$$

where I *is the* $n\times n$ *identity matrix. Prove that with* A *defined in* (5) *above that:*

$$\frac{1}{\sigma^n(\Delta t)^{n/2}}\varphi\left(A^{-1}(w-\mu)\right) = (2\pi)^{-n/2}\left[\det C\right]^{-1/2}\exp\left[-\frac{1}{2}(w-\mu)^T C^{-1}(w-\mu)\right],$$

where $C = AA^T$, *the elements* $\{C_{ij}\}_{i,j=1}^n$ *of* C *are given by:*

$$C_{ij} = \min(i,j)\sigma^2\Delta t,$$

and $\det C$ *denotes the determinant of* C.

Thus by (6.6), w *in* (7) *has a multivariate normal distribution with mean vector* μ *and covariance matrix* C *and the proof of 9.86 is complete.*

As was the case for (9.81), while (9.86) provides a nice mathematical result, this integral will not be easily evaluated. However, (6) of the above proof can be expressed:

$$O_0[X_0] = e^{-rT}E_N\left[O_T\left(X_0\exp((A\bar{N}+\mu)_1),....,X_0\exp\left((A\bar{N}+\mu)_n\right)\right)\right]. \tag{9.87}$$

This expectation is defined relative to the multivariate standard normal $\bar{N} \equiv (N_1,...,N_n)$, meaning with independent $\{N_j\}_{j=1}^n$, and with $\{(A\bar{N}+\mu)_j\}_{j=1}^n$ the components of $A\bar{N}+\mu$, with A defined in (5) of the above proof. Thus:

$$\left(A\bar{N}+\mu\right)_j = \left(R(r)-\sigma^2/2\right)j\Delta t + \sum\nolimits_{i=1}^{j} N_i\sigma\sqrt{\Delta t}. \tag{9.88}$$

Defined as an expectation of this settlement function under $\bar{N}$, it makes sense to investigate estimation of $O_0[X_0]$ with samples.

Definition 9.71 (Monte Carlo price of a path-dependent European derivative) *Given a European path-dependent financial derivative with settlement function* $O_T(X_{\Delta t},...,X_{n\Delta t})$ *at time* T, *let* $\{\bar{N}^{(k)}\}_{k=1}^M$ *be an* M*-sample of independent multivariate standard normals, so* $\bar{N}^{(k)} \equiv \left(N_1^{(k)},...,N_n^{(k)}\right)$ *with independent, standard normals* $\left\{N_j^{(k)}\right\}_{j=1}^n$. *The random variable:*

$$O_M(X_0) = \frac{e^{-rT}}{M}\sum\nolimits_{k=1}^{M} O_T\left(X_0\exp\left(\left(A\bar{N}^{(k)}+\mu\right)_1\right),....,X_0\exp\left(\left(A\bar{N}^{(k)}+\mu\right)_n\right)\right), \tag{9.89}$$

with $\left(A\bar{N}^{(k)}+\mu\right)_j$ *defined as in (9.88), is called the* ***Monte Carlo price estimate****.*

As expected, we have the following result.

Proposition 9.72 (Monte Carlo price convergence) *With $O_M(X_0)$ defined in (9.89), assume that the settlement function $O_T(X_{\Delta t}, ..., X_{n\Delta t})$ satisfies the assumptions of Proposition 9.69. Then:*

1. *The expected value of $O_M(X_0)$ equals the option price in (9.86):*

$$E\left[O_M(X_0)\right] = O_0(X_0). \tag{9.90}$$

2. *As $M \to \infty$, $O_M(X_0)$* ***converges with probability*** 1 ***to*** *$O_0(X_0)$:*

$$O_M(X_0) \to_1 O_0(X_0). \tag{9.91}$$

Proof. *The proof is virtually identical to that of Proposition 9.64, but we provide it for completeness.*

First, (9.90) follows from (9.89) and (9.87) by linearity of expectations.

For item 2, assume $\{\bar{N}^{(k)}\}_{k=1}^{\infty}$ are given and define the random variable:

$$Y_k = e^{-rT}O_T\left(X_0 \exp\left(\left(A\bar{N}^{(k)} + \mu\right)_1\right), ..., X_0 \exp\left(\left(A\bar{N}^{(k)} + \mu\right)_n\right)\right).$$

Since O_T is Borel measurable and $\{\bar{N}^{(k)}\}_{k=1}^{\infty}$ are independent random vectors, $\{Y_k\}_{k=1}^{\infty}$ are independent random variables by Proposition II.3.56. In addition, $\{Y_k\}_{k=1}^{\infty}$ are identically distributed.

By boundedness of O_T:

$$Var\left[O_T\left(X_0 \exp\left(\left(A\bar{N}^{(k)} + \mu\right)_1\right), ..., X_0 \exp\left(\left(A\bar{N}^{(k)} + \mu\right)_n\right)\right)\right] < \infty,$$

and so $Var\left[Y_k\right]$ is finite and independent of k, and:

$$\sum\nolimits_{k=1}^{\infty} Var\left[Y_k\right]/k^2 < \infty.$$

Since $E\left[Y_k\right] = O_0(X_0)$ from item 1, the strong law of large numbers in Proposition IV.6.50 obtains:

$$\frac{1}{M}\sum\nolimits_{k=1}^{M} Y_k - O_0^{(n)}(X_0) \to_1 0,$$

and the proof is complete. ■

10

Limits of Binomial Motion

In Propositions 8.46 and 9.29, we were given a fixed time horizon of T, and a binomial return model with a real-world or risk-neutral probability structure. Since asset prices evolved in time steps of $\Delta t \equiv T/n$ for given but arbitrary integer n, these results could only characterize the limiting distributions of the asset price X_t for $t = \frac{k}{n}T$ for $1 \leq k \leq n$. While the collection of all such time points are dense in the interval $[0, T]$, it would be of interest to attempt to push these models a little further to fill in these gaps. In this chapter we introduce such a model, and investigate its properties. Books VII and beyond will formalize and expand this investigation.

Beginning with **binomial paths**, which are defined to interpolate binomial sums defined with $p = \frac{1}{2}$ between successive time points $k\Delta t$ and $(k+1)\,\Delta t$, we investigate two types of limits as $n \to \infty$. **Pathwise binomial motion** is first considered as the uniform limits of these binomial paths. While continuous, the distributional properties of such limiting binomial paths are lost.

A more rewarding investigation into the distributional limits of binomial paths is studied next, defining such limits to be **standard binomial motion**. After deriving a variety of results on properties of such distributional limits, **nonstandard binomial motion** is introduced. The first inquiry is into binomial paths with $p \neq \frac{1}{2}$, and then where $p = p(\Delta t)$ is given by the risk-neutral probabilities or the alternative real-world probabilities of Chapter 9.

The final section considers distributional limits of real-world and risk-neutral asset models defined with respect to the respective binomial paths, extending earlier results.

10.1 Binomial Paths

To generalize the evolution of asset prices from discrete to continuous time, let T be given, and define $b(\Delta t)$ as a binomial variate:

$$b(\Delta t) = \pm\sqrt{\Delta t}, \qquad p = 1/2.$$

as was used in the real-world asset price model. Equivalently, this can be expressed:

$$b(\Delta t) = b\sqrt{\Delta t}, \qquad b = \pm 1, \qquad p = 1/2, \tag{10.1}$$

separating the binomial random variate $b = \pm 1$ from the time interval parameter Δt.

Given Δt, initial asset price $X_0 > 0$, and independent binomial variates $\{b_j(\Delta t)\}_{j=1}^{n}$, the temporal binomial models of Chapter 8 modeled asset prices at time $j\Delta t$ by:

1. **Additive temporal model**:

$$X_j = X_0 + \mu j\Delta t + \sigma B_j(\Delta t),$$

DOI: 10.1201/9781003275770-10

2. **Multiplicative temporal model:**

$$\ln [X_j / X_0] = \mu j \Delta t + \sigma B_j(\Delta t),$$

where

$$B_j(\Delta t) \equiv \sum\nolimits_{k=1}^{j} b_k(\Delta t). \tag{10.2}$$

It is easy to extend the definition of $B_j(\Delta t)$ to a continuous function $B_t(\Delta t)$ for $t \in [0, T]$ for any such collection $\{b_j(\Delta t)\}_{j=1}^n$.

Definition 10.1 (Binomial path) *Given $T = n\Delta t > 0$ and $\{b_j(\Delta t)\}_{j=1}^n$, define the piecewise linear **binomial path** $B_t(\Delta t)$ for $t \in [0, T]$ by:*

$$B_t(\Delta t) = (1 - s)B_{j-1}(\Delta t) + sB_j(\Delta t), \tag{10.3}$$

where we define $B_0(\Delta t) = 0$, and $t \equiv (j - 1 + s)\,\Delta t$ for $0 \le s \le 1$. Equivalently:

$$B_t(\Delta t) = B_{j-1}(\Delta t) + sb_j(\Delta t). \tag{10.4}$$

So defined, we can think of $B_t(\Delta t)$ as a random variable for each t, or define a **realization of** $B_t(\Delta t)$ to mean the continuous function defined by (10.3) or (10.4) based on a realization of binomial variates $\{b_j(\Delta t)\}_{j=1}^n$. By a **realization** of a binomial sequence $\{b_j(\Delta t_n)\}_{j=1}^\infty$ is meant any sequence of $\pm\sqrt{\Delta t_n}$, without regard for the associated probability.

In the above parametrization $j - 1 = \lfloor t/\Delta t \rfloor$, where $\lfloor t/\Delta t \rfloor = \max\{k \in \mathbb{N} | k \le t/\Delta t\}$ denotes the **greatest integer function.** Hence $s \equiv t/\Delta t - \lfloor t/\Delta t \rfloor$, and the formula for $B_t(\Delta t)$ can be explicitly expressed in terms of t:

$$B_t(\Delta t) = B_{\lfloor t/\Delta t \rfloor}(\Delta t) + (t/\Delta t - \lfloor t/\Delta t \rfloor)\, b_{\lfloor t/\Delta t \rfloor + 1}(\Delta t).$$

While mathematically concise, this notation is a bit cumbersome for our needs, for which it will be convenient to just express $B_t(\Delta t)$ piecewise on each subinterval $[(j - 1)\,\Delta t, j\Delta t]$.

In the next three sections, we investigate pathwise limits, and then various results on distributional limits of binomial paths as $\Delta t \to 0$. As was the case for Chapter 8, for notational simplicity, and because it is generally adequate in practice, we use only uniform partitionings of $[0, T]$ with $\Delta t = T/n$.

However, the results of this chapter could be generalized to arbitrary partitions:

$$0 \equiv s_0 < s_1 < s_2 < .... < s_n = T < \infty,$$

with binomial paths defined using binomial variates:

$$b_j\left(\Delta t_n^{(j)}\right) \equiv b_j \sqrt{\Delta t_n^{(j)}},$$

where $\Delta t_n^{(j)} \equiv s_j - s_{j-1}$.

In this general model, comparable results are obtained if we replace $\Delta t \to 0$ as $n \to \infty$ with $\max_j\{\Delta t_n^{(j)}\} \to 0$ as $n \to \infty$. The reader may recall that $\delta \equiv \max_j\{\Delta t_n^{(j)}\}$ was called the **mesh size** of the above partition in Books III and V, and this played a role in the integration theories developed there.

10.2 Uniform Limits of $B_t(\Delta t)$

In this section, we investigate uniform limits of binomial paths as $\Delta t_n \to 0$. We begin with a natural idea, defining a **pathwise binomial motion** on $[0, T]$ as a uniform limit of binomial paths, when such a limit exists. When $T < \infty$, the time-steps $\Delta t_n \equiv T/m_n$ for some collection of integers $\{m_n\}_{n=1}^{\infty}$ with $m_n \to \infty$. When $T = \infty$, Δt_n cannot be defined in this way, but are again constant for each binomial path, and $\Delta t_n \to 0$.

Recall that as noted above, a **realization** of a binomial sequence $\{b_j(\Delta t_n)\}_{j=1}^{\infty}$ is any sequence of $\pm\sqrt{\Delta t_n}$, without regard for the associated probability.

Definition 10.2 (Pathwise binomial motion) *A function B_t^U defined on $[0, T]$ for $T \leq \infty$ is called a* ***pathwise binomial motion*** *if there exists $\Delta t_n \to 0$ and binomial realizations $\{\{b_j(\Delta t_n)\}_{j=1}^{\infty}\}_{n=1}^{\infty}$ so that:*

$$\sup_{t\in[0,T]} \left|B_t^U - B_t(\Delta t_n)\right| \to 0, \; n \to \infty, \tag{10.5}$$

where $B_t(\Delta t_n)$ is defined in (10.3).

We denote the class of all pathwise binomial motions on $[0, T]$ by $\mathcal{BIN}([0, T])$, where $[0, T] \equiv [0, \infty)$ if $T = \infty$.

Remark 10.3 (On B_t^U) *For each $B_t^U \in \mathcal{BIN}([0, T])$, there exists realizations of binomial sequences $\{\{b_j(\Delta t_n)\}_{j=1}^{\infty}\}_{n=1}^{\infty}$, or equivalently, realizations of associated binomial paths $\{B_t(\Delta t_n)\}_{n=1}^{\infty}$, so that $B_t(\Delta t_n) \to B_t^U$* ***uniformly in*** *$t \in [0, T]$ as $n \to \infty$.*

Also, for all $B_t^U \in \mathcal{BIN}([0, T])$:

$$B_0^U = 0,$$

since $B_0(\Delta t) = 0$ for all Δt by Definition 10.2.

The first task is to ensure that $\mathcal{BIN}([0, T])$ is not an empty class making the above definition vacuous, and this we do in Example 10.6 and Proposition 10.9. For this, recall the notion that a sequence of functions **satisfies the Cauchy criterion,** named for **Augustin-Louis Cauchy** (1789–1857).

Definition 10.4 (Cauchy criterion) *A function sequence $\{f_n(t)\}_{n=1}^{\infty}$ is said to* ***satisfy the Cauchy criterion*** *on a set A, or, is* ***uniformly Cauchy*** *on A, if for any $\epsilon > 0$ there is an $N \in \mathbb{N}$ so that for $n, m \geq N$:*

$$|f_n(t) - f_m(t)| < \epsilon \text{ for all } t \in A. \tag{10.6}$$

The following proposition states that this criterion is equivalent to the statement that $\{f_n(t)\}_{n=1}^{\infty}$ converges uniformly on A. But in practice this is a very useful reformulation of uniform convergence because it can be verified without identifying the convergent function $f(t)$.

Proposition 10.5 (Cauchy Criterion for Uniform Convergence) *A function sequence $\{f_n(t)\}_{n=1}^{\infty}$ converges uniformly to a function $f(t)$ on a set A if and only if this sequence satisfies the Cauchy criterion on A.*

Proof. *If $f_n(t) \to f(t)$ uniformly on A, then given $\epsilon > 0$ there exists N so that $|f_n(t) - f(t)| < \epsilon/2$ for all $n \geq N$ and all $t \in A$. Thus for $n, m \geq N$ and all $t \in A$:*

$$|f_n(t) - f_m(t)| \leq |f_n(t) - f(t)| + |f(t) - f_m(t)| < \epsilon,$$

and $\{f_n(t)\}_{n=1}^{\infty}$ satisfies the Cauchy criterion on A.

Conversely, if $\{f_n(t)\}_{n=1}^{\infty}$ satisfies the Cauchy criterion on A then for each $t_0 \in A$, $\{f_n(t_0)\}_{n=1}^{\infty}$ is a Cauchy sequence of real numbers. Since $\mathbb{R}$ is complete (see the proof of Proposition V.9.10 for example), there exists a real number which we denote by $f(t_0)$, such that $f_n(t_0) \to f(t_0)$. Now given $\epsilon > 0$, the Cauchy criterion assures the existence of N so that (10.6) is satisfied for all $n, m \geq N$ and all $t \in A$. Thus for all $t \in A$ and $n \geq N$:

$$|f_n(t) - f(t)| = \lim_{m\to\infty} |f_n(t) - f_m(t)| < \epsilon,$$

and $\{f_n(t)\}_{n=1}^{\infty}$ converges uniformly to $f(t)$ on A. ■

Example 10.6 (Existence of pathwise binomial motion) *In this example we complete a construction of binomial motion on $[0, T]$, allowing for the case $T = \infty$.*

Arbitrarily choose $\Delta t_0 > 0$, and define $\Delta t_n \equiv \Delta t_0/m_n$ with $m_n = 4^n$. Thus each successive refinement of a partition quarters all intervals:

$$\Delta t_{n+1} = \Delta t_n/4.$$

Now arbitrarily generate a realization of a binomial sequence to obtain a realization of $B_t(\Delta t_0)$ on $[0, T]$.

We claim that we can then sequentially choose realizations of binomial sequences at each successive step, so that the step $n+1$ binomial path $B_t(\Delta t_{n+1})$ equals the step n binomial path $B_t(\Delta t_n)$ at all $k\Delta t_n$. Since $B_0(\Delta t_n) = 0$ for all n, it is enough to prove that given $B_t(\Delta t_n)$ and any Δt_n-interval $[k\Delta t_n, (k+1)\Delta t_n]$, that if:

$$B_{k\Delta t_n}(\Delta t_n) = B_{k\Delta t_n}(\Delta t_{n+1}), \tag{1}$$

then there are binomial sequences for the associated Δt_{n+1} subintervals such that:

$$B_{(k+1)\Delta t_n}(\Delta t_n) = B_{(k+1)\Delta t_n}(\Delta t_{n+1}). \tag{2}$$

In other words, if these paths agree at the beginning of the interval, they agree at the end, and the construction is complete by induction.

For example, if $b_{k+1}^{(n)} = 1$ and thus:

$$B_{(k+1)\Delta t_n}(\Delta t_n) = B_{k\Delta t_n}(\Delta t_n) + \sqrt{\Delta t_n},$$

let $\left(b_{4k+1}^{(n+1)}, b_{4k+2}^{(n+1)}, b_{4k+3}^{(n+1)}, b_{4k+4}^{(n+1)}\right)$ denote the four binomial variates for this interval associated with the Δt_{n+1} time-steps. Then since $\sqrt{\Delta t_{n+1}} = \sqrt{\Delta t_n}/2$, if these variates equal $\{-1, 1, 1, 1\}$ in any order, then (1) will obtain (2). If $b_{k+1}^{(n)} = -1$, then $\left(b_{4k+1}^{(n+1)}, b_{4k+2}^{(n+1)}, b_{4k+3}^{(n+1)}, b_{4k+4}^{(n+1)}\right)$ equalling $\{-1, -1, -1, 1\}$ in any order obtain the same result.

Thus by construction, each binomial path $B_t(\Delta t_{n+1})$ equals the prior path $B_t(\Delta t_n)$ for all $t = k\Delta t_n$. By elementary geometry we can now conclude that independent of how one chooses the four-tuples of binomial variates in each subinterval, that for all t:

$$|B_t(\Delta t_{n+1}) - B_t(\Delta t_n)| \leq 0.75\sqrt{\Delta t_n} = 0.75\sqrt{\Delta t_0}/2^n.$$

The resulting sequence of binomial paths $\{B_t(\Delta t_n)\}_{n=1}^{\infty}$ thus satisfies the Cauchy criterion. If $n > m \geq N$:

$$|B_t(\Delta t_m) - B_t(\Delta t_n)| \leq 0.75\sqrt{\Delta t_0}2^{-(N-1)}.$$

By Proposition 10.5, $\{B_t(\Delta t_n)\}_{n=1}^{\infty}$ converges uniformly to some function B_t^U, which is then by definition a pathwise binomial motion.

Exercise 10.7 (Pathwise binomial motion with $\Delta t_{n+1} = \Delta t_n/m^2$) *Provide the details of the construction of pathwise binomial motion using the approach of Example 10.6, but where now $\Delta t_{n+1} = \Delta t_n/m^2$ for any positive integer m. Hint: The above construction required only that $m^2 - m$ was even.*

Because the convergence in (10.5) is **uniform** on the interval $[0,T]$, all B_t^U are continuous, and thus by Proposition I.3.11, pathwise binomial motion is also Borel measurable.

Proposition 10.8 ($\mathcal{BIN}([0,T]) \subset \mathcal{C}([0,T])$) *For all intervals $[0,T]$:*

$$\mathcal{BIN}([0,T]) \subset \mathcal{C}([0,T]), \tag{10.7}$$

where $\mathcal{C}([0,T])$ denotes the class of continuous functions on $[0,T]$. When $T < \infty$, the functions in $\mathcal{BIN}([0,T])$ are in fact uniformly continuous.

Proof. *Continuity of $B_t(\Delta t_n)$ and uniform convergence assures continuity of the limit function B_t^U by Exercise III.1.44. When $T < \infty$, B_t^U is uniformly continuous by Exercise III.1.14.* ■

The next result shows that $\mathcal{BIN}([0,T])$ is sequentially closed.

Proposition 10.9 ($\mathcal{BIN}([0,T])$ is sequentially closed) *If f_t is any function on $[0,T]$ with the property that for any $\delta > 0$ there is a $B_t^{U(\delta)} \in \mathcal{BIN}([0,T])$ with $\left|B_t^{U(\delta)} - f_t\right| < \delta$ for all t, then $f_t \in \mathcal{BIN}([0,T])$.*

Proof. *For any $\delta > 0$, let $\{B_t^{(\delta)}(\Delta t_n)\}_{n=1}^{\infty}$ be a sequence of binomial paths which converge uniformly to $B_t^{U(\delta)}$ as $n \to \infty$, where $\{\Delta t_n\}_{n=1}^{\infty}$ may depend on δ. By the triangle inequality, for all t:*

$$\begin{aligned}
\left|B_t^{(\delta)}(\Delta t_n) - f_t\right| &\leq \left|B_t^{(\delta)}(\Delta t_n) - B_t^{U(\delta)}\right| + \left|B_t^{U(\delta)} - f_t\right| \\
&< \left|B_t^{(\delta)}(\Delta t_n) - B_t^{U(\delta)}\right| + \delta.
\end{aligned}$$

Given positive $\epsilon_m \to 0$, let $\delta_m = \epsilon_m/2$ above and choose Δt_n so that $\left|B_t^{(\delta_m)}(\Delta t_n) - B_t^{U(\delta_m)}\right| < \epsilon_m/2$. Then $\left|B_t^{(\delta_m)}(\Delta t_n) - f_t\right| < \epsilon_m$ for all t, and $f_t \in \mathcal{BIN}([0,T])$. ■

It would be interesting to identify if the functions in $\mathcal{BIN}([0,T])$ can be characterized in more detail beyond (10.7). However, even if we could characterize all such functions, a moment of thought reveals that we have lost crucial information.

By focusing on the resulting paths, we have lost all information on the distributional properties of such paths. For example, while it is possible to make probabilistic statements about $B_t(\Delta t_n)$ for any $t = k\Delta t_n$, or any finite collection of such times, there is no way to evaluate such probabilistic statements about B_t^U, the limiting paths of $B_t(\Delta t_n)$ as $\Delta t_n \to 0$.

Thus in the next section we abandon the question of path-based limits, and instead consider distributional limits of $B_t(\Delta t_n)$ as $\Delta t_n \to 0$ for finite collections of time points t.

10.3 Distributional Limits of $B_t(\Delta t)$

We assume as given an interval $[0,T]$ with $T < \infty$, and define $\Delta t_n = T/n$ so $\Delta t_n \to 0$ as $n \to \infty$. In this section, distributional and other limits of binomial paths $B_t(\Delta t_n)$ are investigated for finite collections of time points t as $\Delta t_n \to 0$. The insights derived will be useful to our understanding of what follows in Book VII.

The various results in the following propositions treat $t \in [0,T]$, $\{t_i\}_{i=1}^m \subset [0,T]$, etc., as fixed though arbitrary, and thus the stated limiting distributional results apply to the **random variables** $B_t(\Delta t_n)$, $\{B_{t_i}(\Delta t_n)\}_{i=1}^m$, etc., or the **random vector** $(B_{t_1}(\Delta t_n), ..., B_{t_m}(\Delta t_n))$. While we do not yet have a probability space in which $B_t(\Delta t_n)$ or its convergent B_t exist as a path or function, once time points are fixed, these random variables and vectors can, if needed for context, be inferred to exist on a fixed probability space provided by Propositions 1.20 and 1.24.

Recall the notion of random variable or random vector convergence of Definitions 3.2 and 3.8, denoted $X_n \to_d X$, and called **convergence in distribution** or **convergence in law**.

Proposition 10.10 (Convergence in distribution of $B_t(\Delta t_n)$) *Let $B_t(\Delta t_n)$ be defined as in Definition 10.1, and let $N(0,t)$ denote a normally distributed random variable with mean 0 and variance t. Then:*

1. *For every $t \in (0,T]$, as $\Delta t_n \to 0$:*
$$B_t(\Delta t_n) \to_d N(0,t).$$
2. *For $0 \leq s < t \leq T$, as $\Delta t_n \to 0$:*
$$B_t(\Delta t_n) - B_s(\Delta t_n) \to_d N(0,t-s).$$
3. *For $0 \leq s < t \leq T$, if $\Delta t_n < (t-s)/2$:*
$$E\left[B_s(\Delta t_n)B_t(\Delta t_n)\right] = \min[s,t].$$

Proof. *For item 1 fix $t \in (0,T]$, and for each n let $t = [j(n) + e(n)]\,\Delta t_n$ with integer $j(n)$ and $0 \leq e(n) < 1$. Then by (10.4):*

$$B_t(\Delta t_n) = B_{j(n)}(\Delta t_n) + e(n)b_{j(n)+1}(\Delta t_n).$$

Independence of $\{b_j(\Delta t_n)\}_{j=1}^\infty$ obtains independence of $B_{j(n)}(\Delta t_n)$ and $b_{j(n)+1}(\Delta t_n)$ by Exercise II.3.50 and Proposition II.3.56, and then Proposition 5.23 obtains the identity in characteristic functions:

$$C_{B_t(\Delta t_n)}(y) = C_{B_{j(n)}}(y)C_{e(n)b_{j(n)+1}}(y). \tag{1}$$

Since $b_j(\Delta t_n)$ is binomial with values ± 1 with probability $p = 1/2$, it follows from (5.20):

$$\begin{aligned} C_{e(n)b_{j(n)+1}}(y) &= 0.5\left[\exp\left(iye(n)\sqrt{\Delta t_n}\right) + \exp\left(-iye(n)\sqrt{\Delta t_n}\right)\right] \\ &= \cos\left[ye(n)\sqrt{\Delta t_n}\right]. \end{aligned}$$

Hence for all y,

$$C_{e(n)b_{(n)j+1}}(y) \to 1, \tag{2}$$

as $\Delta t_n \to 0$.

Now $\{b_j(\Delta t_n)\}_{j=1}^{j(n)}$ satisfies Lindeberg's condition of Definition 7.6, since $|Z_j| \equiv |b_j(\Delta t_n)|/\sqrt{\Delta t_n} = |b|$, where $b = \pm 1$ with $p = 1/2$. Hence for any $s > 0$ and $j(n) > 1/s^2$:

$$\int_{|Z_j|>s\sqrt{j(n)}} Z_j^2 d\lambda \equiv \int_{|b|>s\sqrt{j(n)}} 1 d\lambda = 0.$$

Thus since $E\left[B_{j(n)}(\Delta t_n)\right] = 0$ *and* $Var\left[B_{j(n)}(\Delta t_n)\right] = j(n)\Delta t_n \to t$*, Corollary 7.14 of Lindeberg's central limit theorem obtains that* $B_{j(n)}(\Delta t_n) \to_d N(0,t)$*. Lévy's continuity theorem of Proposition 5.30 states that this is equivalent to:*

$$C_{B_{j(n)}}(y) \to C_{N(0,t)}(y). \tag{3}$$

Combining results (1)–(3) *obtains that* $C_{B_t(\Delta t_n)}(y) \to C_{N(0,t)}(y)$*, and another application of Lévy's continuity theorem completes the proof of item* 1.

Items 2 *and* 3 *require the same setup, expressing* s *and* t *in terms of* Δt_n*. Since* $s < t$*, let* $s = [j(n) + e(n)]\,\Delta t_n$ *and* $t = [k(n) + f(n)]\,\Delta t_n$*, where* $j(n) \leq k(n)$*, and* $0 \leq e(n)$*,* $f(n) < 1$*. Since* $s < t$*, it follows that for* n *large and defined so that* $\Delta t_n < (t-s)/2$ *say, that* $j(n) < k(n)$ *and this will be assumed to avoid notational complexity.*

To prove item 2*:*

$$B_t(\Delta t_n) - B_s(\Delta t_n) = (1 - e(n))\, b_{j(n)+1} + \sum\nolimits_{i=j(n)+2}^{k(n)} b_i + f(n) b_{k(n)+1},$$

where we suppress Δt_n*. Denoting* $X \equiv \sum_{i=j(n)+2}^{k(n)} b_i$*, the justifications above obtain:*

$$C_{[B_t(\Delta t_n)-B_s(\Delta t_n)]}(y) = C_{(1-e(n))b_{j(n)+1}}(y) C_X(y) C_{f(n)b_{k(n)+1}}.$$

Again as for (2), $C_{(1-e(n))b_{j(n)+1}}(y) \to 1$ *and* $C_{f(n)b_{k(n)+1}} \to 1$ *as* $\Delta t_n \to 0$*. That* $C_X(y) \to C_{N(0,t-s)}(y)$ *follows as before with Lindeberg's central limit theorem since* $E[X] = 0$ *and* $Var[X] = (k(n) - j(n) - 1)\,\Delta t_n \to t - s$.

For item 3*, again suppressing* Δt_n*:*

$$\begin{aligned} B_s(\Delta t_n) B_t(\Delta t_n) &= \left[B_{j(n)}(\Delta t_n) + e(n) b_{j(n)+1}\right]\left[B_{k(n)}(\Delta t_n) + f(n) b_{k(n)+1}\right] \\ &= \sum\nolimits_{i=1}^{j(n)} b_i^2 + e(n) b_{j(n)+1}^2 + \sum\nolimits_{l \neq i} b_l b_i + \textit{Other terms}. \end{aligned}$$

Each term in "other terms" is a mixed binomial $b_l b_i$ *multiplied by* $e(n)$ *and/or* $f(n)$.

By independence, the expectation of the terms in the i, l*-summation and other terms is* 0*, while* $E\left[b_i^2(\Delta t_n)\right] = \Delta t_n$ *for all* i*. Hence:*

$$E\left[B_s(\Delta t_n) B_t(\Delta t_n)\right] = j(n)\Delta t_n + e(n)\Delta t_n = s,$$

and this proves item 3. ∎

Exercise 10.11 *Prove that for* $0 \leq s_1 < t_1 \leq s_2 < t_2 \leq T$*, if* $\Delta t_n < \min(t_j - s_j)/2$*, then:*

$$E\left[\left[B_{t_1}(\Delta t_n) - B_{s_1}(\Delta t_n)\right]\left[B_{t_2}(\Delta t_n) - B_{s_2}(\Delta t_n)\right]\right] = 0.$$

Parts 1 and 2 of the above proposition motivate an investigation into the following limiting random variables.

Definition 10.12 (Standard binomial motion Z_t, and Z'_{t-s}) *For* $t \geq 0$*, define* ***standard binomial motion*** Z_t *as the distributional limit of binomial paths:*

$$B_t(\Delta t_n) \to_d Z_t. \tag{10.8}$$

For $s < t$*, define* ***standard binomial motion increments*** Z'_{t-s} *by:*

$$B_t(\Delta t_n) - B_s(\Delta t_n) \to_d Z'_{t-s}. \tag{10.9}$$

Remark 10.13 ($Z_{t_n} \to_d Z_0$ as $t_n \to 0$) *By Proposition 10.10, $Z'_{t-s} = N(0, t-s)$, and for $t > s$:*

$$Z_t = N(0, t).$$

Since $B_0(\Delta t_n) \equiv 0$ for all n, note that (10.8) defines $Z_0 \equiv 0$, a degenerate random variable. It then also follows that $Z_{t_n} \to_d Z_0$ as $t_n \to 0$, meaning by Definition 3.2 that if $\Phi_{0,t_n}(x)$ denotes the distribution function of $N(0, t_n)$ evaluated at x, then:

$$\Phi_{0,t_n}(x) \to \begin{cases} 0, & x < 0, \\ 1, & x > 0. \end{cases}$$

The limiting result on the right is seen to be the distribution function of degenerate Z_0, evaluated at its points of continuity.

The next results will allow the conclusions of item 3 of Proposition 10.10 and Exercise 10.11 to be applied to $E[Z_s Z_t]$ and $E\left[Z'_{t_1-s_1} Z'_{t_2-s_2}\right]$. Specifically, for $s < t$:

$$E[Z_s Z_t] = \min[s, t], \tag{1}$$

and for $0 \leq s_1 < t_1 \leq s_2 < t_2 \leq T$:

$$E\left[Z'_{t_1-s_1} Z'_{t_2-s_2}\right] = 0. \tag{2}$$

We will also derive the joint distributions of, and the distributional relationships between, the Z_t and Z'_{t-s} variates.

On this latter point, it would be natural to expect that $Z'_{t-s} =_d Z_t - Z_s$, meaning equal in distribution. But this will require a direct proof since as noted in Example II.5.31, $B_t(\Delta t_n) \to_d Z_t$ and $B_s(\Delta t_n) \to_d Z_s$ do not necessarily obtain that $B_t(\Delta t_n) - B_s(\Delta t_n) \to_d Z_t - Z_s$.

Recalling Section V.3.3.5, both covariance results in (1) and (2) above are examples of **integrations to the limit,** the application of which must be justified. The goal of the proofs below is to justify, for example, that:

$$\lim_{n\to\infty} E[B_s(\Delta t_n) B_t(\Delta t_n)] = E\left[\lim_{n\to\infty} B_s(\Delta t_n) B_t(\Delta t_n)\right].$$

Then (1) would follow from item 3 of Proposition 10.10, and similarly for (2).

Remark 10.14 (Independence of $\{Z'_{t_i-s_i}\}_{i=1}^n$) *It is tempting to think that the conclusion in the next result on independence of $\{Z'_{t_i-s_i}\}_{i=1}^n$ would simply be a corollary to a proof of (10.10), that $E\left[Z'_{t_j-s_j} Z'_{t_k-s_k}\right] = 0$ for $j \neq k$. Indeed for normal variates it is "known" that uncorrelated implies independent, and this general result is stated and proved in Proposition 6.50.*

*But recall that there is a detail that must first be confirmed which is illustrated in Example 6.51. To apply Proposition 6.50, it must first be confirmed that these normal variates are **multivariate** normally distributed. Thus we prove that $(Z'_{t_1-s_1}, ..., Z'_{t_n-s_n})$ is indeed multivariate normally distributed, then independence follows from $E\left[Z'_{t_j-s_j} Z'_{t_k-s_k}\right] = 0$ for $j \neq k$.*

Proposition 10.15 ($(Z'_{t_1-s_1}, ..., Z'_{t_n-s_n})$ is multivariate normal) *For $0 \leq s_1 < t_1 \leq s_2 < t_2 \leq ... \leq s_n < t_n \leq T$, the vector of standard binomial motion increments:*

$$(Z'_{t_1-s_1}, ..., Z'_{t_n-s_n}),$$

is multivariate normally distributed with mean n-vector $\mu = 0$, and diagonal covariance matrix C with $C_{ii} = t_i - s_i$, and $C_{ij} = 0$ otherwise.

Thus for $j \neq k$:

$$E\left[Z'_{t_j - s_j} Z'_{t_k - s_k}\right] = 0, \tag{10.10}$$

and:

$$\left\{Z'_{t_j - s_j}\right\}_{j=1}^n,$$

are independent random variables.

Proof. *We prove that $\{Z'_{t_i - s_i}\}_{i=1}^n$ are multivariate normally distributed as specified, and this will be done with the aid of the Cramér-Wold theorem of Proposition 5.42.*

Let $Z = (Z_1, Z_2, ..., Z_n)$ have a multivariate normal density function defined on $\mathbb{R}^n$ as in (6.6) with mean n-vector $\mu = 0$:

$$f_Z(z) = (2\pi)^{-n/2} [\det C]^{-1/2} \exp\left[-\frac{1}{2} z^T C^{-1} z\right], \tag{10.11}$$

and where C is a diagonal covariance matrix with $C_{ii} = t_i - s_i$ and $C_{ij} = 0$ otherwise. Thus the determinant of C is given by $\det C = \prod_{i=1}^n (t_i - s_i)$. By Proposition 6.50, such $\{Z_i\}_{i=1}^n$ are independent since C is diagonal and positive definite.

Define the n-vector $B_m = (B_1(\Delta t_m), B_2(\Delta t_m), ..., B_n(\Delta t_m))$, where $B_i(\Delta t_m) \equiv B_{t_i}(\Delta t_m) - B_{s_i}(\Delta t_m)$, and we show that $B_m \to_d Z$. By the Cramér-Wold device, this is true if and only if $a \cdot B_m \to_d a \cdot Z$ for all constant vectors $a \in \mathbb{R}^n$, where the dot product of n-vectors is defined by $x \cdot y = \sum_{i=1}^n x_i y_i$. By Proposition 6.23, $a \cdot Z$ is normally distributed with mean 0 and variance $\sum_{i=1}^n a_i^2 (t_i - s_i)$, and thus the goal is to show that $a \cdot B_m$ converges in distribution to this variate for all $a \in \mathbb{R}^n$.

To this end, express $s_i = [j_i(m) + e_i(m)] \Delta t_m$ and $t_i = [k_i(m) + f_i(m)] \Delta t_m$, where $j_i(m) \leq k_i(m)$ are integers, and $0 \leq e_i(m), f_i(m) < 1$. As noted in the proof of Proposition 10.10, $s_i < t_i$ assures that $j_i(m) < k_i(m)$ for m large, so we assume this and note that it is possible that $t_i = s_{i+1}$ for any i. Thus, suppressing Δt_m:

$$\begin{aligned}
a \cdot B_m &= \sum_{i=1}^n a_i \left[\sum_{l=j_i(m)+2}^{k_i(m)} b_l + f_i(m) b_{k_i(m)+1} + (1 - e_i(m)) b_{j_i(m)+1}\right] \\
&= \sum_{i=1}^n a_i \sum_{l=j_i(m)+2}^{k_i(m)} b_l \\
&\quad + \sum_{i=1}^n a_i \left[f_i(m) b_{k_i(m)+1} + (1 - e_i(m)) b_{j_i(m)+1}\right] \\
&= X_m + Y_m.
\end{aligned}$$

Since $0 \leq e_i(m), f_i(m) < 1$, and $|b_l| = 1$:

$$|Y_m| \leq 2n \max |a_i| \sqrt{\Delta t_m}.$$

This obtains convergence in probability of Definition 3.22, that $Y_m \to_P 0$ as $m \to \infty$.

For X_m, we first note that this variate equals a sum of independent binomials. While it is possible that $t_i = s_{i+1}$ for any i, and in any such case $k_i(m) = j_{i+1}(m)$, the summation associated with a_i has last term $b_{k_i(m)}$, while the summation associated with a_{i+1} has first term $b_{j_{i+1}(m)+2}$. Thus all binomials are independent, with:

$$\begin{aligned}
Var[X_m] &= \sum_{i=1}^n a_i^2 (k_i(m) - j_i(m) - 1) \Delta t_m \\
&= \sum_{i=1}^n a_i^2 [t_i - s_i - (1 + f_i(m) - e_i(m)) \Delta t_m] \qquad (1) \\
&\to \sum_{i=1}^n a_i^2 (t_i - s_i).
\end{aligned}$$

Further, by definition of these binomials, $E[X_m] = 0$.

To apply Lindeberg's central limit theorem of Proposition 7.12 to the X_m-summation, we must verify that this summation satisfies Lindeberg's condition of Definition 7.6. In the notation of that definition, $s_m^2 \equiv Var[X_m]$, which is bounded as $m \to \infty$ as noted above. Thus (7.6) is satisfied if for every $t > 0$:

$$\sum\nolimits_{i=1}^{n} \sum\nolimits_{l=j_i(m)+2}^{k_i(m)} E\left[(a_i b_l(\Delta t_m))^2 \chi\left[|a_i b_l(\Delta t_m)| \geq t s_m\right]\right] \to 0, \tag{2}$$

as $m \to \infty$. Here χ is the indicator function and is defined by $\chi\left[|a_i b_l(\Delta t_m)| \geq t s_m\right] = 1$ on the set where $|a_i b_l(\Delta t_m)| \geq t s_m$, and is 0 otherwise.

The terms of the sum in (2) can be estimated with Chebyshev's inequality of Proposition IV.4.34 since $(a_i b_l(\Delta t_m))^2 = a_i^2 \Delta t_m$ is constant. Since $Var[a_i b_l(\Delta t_m)] = a_i^2 \Delta t_m$, this obtains:

$$\begin{aligned} E\left[(a_i b_l(\Delta t_m))^2 \chi\left[|a_i b_l(\Delta t_m)| \geq t s_m\right]\right] &= a_i^2 \Delta t_m \Pr\left[|a_i b_l(\Delta t_m)| \geq t s_m\right] \\ &\leq \frac{a_i^4 (\Delta t_m)^2}{(t s_m)^2}. \end{aligned}$$

Hence:

$$\begin{aligned} &\sum\nolimits_{i=1}^{n} \sum\nolimits_{l=j_i(m)+2}^{k_i(m)} E\left[(a_i b_l(\Delta t_m))^2 \chi\left[|a_i b_l(\Delta t_m)| \geq t s_m\right]\right] \\ \leq\ & \sum\nolimits_{i=1}^{n} \sum\nolimits_{l=j_i(m)+2}^{k_i(m)} \frac{a_i^4 (\Delta t_m)^2}{(t s_m)^2} \\ \leq\ & \Delta t_m \frac{\max a_i^2}{(t s_m)^2} \sum\nolimits_{i=1}^{n} a_i^2 \left(k_i(m) - j_i(m) - 1\right) \Delta t_m \\ =\ & \frac{\max a_i^2}{t^2} \Delta t_m, \end{aligned}$$

where this last expression follows because the prior sum is seen in (1) to equal $Var[X_m] = s_m^2$. Thus (2) is satisfied, as is Lindeberg's condition.

By Lindeberg's central limit theorem of Proposition 7.12:

$$\frac{X_m}{\sqrt{\sum_{i=1}^{n} a_i^2 (t_i - s_i)}} \to_d W,$$

where $W \sim N(0,1)$ is a standard normal variate. Thus $X_m \to_d W'$ where W' is normally distributed with mean 0 and variance $\sum_{i=1}^{n} a_i^2 (t_i - s_i)$, and so $W' =_d a \cdot Z$ with Z defined above.

Applying Slutsky's theorem of Proposition II.5.29 to $X_m \to_d a \cdot Z$ and $Y_m \to_P 0$ obtains for all $a \in \mathbb{R}^n$:

$$a \cdot B_m = X_m + Y_m \to_d a \cdot Z. \tag{3}$$

This then proves by the Cramér-Wold theorem that $B_m \to_d Z$.

Now given i, let n-vector a be defined with $a_i = 1$ and $a_j = 0$ for $j \neq i$. Then (3) obtains that $a \cdot B_m \to_d Z_i$, the ith variate of the multivariate normal distribution above. On the other hand, Proposition 10.10 and Definition 10.12 assure that:

$$a \cdot B_m \equiv B_{t_i}(\Delta t_m) - B_{s_i}(\Delta t_m) \to_d Z'_{t_i - s_i},$$

and so $(Z_1, Z_2, ..., Z_n) =_d (Z'_{t_1 - s_1}, ..., Z'_{t_n - s_n})$. This proves that $(Z'_{t_1 - s_1}, ..., Z'_{t_n - s_n})$ is multivariate normal with the given covariance matrix C.

Finally by Proposition 6.50, $\left\{Z'_{t_j - s_j}\right\}_{j=1}^{n}$ are independent since C is diagonal and positive definite, and (10.10) follows from Exercise 6.49. ■

We next derive the joint distribution of $(Z_{t_1}, Z_{t_2}, ..., Z_{t_n})$ by a transformation of the distribution for $(Z'_{t_1-s_1}, ..., Z'_{t_n-s_n})$, choosing $s_1 = 0$ and $s_{j+1} = t_j$ otherwise, and an application of the Mann-Wald theorem of Proposition 3.26.

Proposition 10.16 (**$(Z_{t_1}, Z_{t_2}, ..., Z_{t_n})$ is multivariate normal**) *For $0 < t_1 < t_2 < ... < t_n \leq T$, the variates of standard binomial motion:*

$$(Z_{t_1}, Z_{t_2}, ..., Z_{t_n}),$$

are multivariate normally distributed with mean n-vector $\mu = 0$, and covariance matrix C with $C_{ij} = \min(t_i, t_j)$. Thus:

$$E\left[Z_{t_i} Z_{t_j}\right] = \min(t_i, t_j). \tag{10.12}$$

Proof. *Choosing $s_1 = 0$ and $s_{j+1} = t_j$ otherwise in Proposition 10.15 obtains that as $m \to \infty$:*

$$(B_{t_1}(\Delta t_m), B_{t_2}(\Delta t_m) - B_{t_1}(\Delta t_m), ..., B_{t_n}(\Delta t_m) - B_{t_{n-1}}(\Delta t_m)) \to_d MN_n(0, C'). \tag{1}$$

In the notation of Chapter 6, $MN_n(0, C')$ denotes an n-dimensional multivariate normal vector with mean n-vector $\mu = 0$, and covariance matrix C', here defined by $C'_{ii} = t_i - t_{i-1}$ and $C'_{ij} = 0$ otherwise, where we define $t_0 = 0$.

Define a linear and hence continuous transformation $\mathbb{R}^n \to \mathbb{R}^n$ by the matrix:

$$A \equiv \begin{pmatrix} 1 & 0 & 0 & \cdots & 0 \\ 1 & 1 & 0 & \cdots & 0 \\ \vdots & \vdots & \vdots & \ddots & \vdots \\ 1 & 1 & 1 & \cdots & 0 \\ 1 & 1 & 1 & 1 & 1 \end{pmatrix}.$$

In other words, A is lower diagonal with all nonzero elements equal to 1.

Then:

$$\begin{aligned} A &: (B_{t_1}(\Delta t_m), B_{t_2}(\Delta t_m) - B_{t_1}(\Delta t_m), ..., B_{t_n}(\Delta t_m) - B_{t_{n-1}}(\Delta t_m)) \\ &= (B_{t_1}(\Delta t_m), B_{t_2}(\Delta t_m), ..., B_{t_n}(\Delta t_m)), \end{aligned}$$

and the Mann-Wald theorem and (1) obtain:

$$(B_{t_1}(\Delta t_m), B_{t_2}(\Delta t_m), ..., B_{t_n}(\Delta t_m)) \to_d AZ,$$

where $Z \sim MN_n(0, C')$.

By Proposition 6.23, $AZ \sim MN_n(0, C)$ is multivariate normal with mean vector 0 and covariance matrix $C \equiv AC'A^T$. Now AC' is seen to be lower triangular:

$$AC = \begin{pmatrix} t_1 & 0 & 0 & \cdots & 0 \\ t_1 & t_2 - t_1 & 0 & \cdots & 0 \\ \vdots & \vdots & \vdots & \ddots & \vdots \\ t_1 & t_2 - t_1 & t_3 - t_2 & \cdots & 0 \\ t_1 & t_2 - t_1 & t_3 - t_2 & \cdots & t_n - t_{n-1} \end{pmatrix},$$

and a final right multiplication by A^T obtains that $\left(AC'A^T\right)_{i,j} = \min(t_i, t_j)$. ■

Example 10.17 ($f(z_1, ..., z_n)$ for $(Z_1, ..., Z_n)$) *One can obtain a convenient factorization of the joint density function of $(Z_{t_1}, Z_{t_2}, ..., Z_{t_n})$ as follows.*

Let $f(z_1, ..., z_n)$ denote this joint density function as defined in (6.6) with $\mu = 0$ and covariance matrix C where $C_{ij} = \min(t_i, t_j)$:

$$f(z_1, ..., z_n) = (2\pi)^{-n/2} [\det C]^{-1/2} \exp\left[-\frac{1}{2} z^T C^{-1} z\right],$$

where $z \equiv (z_1, ..., z_n)$. Then with $z_0 \equiv 0$, we claim that:

$$f(z_1, ..., z_n) = \prod\nolimits_{j=1}^{n} f_j(z_j - z_{j-1}), \tag{10.13}$$

where f_j is a normal density with mean 0 and variance $t_j - t_{j-1}$:

$$f_j(x) = [2\pi (t_j - t_{j-1})]^{-1/2} \exp\left[-\frac{x^2}{2(t_j - t_{j-1})}\right],$$

with $t_0 \equiv 0$.

For $\det C$, subtracting row $j-1$ from row j in order from $j = n$ to $j = 2$ produces an upper triangular matrix with the same determinant as C, and with diagonal elements $\{t_j - t_{j-1}\}_{j=1}^{n}$. This diagonal matrix has the determinant equal to the product of the diagonal elements, and thus:

$$\det C = \prod\nolimits_{j=1}^{n} (t_j - t_{j-1}). \tag{10.14}$$

Comparing the numerical coefficients in (10.13) from the definition of $f_j(x)$:

$$(2\pi)^{-n/2} [\det C]^{-1/2} = \prod\nolimits_{j=1}^{n} [2\pi (t_j - t_{j-1})]^{-1/2}.$$

To complete the derivation, it must be shown that with $z \equiv (z_1, ..., z_n)$ and $z_0 \equiv 0$:

$$\exp\left[-\frac{1}{2} z^T C^{-1} z\right] = \prod\nolimits_{j=1}^{n} \exp\left[-\frac{(z_j - z_{j-1})^2}{2(t_j - t_{j-1})}\right],$$

or equivalently:

$$z^T C^{-1} z = \sum\nolimits_{j=1}^{n} \left[\frac{(z_j - z_{j-1})^2}{(t_j - t_{j-1})}\right]. \tag{1}$$

Let D be the diagonal matrix with $D_{jj} = 1/(t_j - t_{j-1})$, and E the matrix defined by $E_{jj} = 1$, $E_{j,j-1} = -1$ for $j \geq 2$, and $E_{j,k} = 0$ otherwise. In other words, E has main diagonal of 1s, main subdiagonal of -1s, and all other elements 0s. Thus:

$$Ez = (z_1, z_2 - z_1, ..., z_n - z_{n-1}),$$

and with $z_0 \equiv 0$, (1) can be expressed:

$$z^T C^{-1} z = z^T E^T D E z, \tag{2}$$

recalling that $(Ez)^T = z^T E^T$ by Exercise 6.4.

To prove (2) and thus (1), we prove that $C^{-1} = E^T D E$, or multiplying on the right by C:

$$E^T D E C = I, \tag{3}$$

where I is the identity matrix.

A calculation obtains that EC is an upper triangular matrix given by $(EC)_{ij} = t_i - t_{i-1}$ for $j \geq i$, and is 0 otherwise. Since D is diagonal with $D_{jj} = 1/(t_j - t_{j-1})$, it follows that $(DEC)_{ij} = 1$ for $j \geq i$, and is 0 otherwise, and thus $E^T D E C = I$ follows as an exercise.

We are finally ready to address the distributional relationship between the Z and Z' variates, using a change of variable in the above distribution functions.

Proposition 10.18 (Distribution functions of Z vs. Z' variates) *For $0 \equiv t_0 < t_1 < t_2 < ... < t_n \leq T$, and where "$=_d$" denotes equal in distribution:*

$$(Z'_{t_1-t_0}, Z'_{t_2-t_1}, ..., Z'_{t_n-t_{n-1}}) =_d (Z_{t_1}, Z_{t_2} - Z_{t_1}, ..., Z_{t_n} - Z_{t_{n-1}}), \tag{10.15}$$

and:

$$(Z_{t_1}, Z_{t_2}, ..., Z_{t_n}) =_d \left(Z'_{t_1-t_0}, \sum\nolimits_{j=1}^{2} Z'_{t_j-t_{j-1}}, ..., \sum\nolimits_{j=1}^{n} Z'_{t_j-t_{j-1}} \right). \tag{10.16}$$

Proof. *From the distribution function for $Z \equiv (Z_{t_1}, Z_{t_2}, ..., Z_{t_n})$, we can identify the distribution for:*

$$Z' \equiv (Z_{t_1}, Z_{t_2} - Z_{t_1}, ..., Z_{t_n} - Z_{t_{n-1}}),$$

using Proposition 6.23.

Define the $n \times n$ matrix B with main diagonal $B_{ii} = 1$ and main subdiagonal $B_{i,i-1} = -1$. Since $Z' = BZ$, it follows that Z' is multivariate normally distributed with mean vector $\mu = 0$ and covariance matrix:

$$C_{Z'} = BC_Z B^T,$$

where C_Z is the covariance matrix of Z given in Proposition 10.16. A calculation confirms that $C_{Z'}$ is the covariance matrix of Proposition 10.15 for $(Z'_{t_1-t_0}, Z'_{t_2-t_1}, ..., Z'_{t_n-t_{n-1}})$, and (10.15) follows.

Similarly, from the distribution function for $Z' \equiv (Z'_{t_1-t_0}, Z'_{t_2-t_1}, ..., Z'_{t_n-t_{n-1}})$ we can derive the distribution function for:

$$Z \equiv \left(Z'_{t_1-t_0}, \sum\nolimits_{j=1}^{2} Z'_{t_j-t_{j-1}}, ..., \sum\nolimits_{j=1}^{n} Z'_{t_j-t_{j-1}} \right).$$

Defining the lower triangular matrix B' by $B'_{ij} = 1$ for $j \leq i$, and 0 otherwise, obtains $Z = B'Z'$. Thus Z is multivariate normally distributed with mean vector $\mu = 0$ and covariance matrix:

$$C_Z = B'C_{Z'} (B')^T,$$

where $C_{Z'}$ is the covariance matrix of Z' given in Proposition 10.15. Again a calculation confirms that C_Z is the covariance matrix of Proposition 10.16 for $(Z_{t_1}, Z_{t_2}, ..., Z_{t_n})$, and (10.16) follows. ■

We next investigate other properties of binomial paths $B_t(\Delta t)$ in the limit as $\Delta t \to 0$. As above, we are given an interval $[0, T]$ with $T < \infty$ and define $\Delta t_n = T/n$, or $\Delta t_n = T/mn$ for fixed m, so that $\Delta t_n \to 0$ as $n \to \infty$.

We begin with an exercise.

Exercise 10.19 (Convergence of moments) *As noted in Example IV.4.64, convergence in distribution does not in general assure convergence of moments. However, show that for $k = 1, 2$:*

$$E\left[B_t^k(\Delta t_n)\right] \to E\left[Z_t^k\right], \ t \geq 0, \tag{10.17}$$

and:

$$E\left[(B_t(\Delta t_n) - B_s(\Delta t_n))^k\right] \to E\left[(Z'_{t-s})^k\right], \ t > s. \tag{10.18}$$

Hint: This is not a theoretical derivation, since all expectations can be explicitly evaluated.

The next result shows that the distributional convergence of $B_t(\Delta t_n)-B_s(\Delta t_n) \to_d Z'_{t-s}$ is quite robust, and (10.18) generalizes significantly. For the proof we need a result on uniform integrability from the following exercise. Recall from Definition 4.21 that a sequence of random variables $\{X_n\}_{n=1}^{\infty}$ is uniformly integrable if:

$$\lim_{N\to\infty} \sup_n \int_{|X_n|\geq N} |X_n(s)|\, d\lambda = 0.$$

Exercise 10.20 ($\{X_n\}_{n=1}^{\infty}$ U.I. and $|Y_n| \leq |X_n|$) *Prove that if $\{X_n\}_{n=1}^{\infty}$ are defined on a probability space $(\mathcal{S}, \mathcal{E}, \lambda)$ and uniformly integrable, and $\{Y_n\}_{n=1}^{\infty}$ are defined on this space with $|Y_n| \leq |X_n|$ for all n, then $\{Y_n\}_{n=1}^{\infty}$ are uniformly integrable. Then generalize that the same result holds if there exists N' so that $|Y_n| \leq |X_n|$ on $\{|X_n| \geq N\}$ for all n and $N \geq N'$.*

Proposition 10.21 (On limits of $|B_t(\Delta t_n) - B_s(\Delta t_n)|^{\alpha}$) *For $0 \leq s < t \leq T$, and any $\alpha > 0$:*

$$|B_t(\Delta t_n) - B_s(\Delta t_n)|^{\alpha} \to_d \left|Z'_{t-s}\right|^{\alpha}, \tag{10.19}$$

as $n \to \infty$, where $Z'_{t-s} \sim N(0, t-s)$ by Proposition 10.10 and Definition 10.12.

In addition:

$$E\left[|B_t(\Delta t_n) - B_s(\Delta t_n)|^{\alpha}\right] \to E\left[\left|Z'_{t-s}\right|^{\alpha}\right]. \tag{10.20}$$

Proof. *By item 2 of Proposition 10.10 and Definition 10.12, $B_t(\Delta t_n)-B_s(\Delta t_n) \to_d Z'_{t-s}$ as $n \to \infty$. The Mann-Wald theorem on $\mathbb{R}$ of Proposition II.8.37 states that if $X^{(n)} \to_d X$, then $g\left[X^{(n)}\right] \to_d g[X]$ for any continuous function g. Since $g(x) = |x|^{\alpha}$ is continuous, (10.19) follows.*

From this and Proposition 4.23, the convergence of moments in (10.20) will be justified if $\{|B_{t'}(\Delta t_n) - B_t(\Delta t_n)|^{\alpha}\}_{n=1}^{\infty}$ is uniformly integrable. Recalling Remark 4.22, the uniform integrability criterion can be restated in terms of Riemann-Stieltjes integrals as in (4.43). In detail, the sequence of random variables $\{|X_n|^{\alpha}\}_{n=1}^{\infty}$ is uniformly integrable if as $N \to \infty$:

$$\lim_{N\to\infty} \sup_n \int_{x\geq N} x dG_n = 0, \tag{1}$$

with G_n the distribution function of $|X_n|^{\alpha}$. We apply this representation with:

$$\{|X_n|^{\alpha}\}_{n=1}^{\infty} \equiv \{|B_t(\Delta t_n) - B_s(\Delta t_n)|^{\alpha}\}_{n=1}^{\infty}.$$

To simplify calculations, we demonstrate uniform integrability of $\{|X_n|^{2M}\}_{n=1}^{\infty}$ for any integer M. The result then follows from the second criterion in Exercise 10.20. Specifically, if $N' = 1$ and $\alpha \leq 2M$, then $|X_n|^{\alpha} \leq |X_n|^{2M}$ if $|X_n|^{2M} \geq N \geq 1$.

As in the proof of Proposition 10.10, let $s = [j(n) + e(n)]\,\Delta t_n$ and $t = [k(n) + f(n)]\,\Delta t_n$, where $j(n) \leq k(n)$, and $0 \leq e(n), f(n) < 1$. For n large, it follows from $s < t$ that $j(n) < k(n)$, and we assume this for notational convenience. Then:

$$\begin{aligned} X_n^{2M} &\equiv [B_t(\Delta t_n) - B_s(\Delta t_n)]^{2M} \\ &= \left[(1-e(n))\, b_{j(n)+1}(\Delta t_n) + \sum\nolimits_{i=j(n)+2}^{k(n)} b_i(\Delta t_n) + f(n) b_{k(n)+1}(\Delta t_n)\right]^{2M} \\ &= \left[\sum\nolimits_{i=1}^{k(n)-j(n)+1} a_i^{(n)} b_i(\Delta t_n)\right]^{2M}, \end{aligned} \tag{2}$$

where $0 \leq a_i^{(n)} \leq 1$ for all i, and we reparametrize for notational simplicity.

We now proceed in two steps to prove uniform integrability of X_n^{2M}.

1. ***Tail Estimate of*** X_n^{2M}: *Define* $b_i' = b_i(\Delta t_n)/\sqrt{\Delta t_n}$, $b_i'' = (b_i'+1)/2$, *and* $l(n) = k(n)-j(n)+1$. *Then* $\{b_i''\}_{i=1}^{l(n)}$ *are independent, standard binomials, meaning with range* $0,1$ *and* $p=1/2$. *By Bernstein's inequality in II.(5.5), for a collection of any* m *such variates:*

$$\Pr\left[\left|\frac{1}{m}\sum\nolimits_{i=1}^{m} b_i'' - \frac{1}{2}\right| \geq \epsilon\right] \leq 2e^{-m\epsilon^2/4},$$

and thus:

$$\Pr\left[\left|\sum\nolimits_{i=1}^{m} b_i'\right| \geq 2m\epsilon\right] \leq 2e^{-m\epsilon^2/4}. \tag{3}$$

Now $X_n^{2M} = (\Delta t_n)^M \left(\sum\nolimits_{i=1}^{l(n)} \alpha_i^{(n)} b_i'\right)^{2M}$ *obtains:*

$$\Pr\left[X_n^{2M} \geq N\right] = \Pr\left[\left|\sum\nolimits_{i=1}^{l(n)} \alpha_i^{(n)} b_i'\right| \geq N^{1/2M}/\sqrt{\Delta t_n}\right].$$

Recalling the definition of $\alpha_i^{(n)}$ *and applying the triangle inequality:*

$$\begin{aligned}\left|\sum\nolimits_{i=1}^{l(n)} \alpha_i^{(n)} b_i'\right| &\leq \left|\sum\nolimits_{i=2}^{l(n)-1} b_i'\right| + \left|(1-e(n))\,b_{j(n)+1} + f(n)b_{k(n)+1}\right| \\ &\leq \left|\sum\nolimits_{i=2}^{l(n)-1} b_i'\right| + 2.\end{aligned}$$

Thus:

$$\Pr\left[X_n^{2M} \geq N\right] \leq \Pr\left[\left|\sum\nolimits_{i=1}^{m} b_i'\right| \geq \frac{N^{1/2M}}{\sqrt{\Delta t_n}} - 2\right],$$

where $m = l(n)-2 = k(n)-j(n)-1$. *Applying* (3) *with* $2m\epsilon = N^{1/2M}/\sqrt{\Delta t_n}-2$ *produces:*

$$\Pr\left[X_n^{2M} \geq N\right] \leq 2\exp\left[-\frac{1}{16m\Delta t_n}\left(N^{1/2M} - 2\sqrt{\Delta t_n}\right)^2\right].$$

To simplify, for n *large,* $\left(N^{1/2M} - 2\sqrt{\Delta t_n}\right)^2 \geq \left(N^{1/2M}-1\right)^2 \geq .5N^{1/M}$, *for example, if* $N^{1/2M} \geq 4$, *and thus for* N *large:*

$$\Pr\left[X_n^{2M} \geq N\right] \leq 2\exp\left[-\frac{N^{1/M}}{32m\Delta t_n}\right].$$

Finally, by definition of m *above:*

$$t - s - 2\Delta t_n < m\Delta t_n < t - s,$$

and this obtains the tail estimate:

$$\Pr\left[X_n^{2M} \geq N\right] \leq 2\exp\left[-\frac{N^{1/M}}{32\,(t-s)}\right]. \tag{4}$$

2. ***Uniform Integrability of*** X_n^{2M}: *The distribution function* G_n *of* X_n^{2M} *has finitely many discontinuities by definition, and thus outside of this collection,* (4) *obtains:*

$$1 - G_n(x) \leq 2\exp\left[-\frac{x^{1/M}}{32\,(t-s)}\right], \tag{5}$$

and so $1 - G_n(x)$ *is Lebesgue integrable over* $[N,\infty)$ *for any* $N > 0$.

Recall Lebesgue-Stieltjes integration by parts of Proposition V.6.10:

$$\int_{(a,b]} g d\mu_f = f(b)g(b) - f(a)g(a) - \int_{(a,b]} f d\mu_g,$$

which we apply with right continuous $g = 1 - G_n$, and with $d\mu_f$ denoting Lebesgue measure, defined with continuous $f(x) = x$.

Let $(a,b] = (N, X_n^{\max}]$, where $X_n^{\max}$ exceeds the maximum value of X_n^{2M}. By (2), each $X_n^{\max}$ can be no more than $(k(n) - j(n) + 1)^{2M} (\Delta t_n)^M \leq (k(n) - j(n) + 1)^M (t - s + 2\Delta t_n)^M$. Then since $G_n(X_n^{\max}) = 1$ by construction, Proposition V.6.10 obtains (see Remark 10.22):

$$\begin{aligned} &\int_{(N,X_n^{\max}]} (1 - G_n(x)) dx \\ &= X_n^{\max} (1 - G_n(X_n^{\max})) - N (1 - G_n(N)) + \int_{(N,X_n^{\max}]} x dG_n \qquad (6) \\ &= -N (1 - G_n(N)) + \int_{(N,\infty)} x dG_n. \end{aligned}$$

Using the bound in (5) for $1 - G_n(x)$, it follows that $\sup_n N (1 - G_n(N))$ is bounded, and:

$$\sup_n \int_{(N,\infty)} (1 - G_n(x)) dx < \infty.$$

So by (6) it follows that for any N:

$$\sup_n \int_{(N,\infty)} x dG_n < \infty.$$

Thus:

$$\sup_n \int_{(N,\infty)} (1 - G_n(x)) dx + \sup_n N (1 - G_n(N)) = \sup_n \int_{(N,\infty)} x dG_n. \qquad (7)$$

But again from (5):

$$\lim_{N \to \infty} \sup_n N (1 - G_n(N)) = 0,$$

and:

$$\lim_{N \to \infty} \sup_n \int_{(N,\infty)} (1 - G_n(x)) dx = 0,$$

and so from (7):

$$\lim_{N \to \infty} \sup_n \int_{(N,\infty)} x dG_n = 0.$$

This proves that $\{X_n^{2M}\}_{n=1}^{\infty}$ is uniformly integrable by (1), and the proof is complete.

■

Remark 10.22 (On $\int_{(N,X_n^{\max}]} x dG_n$ in (6)) *There is an important detail in line (6) of the above proof, where $g \equiv 1 - G_n$ and Lebesgue-Stieltjes integration by parts required $\int_{(N,X_n^{\max}]} x d\mu_g$.*

First, $g(x)$ is monotonic and thus of bounded variation, and so the measure μ_g is well-defined by Proposition V.6.5 on **all bounded** *$(a,b] \in \mathcal{A}'$ by:*

$$\mu_g[(a,b]] = g(b) - g(a) = -[G_n(b) - G_n(a)],$$

where $\mathcal{A}'$ is the semi-algebra of right semi-closed intervals. Since G_n is a distribution function, it follows from Proposition I.5.23 that there exists a Borel measure μ_{G_n} so that for **all** *$(a,b] \in \mathcal{A}'$:*

$$\mu_{G_n}[(a,b]] = G_n(b) - G_n(a),$$

and thus for all bounded $(a,b] \in \mathcal{A}'$:

$$\mu_g = -\mu_{G_n}. \tag{1}$$

Taking limits, (1) is satisfied for all $(a,b] \in \mathcal{A}'$, and then by finite additivity, (1) is satisfied on the associated algebra $\mathcal{A}$ of all finite unions of $\mathcal{A}'$-sets (Exercise I.6.10).

The uniqueness theorem of Proposition I.6.14 obtains that (1) is then satisfied for all $A \in \sigma(\mathcal{A})$, the smallest sigma algebra that contains $\mathcal{A}$. By Proposition I.8.1, $\sigma(\mathcal{A}) = \mathcal{B}(\mathbb{R})$, the Borel sigma algebra, and so (1) is satisfied on $\mathcal{B}(\mathbb{R})$.

Finally, Proposition V.6.8 obtains that:

$$\int_{(N,X_n^{\max}]} x d\mu_g = -\int_{(N,X_n^{\max}]} x d\mu_{G_n},$$

and line (6) of the above proof follows by converting this Lebesgue-Stieltjes integral into a Riemann-Stieltjes integral by Proposition V.3.63.

In the final proposition of this section, we summarize for completeness two key results on standard binomial motion Z_t, and provide preliminary insights on t-continuity of Z_t related to sequential continuity. This time-dependent random variable Z_t will in Book VII be seen to be an example of a **stochastic process.**

In essence, a stochastic process is simply an indexed collection of random variables defined on some probability space $(\mathcal{S}, \mathcal{E}, \lambda)$. Sometimes this indexing is discrete, such as $\{Z_n\}_{n=1}^{\infty}$, and sometimes the indexing is continuous, such as $\{Z_t\}_{t \in I}$ where I is an interval. In virtually all cases, this indexing set reflects time, and thus one understands a stochastic process by understanding how these random variables "evolve" through time. For a stochastic process with continuous indexing like Z_t, continuity in t is of critical interest in many applications.

But as random variables, what does continuity even mean?

Recall that a function f is called **sequentially continuous at** t_0 if $f(t_n) \to f(t_0)$ for any sequence $t_n \to t_0$. A function f is called **sequentially Hölder continuous at** t_0 **with exponent** $\gamma > 0$ if there exists $C > 0$ so that for all such sequences:

$$|f(t_n) - f(t_0)| \leq C\,|t_n - t_0|^{\gamma}.$$

In the big-O notation of Definition 7.2, this can be expressed:

$$f(t_n) - f(t_0) = O\left[|t_n - t_0|^{\gamma}\right].$$

One can similarly define sequential continuity in t for a stochastic process X_t, but for Hölder continuity it is convenient to strengthen this condition to:

$$f(t_n) - f(t_0) = o\left[|t_n - t_0|^{\gamma}\right].$$

We will call this **sequentially strong Hölder continuity,** and this means by Definition V.7.13:

$$\frac{|f(t_n) - f(t_0)|}{|t_n - t_0|^{\gamma}} \to 0.$$

These Hölder criteria are closely related, in that if a function is sequentially Hölder continuous at t_0 with exponent γ, then it is sequentially strong Hölder continuous at t_0 with exponent γ' for all $\gamma' < \gamma$.

As is common in measure theory, the following definition will only require the above criteria for almost all $\omega \in \mathcal{S}$, or equivalently, with (λ-)probability 1.

Definition 10.23 (Sequential continuity of a stochastic process) *A stochastic process $\{X_t\}_{t\in I}$ defined on a probability space $(\mathcal{S}, \mathcal{E}, \lambda)$ for interval I is:*

1. ***Sequentially continuous at** $t \in I$ if given $t_n \to t$, then $X_{t_n}(\omega) \to X_t(\omega)$ for almost all ω.*

2. ***Sequentially strong Hölder continuous at** $t \in I$ **with exponent** γ if given $t_n \to t$, then for almost all ω:*

$$\lim_{n\to\infty} \frac{|X_{t_n} - X_t|}{|t_n - t|^{\gamma}} = 0.$$

Remark 10.24 (On continuity of Z_t; On $(\mathcal{S}, \mathcal{E}, \lambda)$) *For the result below, we will "come close" to proving that binomial motion Z_t is sequentially continuous, and sequentially strong Hölder continuous with exponent $\gamma < \frac{1}{2}$. In detail we will prove that given any sequence, there is a subsequence for which the above criteria are met. So we qualify the above terminology with "nearly."*

We will also prove that if $\gamma > \frac{1}{2}$, then the strong Hölder criterion fails for all subsequences. See also Remark 10.26.

Below we will assume the existence of a probability space $(\mathcal{S}, \mathcal{E}, \lambda)$ on which the various random variables are defined. For any given random variable or finite collection of such, this is justified by Propositions 1.20 and 1.24, where it is seen that this probability space is independent of the random variables.

For item 3, $s_n \to t_0$ implies that $Z_{s_n} \to_d Z_{t_0}$ since $Z_t \sim N(0,t)$ for all t. This by definition implies weak convergence $\mu_{s_n} \Rightarrow \mu_{t_0}$ for the associated probability measures on $\mathbb{R}$. Thus Skorokhod's representation theorem of Proposition II.8.30 can also be applied to obtain such a space.

But outside these special situations, we have not identified such a probability space, and indeed will not do so until Book VII.

Proposition 10.25 (On standard binomial motion Z_t) *Let Z_t be standard binomial motion of Definition 10.12, and:*

$$0 \le s_1 < t_1 \le s_2 < t_2 \le ... \le s_n < t_n \le T.$$

Then:

1. *$(Z_{t_1}, Z_{t_2}, ..., Z_{t_n})$ is multivariate normally distributed with mean n-vector $\mu = 0$, and covariance matrix C with $C_{ij} = \min(t_i, t_j)$. Thus:*

$$E\left[Z_{t_i} Z_{t_j}\right] = \min(t_i, t_j).$$

2. $(Z_{t_1} - Z_{s_1}, Z_{t_2} - Z_{s_2}, ..., Z_{t_n} - Z_{s_n})$ *is multivariate normally distributed with mean n-vector* $\mu = 0$ *and diagonal covariance matrix* C *with* $C_{ii} = t_i - s_i$ *and* $C_{ij} = 0$ *otherwise. Thus for* $j \neq k$:
$$E\left[\left(Z_{t_j} - Z_{s_j}\right)\left(Z_{t_k} - Z_{s_k}\right)\right] = 0,$$
and these component variates are independent.

3. *For any* $t_0 \geq 0$, Z_t *is* ***"nearly" sequentially continuous*** *at* t_0. *Given* $s_n \to 0$ *with* $t_0 + s_n \geq 0$ *for all n, there exists a subsequence* s_{n_k} *such that:*
$$\Pr\left[\lim_{k\to\infty} Z_{t_0+s_{n_k}} = Z_{t_0}\right] = 1. \tag{10.21}$$

4. *For any* $t_0 \geq 0$ *and* $0 < \gamma < 1/2$, Z_t *is* ***"nearly" sequentially strong Hölder continuous at*** t_0 ***with exponent*** γ. *Given* $s_n \to 0$ *with* $t_0 + s_n \geq 0$ *for all n, there exists a subsequence* s_{n_k} *such that:*
$$\Pr\left[\lim_{k\to\infty} \frac{\left|Z_{t_0+s_{n_k}} - Z_{t_0}\right|}{\left|s_{n_k}\right|^{\gamma}} = 0\right] = 1. \tag{10.22}$$

5. *Given* $t_0 \geq 0$, $1/2 < \gamma$, $N > 0$, *and sequence* $s_n \to 0$ *with* $t_0 + s_n \geq 0$ *for all n, then there exists a subsequence* s_{n_k} *so that:*
$$\Pr\left[\limsup_{s_{n_k}\to 0} \frac{\left|Z_{t_0+s_{n_k}} - Z_{t_0}\right|}{\left|s_{n_k}\right|^{\gamma}} > N\right] = 1. \tag{10.23}$$
Thus for any $t_0 \geq 0$ *and* $1/2 < \gamma$, Z_t *is* ***not "nearly" sequentially strong Hölder continuous at*** t_0 ***with exponent*** γ.

Proof. *For completeness of this result, item 1 is a restatement of Proposition 10.16, while item 2 is a restatement of Proposition 10.15 made possible by Proposition 10.18.*

For (10.21) of item 3, let $t \equiv t_0 \geq 0$ *and* $\epsilon > 0$ *be given. Given* $s_n \to 0$, *define:*
$$C_n = \{|Z_{t+s_n} - Z_t| > \epsilon\}.$$
Since $Z_{t+s_n} - Z_t \sim N(0, |s_n|)$ *by item 2, II.(9.49) obtains with* $z \sim N(0,1)$ *for notational convenience:*
$$\begin{aligned}\Pr[C_n] &= 2\Pr\left[z \geq \epsilon/\sqrt{|s_n|}\right] \\ &\leq \sqrt{\frac{2}{\pi}}\frac{1}{r_n}e^{-r_n^2/2},\end{aligned}$$
where $r_n \equiv \epsilon/\sqrt{|s_n|} \to \infty$.

Define n_m *as the number of* r_n *terms with* $m \leq r_n < m+1$, *then:*
$$\sum\nolimits_{n=1}^{\infty} \Pr[C_n] \leq \sqrt{\frac{2}{\pi}} \sum\nolimits_{m=1}^{\infty} \frac{n_m}{m} e^{-m^2/2}. \tag{1}$$
This sum need not converge, for example if $n_m = O\left(e^{m^2/2}\right)$.

So choose a subsequence $s_{n_{k,1}} \to 0$ such that $r_{n_{k,1}} \to \infty$ fast enough to ensure that the summation in (1) *converges. By the Borel-Cantelli lemma of Proposition II.2.6,* $\Pr[\limsup C_n] = 0$, *and thus by Definition II.2.1:*

$$\Pr\left[\left|Z_{t+s_{n_{k,1}}} - Z_t\right| > \epsilon \text{ for infinitely many } s_{n_{k,1}}\right] = 0.$$

Equivalently:

$$\Pr\left[\left|Z_{t+s_{n_{k,1}}} - Z_t\right| \leq \epsilon \text{ for all but finitely many } s_{n_{k,1}}\right] = 1.$$

We omit these finitely many $s_{n_{k,1}}$, renumbering the resulting subsequence $s_{n_{k,1}}$, and define:

$$A_1 = \{\left|Z_{t+s_{n_{k,1}}} - Z_t\right| \leq \epsilon \text{ for all } s_{n_{k,1}}\}.$$

Thus $\Pr[A_1] = 1$.

Now with $\epsilon/2$ and this subsequence $s_{n_{k,1}}$, repeat the above argument and choose the further subsequence $\{s_{n_{k,2}}\} \subset \{s_{n_{k,1}}\}$ with $s_{n_{1,2}} > s_{n_{1,1}}$ so that $\sum_{n=1}^{\infty} \Pr[C_n]$ converges with C_n defined in terms of $\epsilon/2$. The Borel-Cantelli lemma again applies to obtain after discarding finitely many terms:

$$\Pr[A_2] \equiv \Pr\left[\left|Z_{t+s_{n_{k,2}}} - Z_t\right| \leq \epsilon/2 \text{ for all } s_{n_{k,2}}\right] = 1.$$

Continuing this way obtains a subsequence $s_{n_{k,j}}$ for each j with $\{s_{n_{k,j}}\} \subset \{s_{n_{k,j-1}}\}$, $s_{n_{1,j}} > s_{n_{1,j-1}}$, and:

$$\Pr[A_j] \equiv \Pr\left[\left|Z_{t+s_{n_{k,j}}} - Z_t\right| \leq \epsilon/j \text{ for all } s_{n_{k,j}}\right] = 1.$$

Finally define $s_{n_k} \equiv s_{n_{k,k}}$ and $A \equiv \bigcap_{j=1}^{\infty} A_j$, and note that by De Morgan's laws of Exercise I.2.2:

$$\Pr[A] = 1 - \Pr[\tilde{A}] = 1 - \Pr\left[\bigcup_{j=1}^{\infty} \tilde{A}_j\right] = 1.$$

We now prove that:

$$\Pr\left[\lim_{k\to\infty} Z_{t+s_{n_k}} = Z_t\right] = 1, \tag{2}$$

by proving such convergence on A.

To this end, given ϵ' there exists j so that $\epsilon/j < \epsilon'$, and thus by definition of A_j:

$$\Pr\left[\left|Z_{t+s_{n_{k,j}}} - Z_t\right| \leq \epsilon' \text{ for all } s_{n_{k,j}}\right] = 1.$$

This statement is therefore true for all $n_{k,j} \geq n_{j,j} \equiv n_j$, and since all s_{n_l} are in this $s_{n_{k,j}}$-sequence for $l \geq j$, it follows that:

$$\Pr\left[\left|Z_{t+s_{n_l}} - Z_t\right| \leq \epsilon' \text{ for all } n_l \geq n_j\right] = 1. \tag{3}$$

Thus given any $\epsilon' > 0$ there exists A_j on which (3) *is satisfied. Hence on $A \equiv \bigcap_{j=1}^{\infty} A_j$,* (3) *is satisfied for all ϵ', which proves* (2).

The proof of item 4 is the same as that of item 3 except now:

$$\frac{Z_{t+s_n} - Z_t}{|s_n|^{\gamma}} \sim N(0, |s_n|^{1-2\gamma}). \tag{4}$$

Since $0 < \gamma < 1/2$ *it follows that* $|s_n|^{1-2\gamma} \to 0$ *as* $n \to \infty$ *and the same proof works with* $|s_n|^{1-2\gamma}$ *in place of* $|s_n|$.

For item 5, (4) again remains valid but $1/2 < \gamma$ *obtains* $|s_n|^{1-2\gamma} \to \infty$. *Given* $N > 0$, *define:*

$$D_n = \left\{ \frac{|Z_{t+s_n} - Z_{t_0}|}{|s_n|^\gamma} \leq N \right\}.$$

By (4), and the crude estimate that $\Pr[|z| \leq x] \leq 2x\varphi(0) = \sqrt{\frac{2}{\pi}}x$:

$$\begin{aligned} \Pr[D_n] &= \Pr\left[|z| \leq N/\sqrt{|s_n|^{1-2\gamma}}\right] \\ &\leq \sqrt{\frac{2}{\pi}}N_n, \end{aligned}$$

where $N_n \equiv N/\sqrt{|s_n|^{1-2\gamma}}$. *Thus:*

$$\sum\nolimits_{n=1}^{\infty} \Pr[D_n] \leq \sqrt{\frac{2}{\pi}} \sum\nolimits_{m=1}^{\infty} N_n.$$

Now $N_n \to 0$ *since* $|s_n|^{1-2\gamma} \to \infty$, *so choose* $s_{n_k} \to \infty$ *fast enough to ensure that this series converges. Thus* $\Pr[\limsup D_{n_k}] = 0$ *by the Borel-Cantelli lemma:*

$$\Pr\left[\frac{\left|Z_{t+s_{n_k}} - Z_t\right|}{|s_{n_k}|^\gamma} \leq N \textit{ for infinitely many } n_k \right] = 0.$$

Equivalently:

$$\Pr\left[\frac{\left|Z_{t+s_{n_k}} - Z_t\right|}{|s_{n_k}|^\gamma} > N \textit{ for all but finitely many } n_k \right] = 1,$$

which is (10.23).

Thus Z_t *cannot be nearly sequentially strong Hölder continuous, since given any proposed subsequence, there is a further subsequence that satisfies (10.23), obtaining a contradiction.*

■

Remark 10.26 (On continuity) *It is quite tempting to now declare that despite defining* Z_t *pointwise in terms of the distributional limits of binomial motion* $B_t(\Delta t)$ *as* $\Delta t \to 0$, *that we have in some sense arrived at a pathwise result reminiscent of the pathwise limits of the prior section. Unfortunately, this is an over-reach.*

First, the sequential continuity result obtained does not assure sequential continuity. But even if it did assure this in the best case, sequential continuity at each point t_0 *with probability* 1 *by (10.21) would not imply sequential continuity for all* t *with probability* 1. *Specifically, if*

$$A_t \equiv \left\{ \lim_{s \to t} Z_s = Z_t \right\}, \ A \equiv \left\{ \lim_{s \to t} Z_s = Z_t \textit{ for all } t \right\},$$

then $A = \bigcap_t A_t$ *is an intersection of uncountably many sets.*

Since the sigma algebra $\mathcal{E}$ *need only be closed for countable intersections,* A *need not be measurable, and even if it is, we could not conclude that* $\Pr[A] = 1$ *unless the union of complements,* $\bigcup_t \tilde{A}_t$, *has measure* 0. *But indeed it is not true that the union of uncountably many sets of measure* 0 *has measure* 0, *as* $\mathbb{R} = \bigcup_t \{t\}$ *attests with Lebesgue measure.*

If indeed Z_t turned out to be sequentially continuous at each t with probability 1, then it is simultaneously sequentially continuous on any countable collection of such points with probability 1. For example, this countable collection could be taken as the rationals, which are dense in any interval. Unfortunately, continuity on a dense set does not imply continuity, as Thomae's function of Remark I.3.19 demonstrates.

***Andrey Kolmogorov** (1903–1987) had an insight to address this question that will be developed in Book VII. Kolmogorov proved that if probability 1 continuity of Z_t on a dense set of $t \in [0,T]$ with $T < \infty$ can be improved to **Hölder continuity** on such a set (again with probability 1), then Z_t can be redefined on the complement of this set to make Z_t continuous (again with probability 1). Further, this redefinition preserves all of the distributional properties of Z_t noted in Proposition 10.25. Kolmogorov also proved that Hölder continuity on such a dense set will follow from Hölder-like results for moments, which we can prove next for Z_t.*

Exercise 10.27 (On $E\left[|Z_t - Z_s|^{2\alpha}\right]$) *For $0 \leq s < t \leq T$ and any $\alpha \geq 0$, prove that:*

$$E\left[|Z_t - Z_s|^{2\alpha}\right] = c_\alpha |t-s|^\alpha, \tag{10.24}$$

where $c_\alpha = \frac{1}{\sqrt{2\pi}}\int_{-\infty}^{\infty} |x|^{2\alpha} \exp\left(-x^2/2\right) dx$. Hint: The identity in (10.24) follows from Proposition 10.18, that $Z_t - Z_s =_d Z'_{t-s} \sim N(0, t-s)$.

10.4 Nonstandard Binomial Motion

In this section we investigate **nonstandard binomial motion**. This will be defined as in Definition 10.12, but where the binomial path variates $b(\Delta t)$ of (10.1) do not require the assumption that $p = 1/2$.

In the first section we consider $0 < p < 1$ fixed, and then turn to $p = q^{Gen}(\Delta t)$, defined in terms of the risk-neutral probabilities of (9.24) and which vary with Δt. These results will also apply to the real-world probability $p(\Delta t)$ of (9.31).

We will not replicate the development of the prior section, but instead, introduce ideas by investigating the effects of these assumptions on the Proposition 10.10 results.

10.4.1 Nonstandard Binomial Motion with $p \neq 1/2$ Fixed

Given $T > 0$ and $\{b_j(\Delta t)\}_{j=1}^n$ defined as in (10.1) but without the assumption that $p = 1/2$, we can define the **binomial path** $B_t(\Delta t)$ for $t \in [0,T]$ as in Definition 10.1 by:

$$B_t(\Delta t) = (1-s)B_{j-1}(\Delta t) + sB_j(\Delta t).$$

or equivalently:

$$B_t(\Delta t) = B_{j-1}(\Delta t) + sb_j(\Delta t).$$

Here again, $B_0(\Delta t) = 0$ and $t \equiv (j-1+s)\,\Delta t$, $0 \leq s \leq 1$.

For general $p \neq 1/2$, there is no counterpart to Proposition 10.10 for convergence in distribution, and this will be seen to be driven by $E[b_j(\Delta t)] = (2p-1)\sqrt{\Delta t}$, and not $Var[b_j(\Delta t)] = 4p(1-p)\Delta t$.

Proposition 10.28 (No convergence in distribution of $B_t(\Delta t_n)$ for $p \neq 1/2$) *Let $B_t(\Delta t_n)$ be defined as in Definition 10.1 with $p \neq 1/2$. Then:*

1. For every $t \in (0, T]$, as $\Delta t_n \to 0$:

$$E\left[B_t(\Delta t_n)\right] \to \begin{cases} \infty, & p > 1/2, \\ -\infty, & p < 1/2. \end{cases}$$

2. For $0 \leq s < t \leq T$, as $\Delta t_n \to 0$:

$$E\left[B_t(\Delta t_n) - B_s(\Delta t_n)\right] \to \begin{cases} \infty, & p > 1/2, \\ -\infty, & p < 1/2. \end{cases}$$

Thus neither $B_t(\Delta t_n)$ nor $B_t(\Delta t_n) - B_s(\Delta t_n)$ converges in distribution.

Proof. *To prove item 1, fix $t \in (0, T]$ and for each n let $t = [j(n) + e(n)]\,\Delta t_n$ with integer $j(n)$ and $0 \leq e(n) < 1$. Then:*

$$B_t(\Delta t_n) = B_{j(n)}(\Delta t_n) + e(n) b_{j(n)+1}(\Delta t_n),$$

and thus:

$$\begin{aligned} E\left[B_t(\Delta t_n)\right] &= (2p-1)\,(j(n)+e(n))\,\sqrt{\Delta t_n} \\ &= \frac{2p-1}{\sqrt{\Delta t_n}} t \to \begin{cases} \infty, & p > 1/2, \\ -\infty, & p < 1/2. \end{cases} \end{aligned}$$

For item 2, since $s < t$, let $s = [j(n) + e(n)]\,\Delta t_n$ and $t = [k(n) + f(n)]\,\Delta t_n$, where $j(n) \leq k(n)$ and $0 \leq e(n), f(n) < 1$. Since $s < t$, it follows that for n large that $j(n) < k(n)$, and this will be assumed to avoid notational ambiguity. Thus suppressing Δt_n:

$$B_t(\Delta t_n) - B_s(\Delta t_n) = (1 - e(n))\, b_{j(n)+1} + \sum\nolimits_{i=j(n)+2}^{k(n)} b_i + f(n) b_{k(n)+1}.$$

Then $E\left[B_t(\Delta t_n) - B_s(\Delta t_n)\right] \to \pm\infty$ as above.

The failure of convergence in distribution then follows as an exercise using Chebyshev's inequality of Proposition IV.4.34, and Remark 10.29. ■

Remark 10.29 (On $Var\left[B_t(\Delta t_n)\right]$) *It is interesting to note, for example in item 1, that the lack of distributional convergence is caused by a divergence of $E\left[B_t(\Delta t_n)\right]$, while $Var\left[B_t(\Delta t_n)\right]$ converges as in Proposition 10.10.*

Independence of $\{b_j(\Delta t_n)\}_{j=1}^{\infty}$ obtains:

$$\begin{aligned} Var\left[B_t(\Delta t_n)\right] &= 4p(1-p) j(n) \Delta t_n + e(n)^2 \Delta t_n \\ &\to 4p(1-p)t. \end{aligned}$$

Similarly,

$$Var\left[B_t(\Delta t_n) - B_s(\Delta t_n)\right] \to 4p(1-p)\,(t-s)\,.$$

10.4.2 Nonstandard Binomial Motion with $p = f(\Delta t)$

Recall the risk-neutral probability $q\,(\Delta t)$ that emerged in the process of pricing certain financial derivatives on a binomial lattice by replication of such options' settlement values. As noted in (9.24), with $R(r)$ defined in (9.25), the general formulas for risk-neutral probability $q\,(\Delta t) \equiv q^{Gen}\,(\Delta t)$ for various underlying assets can be expressed:

$$q(\Delta t) = \frac{e^{R(r)\Delta t} - e^{d(\Delta t)}}{e^{u(\Delta t)} - e^{d(\Delta t)}},$$

where the asset continuous returns $u(\Delta t)$ and $d(\Delta t)$ are given in (9.2) by:

$$u(\Delta t)=\mu\Delta t+\sigma\sqrt{\Delta t}, \qquad d(\Delta t)=\mu\Delta t-\sigma\sqrt{\Delta t}.$$

Analogously, as seen in Section 9.3, we can set $\mu=0$ in the definitions of $u(\Delta t)$ and $d(\Delta t)$ and build a real-world model with the probability function in (9.31):

$$p(\Delta t)=\frac{\exp\left(\left[\mu+\frac{\sigma^2}{2}\right]\Delta t\right)-e^{-\sigma\sqrt{\Delta t}}}{e^{\sigma\sqrt{\Delta t}}-e^{-\sigma\sqrt{\Delta t}}}.$$

Given the negative results of the prior section with fixed $p\neq 1/2$, it is a remarkable fact that binomial paths $B_t(\Delta t)$ for $t\in[0,T]$ converge in distribution as $\Delta t\to 0$, using $p=q(\Delta t)$ or $p=p(\Delta t)$. And indeed, these paths converge weakly at each $t>0$ to a normal random variable, though not a standard normal.

That this is so reflects the approximation formula for $q(\Delta t)$ derived in (9.26):

$$q(\Delta t)=\frac{1}{2}+\frac{1}{2\sigma}\left(R(r)-\mu-\sigma^2/2\right)\sqrt{\Delta t}+O(\Delta t^{3/2}). \tag{10.25}$$

By a change of notation, setting $\mu=0$ to reflect the redefinition of $u(\Delta t)$ and $d(\Delta t)$, and replacing $R(r)$ with $\mu+\frac{\sigma^2}{2}$, obtains the approximation:

$$p(\Delta t)=\frac{1}{2}+\frac{1}{2\sigma}\mu\sqrt{\Delta t}+O(\Delta t^{3/2}).$$

While it is apparent from this that $q(\Delta t),p(\Delta t)\to 1/2$ as $\Delta t\to 0$, what makes the next results possible is that the rate of convergence here is fast enough to prevent the divergence of the means seen in the previous section.

Proposition 10.30 (Convergence in distribution of $B_t(\Delta t_n)$ for $p=q(\Delta t)$) *Let $B_t(\Delta t_n)$ be defined as in Definition 10.1 with $p=q^{Gen}(\Delta t_n)$ of (9.24), and let $N(a,b^2)$ denote a normally distributed random variable with mean a and variance b^2. Then:*

1. *For every $t\in(0,T]$, as $\Delta t_n\to 0$:*

$$B_t(\Delta t_n)\to_d N\left(\frac{1}{\sigma}\left(R(r)-\mu-\frac{\sigma^2}{2}\right)t,t\right).$$

2. *For $0\leq s<t\leq T$, as $\Delta t_n\to 0$:*

$$B_t(\Delta t_n)-B_s(\Delta t_n)\to_d N\left(\frac{1}{\sigma}\left(R(r)-\mu-\frac{\sigma^2}{2}\right)(t-s),t-s\right).$$

Proof. *To prove item 1, fix $t\in(0,T]$. For each n let $t=[j(n)+e(n)]\,\Delta t_n$ with integer $j(n)$ and $0\leq e(n)<1$, and so:*

$$B_t(\Delta t_n)=B_{j(n)}(\Delta t_n)+e(n)b_{j(n)+1}(\Delta t_n).$$

Independence of $\{b_j(\Delta t_n)\}_{j=1}^{\infty}$ obtains independence of $B_{j(n)}(\Delta t_n)$ and $b_{j(n)+1}(\Delta t_n)$ by Exercise II.3.50 and Proposition II.3.56, and then Proposition 5.23 obtains the identity in characteristic functions:

$$C_{B_t(\Delta t_n)}(y)=C_{B_{j(n)}}(y)C_{e(n)b_{j(n)+1}}(y). \tag{1}$$

Since $b_j(\Delta t_n)$ *is binomial with values* ± 1 *with respective probabilities* $q(\Delta t_n)$ *and* $1 - q(\Delta t_n)$*, it follows from (5.20):*

$$\begin{aligned} C_{e(n)b_{j(n)+1}}(y) &= q(\Delta t_n)\exp\left(iye(n)\sqrt{\Delta t_n}\right) + (1-q(\Delta t_n))\exp\left(-iye(n)\sqrt{\Delta t_n}\right) \\ &= 2iq(\Delta t_n)\sin\left[ye(n)\sqrt{\Delta t_n}\right] + \exp\left(-iye(n)\sqrt{\Delta t_n}\right). \end{aligned}$$

Since $q(\Delta t_n) < 1.0$ *for* Δt_n *small by (10.25), it follows that for all* y*,*

$$C_{e(n)b_{(n)j+1}}(y) \to 1, \tag{2}$$

as $\Delta t_n \to 0$.

Next, we claim that $\{b_j(\Delta t_n)\}_{j=1}^{j(n)}$ *satisfies Lindeberg's condition of Definition 7.6. In detail, since* $E[b_j(\Delta t)] = (2q(\Delta t_n) - 1)\sqrt{\Delta t_n}$ *and* $Var[b_j(\Delta t)] = 4q(\Delta t_n)(1 - q(\Delta t_n))\Delta t_n$*:*

$$\begin{aligned} Z_j &= \frac{b_j(\Delta t) - (2q(\Delta t_n)-1)\sqrt{\Delta t_n}}{2\sqrt{q(\Delta t_n)(1-q(\Delta t_n))}\sqrt{\Delta t_n}} \\ &= \frac{b_j - 2q(\Delta t_n) + 1}{2\sqrt{q(\Delta t_n)(1-q(\Delta t_n))}}, \end{aligned}$$

where $b_j = \pm 1$.

Hence:

$$-\sqrt{\frac{q(\Delta t_n)}{1-q(\Delta t_n)}} \leq Z_j \leq \sqrt{\frac{1-q(\Delta t_n)}{q(\Delta t_n)}},$$

and it follows from (10.25) that $|Z_j| \leq 2$*, say, for* Δt_n *small. Then for any* $s > 0$*, and* $j(n) > 4/s^2$*:*

$$\int_{|Z_j| > s\sqrt{j(n)}} Z_j^2 d\lambda = 0,$$

verifying Lindeberg's condition.

Again by (10.25):

$$\begin{aligned} E\left[B_{j(n)}(\Delta t_n)\right] &= \frac{2q(\Delta t_n)-1}{\sqrt{\Delta t_n}}t \to \frac{1}{\sigma}\left(R(r) - \mu - \frac{\sigma^2}{2}\right)t, \\ Var\left[B_{j(n)}(\Delta t_n)\right] &= 4q(\Delta t_n)(1-q(\Delta t_n))t \to t, \end{aligned}$$

and Corollary 7.14 of Lindeberg's central limit theorem obtains:

$$B_{j(n)}(\Delta t_n) \to N\left(\frac{1}{\sigma}\left(R(r) - \mu - \frac{\sigma^2}{2}\right)t, t\right).$$

Lévy's continuity theorem of Proposition 5.30 states that this is equivalent to:

$$C_{B_{j(n)}}(y) \to C_{N\left(\frac{1}{\sigma}\left(R(r)-\mu-\frac{\sigma^2}{2}\right)t,t\right)}(y). \tag{3}$$

Combining results of (1)–(3) obtains:

$$C_{B_t(\Delta t_n)}(y) \to C_{N\left(\frac{1}{\sigma}\left(R(r)-\mu-\frac{\sigma^2}{2}\right)t,t\right)},$$

and another application of Lévy's continuity theorem proves item 1.

For item 2, since $s < t$, let $s = [j(n) + e(n)]\,\Delta t_n$ and $t = [k(n) + f(n)]\,\Delta t_n$, where $j(n) \leq k(n)$ and $0 \leq e(n),\ f(n) < 1$. Since $s < t$, it follows that for n large and defined so that $\Delta t_n < (t-s)/2$ say, that $j(n) < k(n)$ and this will be assumed to avoid notational ambiguity.

Then suppressing Δt_n:

$$B_t(\Delta t_n) - B_s(\Delta t_n) = (1 - e(n))\, b_{j(n)+1} + \sum\nolimits_{i=j(n)+2}^{k(n)} b_i + f(n) b_{k(n)+1}.$$

Denoting $X \equiv \sum\nolimits_{i=j(n)+2}^{k(n)} b_i$, the justifications above obtain:

$$C_{[B_t(\Delta t_n) - B_s(\Delta t_n)]}(y) = C_{(1-e(n))b_{j(n)+1}}(y) C_X(y) C_{f(n)b_{k(n)+1}}. \tag{4}$$

As in (2), $C_{(1-e(n))b_{j(n)+1}}(y) \to 1$ and $C_{f(n)b_{k(n)+1}} \to 1$ as $\Delta t_n \to 0$, while as in (3), since $E[X] \to \frac{1}{2\sigma}\left(R(r) - \mu - \frac{\sigma^2}{2}\right)(t-s)$ and $Var[X] \to t - s$:

$$C_X(y) \to C_{N\left(\frac{1}{\sigma}\left(R(r)-\mu-\frac{\sigma^2}{2}\right)(t-s),\, t-s\right)}(y).$$

Combining results into (4) proves item 2 with a final application of Lévy's continuity theorem. ∎

Proposition 10.31 (Convergence in distribution of $B_t(\Delta t_n)$ for $p = p(\Delta t)$) *Let $B_t(\Delta t_n)$ be defined as in Definition 10.1 with $p = p(\Delta t_n)$ of (9.31), and let $N(a, b^2)$ denote a normally distributed random variable with mean a and variance b^2. Then:*

1. *For every $t \in (0, T]$, as $\Delta t_n \to 0$:*

$$B_t(\Delta t_n) \to_d N\left(\frac{1}{\sigma}\mu t, t\right).$$

2. *For $0 \leq s < t \leq T$, as $\Delta t_n \to 0$:*

$$B_t(\Delta t_n) - B_s(\Delta t_n) \to_d N\left(\frac{1}{\sigma}\mu\,(t-s)\,, t-s\right).$$

Proof. *As noted for the approximation formula for $p(\Delta t)$, these results follow from Proposition 10.30 by setting $\mu = 0$, to reflect the redefinitions of $u(\Delta t)$ and $d(\Delta t)$, and then replacing $R(r)$ with $\mu + \frac{\sigma^2}{2}$.* ∎

10.5 Limits of Binomial Asset Models

In this final section, we revisit the multiplicative binomial asset model and conclusions of various propositions from the point of view of binomial motion.

Recall that given initial asset price X_0, return parameters μ and $\sigma > 0$, fixed time horizon $T > 0$, and $\Delta t \equiv T/n$ for fixed n, we defined the price variate $X_j^{(n)}$ at time $t = j\Delta t$ for $1 \leq j \leq n$:

$$X_j^{(n)} \equiv X_0 \exp\left[\sum\nolimits_{i=1}^{j} Z_i^{(n)}\right],$$

where:

$$Z_i^{(n)} = \mu\Delta t + \sigma\sqrt{\Delta t} b_i.$$

Here $\{b_i\}_{i=1}^n$ are independent and binomially distributed to equal 1 or -1 with probability:

- **Real-world model of Proposition 8.46:** $p = 1/2$.
- **Risk-neutral model of Proposition 9.29:** $q(\Delta t)$ and $1 - q(\Delta t)$, respectively, with $q(\Delta t) \equiv q^{Gen}(\Delta t)$ of (9.24).
- **Real-world model of Proposition 9.31:** $p(\Delta t)$ and $1 - p(\Delta t)$, respectively, with $p(\Delta t)$ of (9.31), and with $\mu = 0$ in the definition of $Z_i^{(n)}$.

Within this framework, $\{X_j^{(n)}\}_{j=1}^n$ could be modeled with any n, obtaining asset prices $X_t^{(n)}$ for $t = j\Delta t$ for $1 \leq j \leq n$. To investigate limiting distributions of such prices as $\Delta t \to 0$, it was necessary to then further "refine" the modeling of these prices as follows.

Given integer m and $\{b_i\}_{i=1}^{mn}$ defined as above, the refined price variate at time $t = j\Delta t$ for $1 \leq j \leq n$ was denoted:

$$X_{mj}^{(mn)} \equiv X_0 \exp\left[\sum\nolimits_{i=1}^{mj} \left(\mu T/mn + \sigma\sqrt{T/mn}b_i\right)\right].$$

Then with $Z \sim N(0,1)$, a standard normal variate, as $m \to \infty$:

- **Real-world model of Proposition 8.46:**

  $X_{mj}^{(mn)}$ converges in distribution to a lognormal distribution with parameters $\mu j\Delta t + \ln X_0$ and $\sigma^2 j\Delta t$, and with $t = j\Delta t$:

$$X_{mj}^{(mn)} \to_d X_t = X_0 \exp\left[\mu t + \sigma\sqrt{t}Z\right].$$

- **Risk-neutral model of Proposition 9.29:**

  $X_{mj}^{(mn)}$ converges in distribution to a lognormal distribution with parameters $\ln X_0 + (R(r) - \sigma^2/2)t$ and $\sigma^2 t$, where $R(r)$ is defined in (9.25), and with $t = j\Delta t$:

$$X_{mj}^{(mn)} \to_d X_t = X_0 \exp\left[(R(r) - \sigma^2/2)t + \sigma\sqrt{t}Z\right].$$

- **Real-world model of Proposition 9.31:**

  $X_{mj}^{(mn)}$ converges in distribution to a lognormal distribution with parameters $\mu j\Delta t + \ln X_0$ and $\sigma^2 j\Delta t$, and with $t = j\Delta t$:

$$X_{mj}^{(mn)} \to_d X_t = X_0 \exp\left[\mu t + \sigma\sqrt{t}Z\right].$$

With above results on binomial paths and binomial motion, we can derive more general conclusions on convergence in distribution of asset prices within these models. Rewriting the above formula for $X_j^{(n)}$, the asset price at time $t = j\Delta t$:

$$X_j^{(n)} \equiv X_0 \exp\left[\mu j\Delta t + \sigma B_{j\Delta t}(\Delta t)\right],$$

where

$$B_{j\Delta t}(\Delta t) \equiv \sum\nolimits_{i=1}^{j} b_i\sqrt{\Delta t}.$$

Thus $B_{j\Delta t}(\Delta t)$ agrees with the binomial path $B_t(\Delta t)$ of Definition 10.1 for $t = j\Delta t$. This motivates the following:

Definition 10.32 (Asset price paths) *Given initial asset price X_0, return parameters μ and $\sigma > 0$, fixed time horizon $T > 0$ and $\Delta t \equiv T/n$ for fixed n, we defined the* **asset price path** *$X_t(\Delta t)$ for $0 \leq t \leq T$:*

$$X_t(\Delta t) = X_0 \exp\left[\mu t + \sigma B_t(\Delta t)\right]. \tag{10.26}$$

Here $B_t(\Delta t)$ is given in Definition 10.1 with respect to independent and binomially distributed $\{b_i\}_{i=1}^n$, equal 1 or -1 with probabilities to be specified.

Proposition 10.33 (Distributional limits of $X_t(\Delta t)$ as $\Delta t \to 0$) *With $X_t(\Delta t)$ defined in (10.26), then for all $0 \leq t \leq T$, $X_t(\Delta t)$ converges in distribution to a lognormal variate as $\Delta t \to 0$, the parameters of which reflect the probability model of the driving binomial variates $\{b_i\}_{i=1}^n$.*

With $Z \sim N(0,1)$, a standard normal variate:

- ***Real-world model:*** *$p = 1/2$:*

$$X_t(\Delta t) \to_d X_t = X_0 \exp\left[\mu t + \sigma\sqrt{t}Z\right]. \tag{10.27}$$

- ***Risk-neutral model:*** *$p = q(\Delta t)$ of (9.24):*

$$X_t(\Delta t) \to_d X_t = X_0 \exp\left[(R(r) - \sigma^2/2)t + \sigma\sqrt{t}Z\right], \tag{10.28}$$

where $R(r)$ is defined in (9.25).

- ***Real-world model:*** *$p = p(\Delta t)$ of (9.31) and $\mu = 0$ in (10.26):*

$$X_t(\Delta t) \to_d X_t = X_0 \exp\left[\mu t + \sigma\sqrt{t}Z\right]. \tag{10.29}$$

Proof. *For (10.27), Proposition 10.10 obtains that for all t, $B_t(\Delta t) \to_d N(0,t)$, a normal variate with mean 0 and variance t, which we express as $\sqrt{t}Z$ with $Z \sim N(0,1)$. Since $g(x) = X_0 \exp[\mu t + \sigma x]$ is continuous, the Mann-Wald theorem of Proposition 3.26 obtains that:*

$$X_t(\Delta t) = g(B_t(\Delta t)) \to_d g(\sqrt{t}Z),$$

which is (10.27).

Similarly for (10.28), Proposition 10.30 obtains that for all t:

$$B_t(\Delta t) \to_d N\left(\frac{1}{\sigma}\left(R(r) - \mu - \frac{\sigma^2}{2}\right)t, t\right),$$

which can be expressed as $\frac{1}{\sigma}\left(R(r) - \mu - \frac{\sigma^2}{2}\right)t + \sqrt{t}Z$ with $Z \sim N(0,1)$. Again by Mann-Wald:

$$X_t(\Delta t) = g(B_t(\Delta t)) \to_d g\left(\frac{1}{\sigma}\left(R(r) - \mu - \frac{\sigma^2}{2}\right)t + \sqrt{t}Z\right),$$

which is (10.28).

For (10.29), Proposition 10.31 states that for all t:

$$B_t(\Delta t) \to_d N\left(\frac{1}{\sigma}\mu t, t\right),$$

which can be expressed as $\frac{1}{\sigma}\mu t + \sqrt{t}Z$ *with* $Z \sim N(0,1)$*. Using continuous* $\tilde{g}(x) = X_0 \exp[\sigma x]$*, again by Mann-Wald:*

$$X_t(\Delta t) = \tilde{g}(B_t(\Delta t)) \to_d \tilde{g}\left(\frac{1}{\sigma}\mu t + \sqrt{t}Z\right),$$

which is (10.29). ■

The final exercise creates a multivariate version of the above result for the real-world model with $p = \frac{1}{2}$. It has a parallel statement for the other models, but we have not developed the multivariate counterpart to Proposition 10.16 for the risk-neutral model with $q(\Delta t)$, or the real-world model with $p(\Delta t)$.

Exercise 10.34 (Multivariate (10.27)) *Given* $0 < t_1 < t_2 < ... < t_n \leq T$*, define the random vector* $(X_{t_1}(\Delta t), X_{t_2}(\Delta t), ..., X_{t_n}(\Delta t))$ *componentwise as in (10.26), in terms of* $(B_{t_1}(\Delta t), B_{t_2}(\Delta t), ..., B_{t_n}(\Delta t))$ *and* $p = \frac{1}{2}$*. Using the Mann-Wald theorem of Proposition 3.26, state and prove the multivariate version of (10.27), on the limiting random vector as* $\Delta t \to 0$.

References

I have listed below a number of textbook references for the mathematics and finance presented in this series of books. All provide both theoretical and applied materials in their respective areas that are beyond those developed here and are worth pursuing by those interested in gaining a greater depth or breadth of knowledge. This list is by no means complete and is intended only as a guide to further study. In addition, various published research papers have been identified in some chapters where these results were discussed.

The reader will no doubt observe that the mathematics references are somewhat older than the finance references and upon web searching will find that some older texts have been updated to newer editions, sometimes with additional authors. Since I own and use the editions below, I decided to present these editions rather than reference the newer editions which I have not reviewed. As many of these older texts are considered "classics," they are also likely to be found in university and other libraries.

That said, there are undoubtedly many very good new texts by both new and established authors with similar titles that are also worth investigating. One that I will at the risk of immodesty recommend for more introductory materials on mathematics, probability theory and finance is:

1. Reitano, Robert, R. *Introduction to Quantitative Finance: A Math Tool Kit.* Cambridge, MA: The MIT Press, 2010.

Linear Algebra, Topology, Measure and Integration

2. Doob, J. L. *Measure Theory.* New York, NY: Springer-Verlag, 1994.
3. Dugundji, James. *Topology.* Boston, MA: Allyn and Bacon, 1970.
4. Edwards, Jr., C. H. *Advanced Calculus of Several Variables.* New York, NY: Academic Press, 1973.
5. Gemignani, M. C. *Elementary Topology.* Reading, MA: Addison-Wesley Publishing, 1967.
6. Halmos, Paul R. *Measure Theory.* New York, NY: D. Van Nostrand, 1950.
7. Hewitt, Edwin, and Karl Stromberg. *Real and Abstract Analysis.* New York, NY: Springer-Verlag, 1965.
8. Hoffman, Kenneth, and Ray Kunze. *Linear Algebra,* 2nd Edition. Englewood Cliffs, NJ: Prentice-Hall, 1971.
9. Royden, H. L. *Real Analysis,* 2nd Edition. New York, NY: The MacMillan Company, 1971.
10. Rudin, Walter. *Principals of Mathematical Analysis,* 3rd Edition. New York, NY: McGraw-Hill, 1976.

11. Rudin, Walter. *Real and Complex Analysis,* 2nd Edition. New York, NY: McGraw-Hill, 1974.

12. Shilov, G. E., and B. L. Gurevich. *Integral, Measure & Derivative: A Unified Approach.* New York, NY: Dover Publications, 1977.

13. Stein, Elias M., and Guido Weiss. *Introduction to Fourier Analysis on Euclidean Spaces.* Princeton, NJ: Princeton University Press, 1971.

14. Strang, Gilbert. *Introduction to Linear Algebra,* 4th Edition. Wellesley, MA: Cambridge Press, 2009.

Probability Theory & Stochastic Processes

15. Billingsley, Patrick. *Probability and Measure,* 3rd Edition. New York, NY: John Wiley & Sons, 1995.

16. Chung, K. L., and R. J. Williams. *Introduction to Stochastic Integration.* Boston, MA: Birkhäuser, 1983.

17. Davidson, James. *Stochastic Limit Theory.* New York, NY: Oxford University Press, 1997.

18. de Haan, Laurens, and Ana Ferreira. *Extreme Value Theory: An Introduction.* New York, NY: Springer Science, 2006.

19. Durrett, Richard. *Probability: Theory and Examples,* 2nd Edition. Belmont, CA: Wadsworth Publishing, 1996.

20. Durrett, Richard. *Stochastic Calculus: A Practical Introduction.* Boca Raton, FL: CRC Press, 1996.

21. Feller, William. *An Introduction to Probability Theory and Its Applications,* Volume I. New York, NY: John Wiley & Sons, 1968.

22. Feller, William. *An Introduction to Probability Theory and Its Applications,* Volume II, 2nd Edition. New York, NY: John Wiley & Sons, 1971.

23. Friedman, Avner. *Stochastic Differential Equations and Applications, Volume 1 and 2.* New York, NY: Academic Press, 1975.

24. Ikeda, Nobuyuki, and Shinzo Watanabe. *Stochastic Differential Equations and Diffusion Processes.* Tokyo: Kodansha Scientific, 1981.

25. Karatzas, Ioannis, and Steven E. Shreve. *Brownian Motion and Stochastic Calculus.* New York, NY: Springer-Verlag, 1988.

26. Kloeden, Peter E., and Eckhard Platen. *Numerical Solution of Stochastic Differential Equations.* New York, NY: Springer-Verlag, 1992.

27. Lowther, George, *Almost Sure: A Random Mathematical Blog,* https://almostsuremath.com/about/

28. Lukacs, Eugene. *Characteristic Functions.* New York, NY: Hafner Publishing, 1960.

29. Nelson, Roger B. *An Introduction to Copulas,* 2nd Edition. New York, NY: Springer Science, 2006.

30. Øksendal, Bernt. *Stochastic Differential Equations: An Introduction with Applications,* 5th Edition. New York, NY: Springer-Verlag, 1998.

31. Protter, Phillip. *Stochastic Integration and Differential Equations: A New Approach.* New York, NY: Springer-Verlag, 1992.

32. Revuz, Daniel, and Marc Yor. *Continuous Martingales and Brownian Motion,* 3rd Edition. New York, NY: Springer-Verlag, 1991.

33. Rogers, L. C. G., and D. Williams. *Diffusions, Markov Processes and Martingales,* Volume 1, Foundations, 2nd Edition. Cambridge, UK: Cambridge University Press, 2000.

34. Rogers, L. C. G., and D. Williams. *Diffusions, Markov Processes and Martingales,* Volume 2, Itô Calculus, 2nd Edition. Cambridge, UK: Cambridge University Press, 2000.

35. Sato, Ken-Iti. *Lévy Processes and Infinitely Divisible Distributions.* Cambridge, UK: Cambridge University Press, 1999.

36. Schilling, René L. and Lothar Partzsch. *Brownian Motion: An Introduction to Stochastic Processes,* 2nd Edition. Berlin/Boston: Walter de Gruyter GmbH, 2014.

37. Schuss, Zeev. *Theory and Applications of Stochastic Differential Equations.* New York, NY: John Wiley and Sons, 1980.

Finance Applications

38. Etheridge, Alison. *A Course in Financial Calculus.* Cambridge, UK: Cambridge University Press, 2002.

39. Embrechts, Paul, Claudia Klüppelberg, and Thomas Mikosch. *Modelling Extremal Events for Insurance and Finance.* New York, NY: Springer-Verlag, 1997.

40. Hull, John C. *Options, Furues, and Other Derivatives,* 11th Edition. New York, NY: Pearson, 2022.

41. Hunt, P. J., and J. E. Kennedy. *Financial Derivatives in Theory and Practice,* Revised Edition. Chichester, UK: John Wiley & Sons, 2004.

42. McDonald, Robert L. *Derivatives Markets,* 2nd Edition. Boston, MA: Addison, Wesley, 2006.

43. McLeish, Don L. *Monte Carlo Simulation and Finance.* New York, NY: John Wiley, 2005.

44. McNeil, Alexander J., Rüdiger Frey, and Paul Embrechts. *Quantitative Risk Management: Concepts, Techniques, and Tools.* Princeton, NJ: Princeton University Press, 2005.

45. Shreve, Steven E. *Stochastic Calculus for Finance I: The Binomial Asset Procing Model.* New York, NY: Springer Finance, 2004.

Research Papers for Book VI

46. Amin, K. and A. Khanna. "Convergence of American Option Values from Discrete to Continuous-Time Financial Models." *Mathematical Finance* 4, 289, 1994.

47. Black, Fisher S. and Myron S. Scholes. "The Pricing of Options and Corporate Liabilities." *Journal of Political Economy,* Vol. 81, No. 3, 637-654. 1973.

48. Cox, John C., Stephen A. Ross,and Mark Rubinstein. "Option Pricing: A Simplified Approach." *Journal of Financial Economics,* Vol. 7, Issue 3, 229-263, 1979.

49. Jiang, Lishang and Min Dai. "Convergence of Binomial Tree Method for American Options." *Partial Differential Equations and Their Applications,* 106-118, 1999.

50. Leisen, Dietmar P.J. "Pricing the American Put Option: A Detailed Convergence Analysis for Binomial Models." *Journal of Economic Dynamics and Control,* 22, 1419-1444, 1998.

51. Merton, Robert C. "Theory of Rational Option Pricing." *The Bell Journal of Economics and Management Science,* Vol. 4, No. 1, 141-183, 1973.

52. Thorin, Olof. "On the Infinite Divisibility of the Lognormal Distribution." *Scandinavian Actuarial Journal,* 3, 121-148, 1977.

Index

Printed in the United States
by Baker & Taylor Publisher Services

Printed in the United States
by Baker & Taylor Publisher Services